T0289748

Coastal Ocean Observing Systems

Coastal Ocean Observing Systems

Edited by

Yonggang Liu

College of Marine Science, University of South Florida, St. Petersburg, FL, USA

Heather Kerkering

Pacific Islands Ocean Observing System, University of Hawaii at Manoa, Honolulu, HI, USA; Sea Connections Consulting, Virginia, USA

Dr. Robert H. Weisberg

College of Marine Science, University of South Florida, St. Petersburg, FL, USA

AMSTERDAM • BOSTON • HEIDELBERG • LONDON
NEW YORK • OXFORD • PARIS • SAN DIEGO
SAN FRANCISCO • SINGAPORE • SYDNEY • TOKYO

Academic Press is an imprint of Elsevier

Academic Press is an imprint of Elsevier
32 Jamestown Road, London NW1 7BY, UK
525 B Street, Suite 1800, San Diego, CA 92101-4495, USA
225 Wyman Street, Waltham, MA 02451, USA
The Boulevard, Langford Lane, Kidlington, Oxford OX5 1GB, UK

Notices

Knowledge and best practice in this field are constantly changing. As new research and experience broaden our understanding, changes in research methods, professional practices, or medical treatment may become necessary.

Practitioners and researchers must always rely on their own experience and knowledge in evaluating and using any information, methods, compounds, or experiments described herein. In using such information or methods they should be mindful of their own safety and the safety of others, including parties for whom they have a professional responsibility.

To the fullest extent of the law, neither the Publisher nor the authors, contributors, or editors, assume any liability for any injury and/or damage to persons or property as a matter of products liability, negligence or otherwise, or from any use or operation of any methods, products, instructions, or ideas contained in the material herein.

ISBN: 978-0-12-802022-7

British Library Cataloguing in Publication Data
A catalogue record for this book is available from the British Library

Library of Congress Cataloging-in-Publication Data
A catalog record for this book is available from the Library of Congress

For Information on all Academic Press publications
visit our website at http://store.elsevier.com

Cover design provided by Amanda Toperoff of Big Blue Nowhere (www.bigbluenowhere.com).

Contents

Contributors

Eric J. Anderson
Great Lake Environmental Research Laboratory, Ann Arbor, MI, USA

Matthew R. Archer
Department of Ocean Sciences, Rosenstiel School of Marine and Atmospheric Science, University of Miami, FL, USA

Luiz Paulo de Freitas Assad
Universidade Federal do Rio de Janeiro, Rio de Janeiro, Brazil

Becky Baltes
U.S. IOOS Program Office, NOAA, Silver Spring, MD, USA

Cecília Bergman
PROOCEANO Serviço Oceanográfico, Rio de Janeiro, Brazil

Aric Bickel
Central and Northern California Ocean Observing System, Moss Landing, CA, USA

Eric P. Bjorkstedt
NOAA Fisheries, Southwest Fisheries Science Center, Santa Cruz, CA, USA

Ana Carolina Boechat
PROOCEANO Serviço Oceanográfico, Rio de Janeiro, Brazil

Jose Carlos Nieto Borge
Universidad de Alcalá, Spain

Marie Bundy
Office for Coastal Management/NOS/NOAA, Silver Spring, MD, USA

Edward J. Buskey
Marine Science Institute, University of Texas, Port Aransas, TX, USA

Marcelo Montenegro Cabral
PROOCEANO Serviço Oceanográfico, Rio de Janeiro, Brazil

Russell Callender
National Ocean Service, NOAA, Silver Spring, MD, USA

Ruben Carrasco
Institute of Coastal Research, Helmhotlz-Zentrum Geesthacht, Germany

Eugenio Pugliese Carratelli
Maritime Engineering Division University of Salerno (MEDUS), University of Salerno, Fisciano, Italy; CUGRI—University Centre for Research on Major Hazard, Fisciano, Italy

Matthew Carrier
Naval Research Laboratory, Stennis Space Center, Mississippi, USA

Melissa L. Carter
Scripps Institution of Oceanography, University of California, San Diego, CA, USA

Gabriel Vieira de Carvalho
PROOCEANO Serviço Oceanográfico, Rio de Janeiro, Brazil

Rubén Castro
Facultad de Ciencias Marinas, Universidad Autónoma de Baja California, Ensenada, Baja California, México

Jeremy Cothran
Arnold School of Public Health and the Baruch Institute for Marine and Coastal Sciences, University of South Carolina, Columbia, SC, USA

Leonardo Maturo Marques da Cruz
PROOCEANO Serviço Oceanográfico, Rio de Janeiro, Brazil

Fabio Dentale
Maritime Engineering Division University of Salerno (MEDUS), University of Salerno, Fisciano, Italy

L. Kellie Dixon
Mote Marine Laboratory, Sarasota, FL, USA

Feliciano Dominguez
Instituto de Investigaciones Oceanológicas, Universidad Autónoma de Baja California, Ensenada, Baja California, México

Jennifer Dorton
Center for Marine Science, University of North Carolina—Wilmington, Wilmington, NC, USA

Reginaldo Durazo
Facultad de Ciencias Marinas, Universidad Autónoma de Baja California, Ensenada, Baja California, México

Christopher A. Edwards
Department of Ocean Sciences, University of California Santa Cruz, Santa Cruz, CA, USA

Todd Fake
University of Connecticut, Storrs, CT, USA

Matthew C. Ferner
San Francisco Bay NERR, San Francisco State University, Tiburon, CA, USA

Jerome Fiechter
Department of Ocean Sciences, University of California Santa Cruz, Santa Cruz, CA, USA

Xavier Flores-Vidal
Instituto de Investigaciones Oceanológicas, Universidad Autónoma de Baja California, Ensenada, Baja California, México

Maurício da Rocha Fragoso
PROOCEANO Serviço Oceanográfico, Rio de Janeiro, Brazil

Charlton Galvarino
Second Creek Consulting, Columbia, SC, USA

Henery Ferreira Garção
PROOCEANO Serviço Oceanográfico, Rio de Janeiro, Brazil

Eduardo Gil
Instituto de Investigaciones Oceanológicas, Universidad Autónoma de Baja California, Ensenada, Baja California, México

Eric W. Gill
Faculty of Engineering and Applied Science, Memorial University, NL, Canada

Klaus-Werner Gurgel
Institute of Oceanography, University of Hamburg, Hamburg, Germany

Jack Harlan
NOAA/U.S. Integrated Ocean Observing System, Silver Spring, MD, USA

Lisa Hazard
Scripps Institution of Oceanography, La Jolla, CA, USA

Debra Hernandez
Southeast Coastal Ocean Observing Regional Association, Charleston, SC, USA

Jochen Horstmann
Institute of Coastal Research, Helmhotlz-Zentrum Geesthacht, Germany

Matthew K. Howard
Texas A&M University, College Station, TX, USA

Meredith D.A. Howard
Southern California Coastal Water Research Project, Costa Mesa, CA, USA

Weimin Huang
Faculty of Engineering and Applied Science, Memorial University, NL, Canada

Michael G. Jacox
Department of Ocean Sciences, University of California Santa Cruz, Santa Cruz, CA, USA

Benjamin Jaimes
Department of Ocean Sciences, Rosenstiel School of Marine and Atmospheric Science, University of Miami, FL, USA

Robert E. Jensen
US Army Corps of Engineers, Environmental Research and Development Center, Washington, DC, USA

Ann E. Jochens
Texas A&M University, College Station, TX, USA

Adrian Jones
University of Maryland Center for Environmental Science, Cambridge, MD, USA

Carolyn Keen
Scripps Institution of Oceanography, La Jolla, CA, USA

Heath Kelsey
University of Maryland Center for Environmental Science, Cambridge, MD, USA

Heather Kerkering
Pacific Islands Ocean Observing System, University of Hawaii at Manoa, Honolulu, HI, USA; Sea Connections Consulting, Virginia, USA

Barbara Kirkpatrick
Gulf of Mexico Coastal Ocean Observing System, College Station, TX, USA; Mote Marine Laboratory, Sarasota, FL, USA

Gary J. Kirkpatrick
Mote Marine Laboratory, Sarasota, FL, USA

Shinichi Kobara
Texas A&M University, College Station, TX, USA

Josh Kohut
Mid-Atlantic Regional Association Coastal Ocean Observing System and Rutgers University, Newport, NJ, USA

Raphael M. Kudela
Ocean Sciences Department, Institute for Marine Sciences, University of California, Santa Cruz, CA, USA

Luiz Landau
Universidade Federal do Rio de Janeiro, Rio de Janeiro, Brazil

Chad Lembke
College of Marine Science, University of South Florida, St. Petersburg, FL, USA

Lynn Leonard
Department of Geography and Geology, University of North Carolina—Wilmington, Wilmington, NC, USA

Yonggang Liu
College of Marine Science, University of South Florida, St. Petersburg, FL, USA

Giovanni Ludeno
Institute for Electromagnetic Sensing of the Environment (IREA), Italian National Research Council (CNR), Napoli, Italy; Department of the Industrial and Information Engineering, Second University of Naples, Aversa, Italy

Rick Luettich
University of North Carolina, Chapel Hill, NC, USA

Bjoern Lund
Rosenstiel School of Marine and Atmospheric Science, University of Miami, Coral Gables, FL, USA

John Manderson
National Marine Fisheries Service, NOAA, Silver Spring, MD, USA

Lívia Sant'Angelo Mariano
PROOCEANO Serviço Oceanográfico, Rio de Janeiro, Brazil

Jorge Martinez-Pedraja
Department of Ocean Sciences, Rosenstiel School of Marine and Atmospheric Science, University of Miami, FL, USA

Molly McCammon
Alaska Ocean Observing System, Anchorage, AK, USA

Clifford R. Merz
College of Marine Science, University of South Florida, St. Petersburg, FL, USA

Tiago Cardoso de Miranda
PROOCEANO Serviço Oceanográfico, Rio de Janeiro, Brazil

Andrew M. Moore
Department of Ocean Sciences, University of California Santa Cruz, Santa Cruz, CA, USA

Philip Muscarella
Naval Research Laboratory, Stennis Space Center, Mississippi, USA

Antonio Natale
Institute for Electromagnetic Sensing of the Environment (IREA), Italian National Research Council (CNR), Napoli, Italy

Luis F. Navarro
Instituto de Investigaciones Oceanológicas, Universidad Autónoma de Baja California, Ensenada, Baja California, México

Hans Ngodock
Naval Research Laboratory, Stennis Space Center, Mississippi, USA

André Luis Santi Coimbra de Oliveira
PROOCEANO Serviço Oceanográfico, Rio de Janeiro, Brazil

Mark Otero
Scripps Institution of Oceanography, La Jolla, CA, USA

Júlio Augusto de Castro Pellegrini
PROOCEANO Serviço Oceanográfico, Rio de Janeiro, Brazil

Leif Petersen
Helzel Messtechnik GmbH, Kaltenkirchen, Germany

William T. Peterson
NOAA Fisheries, Northwest Fisheries Science Center, Seattle, WA, USA

Flávia Pozzi Pimentel
PROOCEANO Serviço Oceanográfico, Rio de Janeiro, Brazil

Dwayne E. Porter
Arnold School of Public Health and the Baruch Institute for Marine and Coastal Sciences, University of South Carolina, Columbia, SC, USA

Josie Quintrell
IOOS Association, Harpswell, ME, USA

Dan Ramage
Arnold School of Public Health and the Baruch Institute for Marine and Coastal Sciences, University of South Carolina, Columbia, SC, USA

Jennifer Read
University of Michigan Water Center, Ann Arbor, MI, USA; Great Lakes Observing System, Ann Arbor, MI, USA

Ferdinando Reale
Maritime Engineering Division University of Salerno (MEDUS), University of Salerno, Fisciano, Italy

William G. Reay
Virginia Institute of Marine Science, Gloucester Point, VA, USA

Frederico Luna Rinaldi
PROOCEANO Serviço Oceanográfico, Rio de Janeiro, Brazil

Leslie Rosenfeld
Central and Northern California Ocean Observing System, Moss Landing, CA, USA

Moninya Roughan
Coastal and Regional Oceanography Lab, School of Mathematics and Statistics, UNSW Australia, UNSW, Sydney, NSW, Australia

Francisco Alves dos Santos
PROOCEANO Serviço Oceanográfico, Rio de Janeiro, Brazil

Natalia Gomes dos Santos
PROOCEANO Serviço Oceanográfico, Rio de Janeiro, Brazil

Amandine Schaeffer
Coastal and Regional Oceanography Lab, School of Mathematics and Statistics, UNSW Australia, UNSW, Sydney, NSW, Australia

Oscar M. Schofield
Institute of Marine and Coastal Sciences, Rutgers University, New Brunswick, NJ, USA

David J. Schwab
University of Michigan Water Center, Ann Arbor, MI, USA

Jorg Seemann
Institute of Coastal Research, Helmhotlz-Zentrum Geesthacht, Germany

Francesco Serafino
Institute for Electromagnetic Sensing of the Environment (IREA), Italian National Research Council (CNR), Napoli, Italy

Justin Shapiro
Mote Marine Laboratory, Sarasota, FL, USA

Lynn K. Shay
Department of Ocean Sciences, Rosenstiel School of Marine and Atmospheric Science, University of Miami, FL, USA

Christina Simoniello
Texas A&M University, Based at University of South Florida, College of Marine Science, St. Petersburg, FL, USA

Erik Smith
North Inlet Winyah Bay NERR, Georgetown, SC, USA; The Baruch Institute for Marine and Coastal Sciences, University of South Carolina, Columbia, SC, USA

Scott Smith
Naval Research Laboratory, Stennis Space Center, Mississippi, USA

Felipe Lobo Mendes Soares
PROOCEANO Serviço Oceanográfico, Rio de Janeiro, Brazil

Innocent Souopgui
Department of Marine Science, University of Southern Mississippi, Stennis Space Center, Mississippi, USA

Michael Spranger
University of Florida, Gainesville, FL, USA

Richard P. Stumpf
National Centers for Coastal Ocean Science, NOAA, Silver Spring, MD, USA

Vembu Subramanian
Southeast Coastal Ocean Observing Regional Association, Charleston, SC, USA

Iain M. Suthers
School of Biological, Earth and Environmental Sciences, UNSW Australia, Sydney, NSW, Australia

Eric Terrill
Scripps Institution of Oceanography, La Jolla, CA, USA

Julie Thomas
Scripps Institution of Oceanography, La Jolla, CA, USA

Pedro Marques São Tiago
PROOCEANO Serviço Oceanográfico, Rio de Janeiro, Brazil

Michelle Tomlinson
University of North Carolina, Chapel Hill, NC, USA

Dwight Trueblood
Office for Coastal Management/NOS/NOAA, University of New Hampshire, Durham, NH, USA

Stephanie Watson
Gulf of Mexico Coastal Ocean Observing System Consultant, Based at Stennis, MS, USA

Robert H. Weisberg
College of Marine Science, University of South Florida, St. Petersburg, FL, USA

Zdenka Willis
U.S. Integrated Ocean Observing System Office, Silver Spring, MD, USA

Lucy R. Wyatt
ACORN, College of Science, Technology and Engineering, Centre for Tropical Water and Aquatic Ecosystem Research, James Cook University, Townsville, QLD, Australia; School of Mathematics and Statistics, University of Sheffield, Sheffield, UK

Lianyuan Zheng
College of Marine Science, University of South Florida, St. Petersburg, FL, USA

Preface

Whether far or near, the ocean plays a role in everyone's life. The ocean drives weather and climate patterns across the globe, hosts an abundance of wildlife that support fishing industries and provide food for the world, serves as a highway for vessels that deliver everyday materials, and supports economies as a tourism destination. A healthy relationship with the oceans requires that we understand it. One way to understand it is through observing.

In 2004, the Pew Ocean Commission and the U.S. Ocean Commission released reports that stressed the need for expanding coastal ocean observing capabilities and improving collaborations among entities collecting ocean information. The recommendation mirrored similar initiatives in other countries, as well as set the stage for other countries to follow suit. In response to these recommendations, there are now regional, national, and global ocean observing systems designed to provide critical coastal and ocean information for decision-making. *Coastal Ocean Observing Systems: Advances and Syntheses* highlights the system development, scientific discoveries, technology advancements, societal benefits, and partnerships made over the last decade of growth and international support of coastal ocean observing entities.

The contents of this book were originally derived from the 2014 Ocean Sciences Meeting Session #009 entitled, "Scientific and Societal Benefits from Integrated Coastal Ocean Observations and Networked Marine Laboratories." The session included 24 presentations from different academic institutions, private sectors, government agencies, and observing programs. The presentations covered various aspects of the development and value of coastal ocean observing systems and provided a state-of-the-art overview of the science and technology in this field. Some of the book chapters are contributed by the participants and are based on their presentation topics and developments since. Other invited chapters cover lessons learned in building effective coastal observing systems and new technologies; improving, analyzing, and sharing acquired data; applying observed data to advance the science of coastal oceanography; creating interagency and interinstitutional partnerships at regional and international scales; and in developing decision-making products that greatly impact our economy, society, and environment.

The book includes worldwide examples of advancement in coastal observing systems. It features a collection from international academics, managements, and industries on the latest developments in several emerging issues of coastal ocean observing systems, with a focus on reporting scientific and technological knowledge gained and the resulting societal benefits.

The audience for this book may be as diverse as the individuals involved in moving a coastal ocean observing system from an idea to an operational, beneficial entity. Government and academic leaders, scientists, oceanographers, and ocean engineers will find value in the coastal ocean observing system developments, scientific syntheses, and technology advancements. Resource managers, students, state and federal agencies, and legislators, for example, will also find value in learning more about the societal benefits and users of coastal ocean observing systems and data. The chapters that highlight lessons learned on partnerships, governance, structure, data management, and stakeholder relationships may act as guidance to parties interested in establishing, restructuring, or improving similar systems. Additional readers may include the media and general public interested in marine science, ocean observing, and the coastal environment.

Yonggang Liu, Heather Kerkering, and Robert H. Weisberg
Editors

Acknowledgment

The editors would like to extend their sincere thanks to Drs Robert A. Weller (Woods Hole Oceanographic Institution, USA), Stefano Vignudelli (Consiglio Nazionale delle Ricerche, Area Ricerca CNR, Italy), Shuqun Cai (South China Sea Institute of Oceanography, Chinese Academy of Science, China), and Thomas Helzel (HELZEL Messtechnik GmbH, Germany) for their kind encouragement, constructive comments, and helpful suggestions during the early stage of the book development. The editors also gratefully thank all the authors for contributing their excellent work to this collection and for their cooperation during the peer-review and revision processes.

Each chapter in the book was peer-reviewed to Elsevier standards by two or more anonymous reviewers, selected by the editors in terms of experiences and knowledge of the topic. The editors sincerely thank the 45 reviewers for their valuable volunteer work and insightful comments that helped improve the quality of the book. The following is a list of those anonymous reviewers who agreed to be acknowledged here:

Alexander, Charles	Office of Marine and Aviation Operations, National Oceanic and Atmospheric Administration, USA
Archer, Matthew	Rosenstiel School of Marine and Atmospheric Science, University of Miami, USA
Arellano, Ana R.	College of Marine Science, University of South Florida, USA
Beard, Russ	National Oceanic and Atmospheric Administration, National Coastal Data Development Center, USA
Beardsley, Robert C.	Woods Hole Oceanographic Institute, USA
Beck, Marcus	U.S. Environmental Protection Agency, USA
Blumberg, Alan F.	Davidson Laboratory, Stevens Institute of Technology, USA
Chen, Zhiqiang	South Florida Water Management District, USA
Cosoli, Simone	Istituto Nazionale di Oceanografia e Geofisica Sperimentale, Italy
Flampouris, Stylianos	Naval Research Laboratory, University of Southern Mississippi, USA
Horstmann, Jochen	Department of Radar Hydrography, Helmholtz-Zentrum Geesthacht, Germany
Huang, Weimin	Department of Electrical Engineering and Computer Engineering, Memorial University, Canada
Kirkpatrick, Barbara	Gulf of Mexico Coastal Ocean Observing System, Mote Marine Laboratory, USA
Le Hanaff, Matthieu	Rosenstiel School of Marine and Atmospheric Science, University of Miami, USA

Li, Zhijin	Jet Propulsion Laboratory, California Institute of Technology, USA
McCabe, Ryan M.	Joint Institute for the Study of the Atmosphere and Ocean, University of Washington, USA
McCammon, Molly	Alaska Ocean Observing System, USA
Merz, Clifford R.	College of Marine Science, University of South Florida, USA
Moltmann, Tim	Integrated Marine Observing System, University of Tasmania, Australia
Morrison, J. Ru	Northeastern Regional Association of Coastal Ocean Observing Systems, USA
Paduan, Jeffrey D.	Naval Postgraduate School, USA
Oliver, Matthew	College of Earth, Ocean and Environment, University of Delaware, USA
Pan, Chudong	Harbor Branch Oceanographic Institute, Florida Atlantic University, USA
Proctor, Roger	Integrated Marine Observing System, University of Tasmania, Australia
Quintrell, Josie	The U.S. Integrated Ocean Observing System (IOOS) Association, USA
Rivas, David A.	Centro de Investigación Científica y de Educación Superior de Ensenada (CICESE), Departamento de Oceanografía Biológica, Mexico
Rubio, Anna	AZTI-Tecnalia, Marine Research Division, Herrera Kaia Portualdea, Spain
Senet, Christian	Bundesamt fuer Seeschifffahrt und Hydrographie, Germany
Solabarrieta, Lohitzune	AZTI-Tecnalia, Marine Research Division, Herrera Kaia Portualdea, Spain
Subramanian, Vembu	Southeast Coastal Ocean Observing Regional Association, USA
Vann, Timi S.	National Oceanic and Atmospheric Administration, USA
Vicen-Bueno, Raúl	Signal Theory and Communications Department, Superior Polytechnic School, University of Alcalá, Spain
Vilibic, Ivica	Institute of Oceanography and Fisheries, Physical Oceanography Laboratory, Croatia
Wee, Brian	National Ecological Observatory Network, Inc., Smithsonian Institution, USA
White, David L.	College of Engineering and Science, Clemson University, USA
Zhang, Aijun	Center for Operational Oceanographic Products & Services, National Ocean Service, National Oceanic and Atmospheric, USA
Zhang, Minwei	College of Marine Science, University of South Florida, USA
Zhao, Jun	Masdar Institute of Science and Technology, United Arab Emirates

CHAPTER

Introduction to Coastal Ocean Observing Systems

1

Yonggang Liu[1,*], Heather Kerkering[2,3], Robert H. Weisberg[1]

College of Marine Science, University of South Florida, St. Petersburg, FL, USA[1]; Sea Connections Consulting, Virginia, USA[2]; Pacific Islands Ocean Observing System, University of Hawaii at Manoa, Honolulu, HI, USA[3]

**Corresponding author: E-mail: yliu@mail.usf.edu*

CHAPTER OUTLINE

1. INTRODUCTION

The ocean plays a role in everyone's life. The ocean affects weather and climate patterns around the globe, hosts an abundance of wildlife that support fishing industries and provide food for the world, serves as a highway for vessels that deliver everyday materials, and supports economies as a tourism destination. The coastal ocean is the part of the earth system where land, water, air, and people meet together. Populations, businesses, and infrastructure are increasing along coastlines, which are all susceptible to changing coastal ocean conditions. Now, more than ever, there is a need for regional to global observing systems that can provide accurate real-time data and forecasts on coastal ocean conditions.

Coastal ocean observing systems (COOS) are necessary for advancing our understanding on the state of the coastal ocean worldwide and its impact on matters of societal importance. These systems integrate a network of people, organizations, technologies, and data to share advances, improve research capabilities, and provide decision-makers with access to information and scientific interpretations. Data, observations, and models integrated into the COOS come from a variety of platforms, including, for example, moorings, high-frequency (HF) radars, underwater gliders and profilers, satellites, and ships. The resulting data are used to better understand, respond to, and prepare for short-term events such as oil spills, harmful algal blooms, and fish kills, longer term changes in our oceans resulting in acidification, hypoxia, and sea level rise, and in everyday decisions related to maritime operations, public health, and management of healthy ecosystems.

COOS advances have benefited from an evolving set of ocean observing efforts. In the United States, for example, the concept of the U.S. Integrated Ocean Observing System (IOOS®) was developed through the establishment of Ocean.US under the auspices of the National Oceanographic Partnership Program (NOPP) to help coordinate emerging activities.[1–3] The envisioned concept was a coordinated national and international network of observations, data management, and analyses that systematically acquired and disseminated data and information on past, present, and future states of the oceans.[4,5] Its coastal component was designed to assess and predict the effects of weather, climate, and human activities on the state of the coastal ocean, its ecosystems and living resources, and on the nation's economy.[6] In 2003 and 2004, the Pew Oceans Commission and the U.S. Commission on Ocean Policy, respectively, released reports[7,8] that further identified the need for a national integrated ocean observing system. The Bush Administration responded with the U.S. Ocean Action Plan,[9] which called for the establishment of a U.S. IOOS and recognized it as a major contribution to the Global Ocean Observing System (GOOS). In March of 2009, President Obama signed the Integrated Coastal and Ocean Observation System Act (ICOOS Act),[10] establishing statutory authority for the development of the U.S. IOOS and mandating the establishment of a national integrated system coordinated at the federal level. Through the ICOOS Act, NOAA became the lead federal agency of the U.S. IOOS. Additionally, the Obama Administration established the Ocean Policy Task Force in 2009 and, in 2010, signed an executive order that adopted the task force's final recommendations, which included strengthening and integrating federal and nonfederal ocean observing systems into a national system and integrating the national system into international observation efforts.[11] These actions highlight the belief in the value of integrated observing systems. Today, U.S. IOOS contributes to the observing strategies implemented around the globe and plays a major role in global coordination and strategic planning. Coastal ocean observing efforts are implemented through the efforts of regional programs distributed around the US coastal regions. The methods, processes, and lessons learned in establishing these regional associations are documented in a number of journal articles.[12–14] Valuable lessons learned from establishing a regional COOS, e.g., the Southeast Coastal Ocean Observing System (SEACOOS, now SECOORA), were reported in a special issue of the Marine Technology Society Journal.[13,14] U.S. IOOS achievements are also found in two other special issues of Marine Technology Society Journal ("United States Integrated Ocean Observing System: Our

Eyes on Our Oceans" and "Coasts and Great Lakes: Part I & Part II"). These special issues, however, focused mainly on the US coastal waters.[15–17] Only a few papers were devoted to coastal ocean observing activities in other countries, e.g., the European and Australian coasts.[18,19]

This book serves as a collection of state-of-the-art information on the advances and syntheses of the COOS in the United States and beyond, including international partners and success stories. The chapters are contributed by a wide range of authors from both research and management communities, and the content includes COOS development, scientific findings, technology advancements, data management, as well as societal benefits of the coastal ocean observing systems. The goal is to offer best practices, lessons learned, and main achievements that could advise and guide stakeholders, business, and science communities in developing and utilizing evolving COOS information resources.

This book is loosely organized as follows: The chapters on the topic of COOS development are arranged in the first part of the book, followed by the chapters on scientific syntheses and technology advancement, which include HF radar and autonomous underwater vehicles (AUV) applications and data analyses, as well as data assimilation experiments. The chapters on the societal benefits of the coastal ocean observing systems are arranged in the last part of the book. Interested readers can quickly find the relevant chapters of specific interest. The main topics of the book chapters are briefly summarized in the next three sections.

2. COASTAL OCEAN OBSERVING SYSTEMS DEVELOPMENT

Most large-scale coastal ocean observing systems are funded through national governments for their own interests, often with different foci, but the world's oceans are connected, therefore partnering is the key to success. Willis[20] discusses how the U.S. IOOS, Australia's Integrated Marine Observing System (IMOS), and Canada's Ocean Tracking Network (OTN) are progressing in their respective regions and are working together to observe and compile ocean information in a way that is easily accessible to scientists and managers.

Using the U.S. IOOS as an example, Quintrell et al.[21] discuss the importance of federal and regional partnerships in coastal ocean observing. The U.S. IOOS was designed by Congress to be a partnership of 17 Federal agencies and enacted through 11 regional systems. Through collaborative projects and shared objectives, these partnerships and regional systems are improving the understanding of the coastal ocean environment, increasing data available to modeling and analysis, improving forecasting capabilities, and improving decision support tools.

For a specific example in the eastern Gulf of Mexico, Weisberg et al.[22] present a comprehensive coastal ocean observing and modeling system for the west coast of Florida based on the lessons learned through their sustained long-term coastal ocean observing and modeling efforts over the last two decades. They develop a rationale, offer a system design, and argue that describing and understanding how the coastal ocean works is a prerequisite to predicting the outcomes of either natural or anthropogenic occurrences, and they provide a set of representative examples. The proposed comprehensive coastal ocean observing system may also serve as a guideline for similar systems elsewhere.

For the US West Coast, Kudela et al.[23] report on the development of the California Harmful Algal Bloom Monitoring and Alert Program (CalHABMAP) through ocean observing. The program is an integrated, statewide, harmful algal bloom monitoring and alert network coordinated by organizations and researchers currently collecting HAB data and developing a centralized portal for the dissemination of this information. With the main goal of implementing a statewide HAB network and forecasting system for California, and potentially the US West Coast, CalHABMAP has succeeded in highlighting the need for a coordinated network and serves as partner for regional and national efforts led by the NOAA National Ocean Service, the IOOS, and the NASA Applied Sciences Program.

COOS are in development beyond the US coastlines. For example, the Australian IMOS was formed in 2007, with equipment deployed from the next year onward. Scientific "nodes" were formed broadly around state boundaries with both nationally unified overarching science goals and local priorities. Roughan et al.[24] report the NSW-IMOS as an example of a successfully implemented ocean observing system along the coast of southeastern Australia. The current observational array is designed around pertinent science questions, leveraged existing data streams, and opportunities for further oceanographic research.

On the Brazil coast, dos Santos et al.[25] present a newly developed coastal ocean observing system in an active oil and gas area offshore of southeastern Brazil. Since the three-year operational oceanography pilot program for Santos Basin (Projeto Azul) began in 2013, they have collected a variety of data in the Brazil Current meander and Cabo Frio eddy and started data assimilation experiments. Their goal is to build and sustain an industry-oriented coastal ocean observing system.

Bjorkstedt and Peterson[26] report on their observations of zooplankton communities and their environment from monthly (or more frequent) 12-h cross-shelf transects in the northern California Current. The authors argue that the "old-fashioned" approaches to ocean observing—going out to sea on a regular and frequent basis to sample the system—are as holistically as practicable using relatively simple methods. Thus, they enrich the contributions of ocean observing systems to ecosystem-based management in the northern California Current. Challenges and opportunities of the frequently conducted coastal transects in ocean observing systems are also discussed.

3. SCIENCE AND TECHNOLOGY ADVANCEMENT

3.1 HF RADAR APPLICATIONS

Shore-based HF radars are a mainstay of many COOS. Operating within a frequency band of 3 to 30 MHz, HF-radar networks are capable of mapping offshore surface currents out to ranges approaching 200 km and with a horizontal resolution of a few kilometers. They may also be used to estimate ocean surface waves. The ability of HF radars to map ocean surface currents and waves over a two-dimensional area, in an operational long-term deployment, even in severe weather conditions, makes them a unique and powerful tool of coastal ocean observing.

Wyatt[27] reports on the advance of Australian Coastal Ocean Radar Network (ACORN) as a facility of the Australian IMOS for coastal ocean monitoring. ACORN is currently operating 12 radars arranged in six pairs, and the network is expanding. Examples and analysis on the progress of data products integrating HF radar–measured surface currents, waves, and winds are presented.

Thomas et al.[28] discuss the societal benefits of high-resolution wave measurements and HF radar–derived surface currents to maritime operations, emergency responders, and the coastal recreation community. They also highlight the contributions of the Coastal Data Information Program (CDIP) and the Coastal Observing Research and Development Center (CORDC) on those measurements, respectively. CDIP manages and ingests data from over 60 coastal wave buoys, supporting nearshore navigation. CORDC manages the data acquisition and near real-time processing of the U.S. High Frequency Radar network (HFRNet), a network that includes numerous participating organizations. These programs work in partnership with the U.S. IOOS.

Archer et al.[29] present recent results and benefits of using radar to investigate shear-zone instability along the frontal regions of the Florida Current, a rapidly evolving western boundary current. They investigated the flow field kinematics of a cyclonic submesoscale frontal eddy and analyzed a near-inertial velocity signal along the anticyclonic flank of the Florida Current, which would be difficult to capture with only ship and moored point measurements.

Flores-Vidal et al.[30] report on a study of fine-scale tidal and subtidal variability in an upwelling-influenced bay using HF radar and surface drifter data. They found that two main factors influence the drifter trajectory distribution. One factor is the impinging of the California Coastal Current, which develops a barrier, and the other is the wind field that is confined by the surrounding mountain chain and develops a shadow-like zone inside the bay.

Merz et al.[31] discuss the effects of spatial/temporal radio frequency interference (RFI) variations of two nearby HF radar sites deployed along the West Florida coast, initially observed via their uneven data storage fill rates. Their experiments show that the application of WERA's "listen before talk" adaptive algorithm, along with a wide enough bandwidth to operate within, can increase data coverage and signal-to-noise ratio. This topic is of value not only to the WERA user but also as general information to the overall HF radar and integrated coastal and ocean observing communities.

Another type of ocean remote sensing system includes X-band shipborne nautical radar operating at a much higher frequency (around 10 GHz). The radar can be operated from a variety of platforms, such as coastal stations, offshore platforms, as well as moving vessels. The following three chapters are devoted to this topic.

Huang and Gill[32] introduce the applications of the X-band radar along the eastern Canadian coast. They describe the methods for extracting sea surface wave information and wind parameters from the radar images, and they present an algorithm for improving the extraction of wind speed from rain-contaminated radar images.

Ludeno et al.[33] provide a review of the state-of-the-art algorithms of X-band radar in estimating nearshore bathymetry. They discuss the limitations of the algorithms in deep water applications, and adopt a "correlation" procedure to estimate the sea water depth. Examples of X-band radar applications are shown for Italy and Germany. They determined that X-band accuracy is fairly adequate for shallow water, but it decreases significantly as the depth increases. They suggest that the algorithm is applicable to water depths up to 20 m.

Horstmann et al.[34] systematically describe X-band marine radar applications in wind, wave, and current retrieval. Overall, they report that winds, waves, and currents are more accurately measured using X-band radar than traditional sensors. Advantages and limitations of such observations are discussed.

3.2 GLIDER APPLICATIONS

Gliders and other AUVs are another coastal ocean observing system asset. Gliders are operated remotely, travel long distances, and cover a large range of depths. They serve as a convenient platform for a variety of ocean sensors, such as temperature, conductivity, dissolved oxygen, and various bio-optical measures.

Quality control and quality assurance of profiled data are required for oceanographic analyses. For example, glider salinity data may have errors around thermocline if unpumped CTD sensors are used. Although new thermal lag correction methods are powerful for adjusting the mismatches of the downcast and upcast glider salinity profiles in weakly stratified ocean waters or weak thermoclines, Liu et al.[35] found that they were not very effective in correcting the salinity errors in the case of a sharp thermocline. Based on the CTD data collected by an autonomous underwater glider on the West Florida Shelf, Liu et al.[35] propose an improved method of glider salinity error correction that can effectively remove these salinity spikes. They also suggest practical procedures of glider salinity correction that are especially useful for glider applications in waters of strong stratification and sharp thermocline.

New sensors for the AUV platform are under development. Shapiro et al.[36] report a new sensor for AUVs, the Optical Phytoplankton Discriminator (OPD), which can discriminate phytoplankton community structure and light absorption of chromophoric dissolved organic matter. Identification and quantification of phytoplankton are important because this may help to determine causation and possible effects.

3.3 DATA ASSIMILATION EXPERIMENTS

For COOS, forecasting ocean conditions may be improved through formal techniques of data assimilation, i.e., integrating observations with model simulations to continually improve the initial conditions of the forecasts. Two chapters are devoted to this topic.

Moore et al.[37] explore the impact of different observing platforms and control vector elements on four-dimensional variational (4D-Var) analyses of California Current transport using the adjoint Kalman gain matrix to map a transport metric into observation space. They provide a direct quantitative measure of the observing system impact on ocean state estimates spanning three decades and reveal the complex interplay between different observing platforms within the 4D-Var analyses as different observing systems become available.

Ngodock et al.[38] report on their findings of the impact of HF radar observations on constraining and improving model forecasts of the coastal ocean circulation in the Mid-Atlantic region of the US east coast using a very high-resolution Navy Coastal Ocean Model[39] and a 4D-Var data assimilation system. They find that the assimilation system can propagate the influence of these surface velocity measurements through all the model variables in space and time. The benefits and limitations of using high-resolution models are also tested and discussed.

4. SOCIETAL BENEFITS

Deriving societal benefits is at the heart of COOS. System priorities are guided by stakeholder needs. In fact, funding and Congressional support for these systems hinges on the system's ability to demonstrate its value to society. As noted earlier, there are many examples of societal benefits in the previously published special issues on IOOS in the Marine Technology Society Journal (2008, 2010, and 2011). Three chapters are devoted to demonstrating societal value, although as will be apparent upon reading, all previous chapters provide examples of societal benefits.

Buskey et al.[40] discuss benefits resulting from the U.S. National Estuarine Research Reserve System (NERRS) monitoring program. Established in 1995, this monitoring program develops quantitative measurements of short-term variability and long-term changes in abiotic and biotic properties of estuarine ecosystems for the purpose of informing effective coastal management. It also generates a national database on estuarine ecosystems. Buskey et al.[40] demonstrate how these data inform coastal managers on issues such as water quality assessment, habitat mapping and change analysis, establishment of nutrient criteria for estuaries, and understanding the predicted impacts of climate change.

Porter et al.[41] review the state of observing system efforts from the U.S. Southeast Coastal Ocean Observing Regional Association (SECOORA). As one of the 11 regional associations in the U.S. IOOS, the SECOORA implement a cohesive Regional Coastal Ocean Observing System (RCOOS) for the southeast United States, encompassing coastal waters of North Carolina, South Carolina, Georgia, and Florida (including the southeast Atlantic seaboard, the Straits of Florida, and the eastern Gulf of Mexico). Porter et al[41] present case studies demonstrating the value of integrating data from these systems to support marine safety, water quality, and ecosystem management decision making. They also provide valuable recommendations for the path forward for the SECOORA.

Society benefits of a regional coastal ocean observing system are also discussed by Simoniello et al.[42] using the Gulf of Mexico Coastal Ocean Observing System (GCOOS) as an example. Particularly, the economic benefits of the GCOOS ocean monitoring systems are quantified using an economic model, and the return on investment is also assessed. Their results demonstrate that, from an economic perspective, the nominal investment made to date in observing systems results in great value to society.

5. CONCLUDING REMARKS

Albeit based on only a small subset of coastal ocean observing system activities worldwide, the chapters in their composite demonstrate the importance of and the progress made in implementing and utilizing coastal ocean observations for the benefits of society. Effective environmental stewardship must be based on defensible science, which in turn requires observations and hypothesis testing. Coastal ocean observing systems are, therefore, prerequisites for reliably forecasting results from either natural or human-induced perturbations to the coastal ocean. Regardless of the concerns, from damage by severe weather, mishaps from oil and gas operations, management of living marine resources to the simple, aesthetic enjoyment of nature, the advancement of a coastal ocean observing system agenda will make a positive contribution to society.

But the development of a coastal ocean observing system is no small task. Required are partnerships across agencies (federal, state, and local), the private sector, and academia, in addition to international cooperation. The challenges are, therefore, deep and complex, which may help to explain the pace of development. As demonstrated by the contributing authors to this book, success stories come from challenges.

As editors, we hope this book sheds light and provides ample evidence on the value of coastal ocean observing system advances and the importance of building and sustaining them into the future.

ACKNOWLEDGMENTS

Partial salary support to Weisberg and Liu derives from the Southeast Coastal Ocean Observing Regional Association (SECOORA) as a pass through from NOAA Grant # NA11NOS0120033, the NASA Ocean Surface Topography Science Team (OSTST) grant # NNX13AE18G, and the Gulf of Mexico Research Institute through the Florida State University Deep-C Program. This is CPR Contribution 41. Kerkering received salary support from the Pacific Islands Ocean Observing System through Cooperative Agreement #NA11NOS0120039, with NOAA National Ocean Service.

REFERENCES

1. Ocean.US. *An integrated and sustained ocean observing system (IOOS) for the United States: design and implementation*. Ocean.US Publication; May 2002.
2. Lindstrom EJ. Establishing an integrated ocean observing system for the United States. *Mar Technol Soc J* 2003;**37**:28–31.
3. Clark AM. On integrating and sustaining a national ocean observing system. *Mar Technol Soc J* 2003;**37**:3–4.
4. Hankin S, Bernard L, Cornillon P, Grassle F, Legler D, Lever J, et al. Designing the data and communications infrastructure for the U.S. integrated ocean observing System. *Mar Technol Soc J* 2003;**37**:51–4.
5. O'Keefe S. Facing the challenges of an integrated ocean observing system. *Mar Technol Soc J* 2003;**37**:7–8.

6. Seim HE. SEA-COOS: a model for a multi-state, multi-institutional regional observation system. *Mar Technol Soc J* 2003;**37**:92–101.
7. Pew Oceans Commission. *America's living oceans: charting a course for sea change, a report to the nation recommendations for a new ocean policy.* Arlington, Virginia: Pew Oceans Commission; May 2003.
8. U.S. Commission on Ocean Policy. *An ocean blueprint for the 21st century. Final report.* Washington, DC, ISBN 0-9759462-0-X.
9. United States Government. *U.S. Ocean action plan: the Bush administration's response to the U.S. ocean commission on ocean policy.* Washington D.C. Exec. Order No. 13366, 3. C.F.R; December 17, 2004.
10. Interagency Ocean Observation Committee. *The integrated coastal and ocean observation system act of 2009.* Pub. L. No. 111-11, 33. U.S.C; 2009. §3601-3610. Available at:, http://www.iooc.us/ocean-observing-legislation/integrated-coastal-ocean-observation-act-of-2009/.
11. United States Government. *Stewardship of the ocean, our coasts, and the Great Lakes.* Executive Order No. 13547, 3. C.F.R; July 19, 2010. Available at:, http://www.whitehouse.gov/the-press-office/executive-order-stewardship-ocean-our-coasts-and-great-lakes.
12. Briscoe MG, Martin DL, Malone TC. Evolution of regional efforts in international GOOS and U.S. IOOS. *Mar Technol Soc J* 2008;**42**:4–9.
13. Seim HE, Nelson J, Fletcher M, Mooers CNK, Spence L, Weisberg RH, et al. SEACOOS program management. *Mar Technol Soc J* 2008;**42**:11–27.
14. Nelson J, Weisberg RH. In situ observations and satellite remote sensing in SEACOOS: program development and lessons learned. *Mar Technol Soc J* 2008;**42**:41–54.
15. Cowles T, Delaney J, Orcutt J, Weller R. The ocean observatories initiative: sustained ocean observing across a range of spatial scales. *Mar Technol Soc J* 2010;**44**:54–64.
16. Read J, Klump V, Johengen T, Schwab D, Paige K, Eddy S, et al. Working in freshwater: the Great Lakes observing system contributions to regional and national observations, data infrastructure, and decision support. *Mar Technol Soc J* 2010;**44**:84–98.
17. Pettigrew NR, Fikes CP, Beard MK. Advances in the ocean observing system in the gulf of Maine: technical capabilities and scientific results. *Mar Technol Soc J* 2011;**45**: 85–97.
18. Proctor R, Howarth J. Coastal observatories and operational oceanography: a European perspective. *Mar Technol Soc J* 2008;**42**:10–3.
19. Hill K, Moltmann T, Proctor R, Allen S. The Australian Integrated Marine observing system: delivering data streams to address national and international research priorities. *Mar Technol Soc J* 2010;**44**:65–72.
20. Willi Z. *National effort in a global context*; 2015 [this book].
21. Quintrell J, Luettich R, Baltes B, Kirkpatrick B, Stumpf RP, Schwab DJ, et al. *The importance of federal and regional partnerships in coastal observing*; 2015 [this book].
22. Weisberg RH, Zheng L, Liu Y. *Basic tenets for coastal ocean ecosystems monitoring: a west Florida perspective, coastal ocean observing systems: advances and syntheses.* Elsevier; 2015 [this book].
23. Kudela RM, Bickel A, Carter ML, Howard MDA, Rosenfeld L. *The monitoring of harmful algal blooms through ocean observing: the development of CalHABMAP*; 2015 [this book].
24. Roughan M, Schaeffer A, Suthers I. *Sustained ocean observing along the coast of southeastern Australia: NSW-IMOS*; 2015. 2007–2014. [this book].

25. dos Santos FA, Oliveira ALS, Soares FLM, Carvalho GV, São Tiago PM, Santos NG, et al. *Projeto Azul: operational ocean observation and forecasts in an active oil and gas area southeastern Brazil*; 2015 [this book].
26. Bjorkstedt E, Peterson B. *The value of long term time series of high-frequency sampling of hydrography, plankton and krill along multiple transects in shelf and slope waters of the northern California current*; 2015 [this book].
27. Wyatt LR. *The IMOS ocean radar facility.* ACORN; 2015 [this book].
28. Thomas J, Hazard L, Jensen RE, Otero M, Terrill E, Keen C, et al. *How high resolution wave observations and HF radar derived surface currents are critical to decision-making for maritime operations?* 2015 [this book].
29. Archer MR, Shay LK, Jaimes B, Martinez-Pedraja J. *Observing frontal instabilities of the Florida current using high frequency radar*; 2015 [this book].
30. Flores-Vidal X, Castro R, Navarro LF, Dominguez F, Gil E. *Fine scale tidal and subtidal variability on an upwelling influenced bay as measured by the Mexican high frequency radar observing system*; 2015 [this book].
31. Merz CR, Liu Y, Gurgel K-W, Petersen L, Weisberg RH. *Effect of radio frequency interference (RFI) noise energy on WERA performance using the "Listen before Talk" adaptive noise procedure*; 2015 [this book].
32. Huang W, Gill EW. *Ocean remote sensing using X-band shipborne nautical radar - applications in eastern Canada*; 2015 [this book].
33. Ludeno G, Reale F, Dentale F, Pugliese Carratelli E, Natale A, Serafino F. *Estimating nearshore bathymetry from X-band radar data*; 2015 [this book].
34. Horstmann J, Nieto Borge JC, Seemann J, Carrasc R, Lund B. *Wind, wave and current retrieval utilizing X-band marine radars*; 2015 [this book].
35. Liu Y, Weisberg RH, Lembke C. *Glider salinity correction for unpumped CTD sensors across a sharp thermocline*; 2015 [this book].
36. Shapiro J, Dixon LK, Schofield OM, Kirkpatrick BA, Kirkpatrick GJ. *Sensors for ocean observing: the optical phytoplankton discriminator*; 2015 [this book].
37. Moore AM, Edwards CA, Fiechter J, Jacox MG. *Observing system impacts on estimates of California current transport*; 2015 [this book].
38. Ngodock H, Muscarella P, Carrier M, Souopgui I, Smith S. *Assimilation of HF radar observations in the Chesapeake-Delaware Bay using the Navy Coastal Ocean Model (NCOM) and the four dimensional variational (4DVAR) method*; 2015 [this book].
39. Barron CN, Kara AB, Martin PJ, Rhodes RC, Smedstad LF. Formulation, implementation and examination of vertical coordinate choices in the Global Navy Coastal Ocean Model (NCOM). *Ocean Model* 2006;**11**:347–75.
40. Buskey EJ, Bundy M, Ferner MC, Porter D, Reay W, Smith E, et al. *System-wide monitoring program of the national estuarine research reserve system: research to address coastal management issues*; 2015 [this book].
41. Porter D, Dorton J, Leonard L, Kelsey H, Ramage D, Cothran J, et al. *Integrating environmental monitoring and observing systems in support of science to inform decision making: case studies for the southeast*; 2015 [this book].
42. Simoniello C, Watson S, Kirkpatrick B, Spranger M, Jochens AE, Kobara S, et al. *One System, many societal benefits: an efficient, cost-effective ocean observing system for the Gulf of Mexico*; 2015 [this book].

CHAPTER

2 National Ocean Observing Systems in a Global Context

Zdenka Willis

U.S. Integrated Ocean Observing System Office, Silver Spring, MD, USA

E-mail: Zdenka.S.Willis@noaa.gov

CHAPTER OUTLINE

1. WHY DO WE NEED OCEAN OBSERVING?

In 1964, the National Academies of Science report titled the "Economic Benefits of Oceanographic Research" characterized the ocean as being an unlimited resource to be exploited. Subsequent science, including that from national and international observing efforts, has changed what we know about the ocean. Today, we recognize the ocean is a finite and shared resource that we need to manage regionally, nationally, and worldwide for ocean health.

There are a number of significant societal threats and challenges facing humans due to the changing ocean. The increased frequency and intensity of coastal storms and resulting storm surges will affect our coastal communities and disrupt commerce, nationwide. Sea level changes are threatening critical infrastructure worldwide. Harmful algal blooms and oxygen-deficient dead zones threaten water supplies, fisheries, and coastal recreation. Ocean acidification is negatively impacting coral reefs and shellfish harvesting. The increasing size and number of vessels calling on ports present challenges for our already inadequate maritime infrastructure, and pose potential environmental risks as well. The world's growing population will increasingly rely on the ocean for food, but fishing must be done sustainably. Achieving sustainable marine fisheries will take improvements in stock assessment to which sustained observing systems can contribute.

To be able to understand and manage the ocean, we need meaningful measures of the ocean's state.

2. ANSWERING THE CALL—NATIONAL AND GLOBAL OCEAN OBSERVING INFRASTRUCTURES

Observing systems are expensive; the United States invests billions of dollars in civil Earth observations to ensure that the nation's decision-makers and managers have the information they need about climate and weather, disaster events, land-use change, ecosystem health, natural resources, and many other characteristics of the planet.[1] The ocean is a harsh environment in which to operate an observing system, from corrosion due to salinity to bio-fouling. The costs of maintaining instrumentation in the ocean's harsh environment, accessing remote locations, and establishing sufficient communications with deployed technologies are significant. Finally, the fact is that most people are not even aware that the ocean touches their lives every day. Therefore, not only do we need to work together to achieve an adequate understanding of three quarters of our planet, we must work together to articulate a compelling case to sustain the ocean observing systems.

By examining two established enterprise ocean observing systems—the United States Integrated Ocean Observing System (IOOS®) (www.ioos.noaa.gov) and Australia's Integrated Marine Observing System (IMOS) (www.imos.org.au)—along with Canada's Ocean Tracking Network (OTN) (www.oceantrackingnetwork.org), this article explores how these efforts contribute to measuring and understanding what is going on in our oceans, along our coasts, and in the United States Great Lakes.

While the focus and activities of IOOS, IMOS, and OTN are different, these efforts have much in common. While IOOS was established as an operationally focused system, IMOS and OTN were established to answer scientific questions. For IOOS, the infrastructure enables decision-making every day, while concurrently advancing science and technologies. IMOS and OTN had as their prime focus improving scientific understanding of ocean conditions, but the information they generate is increasingly being used by government agencies and other users to inform decisions. All three systems have the following in common:

- Deploying, maintaining, and developing advanced observations technologies
- Providing free and open access to data in support of a wide range of users
- Advancing modeling
- Focusing on education

3. OCEAN OBSERVING TECHNOLOGIES

It is a complex task to measure the ocean in ways that deliver useful products for people. For example, to deliver a five-day weather forecast for any local region, meteorologists must sample the whole planet. Satellites are key tools that provide multispectral images, atmospheric soundings, and sea surface characteristics needed for the forecasts. While satellites enable us to see through the atmosphere, they can only measure the surface of the ocean, and this does not provide the detail needed for accurate weather prediction. Therefore, we also need a complement of in situ measurements that extend our reach to the depths of the oceans at all relevant scales of phenomena. In situ refers to systems that measure on or under the surface of the ocean in continuous and event-driven modes, automatically and by humans.

IOOS and IMOS are global leaders in ocean observation. Both have infrastructures that operate on continental scales, field similar observing technologies, share best practices for a broad suite of variables, and generate masses of data. IOOS and IMOS are both partners with the specialized OTN, a global ocean research and technology development platform, in the area of animal tagging and tracking. To understand animal movements and survival, you also have to understand how the animals respond to changes in environmental conditions. Thus, the animal tracking effort is tightly integrated with the observation of the physical and chemical environment, and all networks depend on common platforms for oceanographic observations. Other observing technologies common to all three programs include Argo, moored buoys, high-frequency radar (HFR), and autonomous underwater vehicles known as gliders.

3.1 ARGO GLOBAL ARRAY

Argo is a successful global observing program composed of automated profiling floats that measure temperature and salinity in the water column. Argo is strongly supported by the United States and Australia, and it has participation from 23 additional countries. There are now more than 3000 floats resulting in more than 100,000 profiles each year. Argo achieved its one-millionth profile in 2013. The Argo program is developing its next generation of floats—units that can descend to 2000 m and are equipped with biological and chemical sensors. A subset of Argo floats also measure dissolved oxygen.

3.2 MOORED BUOYS

Moored buoys are one of the mainstays of all ocean observing networks. Buoy networks such as the global tropical moored arrays are international partnerships that measure the air/sea flux along the equator, critical for understanding global circulation and storm generation. OceanSITES and the United States National Science Foundation's (NSF) Ocean Observatories Initiative (OOI) maintain five global stations to provide interdisciplinary sentinel sites worldwide. In addition to these global observing programs, both IOOS and IMOS have deployed coastal buoys. Within the United States exclusive economic zone, the National Oceanic and Atmospheric Administration (NOAA) has deployed over 100 coastal buoys that support weather forecasting, wave forecasting, and navigation. NOAA also operates tsunami buoys that support the international tsunami warning network. The IOOS Regional Associations (RAs) have deployed and/or leverage data from another 100 coastal buoys. IMOS has deployed deep water moorings and an Australian National Moorings Network comprised of a series of national reference stations and regional moorings designed to monitor particular oceanographic phenomena in Australian coastal ocean waters. OTN maintains a limited number (about five deployed at any given time) of moored subsurface buoys (pods) for oceanographic observations.

3.3 HIGH-FREQUENCY RADAR (HFR)

HFR measures surface currents along the coasts out to a distance of about 150 km. Just as measuring winds in the atmosphere is fundamental to meteorology, measuring ocean currents is fundamental to oceanography. Currents determine the movement of surface waters, knowledge of which is critical to support pollutant tracking, for search and rescue, for harmful algal bloom monitoring, to aid navigation, and to provide ecosystem-based management and coastal and marine spatial planning.

Both IOOS and IMOS operate national HFR networks, and through the Group on Earth Observations (GEO), a global HFR network is being developed. HFR data are being used operationally in both the United States and Australia. Some United States examples include support to the oil spill responses for the M/V Cosco Busan (San Francisco, CA, 2007) and the Deep Water Horizon event (Gulf of Mexico, 2010). The data from the HFR network are also used in the United States Coast Guard Search and Rescue system to reduce potential search areas. HFR data are increasingly being used to better understand hurricane tracks and intensity and in the detection of tsunamis. Radars in California and Japan detected the tsunami resulting from the Tohoku Japan earthquake in 2011. In IMOS, HFR data are advancing the understanding of the mesoscale structure of Australian coastal waters. The IMOS HFR network was one of several IMOS capabilities that assisted in the search for Malaysian Airlines flight MH370. David Griffin, an oceanographer with Australia's national science agency, Commonwealth Scientific and Industrial Research Organization, used advanced models of ocean currents around Australia to determine the likely movement of the wreckage, which allowed search and rescue operations to focus their activities.[2] HFR radar data were used to improve the accuracy of surface velocities in these models. IMOS has been publically recognized for this work as an international leader in ocean observing and is considered the critical observational foundation for much of Australia's marine science.[2]

HFR has also been deployed in both the Arctic and Antarctica. In the Arctic, radars have been deployed by IOOS Alaska Ocean Observing System (AOOS) to understand the circulation of the Chukchi Sea in preparation for safe and secure oil extraction. Through the NSF long-term ecosystem program, in partnership with the University of Alaska and Rutgers University, radars are being deployed in Antarctica to better understand the environmental drivers of food webs and ecosystems.

3.4 PROFILING GLIDERS

The use of steerable, profiling gliders has added significant potential and adaptive sampling to the portfolio of platforms. Over the past 6 years, the IOOS RAs and academic institutions funded by federal, state, and local agencies, private foundations, and industry have completed 33,400 glider days. IMOS has established an Australian National Facility for Ocean Gilders (ANFOG) and operates a fleet of coastal and open ocean gliders conducting regular missions to observe boundary currents and other important continental shelf and coastal processes. OTN has explored using gliders as listening devices to pick up the signals from animal tags. The use of gliders is in large measure funded from grants and short-term contracts. Within the United States, this has resulted in the generation of much knowledge and a promising framework on which to build a nationwide, functional glider network. However, without a source of long-term funding, a true network cannot develop. In August 2014, the United States IOOS Office published the paper "Toward a U.S. IOOS® Underwater Glider Network Plan: Part of a comprehensive subsurface observing system" to provide a conceptual foundation for a true network, http://www.ioos.noaa.gov/glider/strategy/welcome.html. One major step IOOS made toward developing a glider network was establishing the United States Glider Data Assembly Center (DAC) in fall of 2013. The Glider DAC will coordinate data collection, standardize data formats, and aid in the dissemination and preservation of data.

Missions conducted by gliders are as varied as the sensors they can carry. In southern California, gliders were used by the city sanitation district when they needed to perform maintenance to monitor sewage water outflow to assure local residents that these operations were safe, and to observe climate variability in the coastal ocean. In the Gulf of Mexico, gliders have been used to document the evolution of red tide blooms to understand how to forecast and track such harmful algal blooms. In Hawaii, glider data are assimilated into models to improve the understanding of vertical (mixing) and horizontal (current) water movements.

In the western Atlantic, Gliderpalooza was conducted in the fall of 2013 to test our ability to coordinate existing independent assets to operate as an integrated fleet of gliders. This effort brought together 15 gliders from 15 partners in the United States and Canada. Figure 1 shows the coordinated missions of Gliderpalooza 2013. The partners were drawn from government, academia, and industry. Results of Gliderpalooza 2013 included the acquisition of more than 25,000 profiles, the near real-time reporting of data through the Global Telecommunications System (GTS) for assimilation into models, tracking of acoustically tagged fish, and the collection of a comprehensive three-dimensional dataset along the east coast of Canada and the United States.

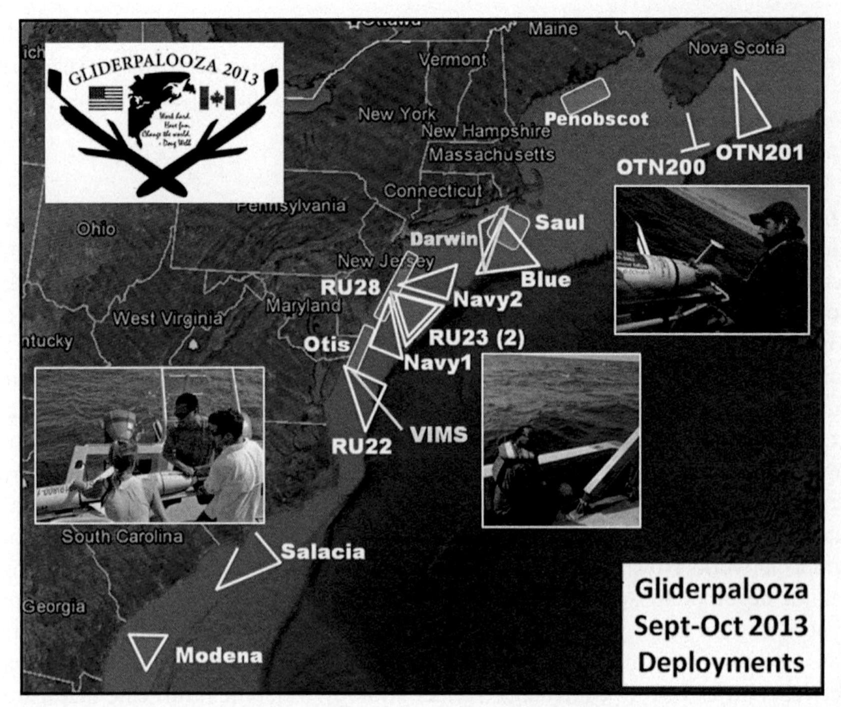

FIGURE 1

The coordinated missions of Gliderpalooza 2013.

3.5 ANIMAL TELEMETRY

Animals are tracked because of the many direct benefits that they provide to coastal communities through fisheries and recreation activities, but also because some species are large enough to carry satellite-linked oceanographic sensor packages and routinely penetrate parts of the ocean (e.g., under pack ice) that our other technologies cannot routinely sample. The Sloan Foundation's Census of Marine Life was the foundation for using such "animal oceanographers" as part of the observing systems.

The use of tagging (satellite, acoustic, and archival) is emerging as a component to both national observing systems like IOOS and IMOS and to global programs such as the OTN. Canada has established OTN (http://oceantrackingnetwork.org), a global ocean research and technology system focused on documenting the movements and survival of aquatic animals primarily through the use of acoustic tags and receivers. OTN is funded by the Canada Foundation for Innovation (CFI) International Joint Ventures Fund and is headquartered at Dalhousie University in Halifax, Canada. Starting in 2008 and beginning full operations in 2010, OTN has been deploying Canadian state-of-the-art acoustic receivers and oceanographic monitoring equipment in key ocean locations around the world.[3] OTN and its partners track many keystone, commercially important, and endangered animals including marine mammals, sea turtles, and fishes including sharks, sturgeon, eels, tuna, salmonids, and cod. Within Canada, OTN has operations in the Atlantic, Pacific, and Arctic. Within the network, natural scientists use their observations of animal activity to work with social scientists to improve governance frameworks for living marine resources, to directly interact with stakeholders, and to suggest ways to weave stronger conservation and management strategies for species. OTN works with programs like IOOS and IMOS to augment national systems, fill gaps in current coverage, and build capacity. Recently, OTN receivers were deployed on the PIRATA South Atlantic Ocean buoys, and OTN is now working to equip the RAMA buoy network in the Indian Ocean with the same capability. Figure 2 shows the OTN partnerships.

IMOS Australian Animal Tagging and Monitoring System (AATAMS) uses acoustic technology, CTD (conductivity-temperature-depth) satellite trackers and bio-loggers to monitor environmental conditions and the movements of marine animals from the Australian mainland to the sub-Antarctic islands and as far south as the Antarctic continent. The AATAMS acoustic telemetry program is strongly supported by the OTN, which has provided equipment for receiver arrays off Perth (Western Australia) and off the east coast of Tasmania. IMOS has found that Southern Ocean elephant seals are not only exhibiting extremely interesting behavior, but are also excellent samplers of the physical environment. In February 2014, IMOS tagged 19 Southern elephant seals in the Indian Ocean and another eight in Antarctica to compare their behavior. The Indian Ocean seals traversed across a wide area of the southern Indian Ocean and into the Southern Pacific Ocean, while the Antarctic seals stayed close to the ice. From February to July of 2014, the seals provided over 6000 conductivity, temperature, and depth profiles that are not only feeding ocean and climate models, but are helping us understand the foraging habits of these seals.[4]

Within the United States, animal tagging is still primarily project based, but telemetrists are developing a strategic plan to establish a U.S. Animal Telemetry Network (ATN) coordinated by IOOS. The United States IOOS Office collaborated with the Office of Naval Research and Stanford University to develop an ATN portal to coordinate the many tagging projects within the United States. One successful effort was the provision of 8138 observations of the Sloan Foundation's Tagging of Pacific Predators (TOPP) Program to the United States Navy. These data were used for model reanalysis and proved that Animal Borne Sensors data are sufficiently accurate and of high enough quality to be useful in filling existing observation gaps in under-sampled ocean regions (i.e., boundary currents, ocean fronts, under sea ice) and to improve operational ocean models.

3.6 WHAT WILL THE NEXT SENSOR BE?

The goal of advancing observing technologies is to extend our observing reach by providing new capabilities to answer new questions and to reduce costs so we can afford to do more. Within IOOS, there are two efforts to support advancement in technologies.

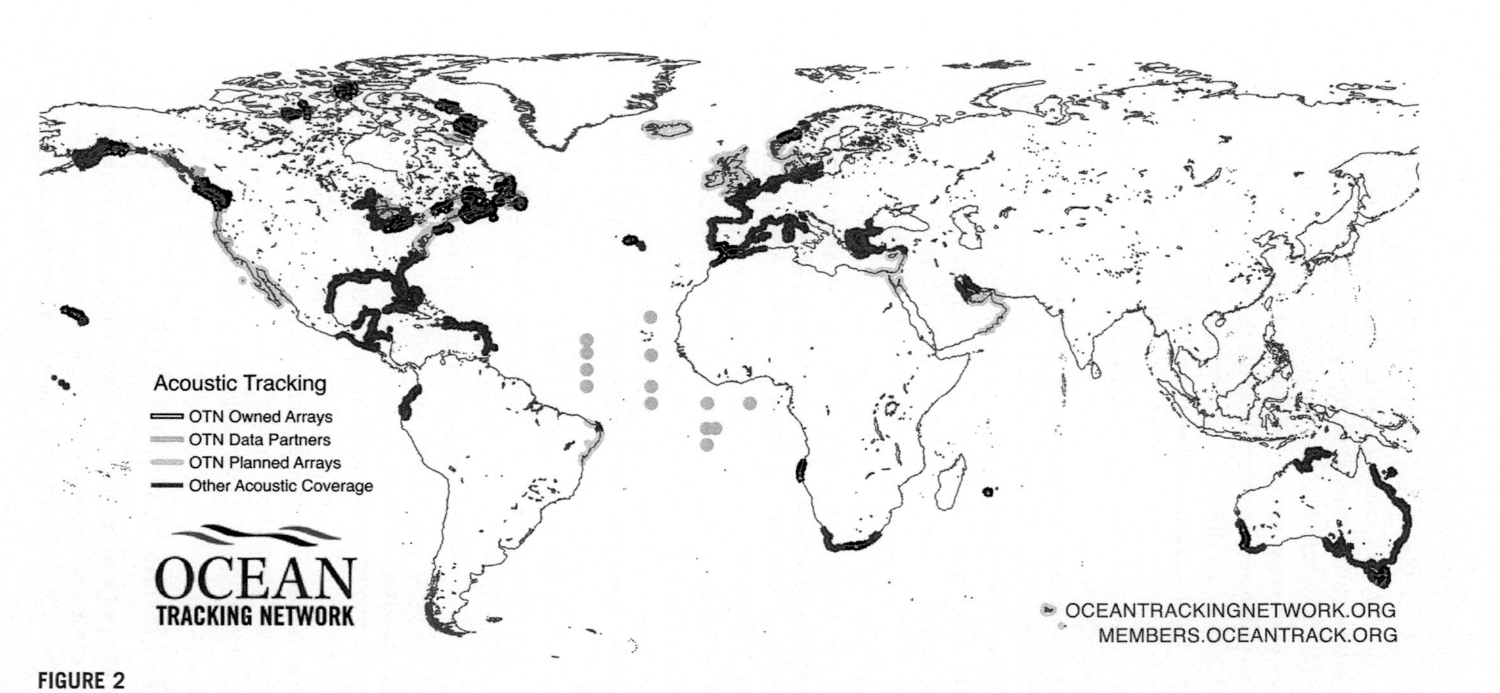

FIGURE 2

OTN network and partnership arrays.

The first component is the Alliance for Coastal Technologies (ACT). ACT provides sensor validation and verification. The latest validation effort involved pH sensors. There is a critical need for reliable, affordable pH sensors in the face of climate-linked ocean acidification. ACT also maintains a comprehensive database of manufacturers and sensors that has been used by the Environmental Protection Agency when doing market surveys for new sensors (http://www.act-us.info/index.php).

In 2013, IOOS initiated the second component, the Ocean Technology Transition project. This project focuses on transitioning sensors and platforms for which there are demonstrated operational end users from the research and development phase to operational mode. Resources are targeted to marine sensors that have moved beyond their proof-of-concept phase, with specific emphasis on transitioning to operating status at affordable lifecycle costs, developing viable data management systems, simplifying operations for ease of use in the field, and controlling maintenance expenses. In the first year (2013), IOOS funded two projects. One project supported the shellfish industry by deploying equipment to oyster hatcheries to measure the total alkalinity of the water. Highly acidic water prevents the settlement and survival of oyster larvae, and it is a threat that could shut down this multimillion dollar industry. The second project was the deployment of the Environmental Sample Processor (ESP) for monitoring harmful algae blooms in the Gulf of Maine along the northeast United States. Scientists deployed three ESPs in the Gulf of Maine between May and June of 2014. It was the first time that multiple ESPs were deployed, and they successfully detected *Alexandrium fundyense*, the primary source of toxic red tide blooms in this region, at levels comparable to traditional labor intensive methods. Five new projects began in late 2014 that include continued support to the shellfish hatcheries, extending the use of the ESP to the northwest coast of the United States, transitioning nutrient sensors into an operational network in the northeast United States, a new buoy design to detect ice freeze-up in the Arctic, and the evaluation of the Imaging Flow CytoBot tool for detection of harmful algal blooms.

Science and technology are dynamic and will inevitably drive the evolution of observational and data management capability into the future. IMOS manages this evolution by routinely assessing its portfolio from the perspective of readiness, as recommended by the Global Ocean Observing System (GOOS).[5] Mature research capability is sustained, with an increasing focus on efficiency over time. Where operational utility begins to outweigh research need, consideration is given to transitioning out of the research environment, noting that this requires close cooperation with relevant operational agencies and a clear framework for transition. Newer capabilities piloted by IMOS are matured if successful, or discontinued if not. Ocean gliders are a good example of newer capability that has been successfully piloted and is now being matured within IMOS. Proof-of-concept activities being undertaken within the research environment are evaluated by IMOS for suitability as next generation pilot capabilities. Marine microbial observing is an example of a technology that is moving through proof-of-concept, using water samples provided by IMOS, and is looking prospective for piloting in the next stage of investment. In this way, IMOS ensures that its capability is continuously evolving to meet current and future needs.

OTN Canada already employs cutting-edge technology, but it is also working closely with its equipment suppliers to provide new sensor and technology capabilities that meet the needs of its researchers. Some acoustic tags already carry sensors that report the depths and temperatures experienced by free-ranging animals. Work is now underway to add new sensors to acoustic tags, for example, by adding/improving accelerometer sensors, which indicate the direction and speed animals are swimming. This information is important for the calculation of the energy budgets of aquatic species and can also be used to indirectly infer the growth/size of animals, as swimming speed frequently correlates with the size of animals. Research is also underway on the development of "predation tags" that following a predation event will signal that a tagged fish has been consumed. Knowledge of the causes of such natural mortality is critical for correctly managing fisheries and for designing recovery strategies for endangered species. Acoustic tracking capabilities are being incorporated into autonomous vehicles, and a transformative advance has just been achieved (December 2014) through a collaboration between United States-based Liquid Robotics, Canada's VEMCO, and the OTN. This development has configured a Liquid Robotics Wave Glider autonomous vehicle to harvest data from subsurface VEMCO acoustic receivers moored by OTN at distances of >100 nautical miles from shore. This capability will hugely reduce the costs of retrieving data, provide all-weather working capabilities, and greatly improve the safety of marine operations for telemetry networks.

4. ACCESS TO THE DATA

All the national systems that participate in GOOS and GEO subscribe to the principal of free and open data. It is policy within the United States that all observational data collected with federal funding support are made freely and openly available, which makes sense from scientific, disaster response, and economic perspectives.

One challenge IOOS and IMOS face is the need to support operational systems, often with unique data formats, while *simultaneously* working to create mutually compatible data access systems and services, and common data formats and metadata standards, in order to facilitate access to this public resource through the internet. Both IOOS and IMOS provide marine data such as temperature, salinity, currents, wind speed/direction, waves, and other primarily physical observations for model assimilation through the GTS. For broader access, IOOS uses three standards to convey the information in an interoperable manner: (1) Open-source Project for a Network Data Access Protocol (OPeNDAP), (2) Sensor Observation Service (SOS) OGC (Open Geospatial Consortium) Standard, and (3) Web Map Services (WMS) OGC Standard. IOOS has developed standard templates for the SOS and developed implementation software. 52 N (http://52north.org/) has developed a software that implements the SOS standard; ncSOS is another software based on the SOS standard that acts like a bridge taking in situ data stored in OPeNDAP-compliant files and delivering SOS-compliant data. In a similar fashion, IMOS makes all of its data openly accessible via a standards-based, open source, marine information infrastructure (https://imos.aodn.org.au/imos123/). Both IMOS and IOOS use a catalog system for data access (Figure 3).

To ensure quality data, IOOS also administers the Quality Assurance of Real Time Ocean Data (QARTOD) project to address standard quality assurance practices for oceanographic data. Six manuals have been published to date covering the following: Ocean Temperature and Salinity; Water Level; Dissolved Oxygen; Waves; Currents; and Wind.

OTN is also working to develop standardized databases and data sharing protocols that will unite all of the global deployments and data systems into a new global observation system that tracks the movements and survival of marine animals, and documents how both are influenced by oceanographic conditions. OTN hosts a data warehouse currently containing more than 95 million animal tracking records, and it is growing. This serves as a publicly accessible repository for data collected by its researchers, and OTN is linking its data systems and formats to the developed animal tracking systems in Australia and to the developing system in the United States and elsewhere. Given the comparative newness of telemetry research, the development of mutually compatible global data systems and formats is relatively easy. OTN is also developing interpretation and visualization tools for analysis of tracking data, which it is making freely available to the scientific community.

5. MODELING AND ANALYSIS

Prediction of future conditions is critical to delivering the full benefits of an ocean observing system. Arguably, within IOOS, the modeling subsystem has been the least coordinated; however, a robust coastal modeling effort exists at both the federal and regional level. The United States ocean, coastal, and Great Lakes modeling community is a federal and regional partnership that is being asked to provide greater resolution models that cover not only traditional physical water circulation, which remains a critical need, but also expand outputs to include inundation forecasting and ecosystem modeling. To focus on the inundation aspect of hurricane forecasting, NOAA established a Storm Surge Roadmap to improve storm surge forecasts and NOAA's ability to communicate the information in a clear way. Other needed forecasting abilities include predicting the following: the prevalence and virulence of pathogens and zoonotics (vector-borne diseases); the distribution, frequency and duration of HABs (water-borne diseases); and the bioaccumulation, transfer, and movement of chemical contaminants in coastal ecosystems and food webs. Contaminants threaten the availability of safe seafood, safe drinking water, safe beach going, the health of ocean sentinel species, the availability of marine natural products, and the provision of ecosystem services. This new forecasting need

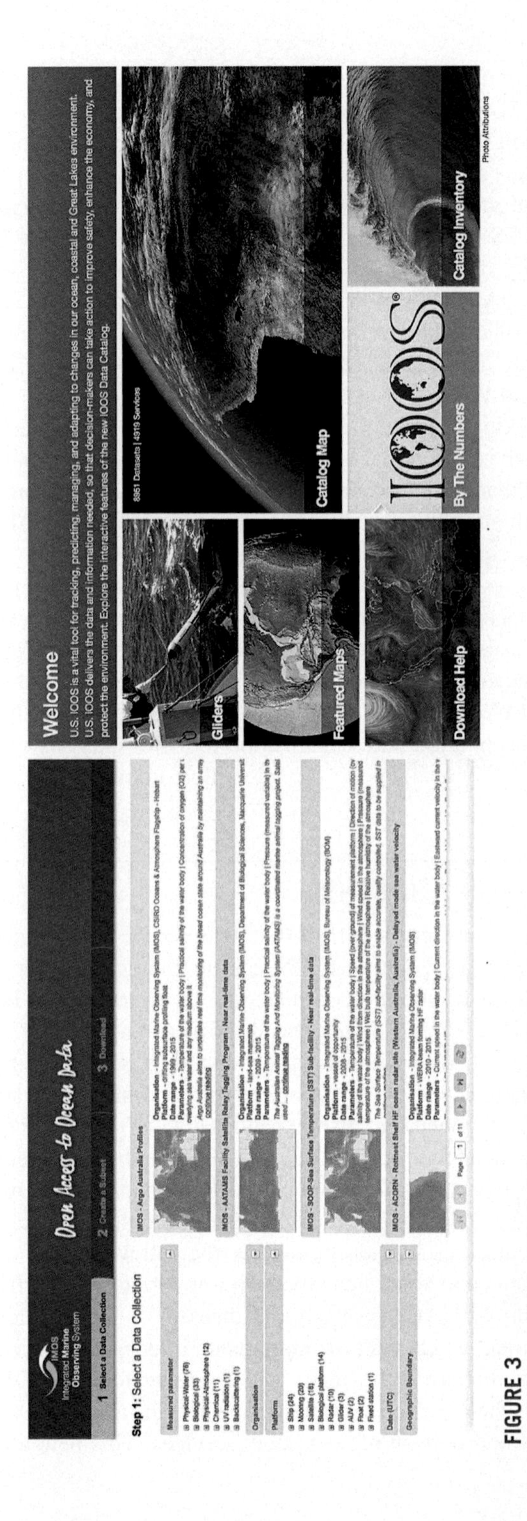

FIGURE 3

IMOS and IOOS use a catalog system for data access.

requires the development of ecological models. Ecological models require an interdisciplinary approach, and in 2014, NOAA published an Ecological Forecasting System strategic plan, http://oceanservice.noaa.gov/ecoforecasting/. A key to coordinating these modeling efforts on a national basis is to transition the models from the research to the operations phase; therefore, IOOS established the IOOS Coastal and Ocean Modeling Testbed (COMT). The vision of COMT is to accelerate the transition of scientific and technical advances from the coastal ocean modeling research community into improved operational ocean products and services (i.e., via feedback loops from research to operations and also from operations to research). One key success of COMT was the inclusion of tides into NOAA's operational inundation model and expanding its horizontal extent.

In May 2014, IMOS published the "IMOS Strategy 2015-25" (http://imos.org.au/plans.html).[6] IMOS has a concerted focus on making data available and seeing that it is used. IMOS has worked with the coastal modeling community to make sure that available data are informing the models. Going forward, IMOS will use its infrastructure to be a coordinating entity to advance the assimilation and further development of coastal modeling. IMOS has begun a structured engagement with the coastal and ocean modeling communities through development of joint products (e.g., in ocean reanalysis), national workshops, and targeted infrastructure investment at the model–data interface (such as virtual laboratories).

The Ocean Tracking Network has three objectives within its modeling and visualization component. The first is to develop visualization tools tailored to the complex marine observations that address OTN research objectives. To arrive at these visualizations, the scientists have to develop the modeling tools necessary to overlay OTN tracking studies (with particular emphasis on estimating migration survival and species' interactions) with environmental conditions. The second objective is to create a repository of documented code and software to be freely shared among OTN researchers and the broader scientific community. The third is to facilitate the exchange of Highly Qualified Personnel (HQP) across Canada and internationally, as resources permit, to foster stronger collaboration on visualization and modeling tools useful to the global scientific community.[3]

6. EDUCATION AND OUTREACH

Those involved in observing the ocean also have an obligation to train the next generation of ocean experts and to communicate to the public the importance of the ocean for the quality and prosperity of our lives. IOOS, IMOS, and OTN have all put in place both formal and informal educational efforts.

To understand the range of these educational efforts, the IOOS has cataloged the accomplishments of the IOOS RAs, http://www.ioos.noaa.gov/education/welcome.html. Products include education, training, and outreach materials ranging from lesson plans for courses to how-to manuals to use real-time data in the classroom. Other examples include IOOS Caribbean Regional Association Coastal Ocean

Observing System (CariCOOS), which teamed up with NOAA's Puerto Rico's Weather Camp to introduce coastal weather and ocean observing tools within the camp. The IOOS Northwest Association of Networked Ocean Observing System (NANOOS), the IOOS RA invited six Ph.D. students from the University of Washington, two classroom teachers, three NOAA volunteers, and one Seattle Aquarium staff/pre-service teacher on a buoy deployment. IOOS Southeast Coastal Observing Regional Association (SECOORA) took advantage of NOAA's Educational Partnership Program by hosting Pedro Matos-Llavona from the University of Puerto Rico at Mayagüez for a summer internship. Rutgers University, a key IOOS partner, developed an undergraduate class based on the deployment of gliders. It was Rutgers' undergraduate students who successfully flew the first glider across the Atlantic Ocean in 2009, known as the Scarlet Knight mission. The class that started in 2008 with 10 students has now grown into a basic and advanced class with an annual enrollment in the two classes of 75 students.

IOOS is also providing informal education opportunities to stimulate the interest of youngsters (and their parents) in the marine environment, leading hopefully to the next generation of scientists. This is being accomplished through the development of the award winning Eco Hero game of the Gulf of Mexico Ocean Observing System (GCOOS). The game is fun, interactive, conservation oriented, teaches the environmental value of the Gulf of Mexico, and "bridges the gap" between research conducted within the Gulf and the relevance of scientists' findings to the public's everyday lives.

IMOS has increasingly focused on broadening its relevance and impact beyond the science, research, and tertiary education sector. In addition to providing observations and data to a large and growing number of research projects, student projects, and academic courses, IMOS is now recognized as an essential partner in large, multi-institutional research programs across multiple sectors. It has contributed to 180 postgraduate projects, over 400 journal publications, and 250 research projects.

Within Canadian research infrastructure programs like the OTN, great value is placed on training of HQP (highly qualified personnel): defined as students, postdoctoral fellows, and other personnel-in-training. OTN researchers generated 77 high-quality peer-reviewed publications and 123 conference or workshop presentations in 2013, most involving students in training. In the same year, the Canadian network engaged 27 principal investigators (PIs), who have formally collaborated with an additional 60 researchers and have jointly trained over 130 HQP.

7. SUMMARY

Ocean observing systems are only as good as our ability to observe and accurately model ocean systems. Finding the resources to sustain these observations is THE challenge for program managers. While the United States federal government is conservatively estimated to annually spend roughly $3.5 billion in civil Earth observations and data analysis, these investments add $30 billion to the United States economy each year by providing Americans with critical data and information about natural resources, climate and weather, disaster events, land-use change, ecosystem health, ocean trends, and many other phenomena.[1] In July 2014, the United States published its first National Plan for Civil Earth Observations (the Plan), laying out a new paradigm for communicating observations and establishing priorities and supporting actions for advancing our civil Earth observing capability. The Plan delineated the need for sustaining observations over relevant time scales. The Plan defined observational systems as either sustained, intended to be deployed for seven years or longer, or experimental, those deployed for less than seven years. It further defined sustained observations as those that support public services and those that support Earth system research. It is exciting to see the priority being placed on sustaining observing systems. As a growing population continues to stress our planet, quality observations will increase in importance. But it is not good enough to measure the planet, those who measure the planet must work to ensure this information comes into play in our communities, our economies, and in management decisions. Working together, IOOS, IMOS, and OTN can help to manage our oceans by measuring them and connecting observations to people.

ACKNOWLEDGMENTS

Thanks to my fellow ocean observer colleagues: Tim Moltmann, Director of IMOS, and Fred Whoriskey, Director of OTN. Both Tim and Fred embrace and demonstrate on a daily basis the commitment to partnerships. Thank you to Tim and Fred for their permission to liberally borrow material from their strategic plans, annual reports, and websites used to write this article. Thank you to Laura Griesbauer, IOOS Office, and Fred Whorsikey for their great assistance in editing.

REFERENCES

1. National Science and Technology Council (NSTC). *Executive office of the President, national plan for civil earth observations.* July 2014. http://www.whitehouse.gov/sites/default/files/microsites/ostp/NSTC/national_plan_for_civil_earth_observations_-_july_2014.pdf.
2. Gunn J, Coffin M, Dittmann S. *Marine science: challenges for a growing 'blue economy'.* May 20, 2014. The conversation, http://theconversation.com/marine-science-challenges-for-a-growing-blue-economy-22845.
3. Ocean Tracking Network. Annual Report 2012–2013. www.oceantrackingnetwork.org, http://oceantrackingnetwork.org/wp-content/uploads/2014/07/AR_2012-2013.pdf.
4. IMOS Marine Matters. July 2014. http://imos.org.au/marine_matters.html.
5. Framework for Ocean Observing. http://www.oceanobs09.net/foo/,
6. IMOS Strategy 2015-25. May 2014 http://imos.org.au/.

CHAPTER 3

The Importance of Federal and Regional Partnerships in Coastal Observing

Josie Quintrell[1,*], Rick Luettich[2], Becky Baltes[3], Barbara Kirkpatrick[4,5], Richard P. Stumpf[6], David J. Schwab[7], Jennifer Read[7,8], Josh Kohut[9], John Manderson[10], Molly McCammon[11], Russell Callender[12], Michelle Tomlinson[2], Gary J. Kirkpatrick[5], Heather Kerkering[14,15], Eric J. Anderson[13]

IOOS Association, Harpswell, ME, USA[1]; University of North Carolina, Chapel Hill, NC, USA[2]; U.S. IOOS Program Office, NOAA, Silver Spring, MD, USA[3]; Gulf of Mexico Coastal Ocean Observing System, College Station, TX, USA[4]; Mote Marine Laboratory, Sarasota, FL, USA[5]; National Centers for Coastal Ocean Science, NOAA, Silver Spring, MD, USA[6]; University of Michigan Water Center, Ann Arbor, MI, USA[7]; Great Lakes Observing System, Ann Arbor, MI, USA[8]; Mid-Atlantic Regional Association Coastal Ocean Observing System and Rutgers University, Newport, NJ, USA[9]; National Marine Fisheries Service, NOAA, Silver Spring, MD, USA[10]; Alaska Ocean Observing System, Anchorage, AK, USA[11]; National Ocean Service, NOAA, Silver Spring, MD, USA[12]; Great Lake Environmental Research Laboratory, Ann Arbor, MI, USA[13]; Pacific Islands Ocean Observing System, University of Hawaii at Manoa, Honolulu, HI, USA[14]; Sea Connections Consulting, Virginia, USA[15]

**Corresponding author: E-mail: jquintrell@comcast.net*

CHAPTER OUTLINE

1. INTRODUCTION

The U.S. IOOS is a coordinated network of federal agencies, regional associations, and private industries created to generate and disseminate information about our coasts, oceans, and Great Lakes. IOOS is a collection of systems, technology, data management, and people working together to enhance the country's ability to collect, deliver, and use coastal information. IOOS provides data and information needed to better characterize, understand, and predict coastal ocean dynamics so decision-makers can take action to improve safety, enhance economy, minimize hazards, and steward the environment. Much of the work that IOOS does is through collaborative partnerships between 17 federal agencies and 11 regional systems.

The federal–regional partnership is a centerpiece of IOOS. Congress created the regional component of IOOS to enhance the ability of federal agencies to provide the scale of information needed to resolve national issues that uniquely manifest themselves at the regional and local level.[1] The complexities of the coastal environment call for partnerships that not only cut across federal agencies but also include regional managers, academia, industry, nongovernmental organizations, and the general public. These partnerships result in a robust network of capabilities that benefit participants, stakeholders, and the general public. Regional systems contribute local expertise, access to high-resolution data needed by local end-users to address national and regional priority issues, tailored products and services, and build relationships with local and regional stakeholders. Federal agencies bring perspectives on national priorities that have regional implications, technical and scientific expertise, operational forecasting capabilities, computational power, and sustainability.

This chapter will describe some of the benefits from these partnerships. The five case studies included in the chapter highlight how partnerships forged by the IOOS system enhance understanding of the coastal ocean environment and improve decision-making at both regional and national levels.

2. WHY A PARTNERSHIP APPROACH TO COASTAL OCEAN OBSERVING?

Coastal ecosystems are complex. The US territorial waters encompass 11 large marine ecosystems (LMEs), as designated by NOAA, that range from the cold waters of the Chukchi Sea in the Arctic to the warm waters of the tropical Pacific.[2] The Great Lakes, with over 10,000 miles of coastline, are the world's largest system of freshwater lakes. Each region is characterized by unique geological, physical, and chemical properties, biological productivity, and human uses.

The complexity of these ecosystems, their large geographic extent, and the differing needs of their users require management responses that are best accomplished in partnership between the multiple federal agencies whose mission is to serve these regions and the national network of 11 IOOS Regional Associations (RAs) that reflect the knowledge and diversity of regional interests. The IOOS RAs provide a forum for convening regional experts, government agencies, private sector companies, nongovernmental organizations (NGOs), and users to identify and prioritize needs, discuss and develop solutions, leverage assets, and share knowledge.

The data and information needs are diverse because they are derived from a broad range of coastal ocean users. Users include mariners who need access to up-to-date sea conditions; fishermen planning their days at sea; resource managers who need data for ecological trends and risk factors; federal agency personnel who need data for modeling and prediction; emergency managers who need forecasts and predictions to protect public health and safety; and the general public who want to plan for coastal activities, recreation, and tourism. The national network of 11 RAs provides services to the entire coastline of the United States, including the Great Lakes, Pacific Islands, and US territories.

3. THE IOOS APPROACH

The Integrated Coastal and Ocean Observation System Act (the Act),[3] enacted in 2009, established a framework for coordinating among the 17 federal agencies that are engaged in coastal and ocean observing and the network of 11 regional systems. The National Oceanic and Atmospheric Administration (NOAA), with its mandate for operational oceanography, is the lead federal agency. The IOOS Program Office, housed within NOAA's National Ocean Service, provides overall coordination and management for the enterprise, including development of data management standards and protocols. An Interagency Ocean Observing Committee (IOOC) is responsible for coordinating the federal agencies.

The Act also established a system of Regional Associations (referred to as Regional Information Coordinating Entities in the Act) to engage stakeholders, foster partnerships at the regional level, and to provide science-based information that addresses priority needs.

The RAs serve as regional forums for experts, researchers, data providers, government agencies, industry, and users to coordinate efforts, leverage assets, and maximize limited resources. Boards of directors comprised of a broad spectrum of regional interests govern each region. As nongovernmental entities (some RAs are nonprofit organizations), they can be flexible and nimble, offering federal agencies the ability to respond quickly to events. As trusted agents, the RAs also link and leverage resources to fulfill critical needs. In the northeast, for example, the RA convened the Northeast Coastal Acidification Network (NECAN) at the request of NOAA's Ocean Acidification Program. NECAN brought together industry, researchers, and the general public to assess the state of the science on ocean acidification in the northeast and to provide resources to researchers, industry, educators, and the general public. This benefited NOAA by convening leading regional experts, and it benefited the public by improving the state of the science and understanding. Similar efforts are underway in other regions as well.

4. BUILDING PARTNERSHIPS THROUGH DATA ACCESSIBILITY

Central to the success of the IOOS partnerships is a robust data management system that works to remove barriers that hinder the sharing of data and information.

Observations are not only collected by the IOOS RAs but also by thousands of other systems operated by industry, academia, government entities, and others.[1] These systems are deployed to collect mission-specific data, and these data are not always accessible or in formats useful for other purposes. The IOOS Data Management and Communication (DMAC) system employs standards, protocols, and methodologies to improve data accessibility and interoperability, thereby facilitating the development of new information products and services.[4] Standardization is essential to easily shared data. IOOS is developing common standards and supporting data transport tools (such as web services or OPeNDAP) and community practices that enable easy access to real-time physical, chemical, and biological data. The RAs are working with regional data providers to expose their data through regional data portals, and as a result, more data are now widely available for public use. The benefits of this are tremendous: Over 50% of the marine data used by the National Weather Service comes from nonfederal sources enabled by DMAC.

5. PRIVATE SECTOR PARTNERSHIPS

The private sector is a vital IOOS partner—both as a provider and user of data and information. Private sector involvement includes IOOS data users such as fishermen preparing for days at sea and mariners guiding ships into congested ports. Private sector industries, such as oil companies, are important data providers. The private sector innovates new technologies, manufactures instruments, and develops software. Private sector partners also add value by utilizing IOOS data to develop tailored information products. Private sector companies are engaged at the federal level through data sharing agreements and at the regional level through involvement with the RAs. The regions serve as important field demonstration sites for companies developing new technologies such as the Liquid Robotics wave glider.

An agreement signed by NOAA and Shell Oil in 2009 set the stage for new and innovative approaches to ocean observing and private sector partnerships. The agreement goes beyond the lease requirements set forth by the Bureau of Ocean Energy Management, and it includes upgrades to existing observation systems, installation of new sensors, and data sharing. Through this partnership, National Weather Service forecasters have access to new data collected through upgrades to oil platform weather stations and transmission of meteorological data to the global telecommunication system. Shell Oil agreed to install new meteorological packages for wind speed and direction, barometric pressure, and air temperature to fill critical gaps along the Louisiana coast. These data help forecasters predict storm intensities, which is especially important for characterizing convective storms that develop over the Gulf of Mexico. In addition, Shell Oil agreed to install thermistor strings on fixed oil platforms at 100-m depth to measure temperature used in modeling and service information. This voluntary partnership benefits Shell by improving forecasts and understanding of the operating environment, and it provides forecasters, researchers, resource managers, and others access to new and important data.

6. CASE STUDIES

The first four case studies highlight how partnerships are making a difference in the areas of data sharing, fisheries, water quality, and forecasting harmful algal blooms. The fifth case study highlights the IOOS Coastal and Ocean Modeling Testbed (COMT) that bridges research modelers and federal operational centers. These case studies are representative of the type of collaborative projects that exist throughout the IOOS enterprise.

Case Study # 1 Agency-Industry-AOOS data sharing agreement in Alaska
Molly McCammon

A historic data sharing agreement between NOAA and three oil companies operating in Alaska's offshore waters, Shell Oil, ConocoPhillips, and StatOil, was signed in August 2011. The agreement laid the groundwork for the Alaska Ocean Observing System (AOOS), the IOOS Regional Association in Alaska, to provide public access to a wealth of oceanographic and environmental data.

The agreement calls for sharing several major data sets in the lease sale areas, including weather and ocean data, environmental study data and results, and eventually, sea floor mapping and other hydrographic data. During the open water season, real-time data is available to NOAA's National Weather Service to include in their weather forecasts, as well as to the AOOS Ocean Data Explorer's real-time sensor application. Historic weather, oceanographic, and environmental studies data are freely available through the AOOS Research Workspace and soon through the AOOS Ocean Data Explorer. All data are being archived at AOOS and at NOAA's National Ocean Data Center (www.aoos.org).

Available data includes Chukchi Sea Environmental Studies Program (CSESP) data between 2008 and 2013 (with 2014 data to be included in fall 2015) and research logistics data (study area boundaries and station locations) from 2013 to 2014. CSESP is a multiyear, interdisciplinary ecological study focused on areas in outer continental shelf oil and gas leases in the northern Chukchi Sea (http://dev.aoos.org/wp-content/uploads/2013/03/assets.jpg).

Industry-funded assets include buoys, research cruises, acoustic recorders, and radars. Real-time data from industry met-ocean buoys stream live through the AOOS Sensor Map during the open water season and are archived at the National Ocean Data Center (NODC). "Despite the wealth of scientific research conducted on the Arctic environment to date, much remains unknown, and no single government agency or entity has the resources or capacity to meet the task alone," Jane Lubchenco, former NOAA administrator and undersecretary of commerce for oceans and atmosphere, stated when the umbrella agreement was signed in 2011. "This innovative partnership will significantly expand NOAA's access to important data, enhance our understanding of the region and improve the United States' ability to manage critical environmental issues efficiently and effectively as climate change continues to impact the Arctic." Under the NOAA/Industry Data Sharing Agreement, NOAA conducts quality control on all industry data before it is incorporated into NOAA products and services.

The information now publicly available through AOOS is raw data from the industry studies and monitoring activities. This collaborative approach delivers many benefits:

- For NOAA: High-quality data enhances NOAA's ability to understand and monitor changes in climate, weather, and physical and biological ocean processes. Increased observations help validate and improve forecast models.
- For industry: By making raw data transparent, industry builds confidence in their research. Data sharing builds a positive relationship between industry and management.
- For the scientific community: It is estimated that the Environmental Studies data collection and quality control alone represents a nearly $100 million research investment. Through the AOOS Website, scientists can now access six years of multiple sets of biological, chemical, and physical data collected in the lease sale areas.
- For the general public: Better understanding of the lease areas could increase response capacity in case of an environmental disaster, such as a hazardous spill. Mariners and coastal residents benefit from improved weather and sea ice forecasts.

Case Study # 2 Fisheries: Cooperative development of dynamic habitat models for integration into formal stock assessments.
Josh Kohut, John Manderson

Long fin squid is a lucrative fishery in the Mid-Atlantic region but has been constrained because of the incidental bycatch of Atlantic butterfish. The two species occur together, and butterfish has been characterized by federal fishery regulators as a "data poor" stock because of the lack of catch data and few butterfish in the trawl surveys. In 2011, the Mid-Atlantic Regional Association Coastal Ocean Observing System (MARACOOS) partnered with NOAA's National Marine Fisheries Service (NMFS) to sponsor a multidisciplinary working group to develop improved habitat models for butterfish.

The working group included experts in marine ecology, physical oceanography, and stock assessment, fishermen, and representatives from government and academia. The goal was to improve understanding of the shifting habitats for butterfish. Models used for fisheries management depend on trawl surveys that are predetermined in where and when they will occur. The working group determined that temperature was the major indicator of where butterfish would be found, suggesting the trawl surveys were missing much of the butterfish and the important environmental data for fisheries management. As ocean conditions change, many fish are shifting their habitats. The trawl surveys are no longer sampling the full range of habitats used by the fish, which can lead to misinterpretations of the population size.

The habitat model developed with input from the working group was coupled with a regional numerical ocean model (the Regional Ocean Modeling System or ROMS) to first determine thermal habitat suitability over the last 40 years. These hindcasts were subsequently compared to the trawl surveys and accounted for the relative change in thermal habitat and the trajectory of sampling on the survey. The group's habitat-based estimate of availability was integrated into the butterfish stock assessment model accepted at the 59th NEFSC stock assessment review. This contribution resulted in a change in stock status and the establishment of a 20,000 metric ton quota for butterfish.

The collaborative approach with private, public, and academic partners increased the level of trust and respect for the expertise of participating partners and led to the development of a product that more accurately captures the dynamic complexity of the mid-Atlantic bight. Such an approach can be applied in the scientific assessments of many other fish stocks. "A lot of people from many different disciplines played an integral part in the success of this, and I think the results speak for themselves," said Greg DiDomenico, Executive Director of the Garden State Seafood Association. "This type of collaboration needs to be applied to other species." Successes like this would not be possible if it were not for the shared approach of our multidisciplinary, multisector study group.[5]

We are grateful for support provided by the NOAA Fisheries habitat assessment improvement team, the Northeast Fisheries Science Center Cooperative Research Program, the US IOOS Mid-Atlantic Regional Association Coastal Ocean Observing System, and the Mid-Atlantic Fishery Management Council.

Case Study #3 Water Quality: Modeling the Huron–Erie Corridor
David J. Schwab, Jennifer Read, and Eric J. Anderson

The Huron–Erie Corridor connects Lakes Huron and Erie and constitutes a section of the international border between Michigan and Ontario. It is an inland waterway made up of the St. Clair River, Lake St. Clair, and the Detroit River, and is the only shipping route between the upper and lower Great Lakes. A steady stream of commercial cargo and recreational boating pass along the corridor. The area is also heavily industrialized—including several major petrochemical plants and docks. The presence of this industry combined with the area's drinking water intake facilities, recreational beaches, and more than 150,000 registered recreational boats make the corridor a high-risk area for contaminant spills and the potential negative impacts a spill could have on the area's industry, recreation, human health, and wildlife.

On March 16–17, 2006, the Great Lakes Observing System (GLOS) and the Southeast Michigan Council of Governments (SEMCOG) co-hosted a technical workshop to develop an implementation strategy to assess the prospects for hydrodynamic model development within the corridor in the 2007–2009 time frame to assist with managing water quality. Nearly 50 individuals, representing a broad cross-section of binational stakeholders engaged in the Huron Erie Corridor (HEC), participated in the two-day workshop. Attendees included representatives of most US and Canadian federal, provincial, county, and municipal governments, regional academic institutions, binational regional agencies, commercial interests, and other stakeholders with strong modeling interest/expertise. The purposes were to clarify the goals, identify user needs, assess types of models, determine data requirements, and discuss potential costs and timelines for implementation.

The major conclusions of this workshop were the following:

- Three-dimensional modeling is needed over the entire geography of the HEC.
- The 3-D model should initially support refinements to real-time monitoring operations, with high detail in critical areas of the system.
- Initially, the 3-D model should be used to predict fate and transport for hazardous spills.
- The 3-D modeling should be public domain.
- Significant care needs to be exercised to ensure that expectations meet reality in terms of model output.

In 2008, GLOS awarded a contract from the Cooperative Institute for Limnology and Ecosystems Research (CILER) at the University of Michigan to develop a hydrodynamic modeling system for providing real-time nowcasts (present conditions) and forecasts (predictions) of physical conditions such as water currents and levels in the HEC. Over the course of the next two years, the Huron–Erie Corridor Waterways Forecasting System (HECWFS) was designed, validated, tested, and made operational in close collaboration with the NOAA Great Lakes Environmental Research Laboratory (GLERL) in Ann Arbor, MI.

The forecasting system provides nowcasts every 3 h, which are made available to the public in real-time using Google Earth, and 48-h forecasts every 12 h. Before 2008, when HECWFS became operational, a real-time prediction system was unavailable for the corridor despite the crucial role it plays in recreation, drinking water, industry, and commercial shipping. Decision-makers, stakeholders, and law enforcement can now use the information provided by the forecast system to aid in navigation, spill response, identifying fish kill sources, search and rescue, forensics, drinking water safety, beach quality forecasting, and more.

In addition to real-time forecasts, HECWFS has been used to simulate contaminant spill scenarios for the area. This information was used to develop the Spill Reference Library, which contains several sets of tables that aim to inform and give decision makers the tools needed to plan for, and react to, a contaminant spill. Each table includes information about how an area might be impacted, like the estimated arrival time of the leading and trailing edge of the spill, the maximum concentration of the contaminant, and whether the contaminant mixed vertically or horizontally in the water column.

Through a continuing partnership between GLERL and GLOS, the nowcast and forecast information from HECWFS is available in real time through a web-based tool that provides detailed maps of currents and water levels. Figure 1 shows the kind of information that is available from the forecasting system. HECWFS has also been used by USGS to optimize the installation of artificial fish-spawning reefs in the HEC.

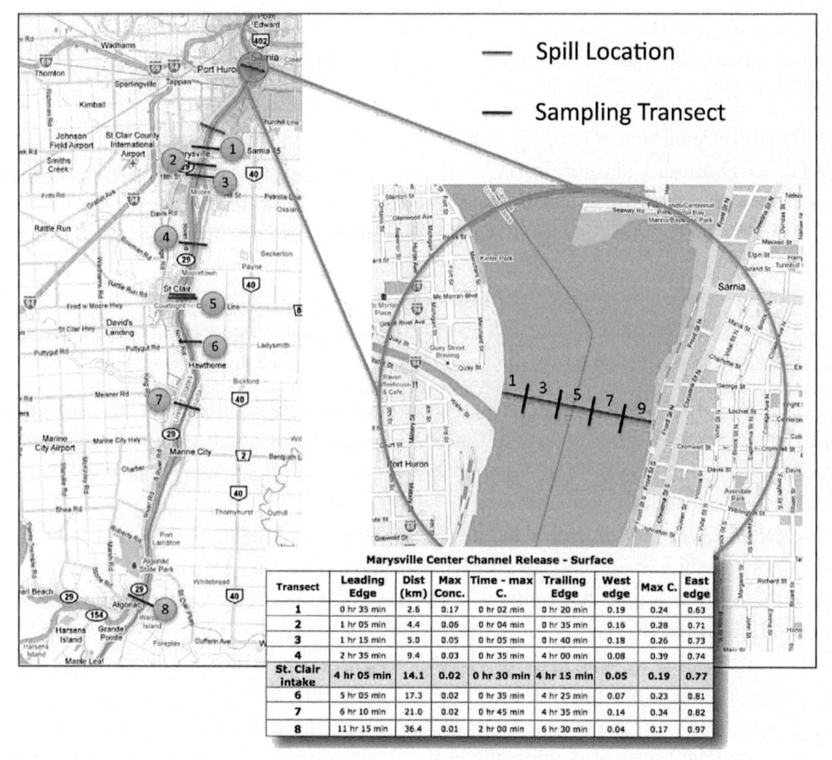

Transect	Leading Edge	Dist (km)	Max Conc.	Time - max C.	Trailing Edge	West edge	Max C.	East edge
1	0 hr 35 min	2.6	0.17	0 hr 02 min	0 hr 20 min	0.19	0.24	0.63
2	1 hr 05 min	4.4	0.06	0 hr 04 min	0 hr 35 min	0.16	0.28	0.71
3	1 hr 15 min	5.0	0.05	0 hr 05 min	0 hr 40 min	0.18	0.26	0.73
4	2 hr 35 min	9.4	0.03	0 hr 35 min	4 hr 00 min	0.08	0.39	0.74
St. Clair intake	**4 hr 05 min**	**14.1**	**0.02**	**0 hr 30 min**	**4 hr 15 min**	**0.05**	**0.19**	**0.77**
6	5 hr 05 min	17.3	0.02	0 hr 35 min	4 hr 25 min	0.07	0.23	0.81
7	6 hr 10 min	21.0	0.02	0 hr 45 min	4 hr 35 min	0.14	0.34	0.82
8	11 hr 15 min	36.4	0.01	2 hr 00 min	6 hr 30 min	0.04	0.17	0.97

FIGURE 1

Screen shot of the Huron–Erie Waterways Forecasting System.

Case Study #4: Harmful Algal Blooms: An Approach to Forecasting Red Tide Impacts on Florida's Beaches

Barbara Kirkpatrick, Richard P. Stumpf, Michelle Tomlinson, and Gary J. Kirkpatrick

An example of a federal–regional partnership that has benefited both partners and their institutions' missions is between the NOAA National Centers for Coastal Ocean Science (NCCOS), the state of Florida, and the Mote Marine Laboratory (Mote) in Sarasota, FL. This one example was selected from the many productive partnerships contributing to the science, forecasting, and management of harmful algal blooms.

The west coast of Florida has almost annual blooms of the toxic dinoflagellate, *Karenia brevis*, which produce the Florida red tide. Identifying the location and movement of blooms is critical for the protection of human health, both with potential consumption of contaminated shellfish and/or the inhalation of toxic aerosols. Prior to 1994, water samples were obtained as an event response action only when dead fish and/or discolored water were reported. In 1994, Mote began a year-round sample schedule that provided early detection of a bloom.

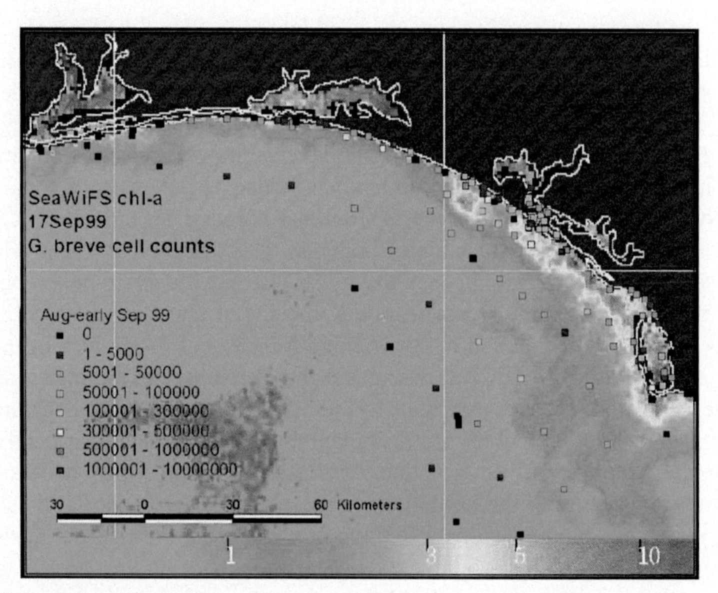

FIGURE 2

Sampling sites along the Florida Panhandle.

Work by Karen Steidinger and Ken Haddad of the Florida Marine Research Institute (FMRI) in the early 1980s showed that data from an ocean color satellite (the Coastal Zone Color Scanner or CZCS) could be useful in finding Florida red tides.[6] However, an operational sensor did not exist until the launch of the satellite Sea-viewing Wide Field of View Sensor (SeaWiFS) in August 1997. At this point, several actions converged. In 1998, Florida Marine Research Institute (FMRI, now the Florida Fish and Wildlife Commission) asked NOAA to help by providing satellite imagery. NOAA through the CoastWatch program (which originated in 1988 after a Florida red tide showed up in North Carolina) arranged to purchase SeaWiFS data. Then, EPA's Gulf of Mexico Program office funded NOAA's NCCOS to collaborate with the Naval Research Lab to determine how to process SeaWiFS data in order to identify the red tide blooms in coastal waters. A dramatically improved atmospheric correction for coastal waters resulted from this work.

By September 1999, the first NCCOS/NRL algorithms were in place, and the Mote Marine Lab led the research cruise. NCCOS produced the first bulletins, providing descriptions of the bloom location to the R/V Pelican at sea off the Florida Panhandle, which led to targeted sampling. These locations are shown in Figure 2.

The system rapidly improved with a bulletin format by late 2001 that included ancillary information, especially cell counts that were pushed to NOAA from Mote and FMRI. The anomaly method of finding blooms was implemented.[7] By 2004, the bulletin was viewed as stable enough for NOAA to move the production from NCCOS, a science office, to operations in the Center for Operational Oceanographic Products and Services (http://tidesandcurrents.noaa.gov/hab/).

With the forecast system in place, the need for validation of the forecasts, both in water samples and respiratory irritation to beach goers became critical.[8] In 2006, Mote initiated the Beach Conditions Reporting System© (BCRS) that provided twice daily beach reports on Sarasota County beaches (http://coolcloud.mote.org/bcrs/). County lifeguards were provided smart phones that allowed real-time reports to be filed from the beach. The data are seamlessly provided on a Website and also pushed to the NOAA Harmful Algal Bloom Operational Forecast System (HAB-OFS) for forecast validation. This partnership of boots on the ground efforts from Mote and remote sensing and forecasting capability by NOAA has proven to be a robust partnership that protects public health from the impacts of these red tides. The BCRS provides important information in determining forecasts of respiratory irritation.

The availability of the BCRS also meant that forecasts of respiratory irritation could be evaluated, a key component to identifying strengths and deficiencies in the forecasts. While the blooms are presumed to be patchy, data on patchiness has been lacking. Stumpf et al.[9] showed that patchiness was quite critical to determining beach impact. Because of the limitations in the data (imagery is not available daily and cannot resolve blooms within 1 to 2 km of the shore), forecasts are made at the resolution of county to half-county at twice a week. The BCRS showed that this worked well, but failed at beach level. Forecasts of respiratory irritation within a county were correct over 70% of the time. If respiratory irritation was forecasted in a county, one of the beaches would report irritation. However, at individual beaches, the forecasts were correct only 20% of the time.

The sustained bloom of 2005 and evidence of health impacts from toxic aerosols has increased need to know impacts at beaches. In addition to the work with the BCRS, Mote has coordinated discussions with the community through public forums and meetings with chambers of commerce, community associations, and visitor and convention bureaus. These discussions influence the character of the forecasts, how they are distributed, and information that is sought. The continued discussions have shaped the need for more targeted information on respiratory impact. The National Weather Service is now being used to distribute warnings on the most extreme conditions, and NOAA and Mote are developing a strategy to improve the resolution of the model, with a goal of achieving "every beach, every day."

Case Study #5: Facilitating Partnership: A modeling testbed for IOOS
Rick Luettich, Becky Baltes

The U.S. IOOS Coastal and Ocean Modeling Testbed (COMT) has been built around partnerships between federal operational centers, academia, and the private sector. Initiated in 2010, the COMT mission is "targeted research and development to accelerate the transition of scientific and technical advances from the coastal ocean modeling research community to improved operational ocean products and services." This mission embraces both "research to operations" via the transition of scientific and technical advances and "operations to research" via targeted research and development. Partnerships between the operational and research communities are critical for successfully completing this mission both in terms of identifying priority research and development areas and achieving transition.

The transition mentioned above can mean adoption of new modeling software within an operational center. However, thus far in the COMT, the focus is on the transition of improved knowledge and understanding about the models currently operated and about prioritizing resource investments to advance modeling capabilities. Knowledge transfer is most effective when researchers and operational end users are working closely together on a project so as to infuse research activities with clear practical objectives and to maximize the likely acceptance of research results within an operational setting.

The initial set of COMT projects covered topics ranging from ecological and hypoxia modeling in the Chesapeake Bay and the Northern Gulf of Mexico to storm surge, wave, and inundation modeling in the Gulf of Maine and the Gulf of Mexico.[10] The group modeling storm surge, waves, and inundation in the Gulf of Mexico was comprised of academic researchers from the University of North Carolina at Chapel Hill, the University of Notre Dame, the University of South Florida, and the Virginia Institute of Marine Sciences along with operational counterparts at the NOAA National Hurricane Center and the NOAA Coast Survey Development Laboratory. The groups worked closely to evaluate the performance of multiple models, physics implementations and parameterizations, grid resolutions, and domain sizes for replicating the observed response along the Louisiana and Texas coastlines to hurricanes Rita and Ike.[11–14] Examples of transition from this group, in terms of the targeted research findings and the operations they impacted, include the following:

- Two-dimensional, vertically integrated models were shown to be able to reproduce combined tide, wind, and wave-driven surge and inundation as accurately and robustly as three-dimensional, vertically resolving models, but with much faster execution speeds. This finding was used by NOAA to pursue development of a coupled runoff, tide, wave, surge, and inundation model using a higher horizontal resolution two-dimensional model versus a lower horizontal resolution three-dimensional model.
- Computational efficiency on multicore, high-performance computing systems varied significantly between the tide/storm surge/inundation models. Due to its accuracy for the testbed storms, advanced physics, and the computational efficiency of its highly scalable solution algorithm, ADCIRC, it was determined to be capable of supporting coupled modeling at a reasonable computational cost and was selected as the basis for further development as NOAA's coupled runoff, tide, wave, surge, and inundation model.
- Model accuracy was shown to depend significantly on the inclusion of tides, on horizontal grid resolution, and on model domain size. In particular, omission of tides, the small size of the model domains, and inadequate grid resolution in the National Hurricane Center's SLOSH model were shown to degrade model accuracy. Since the conclusion of the initial COMT project, the storm surge group at the National Hurricane Center has focused on adding tides to SLOSH and on revising SLOSH basins to expand their coverage area and to increase horizontal grid resolution.

As a result of the success of the first COMT project, a new COMT project has been initiated between the academic community, the NOAA Environmental Modeling Center, the National Hurricane Center, and the U.S. Army Corps of Engineers to evaluate the performance of models for wave-dominated storm surge and inundation in regions of steep bathymetry (e.g., Puerto Rico, US Virgin Islands, and Hawaii). The knowledge and understanding gained from this work is expected to guide decisions determining the future use of models at these operational centers.

7. CONCLUSION

The IOOS partnership structure offers an alternative to large government programs by linking federal agencies with researchers, managers, industry, and stakeholders at the local and regional level to identify needs and to seek solutions. The Regional Associations act as a conduit for the 17 federal agencies to understand how national priorities manifest at the local and regional level and for local and regional partners to share data, expertise, knowledge, and resources. The RA structure is the only regional structure that brings together multiple federal agencies with tribal, state, and local government, industry, nongovernmental organizations, researchers, and the general public to discuss the need for coastal information for multiple purposes. These relations foster innovative efficiencies by leveraging existing resources and expertise. IOOS partnerships are increasing the amount of data available for improving our understanding of coastal oceans and developing information products, enhancing operational forecasts through coordination with researchers, and furthering our understanding of the coastal and Great Lakes environments. The challenge will be to maintain and expand these partnerships so that they become the way of doing business.

REFERENCES

1. U.S. Commission on Ocean Policy. *A blueprint for the 21st century, final report*. 2004 [Washington, DC].
2. Sherman K, Hempel G, editors. *The UNEP Large Marine Ecosystem Report: a perspective on changing conditions in LMEs of the world's Regional Seas. UNEP Regional Seas report and studies No. 182*. Nairobi, Kenya: United Nations Environment Programme; 2008.
3. Interagency Ocean Observation Committee. *Integrated coastal and ocean observing system act of 2009*. 2009. Pub. L. No. 111-11, 33 USC 3601–3610.
4. IOOS. *Data management and communications plan concept of operations*. Version 1.5. Silver Spring, MD: NOAA; 2009.
5. Clark M. *New assessment determines that butterfish are not overfished: council applauds collaborative efforts to determine butterfish stock status*. Dover, DE: Mid-Atlantic Fisheries Management Council Press Release; 2014.

6. Steidinger KA, Haddad K. Biologic and hydrographic aspects of red tides. *BioScience* 1981;**31**:814–9.
7. Stumpf RP, Culver ME, Tester PA, Kirkpatrick GJ, Pederson BA, Tomlinson M, et al. Monitoring *Karenia brevis* blooms in the Gulf of Mexico using satellite ocean color imagery and other data. *Harmful Algae* 2003;**2**:147–60.
8. Kirkpatrick B, Currier R, Nierenberg K, Reich A, Backer L, Stumpf R, et al. Florida red tides and human health: a pilot beach conditions reporting system to minimize human exposure. *Sci Total Environ* 2008;**402**:1–8.
9. Stumpf RP, Tomlinson MC, Calkins JA, Kirkpatrick B, Nierenberg K, Currier R, et al. Skill assessment for an operational algal bloom forecast system. *J Mar Syst* 2009;**76**: 151–61.
10. Luettich Jr RA, Wright LD, Signell R, Friedrichs C, Friedrichs M, Harding J, et al. Introduction to special section on the U.S. IOOS coastal ocean modeling testbed. *J. Geophys Res Oceans* 2013;**118**:6319–28. http://dx.doi.org/10.1002/2013JC008939.
11. Kerr PC, Donahue AS, Westerink JJ, Luettich Jr RA, Zheng LY, Weisberg RH, et al. U.S. IOOS coastal and ocean modeling testbed: inter-model evaluation of tides, waves, and hurricane surge in the Gulf of Mexico. *J. Geophys Res Oceans* 2013;**118**. http://dx.doi.org/10.1002/jgrc.20376.
12. Kerr PC, Martyr RS, Donahue AS, Hope ME, Westerink JJ, Luettich Jr RA, et al. U.S. IOOS coastal and ocean modeling testbed: evaluation of tide, wave, and hurricane surge response sensitivities to mesh resolution and friction in the Gulf of Mexico. *J. Geophys Res Oceans* 2013;**118**. http://dx.doi.org/10.1002/jgrc.20305.
13. Hope ME, Westerink JJ, Kennedy AB, Kerr PC, Dietrich JC, Dawson C, et al. Hindcast and validation of hurricane Ike (2008) waves, forerunner, and storm surge. *J. Geophys Res Oceans* 2013;**118**. http://dx.doi.org/10.1002/jgrc.20314.
14. Zheng L, Weisberg RH, Huang Y, Luettich Jr RA, Westerink JJ, Kerr PC, et al. Implications from the comparisons between two- and three-dimensional model simulations of the hurricane Ike storm surge. *J. Geophys Res Oceans* 2013;**118**. http://dx.doi.org/10.1002/jgrc.20248.

CHAPTER 4

Basic Tenets for Coastal Ocean Ecosystems Monitoring

Robert H. Weisberg*, Lianyuan Zheng, Yonggang Liu
College of Marine Science, University of South Florida, St. Petersburg, FL, USA
**Corresponding author: E-mail: weisberg@usf.edu*

CHAPTER OUTLINE

1. INTRODUCTION

Coastal Ocean Observing Systems (COOS) have been in discussion for over a decade. The need for more comprehensive sets of observations and models became clear during the Deepwater Horizon oil spill in 2010 when the ability of the federal and state agencies to track oil under the unified command structure was brought into question. This was particularly true of the coastal ocean that lacked both adequate ocean observations and high-resolution models. Now, more than a dozen years after an Ocean.US, May 2002 report advanced the rationale and plan for an Integrated Ocean Observing System (IOOS)[1] and more than four years post the Deepwater Horizon event, the oceanographic community remains without sufficient COOS resources.

The Deepwater Horizon event did result in legislation and other vehicles by which resources could be directed. In particular, monetary sources were established through the RESTORE Act, the British Petroleum funded Gulf of Mexico Research Institute, the National Fish and Wildlife Federation, and the National Academy of Sciences. To date, however, none of these entities has advanced the concept of a sustained observing system, either for the Gulf of Mexico as a whole, or for the coastal ocean portion that spans the continental shelf and the estuaries. The coastal ocean is particularly important because that is where society literally meets the sea. Presented herein is a COOS plan for the west coast of Florida that builds upon existing resources, experience, and applications to environmental problems of societal importance. Though specific to the west coast of Florida, its design principles are applicable elsewhere.

The RESTORE Act draws its acronym from The Resources and Ecosystems Sustainability, Tourist Opportunities, and Revived Economies of the Gulf Coast States Act of 2012 (RESTORE Act, P.L. 112–141). Thus, it serves many purposes, from economic restitution to environmental science, and it will define and strongly influence the expenditure of funds for Gulf of Mexico science over the foreseeable future. Given its potential magnitude, other substantial resources will likely not be forthcoming. How RESTORE ACT funds are utilized, in addition to the other Deepwater Horizon–related distributions, will largely define the immediate future for Gulf of Mexico science.

"You cannot fix it if you do not know how it works, and you cannot restore it if you do not know what it was to begin with," opined one of us (Weisberg) in regard to RESTORE Act testimony in December 2011. Determining how something works requires application of the scientific method, which begins with observations. COOS discussions often use observations and monitoring interchangeably. There is an important difference, however, because monitoring is of little value if we do not know what and where to monitor. An automobile temperature gauge provides an illustration. Temperature is monitored because overheating may cause the engine block to crack. But how did we arrive at knowing what and where to monitor? The answer is, we applied principles of materials science to define engine tolerances, and we experimented with temperature gauge placement to ensure adequate performance. In other words, monitoring followed the application of the scientific method, whereby carefully planned observations and scientific and engineering principles were first applied. Studying the coastal ocean must proceed similarly. Otherwise, the environmental aspects of the RESTORE Act will not be realized.

Section 2 lists a set of tenets by which we may study the coastal ocean and how these may be prioritized. Section 3 offers application examples demonstrating the societal utility of a comprehensive and coordinated set of observations and science-based models. Section 4 discusses experimental design, offering a rationale for each data point. Whereas every coastal region is somewhat unique, we believe that the tenets and design are of a general nature, and hence, these may be applicable elsewhere. A summary and set of recommendations (Section 5) concludes this chapter.

2. THE TENETS

Coastal ocean ecology integrates all of the processes that are responsible for organism success. This begins with the coastal ocean circulation that unites nutrients with light, thereby fueling primary productivity and, thence, all higher trophic level interactions. It is not surprising that West Florida Continental Shelf (WFS, Figure 1) hydrographic sections (e.g., Figure 2) generally show the highest levels

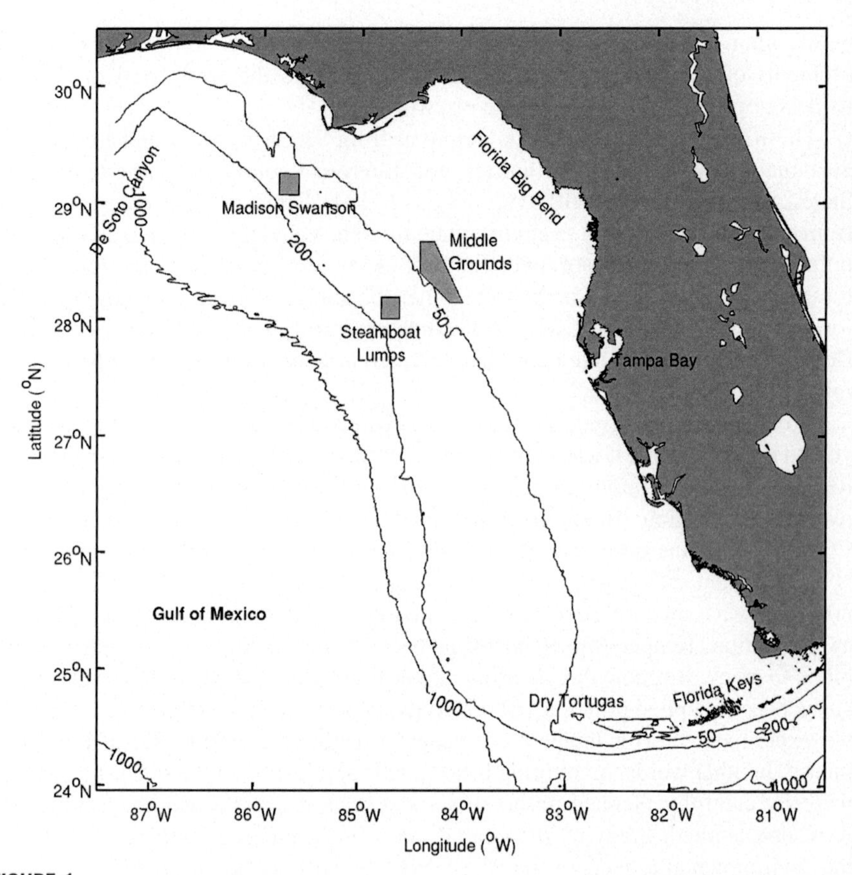

FIGURE 1

A map of the West Florida Continental Shelf with sites identified as referred to in text. Shown are the 50-, 200-, and 1000-m isobaths.

of chlorophyll near the bottom, versus near the surface. Thus, coastal ocean ecology is not merely the purview of biology. It requires a fully multidisciplinary approach.

Progress in understanding WFS coastal ocean ecology accelerated in 1998 when a multidisciplinary team was constituted to study *Karenia brevis* harmful algae blooms (HABs) as part of a NOAA ECOHAB regional field study. Prior to 1998, we were knowledgeable about the organism, and that it tended to preferentially bloom along the shore between the Tampa Bay and Charlotte Harbor estuaries, but we did not have explanations as to why.[2] Such explanations only began to evolve once the phenomenon was viewed in a more systems science manner.[3–6] This, of course, is not unique to the WFS. *Alexandrium* explanations in the Gulf of Maine[7] also relied on a systems science approach to coastal ocean ecology, and similar examples abound elsewhere.

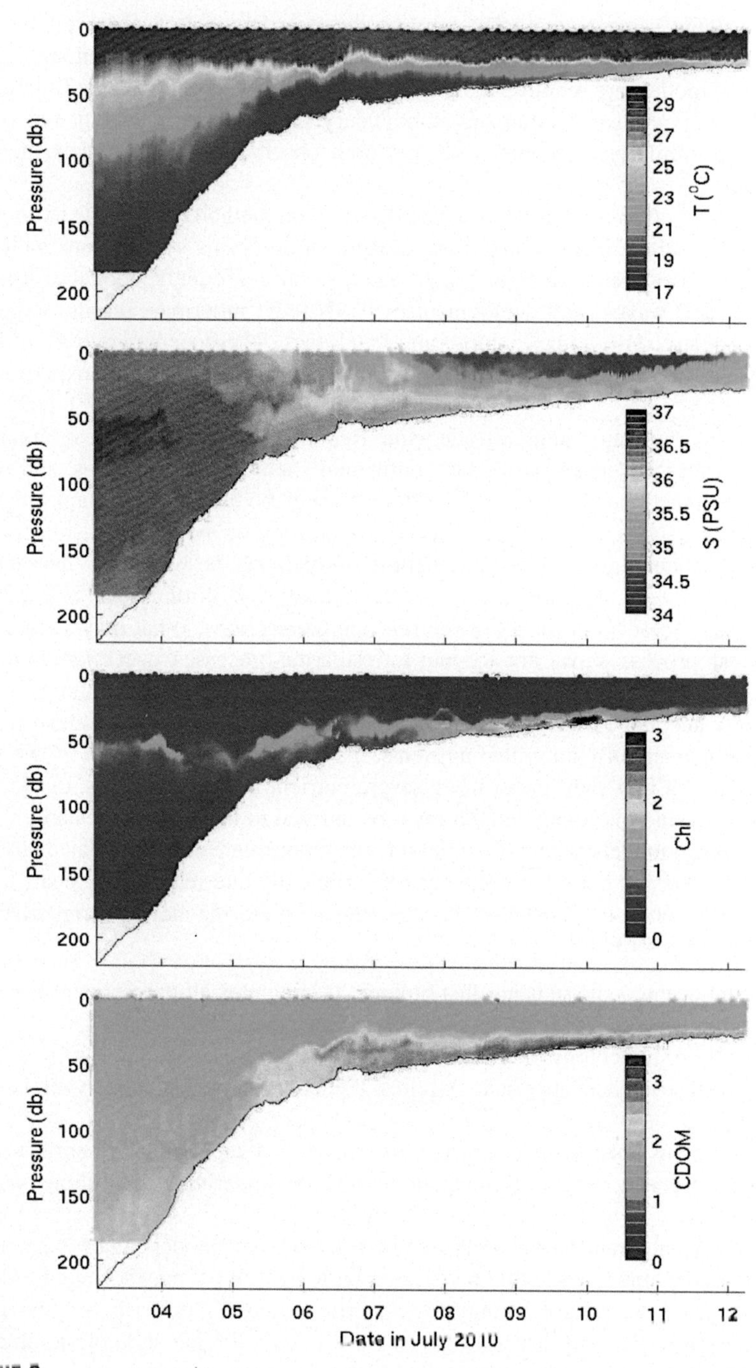

FIGURE 2

Temperature, salinity, chlorophyll, and CDOM sampled on an across-shelf glider transect situated just north of Tampa Bay, FL, during July 2010. *(Adapted from Weisberg et al.[13])* Note the effects of upwelling on the WFS water properties.

A systems science approach requires the coordination of observations with models. Three dimensionality and spatial extent preclude ever having enough observations, so models are required to more fully describe the coastal ocean. But models without observations for initialization, boundary values, veracity testing, and assimilation are similarly incomplete. Hence, both observations and models must be coordinated.

Being that no single sensor or sensor deployment method is adequate to describe spatially vast, three-dimensional phenomenon, an observing system must include a variety of sensors (e.g., velocity, temperature, salinity, nutrients, optical properties, sea level, and waves) and deployment methods (e.g., moorings, profilers, gliders, high-frequency (HF) radars, ships, and satellites). These demands may quickly outstrip any reasonable allocation of resources without a judicious experimental design.

Similarly, no single model is adequate. Requirements for deep ocean modeling may be different from those for the continental shelf and the estuaries, and being that the connections between these different water bodies are important, all must be properly considered. Required then is a hierarchy of models (e.g., for ocean–atmosphere interactions, ocean circulation, waves, and the biological interactions that, together with the circulation, comprise ecology). Within such hierarchy is the necessity to account for all of the relevant connections. What may account for deep ocean processes may not account for estuarine processes, necessitating downscaling from the deep ocean, across the continental shelf, and into the estuaries. What may account for certain biological phenomena, such as diatoms near a river mouth with essentially unlimited nutrients, may not account for slower growing dinoflagellates at mid-shelf under more severe nutrient limitations. Thus, the suite of variables within a biological model must be tailored to the problem at hand. Given that biological models necessarily contain many more state variables than a physical model, their design considerations must be efficiently and realistically dealt with.

The basic tenets for coastal ocean ecosystems monitoring may be summarized by the following set of bullets:

- Coastal ocean ecology is not just biology. It integrates all processes responsible for organism success.
- Coastal ocean ecology, therefore, requires a systems science approach.
- A systems science approach requires the coordination of observations with models.
- Observations must include a variety of sensors and deployment systems, and no single observing method is adequate to describe underlying three-dimensional processes.
- Similarly, hierarchies of models are needed. No single model can be expected to handle all connections, either physical or biological. What may account for deep ocean processes cannot account for estuarine processes, necessitating downscaling from the deep ocean to the estuaries, and a limited suite of variables cannot, in a general way, account for biological phenomena.

- A reduced monitoring array may eventually be possible (from the previous) once we know how the coastal ocean system works and, therefore, what and where to monitor.

3. RECENT APPLICATION EXAMPLES

Three recent examples help to illustrate these tenets. The examples derive from applications of a coordinated program of coastal ocean observing and modeling aimed at addressing WFS ecological matters of societal concern, including the following: *Karenia brevis* HABs, gag grouper recruitment, and the findings of fish lesions subsequent to the Deepwater Horizon oil spill.

Independent from the Deepwater Horizon oil spill,[8] 2010 exhibited a period of anomalously prolonged and intense upwelling that lasted from mid-May through the end of the year. This upwelling was not just wind-induced; instead, it was facilitated by deep ocean interactions with the shelf slope, particularly by the Gulf of Mexico Loop Current contacting the southwest corner of the WFS near the Dry Tortugas. Such contact commenced around May 20, 2010, when the Loop Current shed an eddy and retreated back to the south. Both the eddy and the parent Loop Current stayed in contact with the shelf slope until the eddy eventually sidled westward, leaving only the Loop Current in place.[9] These contacts resulted in relatively high dynamic height perturbations that propagated along the shelf slope toward the north as prescribed by continental shelf wave dynamics.[10,11] The eddy-induced perturbations only extended shoreward by an internal Rossby radius of deformation, whereas the Loop Current perturbations extended across the entire shelf. The reason for these different responses to the eddy and Loop Current interactions is the geometry of the southern WFS, where all shallow water isobaths must wrap around the Dry Tortugas, the westernmost island in the Florida Keys chain.[12] The end result of the Loop Current interaction was a southward-directed geostrophic current that extended (along with the across-shelf pressure gradient force) across the entire WFS. Frictional turning (to the left and, hence, shoreward) across the bottom Ekman layer then drove the anomalous upwelling circulation.

These findings are described in detail,[13] along with how they were observed, using a combination of moorings, HF radar, gliders, and satellites and how the circulation and other water properties were modeled using the West Florida Coastal Ocean Model (WFCOM). The WFCOM[14] downscales from the deep ocean, across the continental shelf, and into the estuaries by nesting the Finite Volume Coastal Ocean Model (FVCOM,[15]) in the Hybrid Coordinate Ocean Model (HYCOM, e.g.,[16]). As a consequence of the anomalously intense and prolonged upwelling, inorganic nutrients of deep ocean origin were transported across the shelf break, thereby favoring faster growing diatoms over the slower growing *Karenia brevis* dinoflagellates. Whereas upwelling is an essential part of a *K. brevis* bloom manifesting along the shoreline, too much upwelling opposes bloom development. From these findings, it was concluded that the ocean circulation physics and the organism biology are each necessary conditions for *K. brevis* bloom development, but neither alone are sufficient conditions.[13]

Using this same logic, we subsequently (and successfully) predicted that there would not be a major *Karenia brevis* bloom in 2013 and that there would be one in 2014. *K. brevis* HABs provide an example of why coastal ocean ecology must be treated in a multidisciplinary manner. Note the deep chlorophyll maximum, indicative of an offshore nutrient source, in Figure 2. These were among the observations (along with velocity profiles from moorings and surface velocity fields by HF radar) that we used for quantitatively gauging the WFCOM performance.

The next example is gag grouper recruitment. Gag are known to spawn near the shelf break and to begin their juvenile growth stage either in near shore or estuarine grass beds.[17] How they transit the shelf from offshore spawning to near shore settlement sites remained unknown. By combining coastal ocean circulation physics with fisheries biology, the observed juvenile observations in 2007 were accounted for using a WFCOM simulation (gauged against coastal ocean circulation observations) and other observed water properties.[18] The circulation physics are similar to those already explained.[13] Loop Current interactions along the shelf slope near the Dry Tortugas are again germane. If these occur in phase with spawning (primarily in late winter to early spring) and last long enough, then a particular year class may be successful. The 2007 year class provided a case in point. We tested both surface and near bottom transport route hypotheses and rejected the failed surface route hypothesis in favor of the successful near bottom route hypothesis. Further support for the near bottom transport route was gained from macro-algae of deep, hard bottom origin that were found to be colocated with the gag juveniles, plus other biochemical evidence. Simulated bottom route trajectories are shown in Figure 3. As with *Karenia brevis* HABs, gag recruitment is intimately tied to the coastal ocean circulation.

Our final example harkens back to the Deepwater Horizon oil spill and the observation of fish lesions detected over the next year. Given the well-established occurrence of anomalously intense and prolonged upwelling, it was hypothesized that any hydrocarbons located within the water column, either dissolved or of sufficiently small particulate size to have effectively been dissolved, would have been carried from the north Florida shelf region, where surface oil was abundant, to the WFS, where no surface oil was detected. Exploring this hypothesis led to the conclusion that hydrocarbons of Deepwater Horizon origin did transit to the WFS sight unseen beneath the surface.[19] We used a WFCOM tracer simulation for which a passive tracer was initialized where satellite imagery defined its presence, and we then modeled the tracer evolution in time and space. The tracer patterns (Figure 4) matched the distribution of observed fish lesions, and limited liver chemistry samples[20] offered additional evidence (albeit not statistically significant due to relatively few samples). By making reasonable assumptions of the initial tracer concentration and the subsequent dilution by advection and diffusion, we arrived at WFS tracer concentrations that may have plausibly affected reef fish.

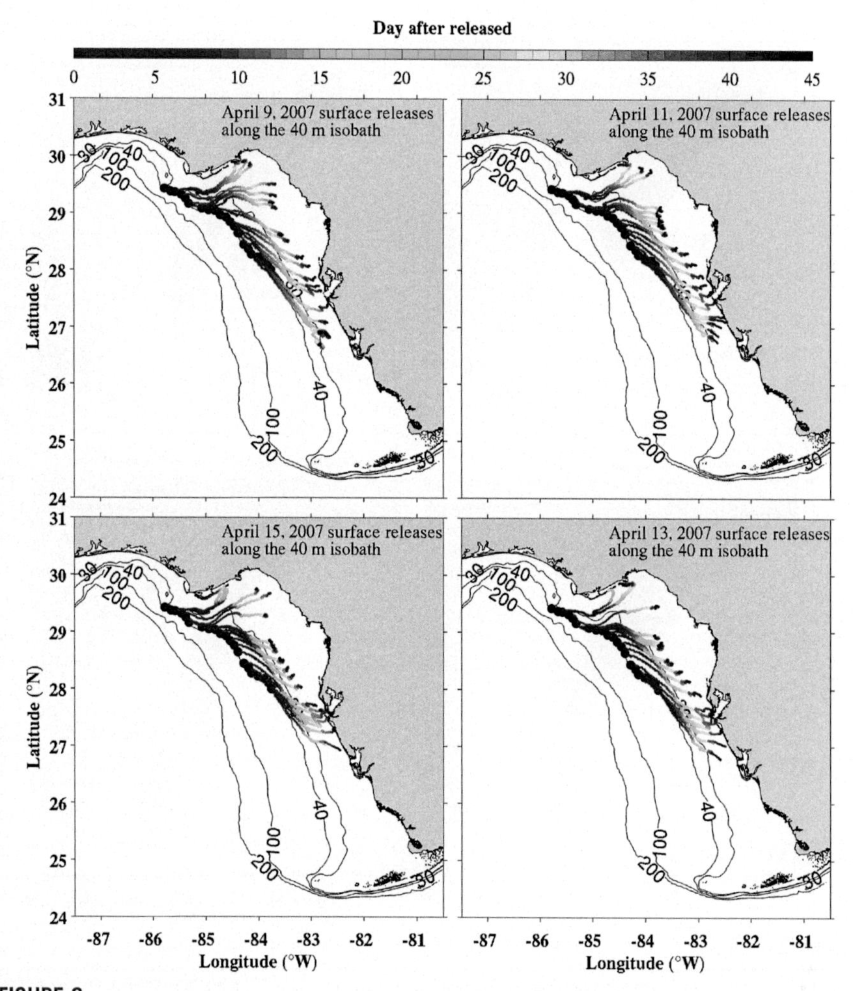

FIGURE 3

Model simulated near bottom particle trajectories for particles initialized on the 40-m isobath, beginning (clockwise from the upper left hand panel) on April 9, April 11, April 13, and April 15, 2007, respectively. *(Adapted from Weisberg et al.[18])* Particles of offshore, Big Bend origin arrived onshore where gag juveniles were observed.

4. EXPERIMENTAL DESIGN

Given the basic tenets of Section 2, how do we prioritize sampling locations to arrive at an array design? The scientific method, which begins with observations, provides guidance. Much is known about the continental shelf. For instance, the shelf spans several dynamical regimes. The outer shelf, where deep ocean interactions occur,

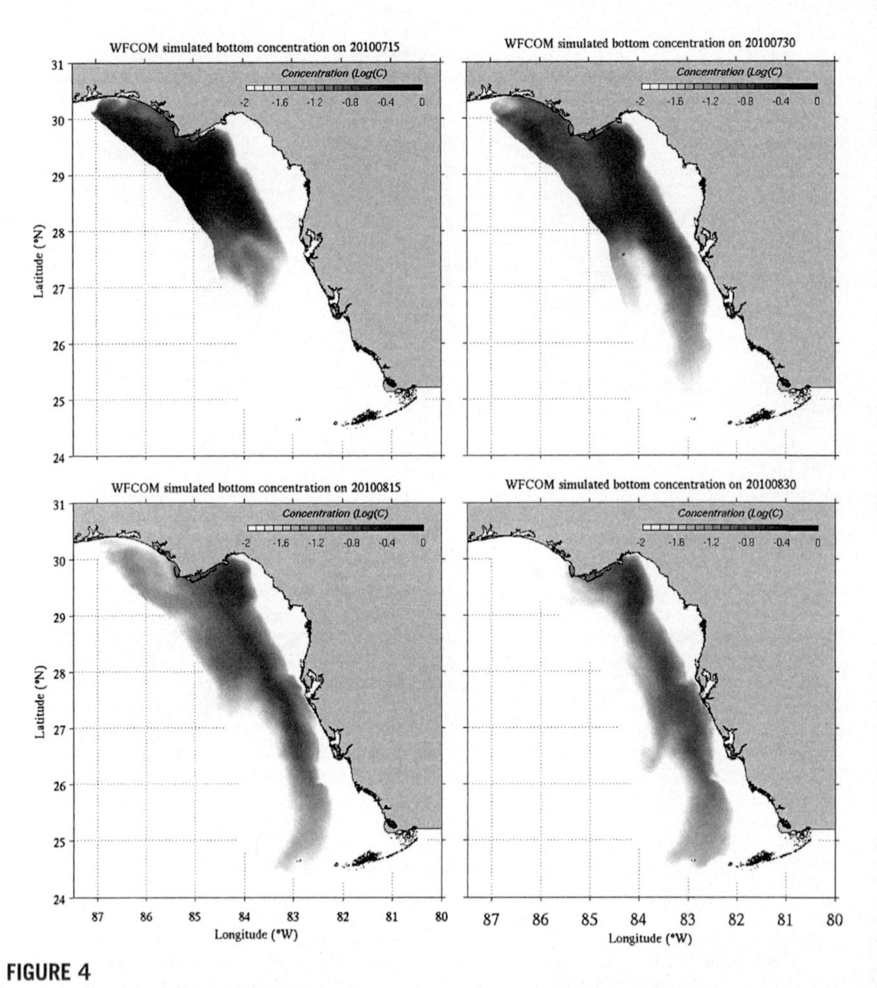

FIGURE 4

Normalized near bottom tracer concentrations on July 15, 2010; July 30, 2010; August 15, 2010; and August 30, 2010. The initial value for the tracer is 1.0, and the tracer concentration gray scale is provided as $\log_{10}$.

Adapted from Weisberg et al.[19]

extends a baroclinic Rossby radius of deformation shoreward from the shelf break. The inner shelf, where surface and bottom Ekman layers interact through divergence, extends across the entire shelf if the shelf is narrow, or only partially across the shelf if the shelf is wide enough. Within the inner shelf is a region over which low-salinity water of estuarine origin may further alter the stratification and baroclinic structure. These dynamical regions provide a physical basis for experimental design.

Recognizing that observations must be coordinated with models, it follows that the models must be driven by the most accurate forcing functions (winds, heat fluxes, rivers, and offshore boundary values) if they are to accurately portray the workings of the coastal ocean, i.e., the continental shelf and estuaries. It has been known for quite some time that winds along the shoreline tend to underestimate winds over the continental shelf,[21] and this is true of the WFS. For instance, when comparing the performance of a coastal ocean model gauged against in situ velocity observations, it was found[22] that the NOAA-modeled winds tended to underestimate the observed currents, and that significant improvement could be achieved by combining the NOAA-modeled winds with in situ observed winds via optimal interpolation (a crude data assimilation technique). In other words, improved ocean state variable estimation requires improved boundary values (winds, in this instance), and this may be achieved by increasing the coastal ocean wind observations available for assimilation into atmosphere models. We may posit that adding more buoys to the coastal ocean may be the most effective way of improving coastal ocean marine weather forecasts and, consequently, coastal ocean state variable estimations and forecasts. We may posit similarly for the open boundary values that must be passed to a coastal ocean model from a larger scale, deep ocean model. Both surface forcing (winds in particular) and deep ocean forcing (open boundary values) improvements require judiciously placed moorings. How might we combine this concept with the dynamical regimes concept previously discussed?

COOS design begins with moorings because, in addition to surface meteorology, these provide the backbone for long-term, full water column measurements. Given the scientific requirements of spanning the different shelf dynamical regimes, increasing the number of surface meteorological data points and assisting with open boundary values, it is clear that arbitrarily chosen buoy lines are not an efficient way to proceed. Instead, we require an array with maximum spatial coverage and attention to the specific dynamical regimes. A hexagonal close-packed array provides for this. For the WFS, with a nearly linearly sloping bottom, we note that the inner shelf extends offshore to about the 50-m isobath[23–27] and that the outer shelf extends in from the shelf break by about 30 km.[25,28] Moorings on the 10-, 25-, 50-, and 75-m isobaths thus tend to both span the dynamical regimes and provide measurement points along the shelf break (at about the 75-m isobath) that are also outside the nesting zone between the outer (HYCOM) and inner (FVCOM) models. Within this context, mooring placement consideration may also be given to specific locations such as marine protected areas. For the WFS, these (Steamboat Lumps, Madison Swanson, and the Florida Middle Grounds; Figure 1) fit nicely within the hexagonal array of moored buoy observing sites shown in Figure 5.

Another justification for moored buoys is their ability to provide sustained observations over a long period of time. It is only in this manner that we may define a seasonal cycle, a long-term mean circulation, and the inter-annual deviations from the long-term mean circulation and the seasonal cycle, all of which are important for shelf ecology. With a decade of data, we may begin to put error bars around the

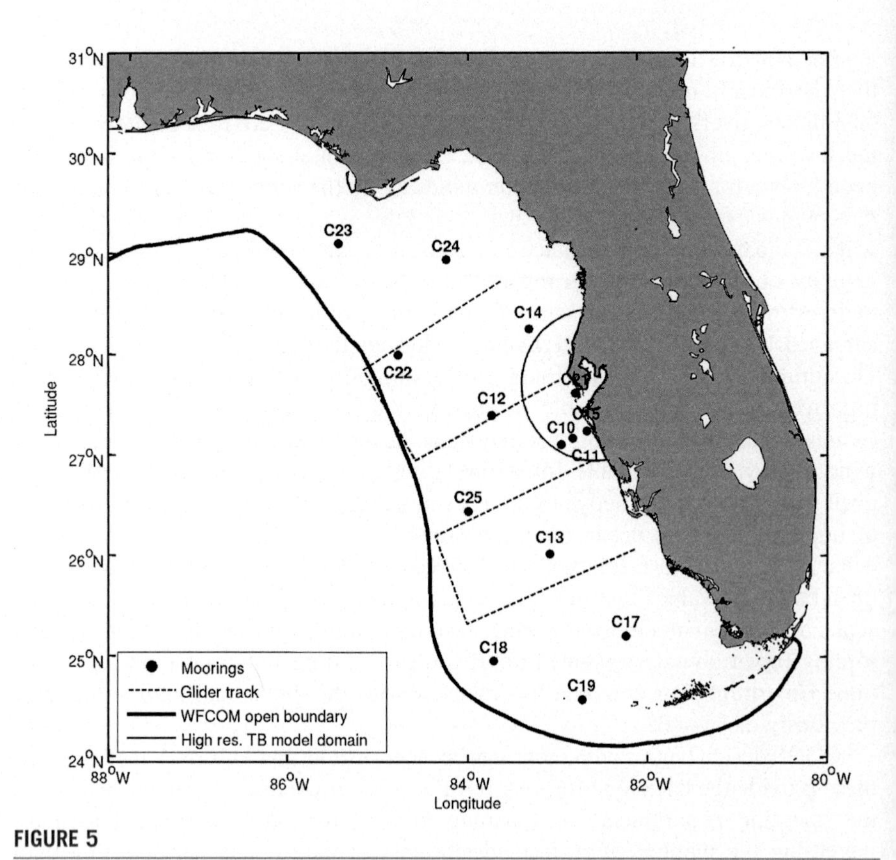

FIGURE 5

West Florida Shelf mooring locations, glider tracks, WFCOM open boundary, and higher resolution Tampa Bay model open boundary.

seasonal variations, and depending on the observed variance, we may also begin to define a long-term mean circulation.[29,30] More importantly, we may then begin to distinguish those more rare events that help to define the ecological workings of the shelf such as the anomalous strong and protracted upwelling years of 1998[3,4] and 2010[13] and years when phasing and duration of anomalous upwelling facilitate gag recruitment success such as 2007.[18]

Moorings are but one aspect of a COOS. Considerable emphasis has been placed on the role of HF radars in mapping surface currents (and waves), especially for search and rescue and the tracking of buoyant hazardous spills (e.g.,[31,32]). Presently, both the west and northeast coasts of the United States are more densely sampled by HF radars than the southeast and the Gulf of Mexico coasts of the United States, where the coverage is relatively sparse. By virtue of producing surface current fields, HF radar data are very useful for assimilation into models (e.g.,[33,34]). Nonetheless, HF radars are limited to surface currents, whereas ecology also depends on what happens near the bottom. Thus HF radars, like moorings, are but one aspect of a COOS.

In a quest for three-dimensional coverage, the next sensor delivery system consists of gliders and profilers. Gliders use buoyancy engines to change their vertical position, coupled with wings to provide forward propulsion. Outfitted with a variety of sensors from velocity, temperature, and salinity to bio-optics and acoustics sensors, gliders provide an effective means for mapping three-dimensional fields on the continental shelf (e.g., Figure 1) and elsewhere. A limitation is nonsynoptic coverage because of their slow transit speed; but, when combined with profiling floats, both three-dimensional and synoptic coverage may be realized.

In addition to the observing assets already mentioned is the obvious importance of satellites for remotely sensing sea surface temperature, height, color, and salinity and for tracking surface drifters. The first two of these variables are essential inputs to ocean models (although extending the use of altimetry onto the continental shelf still requires assessments against in situ data, e.g.,[35,36]), the third yields interpretations of chlorophyll and harmful substance spills, and the fourth may be useful in regions with fairly large salinity contrasts, as occurs in the coastal ocean. Satellite tracked drifters offer a Lagrangian approach to oil spills and HABs as well as evaluating Lagrangian trajectory models (e.g.,[36,37]). Like HF radars, however, they are limited to the near surface. When taken together, these four sensor delivery systems (moored buoys, HF radars, gliders/profilers, and satellites), along with conventional tide and wave gauges, form the basis for a COOS. The ocean state variables that are most readily sampled are velocity, temperature, salinity, sea surface height, bio-optics, and bio-acoustics, as all of these variables have associated robust sensors. Other variables are equally important, such as the various nutrients, oxygen, carbon compounds and alkalinity, toxins, and plankton species discriminators; but, sensors for these variables still remain largely in development. Atmosphere state variables are also readily sampled with robust sensors. It is important to begin with what we know we can do well and add other sensors as engineering developments allow.

When taken together, we may offer an array design for the WFS that may also be applicable elsewhere. Figures 5 and 6 show our instrumentation goal for the WFS. The underlying elements are moored buoys, and these are supplemented by HF radars, regularly sampled glider transects, and available satellite sensors. All of these observing assets are coordinated with models. The WFCOM domain is shown in Figure 7 with an example of surface velocity vectors superimposed on surface salinity for June 19, 2010, a time when Deepwater Horizon surface oil was translating eastward along the north Florida coastline. Whereas WFCOM contains all of the major estuaries of the region with horizontal resolution as fine as 150 m, there are instances when even higher resolution may be needed. For this, we may nest a more finely resolved estuary specific model in WFCOM. Application examples for Tampa Bay[38,39] employ resolutions as fine as 20 m. Additional model applications are for hurricane storm surge and waves, but these require larger domains (essentially the entire GOM). Hence, while we engage in such activities (e.g.,[40,41]), these are not included herein.

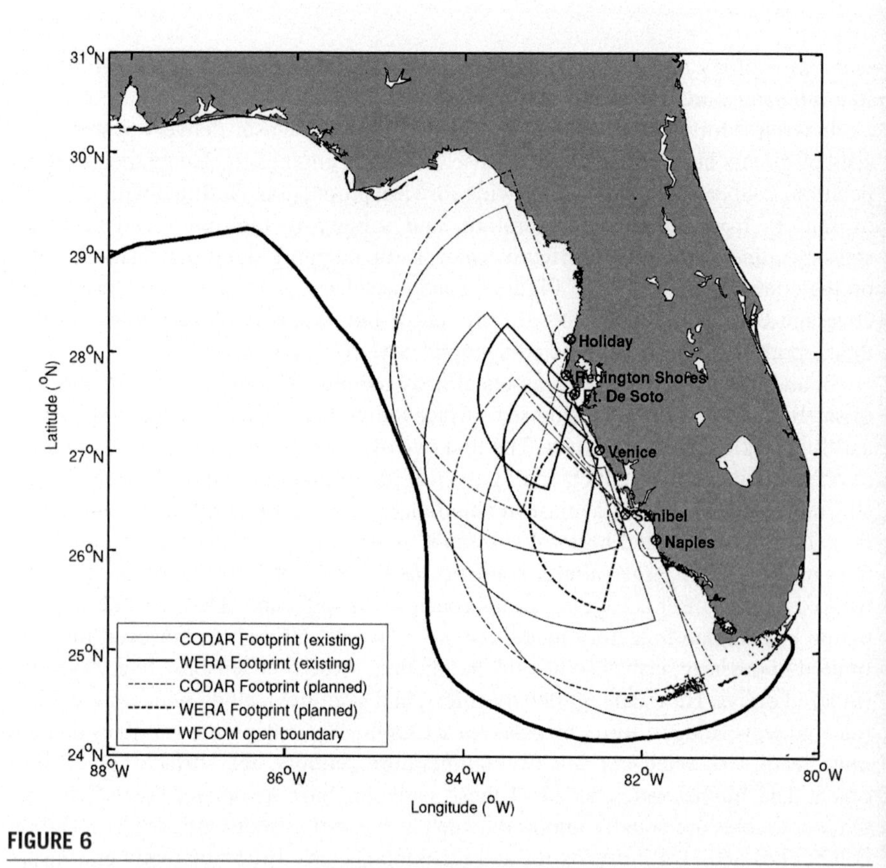

FIGURE 6

West Florida Shelf HF radar footprints and the WFCOM open boundary.

The moored buoy locations of Figure 5 each have specific scientific justification. The overall array is aimed at maximizing wind field coverage (the hexagonal pattern, versus a less efficient set of across shelf lines), and the spacing is at about half the de-correlation, e-folding scale for winds, thereby allowing for wind field constructions and for assimilation into atmosphere models. By locating moorings on the 10-, 25-, 50-, and 75-m isobaths, we span the dynamically distinct inner and outer shelf regions. The shelf break, 75-m isobath locations are particularly important for determining open boundary value veracity and for determining deep ocean water properties that may upwell across the shelf break. The southwest corner location has additional dynamical significance as explained previously, and some of the other sites are colocated with marine protected areas (Figure 1) without deviating from the experimental design criteria.

The HF radar distribution, using two different frequencies with overlapping coverage,[42,43] is meant to span most of the shelf with at least some redundancy for improved data returns. Within the combined HF radar and moored array are

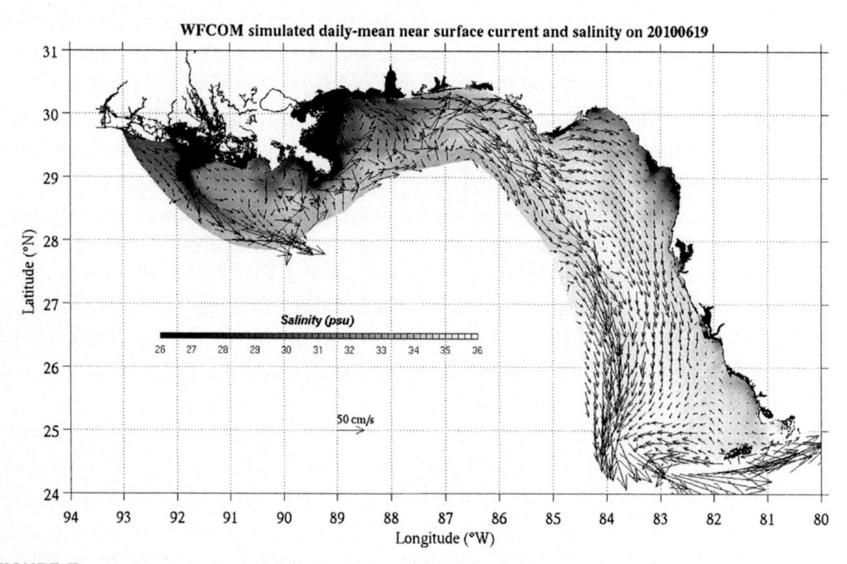

FIGURE 7

West Florida Coastal Ocean Model (WFCOM) simulation of surface velocity vectors superimposed on surface salinity, sampled on June 19, 2010, when oil of Deepwater Horizon origin was observed to be moving swiftly eastward along the northern Florida coastline.

a set of glider transects to be repeated on a quarterly basis. The gliders are particularly important for ecological applications. For instance, we know that *Karenia brevis* HABs originate offshore[44] and are carried to the near shore within the bottom Ekman layer.[6] With advanced information on near bottom chlorophyll, we can provide forecasts of potential blooms near the beach. The addition of profiling floats would further facilitate this objective. Assimilation of these observations in models would also improve upon their forecasting capabilities, and[45] provides a recent WFS example, albeit with very limited observations.

5. SUMMARY AND RECOMMENDATIONS

We present a rationale for ecosystem monitoring with application to the West Florida Shelf (WFS). The basic premise is that monitoring can only commence once the system workings are understood, and for this we require a coordinated and sustained set of coastal ocean observations and models. Thus, the basic COOS tenets advanced herein begin with the realization that coastal ocean ecology integrates all processes responsible for organism success. In other words, ecology is not just biology. Coastal ocean ecological understanding requires a fully multidisciplinary, systems science approach beginning with the coastal ocean circulation that determines the water properties within which organisms reside. Systems science requires the coordination between observations and models; the sampling problem precludes ever defining the system by observations alone, and models without supporting observations are similarly deficient. Observations must include a variety of sensors and deployment systems because no single method is adequate to describe three-dimensional processes. Similarly, hierarchies of models are needed, and no single model can be expected to handle all connections, either physical or biological. Finally, no single focus or theme is adequate. For instance, we will never understand fisheries or harmful algal blooms by simply focusing of fish or harmful algae; coastal ocean scientific matters are all connected. Only through a quantitative understanding of these connections can we predict the outcomes of either natural or human-induced changes to the coastal ocean environment. In view of these tenets, we further note that stakeholder needs cannot be met without the application of defensible science.

In view of the resources either already or potentially available as a consequence of the Deepwater Horizon oil spill (through the Gulf of Mexico Research Institute, the National Fish and Wildlife Federation, the RESTORE Act, and the National Academy of Sciences), the time for implementing COOS for the Gulf of Mexico is now, already some five years after the incident. Experimental design criteria offered herein provide for an efficient pathway toward a scientifically justified observing and modeling system. We may begin with existing technology, adding additional sensors as they become available. There is no reason to wait for additional design guidance from models by way of Observing System Simulation Experiments (OSSEs) or Observing System Experiments (OSEs), both of which are important assessment tools for data assimilation. Whereas the utility of OSSEs and OSEs is demonstrated for deep ocean regions (e.g.,[46]) and they may also be useful for the coastal ocean, the coastal ocean is dynamically more constrained and demonstrated progress on matters of coastal ocean ecology already exist. The window of resource availability is presently open, and to paraphrase Hillel: if not now, when?

ACKNOWLEDGMENTS

The genesis of these thoughts and opinions precedes the formal introduction of IOOS, and a version of these (with colleagues W. Boicourt, A. Jochens, and J. Virmani) was submitted as a community white paper to the 2012 IOOS Summit (http://www.iooc.us/summit/white-paper-submissions/community-white-paper-submissions/). Coastal ocean observing within the College of Marine Science, University of South Florida began in 1993 through a cooperative agreement with the United States Geological Survey. State of Florida support for a Coastal Ocean Monitoring and Prediction System (COMPS) was obtained in 1998. Various awards were sustaining through the present time when external support derives from the Southeast Coastal Ocean Observing Regional Association (SECOORA) as a pass through from NOAA Grant # NA11NOS0120033 for both moored buoy and HF radar operations. Modeling support (now ended) was from the Gulf of Mexico Research Institute through the Florida State University Deep-C Program. We are particularly grateful for the excellent staff support provided through our Ocean Circulation Group. J. Donovan is responsible for all computer and data organization matters. D. Mayer provides QA/QC and scientific guidance. C. Merz maintains the HF radar systems, and J. Law is responsible for all sea-going operations. Without such excellent staff support, none of our work would be possible. This is CPR Contribution 39.

REFERENCES

1. Ocean.US. *An integrated and sustained ocean observing system (IOOS) for the United States: design and implementation*. Ocean.US Publication; May 2002.
2. Steidinger KA, Vargo GA, Tester PA, Tomas CR. Bloom dynamics and physiology of Gymnodinium breve with emphasis on the Gulf of Mexico. In: Anderson DM, Cembella AM, Hallegraeff GM, editors. *Physiological ecology of harmful algal blooms*. New York: Springer; 1998. p. 135–53.
3. Weisberg RH, He R. Local and deep-ocean forcing contributions to anomalous water properties on the West Florida Shelf. *J Geophys Res* 2003;**108**(C6):15. http://dx.doi.org/10.1029/2002JC001407.
4. Walsh JJ, Weisberg RH, Dieterle DA, He R, Darrow BP, Jolliff JK, et al. The phytoplankton response to intrusions of slope water on the West Florida Shelf: models and observations. *J Geophys Res* 2003;**108**(C6):15. http://dx.doi.org/10.1029/2002JC001406.
5. Walsh JJ, Jolliff JK, Darrow BP, Lenes JM, Milroy SP, Remsen D, et al. Red tides in the Gulf of Mexico: where, when, and why. *J Geophys Res* 2006;**111**(C11003). http://dx.doi.org/10.1029/2004JC002813.
6. Weisberg RH, Barth A, Alvera-Azcárate A, Zheng L. A coordinated coastal ocean observing and modeling system for the West Florida Shelf. *Harmful Algae* 2009;**8**: 585–98.
7. McGillicuddy DJ, Anderson DL, Lynch DR, Townsend DW. Mechanisms regulating large scale seasonal fluctuations of *Alexandrium fundyense* populations in the Gulf of Maine. Results froma physical biological model. *Deep-Sea Res* 2005;**II**(52): 2698–714.
8. Liu Y, MacFadyen A, Ji Z-G, Weisberg RH. Monitoring and modeling the Deepwater Horizon oil spill: a record-breaking enterprise. In: *Geophys. monogr. ser.*, vol. 195. Washington, DC: AGU; 2011. p. 271.
9. Liu Y, Weisberg RH, Hu C, Kovach C, Riethmüller R. Evolution of the loop current system during the deepwater Horizon oil spill event as observed with drifters and satellites. In: Liu Y, et al., editors. *Monitoring and modeling the Deepwater Horizon oil spill: a record-breaking enterprise. Geophys. monogr. ser.*, vol. 195. Washington, DC: AGU; 2011. p. 91–101. http://dx.doi.org/10.1029/2011GM001127.
10. Gill AE. *Atmosphere-ocean dynamics*. San Diego, Calif: Academic; 1982. 408 pp.
11. Brink KH. Deep-sea forcing and exchange processes. In: Brink KH, Robinson AR, editors. *The sea*, vol. 10. New York: John Wiley; 1998. p. 21–62. 1998.
12. Hetland RD, Hsueh Y, Leben RR, Niiler PP. A Loop Currentinduced jet along the edge of the West Florida Shelf. *Geophys Res Lett* 1999;**26**:2239–42.
13. Weisberg RH, Zheng L, Liu Y, Lembke C, Lenes JM, Walsh JJ. Why a red tide was not observed on the West Florida Continental Shelf in 2010. *Harmful Algae* 2014;**38**: 119–26. http://dx.doi.org/10.1016/j.hal.2014.04.010.

14. Zheng L, Weisberg RH. Modeling the west Florida coastal ocean by downscaling from the deep ocean, across the continental shelf and into estuaries. *Ocean Model* 2012;**48**:10–29.
15. Chen CS, Liu H, Breadsley RC. An unstructured, finite-volume, three-dimensional, primitive equation ocean model: application to coastal ocean and estuaries. *J Atmos Ocean Technol* 2003;**20**:159–86.
16. Chassignet EP, Hurlburt HE, Metzger EJ, Smedstad OM, Cummings J, Halliwell GR, et al. U.S.GODAE:global ocean prediction with the HYbrid Coordinate Ocean Model (HYCOM). *Oceanography* 2009;**22**:48–59.
17. Fitzhugh GR, Koenig CC, Coleman FC, Grimes CB, Sturges W. Spatial and temporal patterns in fertilization and settlement of young gag (Mycteroperca microlepis) along the West Florida Shelf. *Bull Mar Sci* 2005;**77**:377–96.
18. Weisberg RH, Zheng L, Peebles E. Gag grouper larvae pathways on the West Florida Shelf. *Cont Shelf Res* 2014;**88**:11–23. http://dx.doi.org/10.1016/j.csr.2014.06.003.
19. Weisberg RH, Zheng L, Liu Y, Murawski S, Hu C, Paul J. Did Deepwater horizon hydrocarbons transit to the west Florida Continental shelf? *Deep-Sea Res* 2014;**II**. http://dx.doi.org/10.1016/j.dsr2.2014.02.002.
20. Murawski SA, Hogarth WT, Peebles EB, Barbeiri L. Prevalence of external skin lesions and polycyclic aromatic hydrocarbon concentrations in the Gulf of Mexico, Post-Deepwater Horizon. *Trans Am Fish Soc* 2015;**143**:1084–97. http://dx.doi.org/10.1080/00028487.2014.911205.
21. Weisberg RH, Pietrafesa LJ. Kinematics and correlation of the surface wind field in the South Atlantic Bight. *J Geophys Res* 1983;**88**:4592–610.
22. He R, Liu Y, Weisberg RH. Coastal ocean wind fields gauged against the performance of an ocean circulation model. *Geophys Res Lett* 2004;**31**:L14303. http://dx.doi.org/10.1029/2003GL019261.
23. Li Z, Weisberg RH. West Florida Shelf response to upwelling favorable wind forcing: kinematics. *J Geophys Res* 1999;**104**:13507–27.
24. Li Z, Weisberg RH. West Florida continental shelf response to upwelling favorable wind forcing, 2: dynamics. *J Geophys Res* 1999;**104**:23427–42.
25. Weisberg RH, Li Z, Muller-Karger FE. West Florida shelf response to local wind forcing: April 1998. *J Geophys Res* 2001;**106**:31239–62.
26. Weisberg RH, He R, Liu Y, Virmani JI. West Florida shelf circulation on synoptic, seasonal, and inter-annual time scales. In: Sturges W, Lugo-Fernandez A, editors. *Circulation in the Gulf of Mexico. AGU monograph series, geophysical monograph*, vol. 161; 2005. p. 325–47.
27. Liu Y, Weisberg RH. Ocean currents and sea surface heights estimated across the West Florida Shelf. *J Phys Oceanogr* 2007;**37**(6):1697–713. http://dx.doi.org/10.1175/JPO3083.1.
28. He R, Weisberg RH. A loop current intrusion case study on the West Florida Shelf. *J Phys Oceanogr* 2003;**33**:465–77.
29. Weisberg RH, Liu Y, Mayer D. West Florida Shelf mean circulation observed with long-term moorings. *Geophys Res Lett* 2009;**36**:L19610. http://dx.doi.org/10.1029/2009GL040028.
30. Liu Y, Weisberg RH. Seasonal variability on the West Florida Shelf. *Prog Oceanogr* 2012;**104**:80–98. http://dx.doi.org/10.1016/j.pocean.2012.06.001.
31. Harlan J, Terrill E, Hazard L, Keen C, Barrick D, Whelan C, et al. The integrated ocean observing system high frequency radar network: status and local, regional and national applications. *Mar Tech Soc J* 2010;**44**(6):122–32.

32. Paduan JD, Washburn L. High-frequency radar observations of ocean surface currents. *Ann Rev Mar Sci* 2013;**5**:115–36. http://dx.doi.org/10.1146/annurev-marine-121211-172315.
33. Barth A, Alvera-Azcarate A, Weisberg RH. Assimilation of high-frequency radar currents in a nested model of the West Florida Shelf. *J Geophys Res* 2008;**113**:C08033.
34. Wilkin JL, Zhang WG, Cahill B, Chant RC. Integrating coastal models and observations for studies of ocean dynamics, observing systems and forecasting. In: Schiller A, Brassington G, editors. *Operational oceanography in the 21st century.* Dordrecht: Springer; 2011. p. 487–512.
35. Liu Y, Weisberg RH, Vignudelli S, Roblou L, Merz CR. Comparison of the X-TRACK altimetry estimated currents with moored ADCP and HF radar observations on the West Florida Shelf. *Adv Space Res* 2012;**50**:1085–98. http://dx.doi.org/10.1016/j.asr.2011.09.012.
36. Liu Y, Weisberg RH, Vignudelli S, Mitchum GT. Evaluation of altimetry-derived surface current products using Lagrangian drifter trajectories in the eastern Gulf of Mexico. *J Geophys Res* 2014;**119**:2827–42. http://dx.doi.org/10.1002/2013JC009710.
37. Liu Y, Weisberg RH. Evaluation of trajectory modeling in different dynamic regions using normalized cumulative Lagrangian separation. *J Geophys Res* 2011;**116**:C09013. http://dx.doi.org/10.1029/2010JC006837.
38. Zhu J, Weisberg RH, Zheng L, Han S. On the flushing of Tampa Bay. *Estuaries Coasts* 2015;**38**:118–31. http://dx.doi.org/10.1007/s12237-014-9793-6.
39. Zhu J, Weisberg RH, Zheng L, Han S. Influences of channel deepening and widening on the tidal and non-tidal circulation of Tampa Bay. *Estuaries Coasts* 2015;**38**:132–50. http://dx.doi.org/10.1007/s12237-014-9815-4.
40. Zheng L, Weisberg RH, Huang Y, Luettich RA, Westerink JJ, Kerr PC, et al. Implication from the comparisons between two- and three-dimensional model simulations of the Hurricane Ike storm surge. *J Geophys Res Oceans* 2013;**118**:3350–69. http://dx.doi.org/10.1002/jgrc.20248.
41. Huang Y, Weisberg RH, Zheng L. Gulf of Mexico hurricane wave simulations using SWAN: bulk formula based drag coefficient sensitivity for Hurricane Ike. *J Geophys Res Oceans* 2013;**118**:1–23. http://dx.doi.org/10.1002/jgrc.20283.
42. Merz CR, Weisberg RH, Liu Y. Evolution of the USF/CMS CODAR and WERA HF radar network. *Proc. MTS/IEEE Oceans* 2012;**12**:1–5. http://dx.doi.org/10.1109/OCEANS.2012.6404947.
43. Liu Y, Weisberg RH, Merz CR. Assessment of CODAR Seasonde and WERA HF radars in mapping surface currents on the West Florida Shelf. *J Atmos Ocean Technol* 2014;**31**:1363–82. http://dx.doi.org/10.1175/JTECH-D-13-00107.1.
44. Steidinger KA. Imlications of dinoflagellate life cycleson initiation ofGymnodinium breve red tides. *Environ Lett* 1975;**9**:129–39.
45. Pan C, Zheng L, Weisberg RH, Liu Y, Lembke C. Comparisons of different ensemble schemes for glider data assimilation on West Florida Shelf. *Ocean Model* 2014;**81**:13–24. http://dx.doi.org/10.1016/j.ocemod.2014.06.005.
46. Halliwell Jr GR, Srinivasan A, Kourafalou V, Yang H, Willey D, Le Henaff M, et al. Rigorous evaluation of a fraternal twin ocean OSSE system for the open gulf of Mexico. *J Atmos Ocean Technol* 2014;**31**:105–30. http://dx.doi.org/10.1175/JTECH-D-13-00011.1.

CHAPTER

5

The Monitoring of Harmful Algal Blooms through Ocean Observing: The Development of the California Harmful Algal Bloom Monitoring and Alert Program

Raphael M. Kudela[1,*], Aric Bickel[2], Melissa L. Carter[3], Meredith D.A. Howard[4], Leslie Rosenfeld[2]

Ocean Sciences Department, Institute for Marine Sciences, University of California, Santa Cruz, CA, USA[1]; Central and Northern California Ocean Observing System, Moss Landing, CA, USA[2]; Scripps Institution of Oceanography, University of California, San Diego, CA, USA[3]; Southern California Coastal Water Research Project, Costa Mesa, CA, USA[4]

**Corresponding author: E-mail: kudela@ucsc.edu*

CHAPTER OUTLINE

1. INTRODUCTION AND BACKGROUND

Phytoplankton are at the base of the food chain in most freshwater and marine systems. These single-celled photosynthetic algae provide many positive benefits including production of about half the oxygen on the planet, and transformation of sunlight and inorganic elements into the organic material and energy that drives our productive aquatic ecosystems. The dominance of cells of one or several species of phytoplankton can form an algal bloom, either through rapid growth or minimal loss (e.g., grazing, physical advection, or sinking). Though blooms are often harmless or beneficial to the functioning of marine and freshwater ecosystems, there is increasing awareness that blooms can also be indicative of eutrophication, ecosystem disruption, or altered environmental states.[1–3]

Some algal blooms can have negative impacts to the environment, human and aquatic health, and the economy (such as aquaculture, fisheries, and tourism) and are thus termed harmful algal blooms (HABs). HABs negatively affect many freshwater systems and the majority of coastal regions worldwide. There is consensus that the impact of such blooms has grown over the last few decades, causing harm to public health, ecosystem function, fisheries and aquaculture, and recreation/tourism industries.[4] Despite the broad consensus that HAB impacts are increasing, it is challenging to describe the research, management, and mitigation strategies coherently because the term HAB encompasses a wide, and sometimes bewildering, array of species, life histories, ecosystems, and impacts. There is no unifying ecological or evolutionary theme in the organisms considered to form HABs. They span the majority of algal taxonomic groups, including both eukaryotes and prokaryotes; some produce potent toxins, and others cause harm through a variety of other mechanisms. The dinoflagellates account for the majority (75%) of HAB species, but there are HAB representatives from nearly every algal taxon.[5]

To categorize the many groupings of potentially harmful algae, HABs are often first divided into toxic versus high-biomass blooms. Although they comprise a minority of the phytoplankton assemblage, toxic species can and do have significant impacts on human, marine life, and ecosystem health.[6] In California and along the US West Coast, toxic species are the most frequently encountered HABs; they consist primarily of the dinoflagellate genus *Alexandrium* (Paralytic Shellfish Poisoning) and the diatom genus *Pseudo-nitzschia* (Amnesic Shellfish Poisoning and Domoic Acid Poisoning).[7] Less frequently observed, but emerging as serious potential threats, are the dinoflagellates *Dinophysis* (which can cause Diarrhetic Shellfish Poisoning[8,9]) and *Gonyaulax* and *Lingulodinium* (Yessotoxin producers[10,11]), whereas the freshwater cyanobacteria *Microcystis* (a Microcystin producer) threatens not only the terrestrial environment but also estuarine and coastal waters.[12,13] Other toxic species have been encountered in California, but this relatively short list comprises the majority of significant toxic HAB issues seen to date.

In contrast to blooms of toxic species, high-biomass blooms cause negative impacts through the sheer abundance of cells.[3] This can lead to physical disruption of other organisms via gill irritation, viscosity, and gelatinous barriers that lead to gill clogging, production of allelochemicals, and anoxia or hypoxia following the decay of large blooms. Impacts to humans include disruption of desalination systems and drinking water supplies, as well as impacts to tourism due to accumulation of dead organisms, foul smells, foam production, and discoloration of the water. Many of these HABs fall under the general term ecosystem disruptive algal blooms (EDABs[14,15]). This group includes emergence of organisms previously thought to be harmless such as the dinoflagellate *Akashiwo sanguinea*, which can produce foam with surfactant properties resulting in the large-scale mortality of marine birds,[16] as well as organisms such as the dinoflagellate *Cochlodinium* (fish killer[17–19]) and the raphidophyte *Heterosigma akashiwo* (fish killer[20,21]), which until recently have not been as problematic in California compared to other regions.[8]

This considerable diversity in organisms, impacts, mechanisms, and potential solutions led to the realization that there is no single solution to understanding, monitoring, predicting, or mitigating HAB problems within California or along the US West Coast. Though California has been highly successful at minimizing human health impacts through the establishment of the marine Biotoxin Monitoring Program by the California Department of Public Health, this program was not intended to address the wide array of emerging issues that includes impacts to ecosystem and wildlife health, emerging toxins, and HAB dynamics driven by both terrestrial and marine processes. Research efforts funded by agencies such as California Sea Grant and the NOAA competitive HAB programs ECOHAB and MERHAB provided significant funding leading to advances in understanding and monitoring HABs in California, but these programs were necessarily focused on specific regions, organisms, and impacts. Discussions between the NOAA competitive HAB program, disparate phytoplankton monitoring efforts, agencies managing HAB impacted resources, and the Southern California Coastal Water Research Program (SCCWRP) identified a needed statewide capacity for monitoring and HAB alerts that encompasses all of the existing and emerging HAB issues. It is within this context that the NOAA Center for Sponsored Coastal Ocean Research, SCCWRP, and the California Ocean Science Trust convened the first workshop in 2008 that developed the California Harmful Algal Bloom Monitoring and Alert Program, or CalHABMAP.[22]

The 2008 workshop participants included the leading HAB researchers in California and representatives with a wide variety of stakeholders, such as water quality management representatives, public health officials that manage the shellfish industry, animal rescue communities, universities, and state and local agencies. These participants strongly agreed that there was a need for a coordinated HAB alert system. Thus, the goal of CalHABMAP is to implement a proactive HAB alert network that will provide algal bloom forecasts and facilitate information exchange among HAB researchers, managers, and the general public. Crucial decisions on how to respond to HAB events, how to mitigate their impacts, and how to predict them, require knowledge of their occurrence and their impacts along our coast. Instituting a statewide alert network is a critical first step in moving toward an operational forecast and prevention program.

CalHABMAP created an integrated, statewide HAB monitoring and alert network by coordinating organizations and researchers currently collecting HAB data and created a centralized portal for the dissemination of this information. The program also established an ambitious set of goals including studies to normalize the diverse methodologies employed to identify toxin and toxic algal species in coastal waters, and development of an economic analysis of resources along the California coast and the potential impact of HABs on these resources. Thus the main goal of CalHABMAP is to ultimately implement a statewide

HAB network and forecasting system for California, and potentially the US West Coast, by implementing the following specific objectives:

- Design an HAB network that will meet the needs of, and will be accessible to, all HAB stakeholders.
- Create a web portal within the California Ocean Observing System's programs and act as a mechanism to bring these two programs together. The portal would be a centralized location where HAB data and predictive information could be used by many groups throughout the state.
- Conduct a comparison of analytical methods for toxin analysis and harmful algae identification and enumeration, and review and disseminate the results through a workshop. The method comparison information will be critical to establish guidelines for comparing different data sets currently collected, and it will help determine the finalized design for the CalHABMAP network.
- Conduct an economic analysis of the potential impacts of HABs along the California coast.
- Collaborate with the Water Quality Monitoring Council to ensure that HAB information and data are included in and accessible from water quality websites.

2. THE CalHABMAP NETWORK

At the heart of CalHABMAP is the development of a grassroots-driven, voluntary network of HAB researchers, managers, and other volunteers who share and disseminate information about HABs in California. The first step was to develop a CalHABMAP listserv to provide an efficient mechanism for all of the interested parties to communicate. The listserv provides the following types of data and information:

1. Results from weekly phytoplankton monitoring samples are collected and the results are published to the list in a standardized report format. These weekly HAB Monitoring Reports are posted from the following locations (see also Figure 1):
   a. Santa Cruz Municipal Wharf, Santa Cruz
   b. Monterey Wharf, Monterey
   c. Cal Poly Pier, San Luis Obispo
   e. Goleta Pier, Goleta
   f. Stearn's Wharf, Santa Barbara
   g. Santa Monica Pier, Santa Monica
   h. Newport Pier, Newport
   i. Scripps Pier, La Jolla
2. Monthly Biotoxin Report provided by California Department of Public Health, Plankton Monitoring Program
3. General comments and an open forum to discuss current HAB conditions or marine mammal and bird strandings

FIGURE 1

Locations of existing CalHABMAP stations (white circles) and proposed new stations (black circles) overlaid over the background of mean normalized fluorescence line height, a proxy for phytoplankton biomass. New station locations were conditioned on the continued existence of the existing CalHABMAP stations. The addition of the new stations improves probability of identifying an HAB from ~20% to ~70%. The inset shows the spatial map of HAB coverage for the existing stations in Southern California. Warmer colors indicate the region represented by the station information, whereas blue indicates the regions for which the shore station information is inadequate.

An important component of the listserv is that it *integrates* groups across the region, and it can accommodate expansion to include both Mexican observations (already in place) and observations from the other western states. Though the majority of oceanographic stations and sampling locations are maintained and operated by the Southern California Coastal Ocean Observing System (SCCOOS), the Central and Northern California Ocean Observing System (CeNCOOS), and the California Department of Public Health (CDPH), all of the data are collated and presented as part of CalHABMAP.

Whereas the network is based primarily on historical locations where HABs were monitored, a second component of CalHABMAP is the assessment of an "ideal" network. Toward this end, Frolov et al.[23] and Kudela et al.[24] identified the optimal network for monitoring HABs in California. That analysis found that only the CDPH network, with 50 sites routinely sampled, had sufficient spatial and temporal coverage to capture 60% of potential blooms. In contrast, the SCCOOS/CeNCOOS networks capture only 20% of the variability, but do so at economically and ecologically important sites. However, with the inclusion of just 10 additional optimally placed stations (some of which are on the continental shelf rather than at the shore) to complement the eight occupied by CeNCOOS/SCCOOS, 73% of the variability could be captured, including offshore (more than a few km from the coast) sites that are particularly relevant with the permitting of Catalina Sea Ranch offshore aquaculture farm by the California Coastal Commission in 2014. This represents the first large-scale offshore aquaculture facility in California, and will likely not be efficaciously covered by the existing CDPH network. The results of a more quantitative analysis of an optimal HAB monitoring program demonstrate that better coverage can be achieved with fewer stations while simultaneously extending the program to cover nearshore waters.

In addition to the listserv, CalHABMAP proposed to develop a single web portal that would integrate observations from members, providing a single location where end users could acquire information (Figure 2). This has been more challenging than anticipated because many groups are required to or find it desirable to maintain individual web pages and databases. Nonetheless, a significant accomplishment from CalHABMAP has been the development of a data portal that merges the two Ocean Observing System datasets seamlessly. This portal is hosted at http://www.habmap.info, with financial support from CeNCOOS, SCCOOS, and the California Ocean Protection Council. Data are hosted and maintained by SCCOOS using a common format and web-enabled database. This system allows for data exploration and real-time information, as well as supporting peer-reviewed research on the status and trends of HABs within California (e.g.,[7,23,25]). A similar web-based graphical approach has been taken by CDPH for dissemination of the Marine Biotoxin Monitoring Program data, used by the State to regulate marine biotoxins (Figure 3).

Recognition that HABs are a major contributor to poor health and strandings of marine mammals[26–28] has pushed the development of a coordinated monitoring effort for marine mammals as well.[29] There is no national program for monitoring marine animal health, despite the numerous monitoring initiatives maintained by private groups, and federal and state agencies. To facilitate this effort Axiom Consulting & Design, working with CeNCOOS and in collaboration with SCCOOS, the Marine Mammal Center, and other agencies, is developing an interactive Marine Mammal Health Map (Figure 4). A separate but coordinated

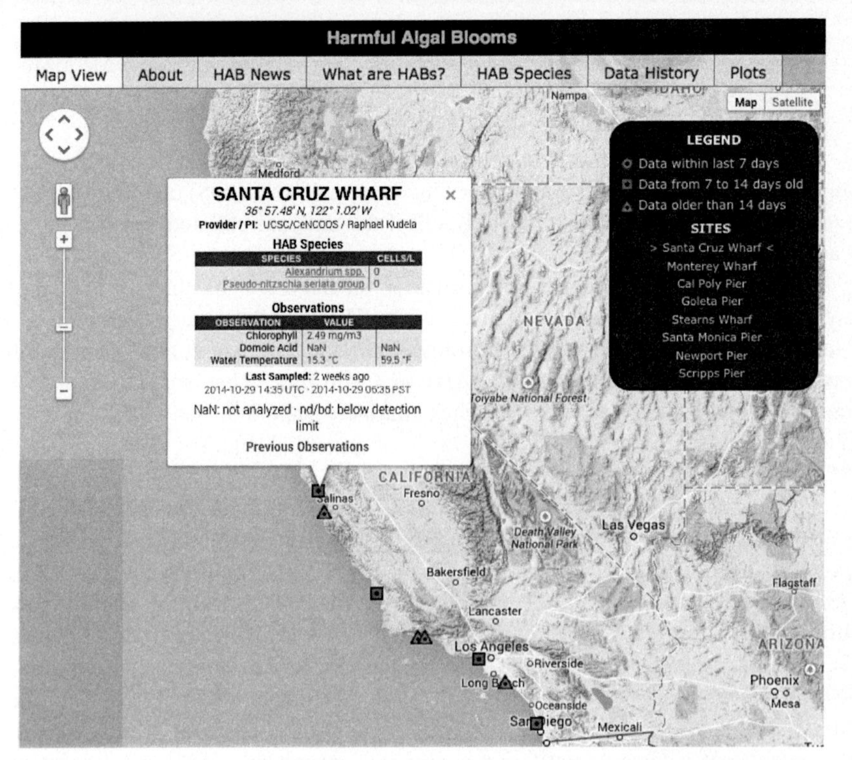

FIGURE 2

The HAB data portal for CalHABMAP, maintained by SCCOOS. The interactive map provides access to the near real-time data from each site and links to historical data. The maps can be accessed at http://www.habmap.info/data.html.

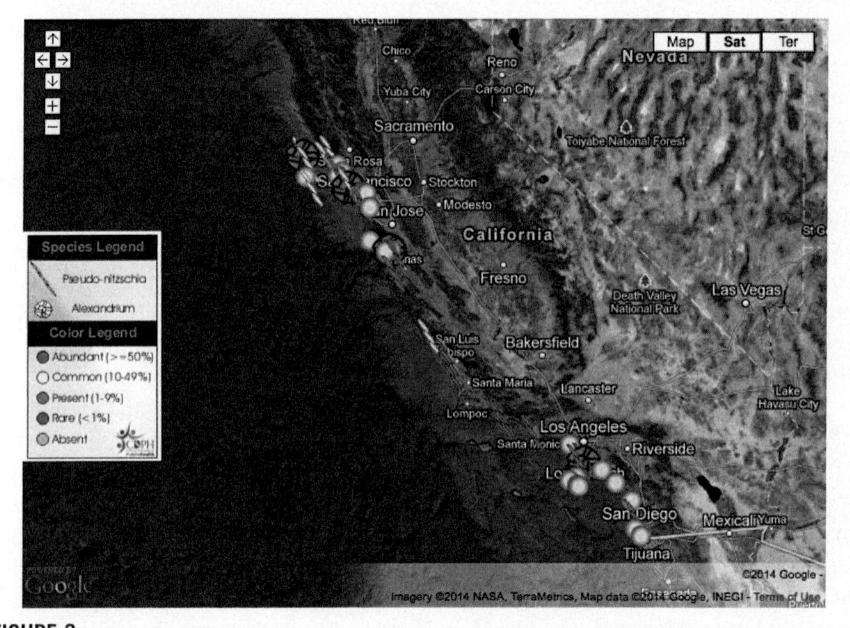

FIGURE 3

HAB location and abundance data provided by the California Department of Public Health for the fourth week of September, 2014. The interactive map can be accessed at http://www.cdph.ca.gov/HealthInfo/environhealth/water/Pages/Toxmap.aspx.

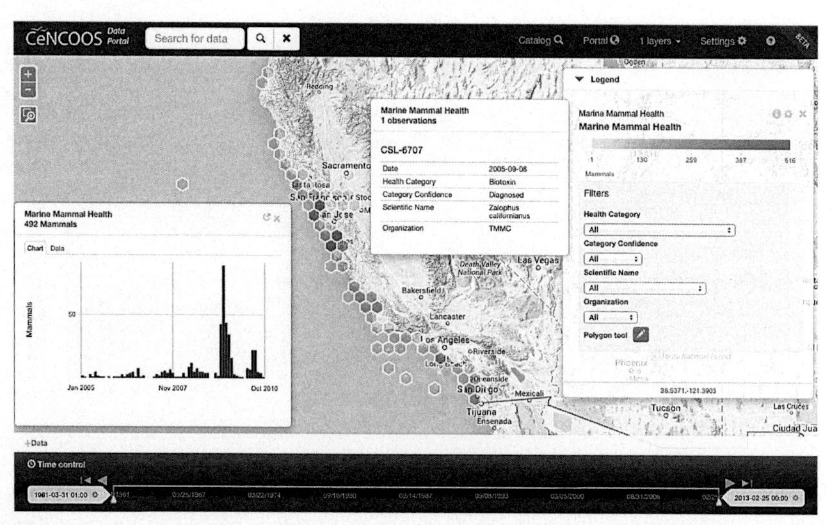

FIGURE 4

A screenshot from the beta version of the Marine Mammal Health Map, hosted by CeNCOOS. The map provides easy visual representation of marine mammal stranding data geographically (polygons) with shading (red in the original) indicating the number of impacted animals. The inset boxes show examples of data available by clicking on polygons. Additional layers, such as numerical model data, satellite imagery, or other oceanographic information can be added to provide context and exploratory data analysis.

program has introduced marine mammal strandings to the Jellywatch Website (http://www.jellywatch.org), creating a crowd-sourcing component that directly engages the public in identifying the impact of HAB events.

3. DEVELOPMENT OF AN HAB FORECASTING CAPABILITY

A significant advancement toward an operational California HAB forecast has been the transition of a research-based forecast system for the toxic diatom *Pseudo-nitzschia*[6,24,30] toward an operational forecasting system. As reviewed by Anderson et al.,[6] there are many ways to implement a regional forecast of HABs based on empirical, statistical, and/or numerical models. Preliminary empirical studies on HABs in the California Current System (CCS) provide the basis to forecast blooms and toxin production using statistical models (Figure 5).

Anderson et al.[30] extended a previous study on the Santa Barbara Channel (SBC), located at the northern extreme of the Southern California Bight,[31,32] and they identified several variables to be good predictors of *Pseudo-nitzschia* abundance and high domoic acid concentrations in surface waters south and east of Pt. Conception. The first study employed stepwise multiple linear regression to

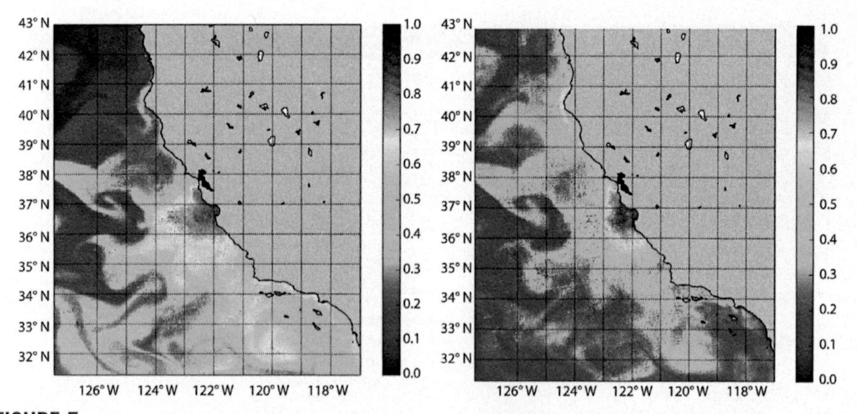

FIGURE 5

Probability of particulate domoic acid (left) and cellular domoic acid (right) for March 26, 2014, from the coupled modeling approach employed by CeNCOOS. Further details are available at http://www.cencoos.org/sections/conditions/blooms/habforecast/.

correlate hydrographic and chemical shipboard data with a broad range of *Pseudo-nitzschia* cell densities and domoic acid loads from inshore and offshore sites within the SBC from 2004–2006.[31] In the follow-on study, Anderson et al.[32] created similar models to predict *Pseudo-nitzschia* spp. cell abundance, particulate domoic acid, and cellular domoic acid from a longer HAB dataset that included new observations from 2009 to 2010 and used the more flexible generalized linear model (GLM) method in place of the optimal least squares approach.[33,34] The significant predictor variables in the new GLMs agree well with previous models,[32] indicating a consistent set of environmental controls on HABs over time, e.g., chlorophyll-a, macronutrient availability (reduced ratios of Si:N and Si:P), and upwelling (negative relationship with temperature and positive relationship with salinity).

Whereas statistical models such as these can be criticized because they are not based on ecological understanding of bloom dynamics and are highly dependent on the training data, a recent analysis demonstrates that similar models[34] developed for the Monterey Bay successfully captured subsequent (2008–2009) bloom events, including an unusual early winter event. The model successfully predicted a mandated shellfish closure (outside the normal time period when the California Department of Public Health recommends not harvesting shellfish) and also captured the physical forcing presumably leading to this anomalous bloom—early and heavy rains leading to enhanced river flow coupled with weak upwelling. Though these results are not a predictor of future model performance, they highlight that these models can successfully capture changes (such as those due to basin-scale or global trends in climate forcing) that may lead to changing phenology of HAB events.

The statistical modeling approach can also successfully identify underlying predictive patterns without requiring absolute fidelity between the model output and reality (required if an assimilative model is used). For example, the Navy Coastal Ocean Model (NCOM[35]) was able to reproduce many observed circulation features of upwelling and relaxation in Central California, and when coupled to a biochemical submodel, outputs were used to drive statistical HAB predictions for Monterey Bay.[34] Initial results using this coupled modeling approach failed completely, with worse predictions than random chance, in part because of subtle problems with the biological parameters (chlorophyll and nutrient fields) from the models. When the statistical models were "re-tuned" to only use the numerical model data, statistical HAB model results improved dramatically, demonstrating that the model had skill but also exhibited significant biases when compared to reality (as determined by comparison of model versus measured fields). This approach takes advantage of this fact, using the models to guide development of statistical relationships while remaining relatively insensitive to errors in the model results.

A long-term goal is to merge this statistical approach with fully coupled physical–biological numerical models to develop short-term (days) forecasts of bloom probabilities and spatial trajectories (Figure 6). However, most existing numerical models are simply not advanced enough (particularly the biology) to predict coastal phytoplankton dynamics at the level of specificity required, and they are often too

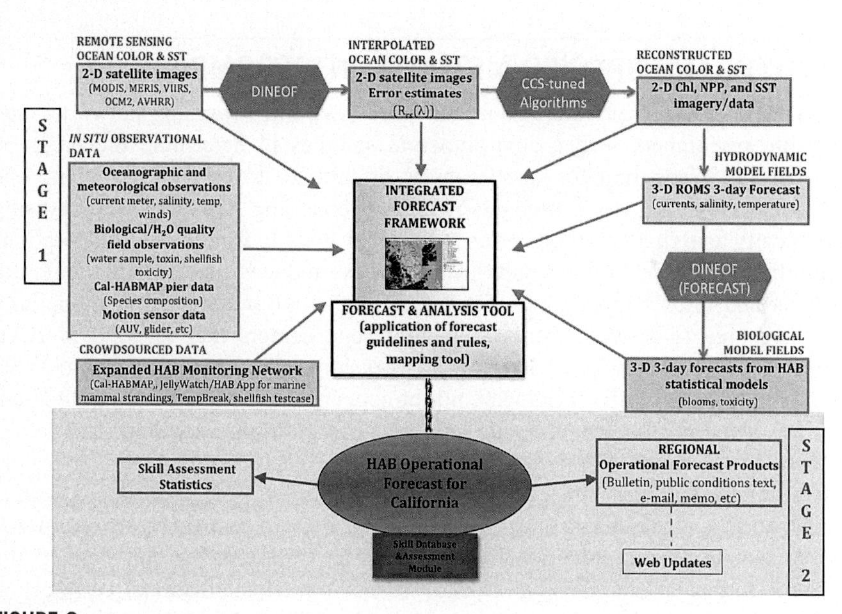

FIGURE 6

Flowchart showing the planned implementation of an operational HAB forecast for California. The effort is currently at Stage 1 (with NASA Applied Sciences Program funding). Partnerships with CeNCOOS and NOAA will transition the effort to Stage 2.

complex and expensive (in terms of development time, computational effort, and operational costs) to maintain indefinitely. In the short term, bio-optical statistical models are being used to create routine HAB forecasts by merging ocean color satellite (MODIS-Aqua) and 3-km resolution Regional Ocean Modeling System (ROMS) model output. Routine forecasts are being run in partnership with the National Ocean Service (NOS) as a testbed for collaboration toward transitioning research results to an operational center, ensuring that a dialogue is established between all stakeholders and that resources are targeted toward sustained operations and maintenance. As coupled physical−biological models improve, so too will the HAB forecasts; this steady progression will ultimately allow for inclusion of other HAB organisms using the same coupled modeling framework.

These forecast efforts highlight the need for and utility of the CalHABMAP framework. The models are integrating across both IOOS regions and utilizing field data provided by the CalHABMAP network. Marine mammal strandings and crowd-sourced data (via Jellywatch) are being used to validate and refine the models. Data are disseminated via CalHABMAP, as well as through partnership with CeNCOOS. And as noted by Frolov et al.,[23] the combination of observations, satellite remote sensing, and integrative modeling provide the only realistic framework for identifying the majority of blooms, including those that may affect offshore development and transport.

4. TOXIN AND SPECIES METHODS INTERCOMPARISON

Regulatory agencies, private, state, and federal wildlife stranding networks, and scientific researchers screen environmental samples to determine the presence of toxins. It is common for small subsets of samples to be screened using one of the less expensive assay methods, with the remaining samples analyzed using more costly analytical methods prior to decision-making for closure of operations or publication of scientific results. This process reduces the cost of regulatory analysis and allows processing of a large number of samples from a spatially extensive area to provide information to rescue centers that respond to HAB events. As part of the initial HABMAP workshop, participants determined that the methodological differences were not of concern for the screening application, as most of the positive samples are analyzed using multiple methods to confirm the presence of toxins. Workshop participants agreed, however, that methodological differences were of substantial concern for the shellfish closure applications, spatial comparisons, and trend assessments. Additionally, the recent emergence of HAB species and toxins not traditionally monitored in California has highlighted the need for further standardization. Outside of the formal international certification process led by AOAC International, there have been limited intercalibration studies to assess methodological differences as well as differences among investigators applying the same methods amongst CalHABMAP collaborators.

Participants concluded that the best way to resolve concerns about method differences affecting these applications was to conduct an interlaboratory intercalibration study. This study would determine the precision of these methods and compatibility among datasets in order to ultimately synchronize and combine data. The interlaboratory intercalibration study would assess the need for standardization to a smaller set of methods in establishing a statewide network and would allow development of a Quality Assurance/Quality Control Plan template to guide any laboratory in minimum acceptable practices to ensure data validity and compatibility.

Workshop participants also agreed that trend analysis and forecasts require improvements in accuracy, consistency, and coordination among phytoplankton species identification methods. Four main identification classes employed within the state were identified (Table 1): microscopy, molecular methods, flow cytometry (including FlowCAM (Figure 7) and Imaging Flow CytoBot), and finally, biomass indicators such as HPLC pigments, chlorophyll detection, optical measurements, and remote sensing. Similar to toxins, a need was identified for an interlaboratory

Table 1 Methods Commonly Used by California Researchers and Agencies for Species Identification, Cell Abundance, and Biomass Detection Identified at the 2008 HABMAP Workshop

Method Class	Information Type
Microscopy	
Traditional microscopy	Qualitative and quantitative; absence/presence, percent composition, cells per unit volume
Electron microscopy	Absence/presence; percent composition
Molecular	
Whole cell (fluorescence in situ hybridization)	Quantitative, absence/presence
Homogenate (sandwich hybridization, quantitative polymerase chain reaction)	Quantitative, absence/presence
Flow cytometry	
FlowCAM	Quantitative
Imaging flow CytoBot	Quantitative
Biomass	
High-performance liquid chromatography	Phytoplankton class
Chlorophyll detection	Biomass indication
Optical measurements	Biomass indication
Remote sensing	Biomass indication, spatial extent

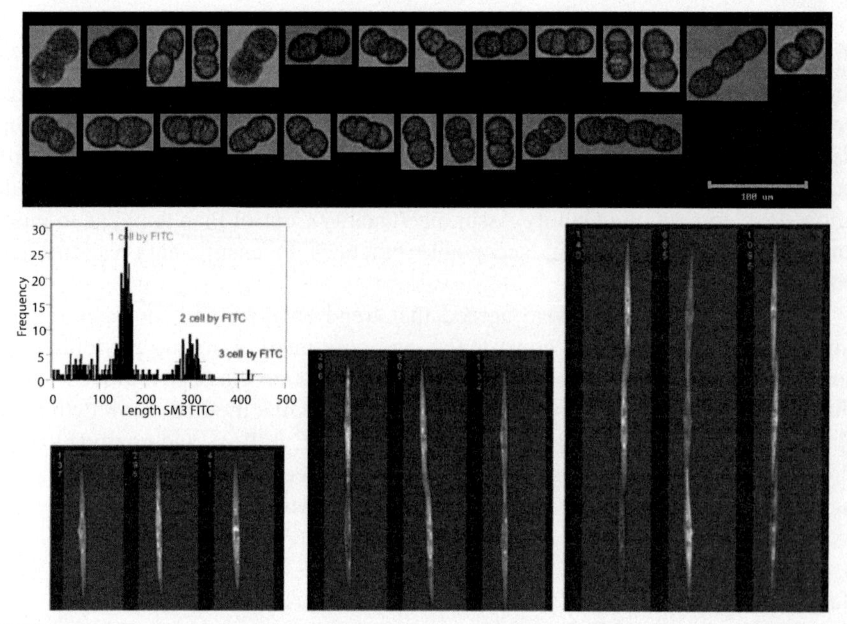

FIGURE 7

Two examples of imaging technologies employed for HAB enumeration. Images collected from FlowCAM analysis (top) of unpreserved seawater from Southern California, November 2012. The FlowCAM's automated image analysis can readily detect/identify *Cochlodinium* cells, whereas traditional microscopy using preserved samples often fails due to cell swelling and bursting. FlowCAM images courtesy of Dave Caron, University of Southern California. A culture of *Pseudo-nitzschia australis* (bottom) was probed with FITC and analyzed on an Amnis ImageStream, with automated detection of chain length (histogram). The ImageStream is capable of handling very long chains without clogging as long as the chain width is less than 250 μm.

intercalibration study to determine data compatibility. This calibration is needed across the four different classes of methods, particularly among molecular methods.

As a grassroots organization, CalHABMAP has limited capabilities and even less funding to implement these comprehensive intercalibration exercises. Instead, endorsing the community participation model, this need was highlighted and Cal-HABMAP members were asked to consider taking on the task by leveraging other funding opportunities and efforts. This was generally very successful. The same year that CalHABMAP made these recommendations, the NOAA Alliance for Coastal Technologies convened a workshop on "Technologies and Methodologies for Detection of Harmful Algae and Their Toxins."[36] Peer-reviewed publications comparing various toxin detection methods were also generated,[37,38] while many participants conducted informal intercalibration efforts between laboratories.

Based on the CalHABMAP recommendations, SCCOOS also centralized toxin detection to ensure consistent data reporting. Though much work remains to be done, particularly more systematic comparisons and intercalibrations for both species and toxin detection, CalHABMAP members have made progress toward consistent measurements, which will ultimately lead to improved trend analysis and forecasts based on accurate, consistent, and coordinated measurements of species and toxins.

5. ECONOMIC ANALYSIS

The final component proposed by CalHABMAP was to develop an economic assessment of the full cost of HABs to the State of California. There is currently no comprehensive global assessment of economic loss due to marine HABs, but it can be conservatively estimated at several billion US$ annually.[39] Using a typical value of information (VOI) estimate of 1% of the "resource" (in this case HAB-related losses),[40] a comprehensive global HAB observing and forecasting information system would represent a value of ±$100 million annually.[39] Downscaling from this global view, marine HAB-related losses in the US are conservatively estimated at ±$95 million annually, adjusted for inflation.[41] In freshwater systems, potential eutrophication-related losses in the United States, primarily due to cyanobacterial blooms, are estimated at up to $4.6 billion annually.[42] In California, the shellfish industry alone was valued at $16.4 million in 2008,[43] representing a large potential economic loss to HAB events.

Despite the frequent request for a comprehensive and up-to-date economic assessment of the true cost of HABs,[22,43,44] little new information has been gained. This is the one outstanding objective for CalHABMAP that has yet to be achieved, and it remains a high priority. The economic value of a truly comprehensive HAB monitoring and forecasting system would increase when the value of long-term monitoring of HAB events and associated environmental drivers is included, vis-à-vis trends in HABs with potential global climate change and increasing anthropogenic pressure on coastal ecosystems.[39]

6. SUMMARY AND RECOMMENDATIONS

In 2008, the newly formed CalHABMAP group set forth an ambitious agenda to establish a coordinated network for harmful algal blooms for the entirety of the State of California. In just a few short years, many of the objectives have been reached, while others remain high-priority items. Concurrently, a state-mandated effort[43] identified a similar set of goals, including the following:

- Assess economic and technical feasibility of a statewide HAB observing system.
- Support technology development by coordinating efforts to improve in situ, real-time detection of algae and toxins and by partnering with the Alliance for Coastal Technologies to evaluate new technology.

- Work toward a statewide HABs observing system by building on the existing observation network, e.g., by expanding pier monitoring in the CeNCOOS region, adding HAB sensors to other monitoring networks, and adding sites in the nearshore zone.
- Develop operational HAB forecasting models based on linking ROMS circulation and nutrient, phytoplankton, zooplankton (NPZ) ecosystem models, building on current pilot projects in Monterey Bay and SBC.
- Build an HAB early warning system by expanding the existing HABMAP system to add participants and information products.
- Improve data management capabilities by adding data sources to the California ocean observing systems' HAB Info System and integrating this system with other state data management initiatives.
- Support core research on effects of nutrient loading from anthropogenic and upwelling sources, focusing primarily on the Southern California Bight.
- Designate a lead entity to coordinate efforts associated with the full range of potential impacts. Plan for the transition from research methods to routinely deployed operational tools.

Not surprisingly, many of the CalHABMAP objectives are mirrored by the California Ocean Protection Council report. It is a testament to the power of a grass-roots effort, supported by state and federal programmatic elements, that nearly all of these objectives have been met or are underway. We identify three key challenges that need to be addressed moving forward. First, despite the repeated recommendation for a comprehensive economic assessment of the impact of HABs on the State of California, no such assessment exists or is planned at this time. Second, there is a clear need for continued integration and support of the nascent CalHABMAP network and potential expansion to include other states. Much of the work done so far has been funded by short-term (a few years) research grants. To capitalize on the successes of CalHABMAP, the monitoring and forecasting efforts must be institutionalized at the regional, state, and federal levels. CalHABMAP should, to the extent possible, champion the integration of the various data sets and data portals (such as the new Marine Mammal Health Map) to achieve the goal of a coordinated, statewide network. Third, there is increasing interest in coordinating the marine HABMAP effort with a complementary freshwater program through development of the California Cyanobacterial Harmful Algal Bloom network (CCHAB). This will require careful consideration about how best to integrate freshwater and marine HAB research, monitoring, and forecasting, as well as how best to coordinate the numerous private, state, and federal groups and agencies who are stakeholders in the process. A similar conversation is taking place in other states, as well as federally, and internationally.[45] It is our hope that CalHABMAP can provide a successful example of how best to implement these strategies.

ACKNOWLEDGMENTS

We thank the many participants of the CalHABMAP program, the volunteer Steering Committee members, and in particular Steve Weisberg, whose initial vision and organization made a statewide HAB monitoring effort a reality. Funding for this synthesis document and associated research efforts was provided by the California Sea Grant and California Ocean Protection Council award R/OPCCONT-12 A 10 (Kudela), National Aeronautics and Space Administration through award NNX13AL28G (Kudela), the NOAA ECOHAB program through award NA11NOS4780030 (Kudela and Howard), the Cal-PReEMPT program funded by NOAA MERHAB (NA04NOS4780239), the Southern California Coastal Ocean Observing System through NOAA award NA11NOS0120029 (McGowan and Carter), and the Central and Northern California Ocean Observing System through NOAA award NA11NOS0120032 (Kudela, Bickel, Rosenfeld). CalHABMAP is supported through the Southern California Ocean Observing System, Central California Ocean Observing System, and the California Ocean Protection Council. This is NOAA ECOHAB Publication #811 and NOAA MERHAB Publication #181, and it is a contribution to the GEOHAB Core Research Program on HABs in Upwelling Systems.

REFERENCES

1. Glibert PM, Anderson DM, Gentien P, Graneli E, Sellner KG. The global, complex phenomena of harmful algal blooms. *Oceanography* 2005;**18**:136–47.
2. Heisler J, Glibert PM, Burkholder JM, Anderson DM, Cochlan W, Dennison WC, et al. Eutrophication and harmful algal blooms: a scientific consensus. *Harmful Algae* 2008;**8**:3–13.
3. GEOHAB. In: Glibert P, Pitcher G, editors. *Global ecology and oceanography of harmful algal blooms*. Science Plan, SCOR and IOC, Baltimore and Paris; 2001. p. 87.
4. Anderson DM, Cembella AD, Hallegraeff GM. Progress in understanding harmful algal blooms: paradigm shifts and new technologies for research, monitoring, and management. *Ann Rev Mar Sci* 2012;**4**:143–76.
5. Smayda TJ. Harmful algal blooms: their ecophysiology and general relevance to phytoplankton blooms in the sea. *Limnol Oceanogr* 1997;**42**:1137–53.
6. Anderson CR, Moore SK, Tomlinson MC, Silke J, Cusack CK. Living with harmful algal blooms in a changing world: strategies for modeling and mitigating their effects in coastal marine ecosystems. In: Ellis J, Sherman D, editors. *Sea and ocean hazards, risks and disasters. Sec. X: harmful algal blooms*, vol. X. Elsevier Publishers; 2014.
7. Lewitus AR, Horner RA, Caron DA, Garcia-Mendoza E, Hickey BM, Hunter M, et al. Harmful algal blooms along the North American west coast region: history, trends, causes, and impacts. *Harmful Algae* 2012;**19**:133–59.
8. Horner RA, Garrison DL, Plumley FG. Harmful algal blooms and red tide problems on the U.S. west coast. *Limnol Oceanogr* 1997;**42**:1076–88.
9. Trainer VL, Moore L, Bill BD, Adams NG, Harrington N, Borchert J, et al. Diarrhetic shellfish toxins and other lipophilic toxins of human health concern in Washington State. *Mar Drugs* 2013;**11**:1815–35.
10. Howard MDA, Silver M, Kudela RM. Yessotoxin detected in mussel (*Mytilus californicus*) and phytoplankton samples from the U.S. west coast. *Harmful Algae* 2008;**7**:646–52.

11. De Wit P, Rogers-Bennett L, Kudela RM, Palumbi SR. Forensic genomics as a novel tool for identifying the causes of mass mortality events. *Nat Comm* 2014;**5**.
12. Miller MA, Kudela RM, Mekebri A, Crane D, Oates SC, Tinker MT, et al. Evidence for a novel marine harmful algal bloom: cyanotoxin (microcystin) transfer from land to sea otters. *PLoS One* 2010;**5**:e12576.
13. Gibble CM, Kudela RM. Detection of persistent microcystin toxins at the land-sea interface in Monterey Bay, California. *Harmful Algae* 2014;**39**:146–53. http://dx.doi.org/10.1016/j.hal.2014.07.004.
14. Sunda WG, Shertzer KW. Modeling ecosystem disruptive algal blooms: positive feedback mechanisms. *Mar Ecol Prog Ser* 2012;**447**:31–47.
15. Sunda WG, Graneli E, Gobler CJ. Positive feedback and the development of persistence of ecosystem disruptive algal blooms. *J Phycol* 2006;**42**(5):963–74.
16. Jessup DA, Miller MA, Ryan JP, Nevins HM, Kerkering HA, Mekebri A, et al. Mass stranding of marine birds caused by a surfactant-producing red tide. *PLoS One* 2009; **4**(2):e4550. http://dx.doi.org/10.1371/journal.pone.0004550.
17. Curtiss CC, Langlois G, Busse LB, Mazzillo F, Silver MW. The emergence of *Cochlodinium* along the California Coast (USA). *Harmful Algae* 2008;**7**:337–46.
18. Kudela R, Ryan J, Blakely M, Lane J, Peterson T. Linking the physiology and ecology of *Cochlodinium* to better understand harmful algal bloom events: a comparative approach. *Harmful Algae* 2008;**7**:278–92.
19. Kudela RM, Gobler CJ. Harmful dinoflagellate blooms caused by *Cochlodinium* sp.: global expansion and ecological strategies facilitating bloom formation. *Harmful Algae* 2012;**14**:71–86. http://dx.doi.org/10.1016/j.hal.2011.10.015.
20. O'Halloran C, Silver MW, Holman TR, Scholin CA. *Heterosigma akashiwo* in central California waters. *Harmful Algae* 2008;**5**:124–32.
21. Herndon J, Cochlan WP. Nitrogen utilization by the raphidophyte *Heterosigma akashiwo*: growth and uptake kinetics in laboratory cultures. *Harmful Algae* 2007;**6**: 260–70.
22. Harmful Algal Bloom Monitoring and Alert Program (HABMAP) Working Group. *The regional workshop for harmful algal blooms (HABs) in California coastal waters.* Report #565. Costa Mesa, CA: Southern California Coastal Water Research Project; 2008.
23. Frolov S, Kudela RM, Bellingham JG. Monitoring of harmful algal blooms in the era of diminishing resources: a case study of the U.S. West Coast. *Harmful Algae* 2013; **21-22**:1–12.
24. Kudela RM, Frolov SA, Anderson CR, Bellingham JG. *Leveraging ocean observatories to monitor and forecast harmful algal blooms: a case study of the U.S. West Coast.* 2013. IOOS Summit. Available at: http://www.iooc.us/summit/white-paper-submissions/community-white-paper-submissions/.
25. Carter M, Hilbern M, Culver C, Mazzillo F, Langlois G. A Southern California perspective on harmful algal blooms. *CalCOFI Rep* 2013;**54**.
26. Moore SE. Marine mammals as ecosystem sentinels. *J Mammal* 2008;**89**:534–40.
27. Gulland FMD, Hall AJ. Is marine mammal health deteriorating? *EcoHealth* 2007;**4**: 135–50.
28. Scholin CA, Gulland F, Doucette GJ, Benson S, Busman M, Chavez FP, et al. Mortality of sea lions along the central California coast linked to a toxic diatom bloom. *Nature* 2000;**403**:80–3.
29. Gulland FMD, Simmons SE, Rowles TK, Moore SE, Sleeman JM, Weise M. *Marine animal health as an ecosystem sentinel.* 2013. IOOS Summit. Available at: http://www.iooc.us/summit/white-paper-submissions/community-white-paper-submissions/.

30. Anderson CR, Kudela RM, Benitez-Nelson C, Sekula-Wood E, Burrell CT, Chao Y, et al. Detecting toxic diatom blooms from ocean color and a regional ocean model. *Geophys Res Lett* 2011;**38**:L04603. http://dx.doi.org/10.1029/2010GL045858.
31. Anderson CR, Brzezinski MA, Washburn L, Kudela R. Mesoscale circulation effects on a toxic diatom bloom in the Santa Barbara Channel, California. *Mar Ecol Prog Ser* 2006; **327**:119–33.
32. Anderson CR, Siegel DA, Kudela RM, Brzezinski MA. Empirical models of toxigenic *Pseudo-nitzschia* blooms: potential use as a remote detection tool in the Santa Barbara Channel. *Harmful Algae* 2009;**8**:478–92.
33. Anderson CR, Sapiano MRP, Prasad MBK, Long W, Tango PJ, Brown CW, et al. Predicting potentially toxigenic diatom blooms in the Chesapeake Bay. *J Mar Sys* 2010;**83**(3–4):127–40. http://dx.doi.org/10.1016/j.jmarsys2010.04.003.
34. Lane JQ, Raimondi P, Kudela RM. The development of a logistic regression model for the prediction of toxigenic *Pseudo-nitzschia* blooms in Monterey Bay, California. *Mar Ecol Prog Ser* 2009;**383**:37–51.
35. Shulman I, Moline MA, Penta B, Anderson S, Oliver M, Haddock SHD. Observed and modeled bio-optical, bioluminescent, and physical properties during a coastal upwelling event in Monterey Bay, California. *J Geophys Res Oceans* 2011;**116**:C1.
36. Alliance for Coastal Technologies. *Technologies and methodologies for the detection of harmful algae and their toxins*. 2008. Ref. No. [UMCES]CBL 08–143. Available online at: http://www.act-us.info/Download/Workshops/2008/USF_HABs/.
37. Litaker RW, Stewart TN, LeEberhardt B-TL, Wekell JC, Trainer VL, Kudela RM, et al. Rapid enzyme-linked immunosorbent assay for detection of the algal toxin domoic acid. *J Shellfish Res* 2008;**27**:1301–10.
38. Seubert EL, Howard MDA, Kudela RM, Stewart TN, Litaker RW, Evans R, et al. Development, comparison and validation using ELISAs for the analysis of domoic acid in California sea lion body fluids. *J AOAC Int* 2014;**97**(2). http://dx.doi.org/10.5740/jaoacint.SGESeubert.
39. Bernard S, Kudela R, Velo-Suarez L. *Developing global capabilities for the observation and prediction of harmful algal blooms. Oceans and society: blue planet*. Cambridge Scholars Publishing; 2014.
40. Macauley M. The value of information: measuring the contribution of space-derived earth science data to resource management. *Space Policy* 2006;**22**(4):274–82.
41. Hoagland P, Scatasta S. The economic effects of harmful algal blooms. In: Graneli E, Turner JT, editors. *Ecological studies 189: ecology of harmful algae*. Berlin: Springer-Verlag; 2006. p. 391–402.
42. Dodds WK, Bouska WW, Eitzmann JL, Pilger TJ, Pitts KL, Riley AJ, Schloesser JT, Thornbrugh DJ. Eutrophication of U.S. freshwaters: analysis of potential economic damages. *Env Sci Tech* 2009;**43**:12–9.
43. Bernstein BB, Buckley E, Price H, Rosenfeld L. *Turning data into information: making better use of California's ocean observing capabilities*. 2011. A Report Prepared for the California Ocean Protection Council. Available online at: http://www.opc.ca.gov/webmaster/ftp/pdf/docs/SCOOP_report_12-20-11.pdf.
44. Trainer VL, Yoshida T, editors. *Proceedings of the workshop on economic impacts of harmful algal blooms on fisheries and aquaculture*; 2014. p. 85. PICES Sci. Rep. No. 47.
45. GEOHAB. Global ecology and oceanography of harmful algal blooms. In: Berdalet E, Bernard S, Burford MA, Enevoldsen H, Kudela RM, Magnien R, et al., editors. *GEO-HAB synthesis open science meeting*. Paris and Newark, Delaware, USA: IOC and SCOR; 2014. p. 78 [alphabetic order].

CHAPTER

6

Sustained Ocean Observing along the Coast of Southeastern Australia: NSW-IMOS 2007–2014

Moninya Roughan[1,*], Amandine Schaeffer[1], Iain M. Suthers[2]

Coastal and Regional Oceanography Lab, School of Mathematics and Statistics, UNSW Australia, UNSW, Sydney, NSW, Australia[1]; School of Biological, Earth and Environmental Sciences, UNSW Australia, Sydney, NSW, Australia[2]

**Corresponding author: E-mail: mroughan@unsw.edu.au*

CHAPTER OUTLINE

1. INTRODUCTION

The East Australian Current (EAC) flows poleward along the coast of southeastern Australia forming the western boundary current of the South Pacific subtropical gyre. This is the most populated region of Australia; hence, the warm current influences the climate and marine economies of nearly half the Australian population. The current transports heat and biota poleward, as well as driving cross-shelf exchange as it flows from the tropics to the Tasman Sea. As the EAC has strengthened in recent decades, the waters off the coast of southeastern Australia have warmed significantly, at a rate of three to four times the global average.[1] Off the coast of New South Wales (NSW),[2] studies have showed that surface waters have warmed at a rate of 0.75 °C per century (~34°S), whereas further to the south (~42.5°S), the surface waters have warmed at a rate of 2.02 °C per century[2]. Increasingly, we are seeing the associated impacts of tropicalization on marine ecosystems, particularly in western boundary currents such as the EAC.[3]

Since 2006, the Australian Federal government has invested over $120 M in Australia's Integrated Marine Observing System (IMOS, www.imos.org.au), which has had matching co-investment from industry, universities, stakeholders, and state and federal agencies. The main goal is to provide a multidisciplinary, multi-institutional approach to enhance the observation and understanding of the oceans around Australia. The funding flows through 10 centrally coordinated "infrastructure" facilities that have the responsibility of deploying ocean observing equipment in the oceans around the Australian continent, covering physical, chemical, and biological variables. (See Ref. 4 this book for an in-depth description of the one such facility, ACORN—the Australian Coastal Ocean Radar Network.) These facilities are supported by the National IMOS office and an eleventh data facility responsible for data management.

Nationally, the scientific community has self-organized into a number of broadly state-based scientific nodes, covering the coastal oceans around Australia. These being NSW-IMOS (southeast coast), Q-IMOS: Queensland (northeast coast and northern Australia), WAIMOS (west coast), SAIMOS (south coast), and Tas-IMOS (Tasmania). In addition, an all-encompassing Blue Water and Climate Node focuses on the deep water and long time scale processes. The IMOS infrastructure is deployed in accordance with the scientific plans (developed by each node) that have each undergone international peer review. In addition to the individual node plans, the national IMOS steering committee has developed a national science plan that encompasses the overarching science themes and objectives of IMOS. Each of these science plans is available on the IMOS Website. IMOS has been remarkably successful to date, and all data collected through IMOS are freely available to the research community (http://imos.aodn.org.au/imos123/home). Here, we focus on the IMOS implementation along the coast of southeastern Australia, as guided by the New South Wales node of IMOS (NSW-IMOS).[5,6]

2. NSW-IMOS IN THE NATIONAL CONTEXT

2.1 SCIENCE OBJECTIVES

Through the evolution of the IMOS program, a set of national science plans has been developed, one for each of the regional science nodes. This process has allowed the marine and climate science community to come together and provide the scientific rationale for a national-scale Integrated Marine Observing System (IMOS). Through this, a number of unifying national overarching goals were identified: (1) multidecadal ocean change, (2) climate variability and weather extremes, (3) major boundary currents and inter-basin flows, (4) continental shelf and coastal processes, and (5) the ecosystem response.[7]

Each of the coastal nodes then formulated their science themes in the context of the national objectives (http://imos.org.au/plans.html); for example, each of the regional nodes is working to understand the dynamics of their respective boundary current and inter-basin flows. In the case of NSW-IMOS, the science themes revolve around the East Australian Current and its role in each of the 5 themes (see Figure 1). For example, under theme 1, multidecadal change, we are addressing questions such as how is the heat transport of the EAC changing, including the offshore recirculation and the poleward eddy transport. As another example, in the context of theme 4, continental shelf processes, we wish to understand the frequency, magnitude, and drivers of upwelling and downwelling processes and slope water intrusions in the coastal waters off NSW and how they influence cross-shelf exchange of properties.

2.2 SOCIO-ECONOMIC DRIVERS

Australia is a vast continent about the size of the mainland United States, with a small population (less than 23 M). More than half these people live within 50 km of the coast that stretches from Brisbane to Melbourne, making this the most populated region of the country. In addition, more than 80% of the population live in coastal regions. Thus, despite the size of the continent, we are primarily a marine nation. Therefore, major coastal issues affect a vast number of the population. These issues include urbanization, water quality, freshwater supply, beach erosion, and severe storms. Severe storms such as East Coast Lows are generally formed over the ocean, bringing high rainfall, high winds, and significant wave height, often causing widespread destruction and coastal erosion. The defense sector collaborates with the research community on operational generation of ocean forecasts, but in general, defense activities are currently not as significant a driver of ocean research in Australia as in other countries, such as the United States.

More than 3500 km^2 of coastal waters along the NSW coastline (to the three-nautical-mile limit) have been designated as multiple use Marine Estates, to conserve marine biodiversity, maintain ecological processes, and provide for

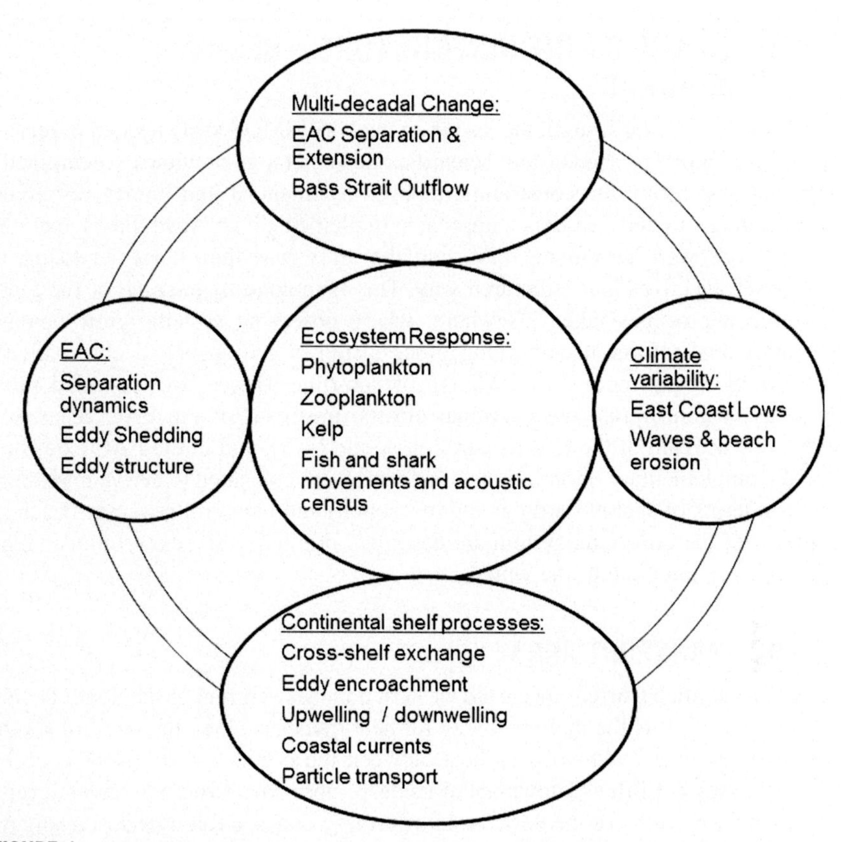

FIGURE 1

Conceptual model of the science goals of NSW-IMOS in the framework of the five national IMOS science themes.

Adapted from Ref. 7.

ecologically sustainable use, public appreciation, and education. In this respect, science needs are associated with a need to understand habitat distributions and connectivity of populations in the context of marine park planning.

Though there are presently no major mining activities off our coastline, the ports associated with the international export of coal are areas of major coastal activities. Along the coast of southeastern Australia, although the fishing effort is modest (the EAC, a western boundary current, is oligotrophic by nature), marine tourism is a major economic driver. Perhaps the biggest threats to the region are associated with the warming and strengthening of the EAC and implications for tropicalization of temperate regions (including habitat loss and invasive species), changes to ocean productivity, and increased storm activity.[7,8]

3. THE NSW-IMOS INFRASTRUCTURE—DESIGN OF THE ARRAY

As with any observing system, the design of the NSW-IMOS array needed to serve a number of competing needs (both scientific and societal as previously mentioned), while working within the constraints imposed by financial and human resources. The advantage to the top-down approach implemented in Australia is that the science needs were articulated first, and the array was then designed to try to meet those objectives in a structured way. This is in contrast to some of the more bottom-up approaches taken elsewhere, where observing systems grew upward from individual efforts at combining process studies and sustained observations. NSW-IMOS covers more than 2000 km of coastline. Hence, we identified three key areas to concentrate observational efforts: upstream of the EAC separation (30°S), downstream of the EAC separation point (34°S), and in the EAC extension (36°S). Complementary observational platforms were colocated to derive maximum benefit in terms of spatial (vertical and horizontal) and temporal coverage. Each of the pieces of the observing system are described briefly in the next sections, along with how they integrate to the whole.

3.1 EAC TRANSPORT ARRAY

The East Australian Current forms the western boundary current of the South Pacific subtropical gyre. It is the main pathway for heat transport along the coast of southeast Australia, and it redistributes heat between the ocean and atmosphere on its journey poleward. Little is known about its deep subsurface structure, undercurrent, and the time variability of the flow. To address these issues, a full transport resolving mooring array was deployed for an 18-month period from April 2012 to August 2013. The array consists of five moorings extending eastward along a line at approximately 26°S, from 154 to 155.5°E, in depths between 1500 and 4750 m (Figure 2(a)). The deep water array is augmented by three shelf moorings, extending the line inshore onto the continental shelf. This data will provide the upstream boundary condition for future modeling studies of the EAC. A data assimilation project is ongoing to assess the utility of this array in predicting EAC separation and eddy shedding.

3.2 HF RADAR

The Coffs Harbour region in northern NSW (~30°S) was chosen as the location of the only high-frequency surface radar pair along the continental shelf of southeastern Australia (Figure 2(a)).[4] The location was chosen for a number or reasons: Coffs Harbour is generally upstream of the EAC separation point, and in this region, submesoscale frontal eddies are frequently formed on the inside edge of the EAC. In addition, both state and federal marine reserves have been designated in the

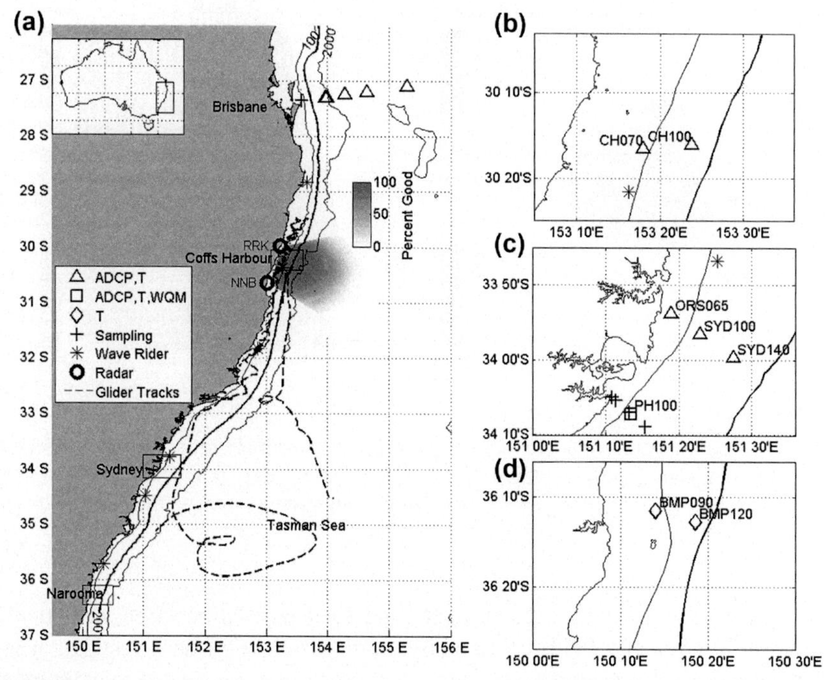

FIGURE 2

(a) Map showing the location of the NSW-IMOS study sites along the east coast of Australia and colocation of equipment. HF radar coverage from the two sites at RRK (Red Rock) and NNB (North Nambucca) is indicated by the shading, with the circles showing the land-based stations. Dashed lines indicate representative glider paths, a star shows the location of wave ride buoys. Right panels show a zoom of the location of the oceanographic moorings at (b) Coffs Harbour, (c) Sydney, and (d) Narooma. Instrumentation is as per the legend—ADCP measures velocity through the water column, T is temperature through the water column, WQM indicates biogeochemical sampling at a fixed depth, and sampling (+) indicates where physical biogeochemical samples are taken.

region, which is known for its biological significance, at the intersection between the northern extent of kelp and southern extent of coral.

Land-based WERA systems were deployed at Red Rock (30°S, Figure 2) and North Nambucca (30.6°S, Figure 2(a)). Each site consists of a transmitter (13.92 initially, now 13.5 MHz, 100 kHz bandwidth) and an array of 12 receivers. This measures surface current velocities at a spatial resolution of 1.5 to 2 km with a maximum range of 100 km offshore.[4,9]

Due to the high spatial resolution of the data, for the first time, we have resolved the surface velocity structure of a number of submesoscale coherent flow structures, including eddies, fronts, and filaments. Increasingly, processes such as coastal cold core eddies are being recognized for their impact in entrainment of shelf waters,

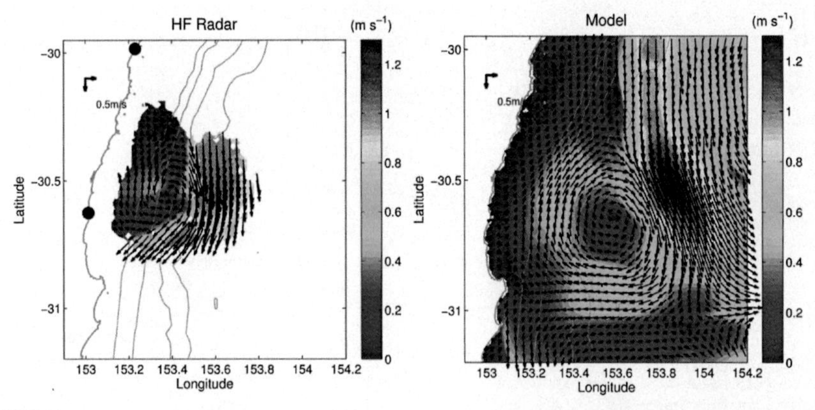

FIGURE 3

Snapshot of surface current vectors and intensity (color) from HF radar observations (left) and ROMS modeling (right) on July 9, 2012. The coastline and 100-, 200-, 1000-, and 2000-m isobaths are shown in gray.[11] Black dots (left) indicate the location of the two HF radar sites.

which are likely enriched with nutrients and seed populations (see Ref. 10 this book). Surface velocity fields captured by the radar allow for an investigation into the evolution of such features, and when combined with numerical modeling, can resolve the dynamical drivers. One such example of a submesoscale cold core eddy is given in Figure 3, where an attempt has been made to reproduce the flow field in a numerical simulation.[11]

3.3 GLIDERS

Under the IMOS arrangement, gliders are managed by a centrally coordinated facility (ANFOG, www.anfog.ecm.uwa.edu.au). This results in a nationally consistent approach to QA/QC, calibration, and data processing, and it reduces maintenance and management costs. There are down sides to this strategy, such as the availability of gliders and limited human resources to manage multiple deployments around a continent the size of Australia; however, these shortcomings are manageable. ANFOG presently maintains a fleet of both Seagliders (deep) and Slocum gliders (shallow).

Gliders have been deployed into the EAC as part of the NSW-IMOS program since 2009. Initially, the program was fairly exploratory in nature while we ascertained what could (and could not) be achieved in a western boundary current (WBC) regime. We deployed a combination of Seagliders and Slocum gliders into EAC eddies to understand the temporal evolution of these mesoscale features that dominate the circulation downstream of the EAC separation point (~31–32°S, more than 50% of the time[12]). This approach resulted in mixed success—some missions proved very successful for understanding eddy dynamics,[13] watermass properties,[14] and entrainment of shelf waters.[15] But, the logistics and cost associated with retrieval of a wayward glider swept offshore in a mesoscale eddy (Figure 2(a)) on top of a number of gear failures (particularly with the Seagliders) resulted in a change of strategy.

Sustained endurance lines that are common in other regions are not feasible in a dynamic WBC—at least not extending onto the shelf and into shallow water where alongshore surface velocities frequently exceed 2 m/s.[16–18] Thus, a strategy was adopted that resulted in gliders being deployed off Yamba (29.5°S) with a zig-zag mission definition extending poleward along the continental shelf. In this manner, the glider is advected poleward with the alongshore currents, and we maximize cross-shelf coverage. The transects generally start at the 25-m isobath, extending to just inshore of the shelf break (~100 m) with a spatial extent of 300 to 400 km (approximately 29–32°S, Figure 2(a)). The final retrieval location (and hence, along-shelf distance) is dependent upon the balance between battery life and weather conditions. From 2009 to 2014, 22 glider missions have been conducted off southeast Australia, and as of May 2014, 13 of these have been repeat cross-shelf (zig-zag) deployments along the continental shelf inshore of the EAC, resulting in a new high-resolution climatology of the shelf waters.[19] This climatology has been used to show the role of along-shelf advection in the momentum balance[20] and the importance of the EAC in driving upwelling as it separates from the coast.[17,18]

3.4 THE SHELF MOORING ARRAY

The longest moored time series on the east coast of Australia is the Ocean Reference Station (ORS065, Figure 2(c), Table 1). It has been maintained (under contract for Sydney Water Corporation) since 1989 and has been included as an IMOS data stream since May 2008. To build on this long history of observations, we created an array of moorings shore normal off Bondi with the addition of two moorings in 100 and 140 m of water (SYD100 and SYD140, respectively), complemented by another mooring ~20 km to the south off Port Hacking (PH100, Figure 2(c)). Presently, these three moorings (Table 1) consist of a bottom-mounted TRDI 300 kHz ADCP housed in a rigid frame with gimbal mount and a line of thermistors extending from the bottom to 15–20 m below the surface (string of Aquatech 520 temperature and temperature/pressure loggers at 8-m intervals through the water column) that is supported by a sub-surface float. In addition, the PH100 mooring is augmented with a Wetlabs water quality meter (WQM) consisting of a SeaBird CTD and measurements of fluorescence, turbidity (Wetlabs FLNTU), and dissolved oxygen, collectively referred to as BGC in Figure 2, Table 1. At various times, this mooring has also had a separate surface float to serve as a warning, but it also allows for measurement of sea surface temperature and to act as a real-time testbed.

Whereas the parameters measured at the CH070 and CH100 moorings (Figure 2(b)) are the same as off Sydney, the ADCP mooring design differs, with the ADCPs suspended 1–2 m above the bottom in floating frame with two acoustic releases hanging directly below the moorings. The final two moorings in the array

Table 1 NSW-IMOS Mooring Deployment Metadata

Name	Latitude (°S)	Water Depth (m)	Distance Offshore (km)	Date Deployed	Parameter Measured	Major Axis	V Bin Depths (m)	T Sensor Depths (m)
CH070	30.27	74	16	August 2009	T, V	20°	10–65	16–72
CH100	30.27	98	25	August 2009	T, V	20°	13–89	11–96
ORS065	33.90	67	2	1989/May 2008	T, V	16°	11–61	16–66
SYD100	33.94	104	10	June 2008	T, V, BGC	19°	12–96	24–102
SYD140	33.99	138	19	June 2008	T, V	24°	23–127	21–137
PH100	34.12	115	6	October 2009	T, V, BGC	33°	14–104	15–111
BMP090	36.19	94	9	March 2011	T	15°		28–92
BMP120	36.21	119	16	March 2011	T	15°		20–116

Parameters measured are T: temperature at 8 m through the water column (4 m at ORS065), V: velocity at 4 m through the water column (8 m at SYD140), BGC: Wetlabs Water quality meter (WQM) consisting of Seabird CTD temperature, salinity, dissolved oxygen, and fluorescence. The ORS065 was first deployed in 1989, with data incorporated into IMOS since May 2008. The major axis orientation is calculated from depth averaged velocity data over a period of Four years from January 2010 to December 2014, except for BMP, where orientations are estimated from bathymetry.

Table adapted from Refs 17,18,23,29.

BMP090 and BMP120 (Figure 2(d)) consist solely of a subsurface temperature/pressure string at 8-m intervals to ~20 m below the surface (Table 1).

Temperature and velocity data are recorded at 5-min intervals, while the WQM records 60 burst samples at a rate of 1 Hz every 15 min. Initial deployment dates are indicated in Table 1, and to date, each mooring continues to return excellent data with a return rate generally between 70 and 90%. An example data set is shown in Figure 4, and additional data can be visualized at www.oceanography.unsw.edu.au.

In general, the NSW-IMOS moored array is fairly low cost by ocean observing standards. Despite the strong surface currents (frequently in excess of 1.5 m/s), the subsurface profile of the moorings means we can limit the size of the weight (vessel) needed; thus, we can deploy from local fishing vessels. The trade-off is that we sacrifice real-time data and surface parameters.

3.5 IN SITU BIOGEOCHEMICAL SAMPLING

The Port Hacking site (Figure 2, Table 1) is one of a series of nine national reference stations (NRS) located around Australia[21] that are instrumented in a similar way and are augmented with in situ biogeochemical sampling monthly.[21,22] Lynch provides a full description of this national network. Each of the national reference stations provides a broad suite of observations that can provide temporal reference for more spatially distributed and intensive shorter term studies.[23] In addition, they provide for the calibration and validation of coastal remote sensing activities. Regionally, individual NRS act as focal points within each of the coastal IMOS nodes.[21]

The two sites at Port Hacking (50 and 100 m) have been occupied nominally monthly since the 1940s, making them some of the longest oceanic temperature records in existence and some of the only long-term records in the southern hemisphere. Initially, temperature was monitored (using reversing thermometers) at PH050 (d = 0, 10, 20, 30, 40, and 50 m) and PH100 (d = 0, 10, 25, 50, 75, and 100 m). Monthly CTD profiles have been taken along a shore normal transect encompassing PH100 since 1997 (Figure 2(c)) at depths of 25, 50, 100, and 125 m. Niskin bottle samples are taken at discrete depths for measurements of total dissolved inorganic carbon, alkalinity, and nutrients. We use combined water column samples from all the Niskin bottles at a site for phytoplankton. Zooplankton measures are taken from a plankton drop net. Zooplankton and phytoplankton samples have also been collected at the 50 and 100 m stations since 1997.

Since the start of IMOS, additional water samples have been included at various times for genetic analyses and investigation of microbial communities. In addition, the monthly sampling off Port Hacking serves to calibrate fluorometric observations obtained from the instrumented moorings and ocean color estimates of chlorophyll obtained from satellites (e.g., MODIS): chlorophyll, total suspended solids, and colored dissolved organic material (CDOM). The methodology is documented in detail in an IMOS operations handbook (http://imos.org.au/facility_manuals.html).

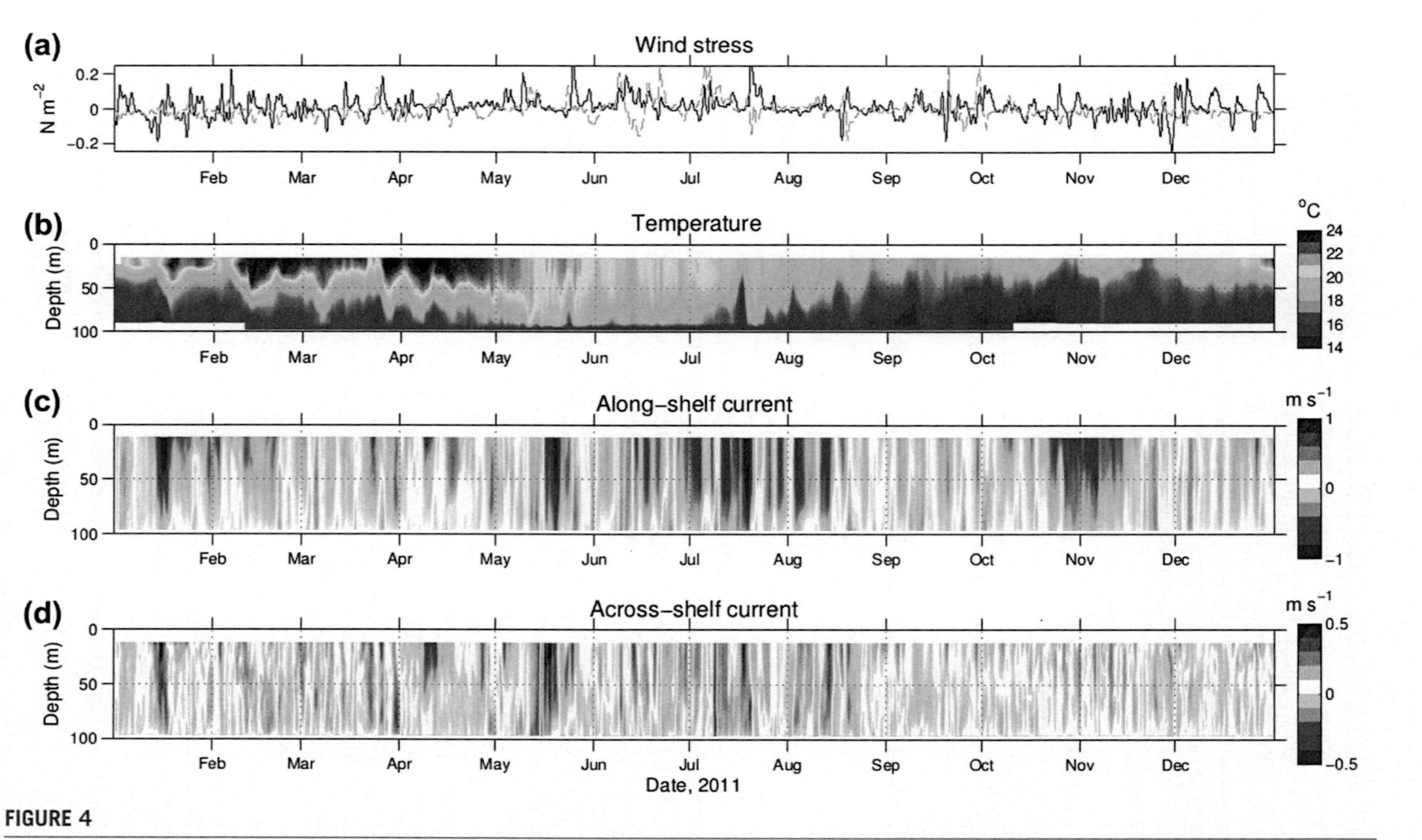

FIGURE 4

Data from the SYD100 mooring during 2011 showing (a) wind stress, where black (gray) is the alongshore (across-shelf) component,

3.6 HIGHER TROPHIC LEVELS

One of the integrating aspects of the IMOS program is the extension of the ocean observing system into plankton and marine animals. Three IMOS facilities, AATAMs, CPR, and AUV, have established a successful long-term observing program of biological responses to oceanographic change. We discuss some of these insights here. AATAMS (Animal Tagging and Monitoring) has two branches, the first focuses on the tagging and tracking of marine mammals (fur seals, sea lion, and elephant seals) and the integration of animal behavior with oceanographic conditions (as measured from CTDs that are glued to the animal's heads, known as bio-logging). This has proved particularly successful in the southern ocean where elephant seals are foraging under ice, thereby collecting new observations of temperature and salinity outside the range of ships and ARGO. This publicly available database provided by IMOS and their French collaborators has approximately 75,000 profiles between the Kerguelen Islands and Pyrdz Bay.[24]

The second branch of AATAMS is the acoustic tagging of fish and sharks in waters around Australia. AATAMS listening stations have been deployed as either arrays or across-shelf curtains around Australia. Specifically, off NSW, arrays that monitor the movement of tagged fish and sharks have been deployed at Coffs Harbour, Sydney, and Bateman's Bay (Figure 2). A national database is now available through the IMOS portal for registered users to access where their tagged fish have unexpectedly turned up. This innovation has revealed some surprisingly large travel distances, such as great white sharks traveling from New Zealand to eastern Australia and bull sharks traveling from Sydney to Townsville (North Queensland).[25] With improved understanding of the actual range of species of interest, we can better incorporate them into biological models and analyses.

Ships of opportunity have been used to deploy a Continuous Plankton Recorder (CPR) between Brisbane, Sydney, and Melbourne, undertaking 33 voyages since 2008 (Figure 5). The CPR is identical to those operated around the world, collecting and preserving plankton on a silk roll, which is converted to abundance. The CPR is complemented by sampling for phyto- and zoo-plankton and microbial diversity at the National Reference Stations, with standardized sampling for larval fish at some NRS commencing in late 2014.

One remarkable discovery by the CPR was the massive abundance of black fungal spores on the CPR silk after the extensive dust storm over eastern Australia in September 2009, after 10 years of drought.[26] The spores were able to be grown on agar from the formalin-fixed material, and they were identified as a terrestrial fungus, *Aspergillus sydowii*. Though the fungus is implicated with coral disease, no particular disease outbreak resulted on this occasion. The significance of more conventional marine phytoplankton blooms (such as the dinoflagellate *Noctiluca*[27]) or swarms of gelatinous zooplankton (such as salps) is starting to be realized.[28]

The Autonomous Underwater Vehicle (AUV) facility has focused on imaging the benthos and automating the characterization of the habitat. Surveys are repeated at very specific locations along the eastern seaboard (and nationally at a number of key locations spread across large latitudinal ranges on both the east and west coasts of

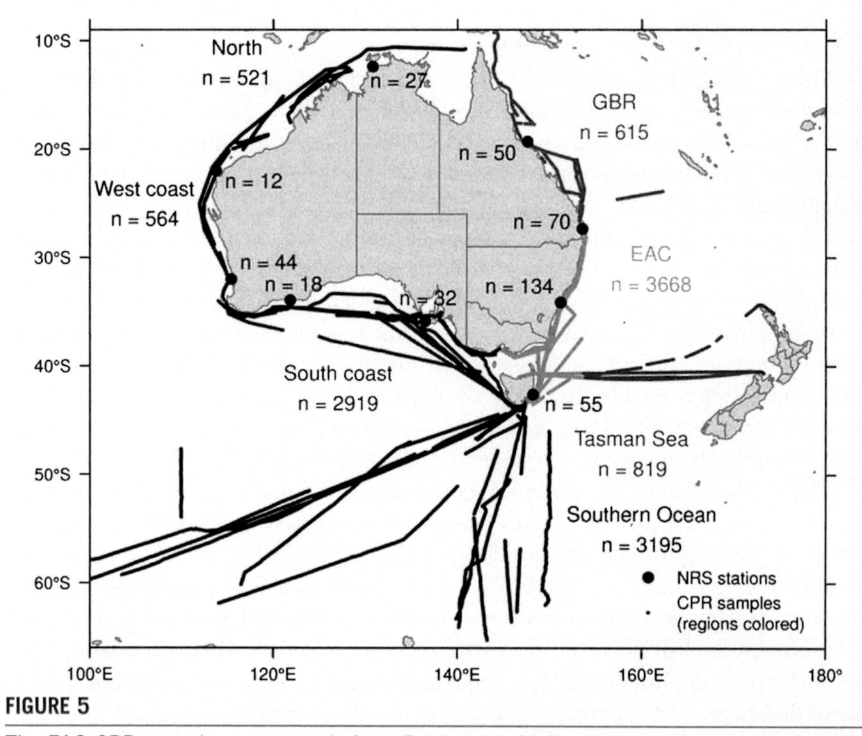

FIGURE 5

The EAC-CPR route is run quarterly from Brisbane to Melbourne on ships of opportunity by the IMOS-SOOP facility, and it is complemented by the quarterly Great Barrier Reef (GBR)-CPR route, the quarterly south coast route from Melbourne to Fremantle, and the annual routes to the Southern Ocean and from New Zealand, as per solid lines.

Adapted from http://imos.org.au/australiancontinuousplanktonr3.html.

Australia). Repeat visits to these sites, stretching from the Solitary Islands (29°S) to Batemans Bay (36°S), have shown, for example, the impact of extreme events on coral bleaching and the subsequent recovery and loss of kelp through marine heat waves. One of the main science questions being addressed by the AUV facility is quantifying kelp and other brown algae at the northern (warmer) boundary of their distribution, while also examining their paucity around major urban centers. Kelp is affected not only by sea urchin grazing as a result of a trophic cascade, but it is also affected by environmental change and pollution, which, therefore, has implications for the location of marine sanctuary zones. Observing by the AUV facility helps provide a science-based case for the contentious questions around marine parks zoning.

4. ASSESSING THE DESIGN OF THE SHELF MOORING ARRAY

In recent years, significant time and financial resources have been invested into ocean observing systems internationally. There is no doubt that these systems as a whole are worthwhile, and they are collecting critical baseline data; however, it is likely that in some instances there is either (1) room for improvement to address data gaps, or (2) room for possible consolidation, where observations are redundant. In the case of NSW-IMOS, the system was designed on the basis of a number of process studies that had been conducted previously, providing a best estimate of a strategy with a view to facilitating as much emerging research in the future as possible. In the case of the shelf mooring array, five years since the deployment of the first moorings, we are now in a position to assess the design of the array and to question if the array is serving our science needs.[29] An objective way of assessing the array is to use both the data and models to understand the spatial and temporal correlations in conjunction with the dominant processes.

4.1 HOW CORRELATED ARE THE DATA?

Spatial correlations of the different parameters measured at the mooring sites provide insight into the relevance of the array design and the variability resolved. For adjacent sites, correlations across the shelf for depth-averaged along-shelf velocities and temperature are high, ranging from 0.84 to 0.89 (lags between 0 to 3 h) and 0.87 to 0.93 (lags between 0 to 6 h), respectively (Table 2). In contrast, for across-shelf velocities, the only correlation coefficients that are greater than 0.25 are CH070–CH100 (0.71) and ORS065–SYD100 (0.57, Table 2). These results suggest the need for at least three moorings at regular intervals across the relatively shelf (<30 km) in order to resolve the cross-shelf dynamics.

Along the coast, the along-shelf velocity exhibits similar variability downstream (ORS065, SYD100, SYD140, and PH100) with correlations greater than 0.62 (Table 2). Between upstream and downstream, correlations for along-shelf velocities are still significant, ranging from 0.24 to 0.34 for lags around 22 to 36 h, and they show the different EAC dynamics along the coast and the potential need for an additional array between 30 and 34°S.

The temperature at Port Hacking and Narooma sites (6, 9, and16 km from the coast, respectively) are most correlated to Sydney inshore moorings (ORS065, 2 km from the coast, Table 3). In contrast, the temperatures at the offshore sites CH100 and SYD140 are the least correlated with moorings at other latitudes, as the dynamics offshore are expected to be mostly driven by the EAC (CH100) and its eddies (SYD140). The temperature variability (Table 3) between inshore and offshore within 25 km of the coastline confirms the relevance of the cross-shelf arrays of two to three moorings.

In terms of temporal variability, the de-correlation time scales at a particular depth are very variable, ranging from 2 to 20 days for temperature, 6 to 50 h and 30 min to 3 h for along- and across-shelf currents, respectively (Table 4). This suggests that high-frequency measurements are necessary to resolve the across-shelf variability; although less so for along-shelf processes. The minimum de-correlation time scales in the water column are driven by local physical processes. Upstream (CH) and in the EAC extension (BMP), the shortest de-correlation time

Table 2 Maximum Correlations and Lags in Hours of the Along- (Black) and Across-Shelf (Gray) Components of the Sub-Inertial Depth-Integrated Velocities

Platform	CH070	CH100	ORS065	SYD100	SYD140	PH100
CH070		0.84 (–1 h)	0.32 (29 h)	0.34 (25 h)	0.30 (28 h)	0.28 (22 h)
CH100	0.71 (3 h)		0.25 (36 h)	0.28 (30 h)	0.29 (31 h)	0.24 (31 h)
ORS065	0.11 (–17 h)	0.11 (214 h)		0.86 (–3 h)	0.71 (–3 h)	0.80 (–2 h)
SYD100	0.11 (55 h)	0.14 (–112 h)	0.29 (1 h)		0.89 (0 h)	0.72 (1 h)
SYD140	0.09 (33 h)	0.04 (–219 h)	0.24 (0 h)	0.57 (–1 h)		0.62 (2 h)
PH100	0.10 (220 h)	0.18 (–203 h)	0.09 (198 h)	0.09 (–102 h)	0.08 (–115 h)	

All correlations are significant at the 95% significance level and are computed using the maximum concomitant time series available for each mooring pair between 2010–2014 (between 564 and 1112 days).

Adapted from Table 2 in Ref. 29.

Table 3 Maximum Correlations and Lags in Hours (h) of the Sub-Inertial Depth-Integrated Temperature

Platform	CH070	CH100	ORS065	SYD100	SYD140	PH100	BMP090	BMP120
CH070		0.88 (6 h)	0.66 (24 h)	0.60 (14 h)	0.44 (2 h)	0.64 (28 h)	0.69 (48 h)	0.71 (36 h)
CH100			0.47 (23 h)	0.43 (13 h)	0.40 (−215 h)	0.44 (−240 h)	0.55 (56 h)	0.59 (81 h)
ORS065				0.89 (2 h)	0.67 (2 h)	0.90 (0 h)	0.76 (11 h)	0.77 (8 h)
SYD100					0.87 (2 h)	0.89 (−1 h)	0.72 (−8 h)	0.58 (240 h)
SYD140						0.72 (−8 h)	0.58 (240 h)	0.51 (226 h)
PH100							0.77 (11 h)	0.66 (151 h)
BMP090								0.93 (0 h)

All correlations are significant at the 95% significance level and are computed using the maximum concomitant time series available for each mooring pair between 2010 and 2014 (between 626 and 1461 days).

Table 4 De-Correlation Time Scales for Each Mooring and Variable Computed as the Maximum Lag Corresponding to Auto-Correlations Greater than a Threshold of 0.7. The Minimum, Maximum, and Their Depths Are Indicated. The Calculations Are Based on 5 min Measurements from 2010 to 2014

Platform	CH070	CH100	ORS065	SYD100	SYD140	PH100	BMP090	BMP120
Temperature	3.3–20 days (bottom–top)	2.3–20 days (bottom–top)	3.3–16 days (44 m–top)	3.6–12.5 days (42 m–top)	3.7–7.4 days (42 m–top)	4.9–15 days (32 m–top)	5.5–10.6 days (92 m–top)	5.8–17 days (114 m–top)
Along-shelf velocity	15–25 h (bottom–17 m)	22–50 h (bottom–16 m)	6–19 h (bottom–19 m)	10–29 h (bottom–24 m)	25–42 h (bottom–31 m)	9–26 h (bottom–20 m)		
Across-shelf velocity	1.2–1.7 h (bottom–17 m)	1.8–3 h (bottom–12 m)	0.6–1 h (bottom–top)	0.8–1.9 h (bottom–16 m)	1.3–2.7 h (bottom–55 m)	0.7–1.3 h (bottom–16 m)		

scale is at the bottom for all variables due to the frequent bottom water uplift driven by the EAC,[9,17,18] whereas in the eddy region (SYD), the most variable temperature is at mid-depth. The design of the mooring array covering the water column seems appropriate for the investigation of both of these highly variable processes.

4.2 MODELING ASSESSMENT

An eddy-resolving model was used to assess the spatial correlations in sea surface temperature and velocity,[30] to understand if the mooring array does an adequate job of resolving inter-annual variability in sea surface temp and velocity along the entire coastline. Results showed that at the inter-annual time scale, temperatures were fairly well correlated along the shelf, and they go some way toward representing the intra-seasonal variability. However, although alongshore correlations were generally high, there was a distinct region of lower correlation at ~32 to 33°S (which was more exacerbated in the velocity fields), suggesting that this region must be instrumented if we are to gain a true understanding of the shelf dynamics in this region.

4.3 MANAGEMENT

Data management is an essential, and in many ways, the most important part of the observing system. To this end, significant investment has been made into an open data repository through the Australian Ocean Data Network (AODN, imos.aodn.org.au).

All IMOS data is made freely available, including enhanced data products and web services in a searchable and interoperable framework. A central data facility (eMII) provides the standards, protocols, and systems to integrate the data and related information into a number of conformal frameworks, and it provides the tools to access and utilize the data. In addition, for some types of data, they provide additional derived data products, as well as web services for processing, integration, and visualization of the data (http://imos.org.au/emii.html). Some IMOS data are also routinely integrated into international programs and databases.

4.4 DATA UPTAKE

Data uptake is an equally important aspect of the program, and this is supported and promoted through an increasing number of data products being made available in addition to regular outreach opportunities to keep the community informed.

NSW-IMOS is one of the IMOS success stories when it comes to data uptake in an integrated manner. We have a vibrant and active science community who use the data in a multitude of ways, and we have actively sought to ensure that the data is being used to maximum benefit. Since 2007, NSW-IMOS data has been integral to the success of more than 50 externally funded research projects that have used IMOS infrastructure as a backbone. In addition, the core NSW-IMOS data have been used by over 45 postgraduate research projects, including 14 Ph.D. completions to date. The vast numbers of conference presentations that acknowledge

NSW-IMOS (more than 200) are likely to increase the number of forthcoming journals articles (presently more than 50). Each of these science outputs contributes to the success and longevity of the program.

One of the ways we have encouraged data uptake and multidisciplinary science is to bring people together regularly in node science meetings (two per year), with a different emphasis at each meeting. Participant numbers range from 30 to 50, with themes including "Model data integration" and "Integrating up the food chain." Alternatively, we hold meetings with a specific regional focus, such as a recent two-day workshop on the 70-year time series at Port Hacking.

In addition, we have designed university courses around the data. For example, a new multi-university master's level subject run through the Sydney Institute of Marine Science (www.sims.org.au) that introduces students to a number of the core NSW-IMOS data streams has proved very successful with up to 40 M.Sc. students enrolling per year. This strategy brings wide exposure to the program as the students come from broad backgrounds nationally and internationally, from industry, nonprofits, and state agencies alike.

5. SHORTCOMINGS AND RECOMMENDATIONS FOR THE FUTURE

5.1 GAPS IN THE ARRAY

The assessment of the array through a combination of models and observations has shown that data gaps exist particularly in the central and southern part of the domain. Unfortunately, due to funding constraints, the mooring array at 36°S has been cut, which leaves a significant gap in monitoring of the EAC extension where the ocean has been shown to be warming rapidly. In addition, this region is a marine protected area, and NSW stakeholders (e.g., NSW Office of Environment and Heritage and Department of Primary Industries) place a priority on understanding coastal processes.

In the short term, these gaps could be filled by an increase in glider coverage. A potentially cost-effective alternative is to run shore normal glider endurance lines in this region, although EAC eddies have proved that this is a difficult environment in which to operate gliders. In addition, an extension of the longshore glider program along the NSW shelf down to the southern part of the domain would provide valuable seasonal and spatial coverage of the shelf hydrography.

The EAC full transport array was removed in Sept 2013 due to funding limitations and problems with shipping schedules. Encouragingly, it is slated to be redeployed in 2015. However, the gap between deployments allows for an important assessment of the array and its utility in predicting EAC dynamics including eddy shedding and separation.

A strategy that has proved cost prohibitive to date is to transition a number of the coastal moorings to real-time data acquisition. This would allow for greater immediate data uptake through the assimilation into coastal circulation models while providing information that meets societal needs (e.g., the commercial and recreational fishing industries). In addition, this would provide a platform for over-ocean wind observations that are critically lacking.[31]

A major gap exists in the biologically critical Stockton Bight area (~32–33.5°S), which is recognized as a fisheries nursery area. The most straightforward way of filling the gap would be an increase in the coastal radar coverage combined with increased southward coverage in glider range. This would provide information on the northward coastal counter current and export from the shelf of coastal waters.

Finally, expanding the reach of the biogeochemical sampling through either regular seasonal or opportunistic sampling at the northern and southern moored arrays (30 and 36°S) would allow more opportunities to validate satellite/remotely sensed products such as Chl-a and provide baseline data for impacts of ocean warming and environmental change.

5.2 MODEL–DATA INTEGRATION

At present, the Australian IMOS program does not have core funding to support model–data integration. This is in contrast to other programs such as the US-IOOS Coastal and Ocean Modeling Testbed, which is focused on operationalizing models out of the research community. To address this issue without creating a significant financial burden, IMOS has taken the lead in coordinating a new workshop: the Australian Coastal and Oceans Modelling and Observations Workshop (ACOMO). Two workshops were held at the Australian Academy of Science in October 2012 and 2014. The goal of these workshops was to bring the somewhat disparate coastal ocean modeling community together to coordinate activities in order to make significant advances in a unified way. The meeting presented an opportunity to detail the state of play and identify future needs and opportunities. Furthermore, a closely related project, the Marine Virtual Laboratory (MARVL) project (http://www.nectar.org.au/marine-virtual-laboratory), is developing software tools to enable model–data integration to happen more efficiently.

Data assimilation techniques are being used increasingly to make assessments of observing systems.[32] Within Australia, collaborators within NSW-IMOS have secured additional federal research funds to conduct data assimilation modeling to assess the design of the observational array. Over the coming years, we envisage that this will provide insight into the types and locations of observations needed to understand the dynamics in the region. Thus, it will help to target new deployments and potentially will provide cost savings.

6. CONCLUSIONS

In Summary, NSW-IMOS is an example of a highly successful implementation of a coastal ocean observing system. The observational array has been built around pertinent science questions, leveraging existing data streams and opportunities.

The science questions are integrated into the national and international context and represent the state of the science today.

The operational aspects are streamlined as they are run through a number of centrally coordinated specialized infrastructure facilities. Data return is increasingly consistent, and data dissemination and uptake is broad. The operations are backed by a team of active and highly successful scientists (the "node") who are using the observing system widely as a backbone to conduct high-quality internationally relevant research.

ACKNOWLEDGMENTS

We acknowledge the enormous contributions made by the national and local IMOS teams, both scientific (e.g., NSW-IMOS node steering committee, National IMOS Office) and technical (e.g., the ANFOG, ACORN, and ANMN technical teams). Projects of this scale require considerable effort from a multitude of people, and the success of NSW-IMOS is a testament to them. IMOS is supported by the Australian Government through the National Collaborative Research Infrastructure Strategy, the Super Science Initiative, and the Education Infrastructure Fund.

REFERENCES

1. Wu L, Cai W, Zhang L, Nakamura H, Timmermann A, Joyce T, et al. Enhanced warming over the global subtropical western boundary currents. *Nat Clim Change* 2012;**2**(3): 161–6.
2. Thompson PA, Baird ME, Ingleton T, Doblin MA. Long-term changes in temperate Australian coastal waters and implications for phytoplankton. *Mar Ecol Prog Ser* 2009;**394**:1–19.
3. Verges A, Steinberg P, Hay M, Poore A, Campbell A, Ballesteros E, et al. The tropicalization of temperate marine ecosystems; climate-mediated changes in herbivory and community phase shifts. *Proc R Soc B* 2014;**281**:20140846. http://dx.doi.org/10.1098/rspb.2014.0846.
4. Wyatt LR. *The IMOS ocean radar facility.* ACORN; 2015 [This book].
5. Roughan M, Morris BD, Suthers IM. NSW-IMOS: an integrated Marine observing system for Southeastern Australia. *IOP Conf Ser Earth Environ Sci* 2010;**11**:012030. http://dx.doi.org/10.1088/1755-1315/11/1/012030.
6. Roughan M, Suthers I, Meyers G. The australian integrated Marine observing system (IMOS) and the regional implementation in New South Wales. In: You J, Henderson-Sellars A, editors. *Climate alert: change monitoring and strategy.* Sydney University Press; 2010, ISBN 9781920899363.
7. Roughan M, Suthers IM, Morris B, Armand L, Baird M, Coleman M, et al. *NSW-IMOS node science and implementation plan July 2009–2013. Tech rep Sydney Institute of Marine Science Mosman, Sydney NSW, 2088, Australia.* 2009.
8. Robertson R, Doblin M, Ingleton T, Roughan M, Suthers IM. *NSW-IMOS node science and implementation plan 2015–2020. Tech rep Sydney Institute of Marine Science Mosman, Sydney NSW, 2088, Australia.* 2014.

9. Schaeffer A, Roughan M, Wood JE. Observed bottom boundary layer transport and uplift on the continental shelf adjacent to a western boundary current. *J Geophys Res Oceans* 2014;**119**. http://dx.doi.org/10.1002/2013JC009735.
10. Archer MR, Shay LK, Jaimes B, Martinez-Pedraja J. *Observing frontal instabilities of the Florida current using high frequency radar.* 2015.
11. Schaeffer A, Roughan M. *MARine virtual laboratory (MARVL) – high-resolution ROMS test: the east Australian current at 30°S in Solitary Island Marine Park.* 2014.
12. Cetina Heredia P, Roughan M, Van Sebille E, Coleman MA. Long-term trends in the East Australian Current separation latitude and eddy driven transport. *J Geophys Res Oceans* 2014;**119**. http://dx.doi.org/10.1002/2014JC010071.
13. Baird ME, Suthers IM, Griffin DA, Hollings B, Pattiaratchi C, Everett JD, et al. Physical-biogeochemical dynamics of a surface flooded warm-core eddy off southeast Australia. *Deep-Sea Res II* 2011;**58**(5):592–605. http://dx.doi.org/10.1016/j.dsr2.2010.10.002.
14. Baird ME, Ridgway K. The southward transport of sub-mesoscale lenses of Bass Strait Water in the centre of anti-cyclonic mesoscale eddies. *Geophys Res Letts* 2012;**39**:L02603.
15. Everett JD, Macdonald HS, Baird ME, Humphries J, Roughan M, Suthers IM. Cyclonic entrainment of preconditioned shelf waters into a Frontal Eddy. *J Geophys Res Oceans* 2015;**120**:677–91. http://dx.doi.org/10.1002/2014JC010301.
16. Roughan M, Middleton JH. On the east australian current; variability, encroachment and upwelling. *J Geophys Res Oceans* 2004;**109**:C07003. http://dx.doi.org/10.1029/2003JC001833.
17. Schaeffer A, Roughan M, Morris B. Cross-shelf dynamics in a western boundary current. Implications for upwelling. *J Phys Oceanogr* 2013;**43**:1042–59.
18. Schaeffer A, Roughan M, Morris B. Corrigendum: cross-shelf dynamics in a western boundary current. Implications for upwelling. *J Phys Oceanogr* 2014;**44**:2812–3. http://dx.doi.org/10.1175/JPO-D-14-0091.1.
19. Schaeffer A, Roughan M. Influence of a western boundary current on shelf dynamics and upwelling from repeat glider deployments. *Geophys Res Lett* 2015;**42**:121–8. http://dx.doi.org/10.1002/2014GL062260.
20. Roughan M, Oke PR, Middleton JH. A modeling study of the climatological current field and the trajectories of upwelled particles in the East Australian Current. *J Phys Oceanogr* 2003;**33**(12):2551–64.
21. Lynch TP, Morello EB, Evans K, Richardson AJ, Rochester W, Steinberg CR, et al. IMOS National Reference Stations: a continental-wide physical, chemical and biological coastal observing system. *PLoS One* 2014;**9**:12. http://dx.doi.org/10.1371/journal.pone.0113652.
22. Lynch TP, Roughan M, Mclaughlan D, Hughes D, Cherry D, Critchley G, et al. A National Reference Station Infrastructure for Australia – using telemetry and central processing to report multidisciplinary data streams for monitoring marine ecosystem response to climate change. In: *MTS/IEEE oceans Canada 2008*; 2008.
23. Roughan M, Morris BD. Using high-resolution ocean timeseries data to give context to long term hydrographic sampling off Port Hacking, NSW, Australia. In: *Proceedings of MTS/IEEE oceans 2011 Kona USA*; 2011.
24. Roquet F, Williams G, Hindell MA, Harcourt R, McMahon C, Guinet C, et al. A Southern Indian Ocean database of hydrographic profiles obtained with instrumented elephant seals. *Sci Data* 2014;**1**:140028. http://dx.doi.org/10.1038/sdata.2014.28.

25. Heupel MR, Simpfendorfer CA, Espinoza M, Smoothey AF, Tobin A, Peddemors V. Conservation challenges of sharks with continental scale migrations. *Front Mar Sci* 2015;**2**:12. http://dx.doi.org/10.3389/fmars.2015.00012.
26. Hallegraeff GM, Coman F, Davies C, Hayashi A, McLeod D, Slotwinski A, et al. Australian duststorm associated with extensive *Aspergillus sydowii* fungal bloom in coastal waters. *Appl Environ Microbiol* 2014;**80**(11):3315–20. ISSN 0099-2240. http://dx.doi.org/10.1128/AEM.04118-13.
27. McLeod DJ, Hallegraeff GM, Hosie GW, Richardson AJ. Climate-driven range expansion of the red-tide dinoflagellate *Noctiluca scintillans* into the Southern Ocean. *J Plankton Res* 2012;**34**(4):332–7.
28. Henschke N, Everett JD, Doblin MA, Pitt KA, Richardson AJ, Suthers IM. Demography and interannual variability of salp swarms (*Thalia democratica*). *Mar Biol* 2014;**161**: 149–63.
29. Roughan M, Schaeffer A, Kioroglou S. Assessing the design of the NSW-IMOS Moored Observation Array from 2008–2013: recommendations for the future. In: *Proceedings of MTS/IEEE oceans 2013, San Diego USA. Sept 2013*; 2013.
30. Oke PR, Sakov P. Assessing the footprint of a regional ocean observing system. *J Mar Syst* 2012;**105–108**:30–51.
31. Wood J, Roughan M, Tate P. Finding a proxy for wind stress over the coastal ocean off Sydney. *Mar Freshw Res* 2012;**63**(6):528–44. http://dx.doi.org/10.1071/MF11250.
32. Xue P, Chen C, Beardsley RC, Limeburner R. Observing system simulation experiments with ensemble Kalman filters in Nantucket Sound, Massachusetts. *J Geophys Res* 2011; **116**:C01011.

CHAPTER

Projeto Azul: Operational Oceanography in an Active Oil and Gas Area Southeastern Brazil

7

Francisco Alves dos Santos[1,*], André Luis Santi Coimbra de Oliveira[1], Felipe Lobo Mendes Soares[1], Gabriel Vieira de Carvalho[1], Pedro Marques São Tiago[1], Natalia Gomes dos Santos[1], Henery Ferreira Garção[1], Flávia Pozzi Pimentel[1], Ana Carolina Boechat[1], Lívia Sant'Angelo Mariano[1], Cecília Bergman[1], Frederico Luna Rinaldi[1], Tiago Cardoso de Miranda[1], Marcelo Montenegro Cabral[1], Leonardo Maturo Marques da Cruz[1], Júlio Augusto de Castro Pellegrini[1], Luiz Paulo de Freitas Assad[2], Luiz Landau[2], Maurício da Rocha Fragoso[1]

PROOCEANO Serviço Oceanográfico, Rio de Janeiro, Brazil[1]; Universidade Federal do Rio de Janeiro, Rio de Janeiro, Brazil[2]

**Corresponding author: E-mail: francisco@prooceano.com.br*

CHAPTER OUTLINE

1. INTRODUCTION

Brazil is the eleventh greatest oil producer in the world, and around 90% of its production comes from the ocean (http://anp.gov.br). It is in the ocean that seismic prospecting, exploration drillings, rig moorings, and all required operations for oil and gas (O&G hereafter) production take place. Consequently, that is where air guns are shot, drill cuttings and produced water are discharged, and oil spills may occur.

Started in the late 1970s in the Campos Basin (southeastern Brazil), the steady growth of offshore O&G operations in Brazil was not followed by a proportional increase in the oceanographic data collection. Without important threats to operations, such as the loop current and hurricanes in the Gulf of Mexico, expensive oceanographic cruises and equipment were not in the order of the day. As a consequence, ocean observations in the area are few and sparse and important aspects of its complex dynamics are still unknown.

The discovery of huge oil reservoirs in the pre-salt layers of Santos Basin (immediately south of Campos Basin) at water columns of up to 3500 m, presented new technological challenges for offshore O&G operations in the area. This fact, added to the increased environmental concern raised in the last decades, boosted private and public investments toward a better understanding and forecasting of the regional dynamics.

In this context, Projeto Azul (translates to Project Blue in English) is a pilot project started in August 2012 for the offshore area of Santos Basin, totally funded by a private O&G operator (BG-Brasil) within the scope of the research and development program of the Brazilian National Petroleum Agency (ANP). It is an attempt to adapt and reproduce in Santos Basin successful ocean observation programs in regional scale like the SCOOS (Southern California Ocean Observing System) and the GCOOS (Gulf of Mexico Coastal Ocean Observing System).

The program's main objectives are the following:

1. To propose and evaluate an operational in situ data collection program capable of efficiently sampling the vertical structure of mesoscale features in Santos Basin
2. To provide meaningful and near real-time data for assimilation in numerical oceanic circulation models responsible for the operational forecasts
3. To create a public database of the collected data and to display both data and analysis in an accessible and useful way for the scientific community, the O&G companies, and the environmental agencies

In this chapter, a brief overview of Santos Basin ocean dynamics and main features influencing offshore operations is presented, followed by a description and first results of the in situ observation program and numerical forecast system proposed.

2. SANTOS BASIN OCEAN DYNAMICS

Santos Basin is located between Cabo Frio (23°S) and Florianopolis (28°S) within the South Brazilian Bight (SBB) and under the influence of the western branch of the

South Atlantic Gyre. A general description of the vertical structure of the current system is usually split into three layers (e.g., Ref. 1): (1) Brazil Current (BC) dominates the upper layers flowing southward over the Continental Shelf Break transporting Tropical Water (TW) and South Atlantic Central Water (SACW). On most of the Santos Basin, BC occupies the upper 500-m layer and presents a strong baroclinic component[2,3] associated to the shear with the (2) Intermediate Western Boundary Current (IWBC). IWBC originates at the Santos Bifurcation around 27°S and transports Antarctic Intermediate Water (AIW) northward, below the BC flow and down to approximately 1300 m.[4–6] Another current inversion occurs toward the bottom layers, where the (3) Western Deep Boundary Current (WDBC) transports the North Atlantic Deep Water (NADW) southward (Figure 1).

An important oceanographic feature in the region is the Cabo Frio Eddy (CFE), a transient cyclonic coherent structure with approximately 100-km diameter frequently observed in the northern part of Santos Basin. Due to the abrupt change in the Brazilian coastline orientation, BC detaches from the shelf break toward deeper waters, increasing potential vorticity and causing the system to meander.[2,8] It is also indicated by Silveira et al.[9] that the meander growth and eventual shed

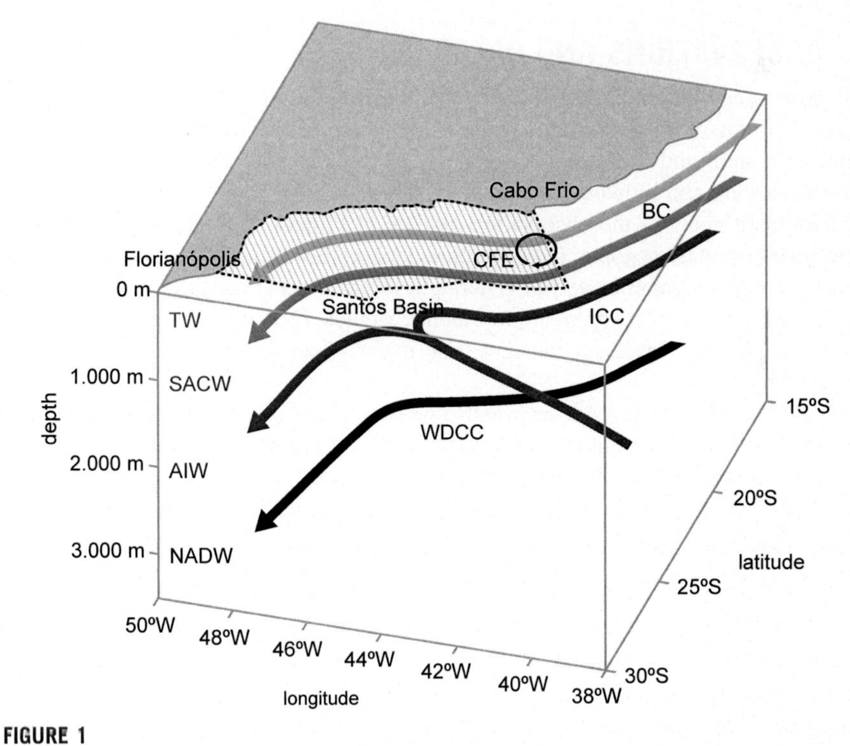

FIGURE 1

Schematic representation of the Santos Basin dynamics.

Adapted from Ref. 7.

of the CFE is linked to specific instability conditions in the baroclinic BC-IWBC layer. The shearing structure (BC-IWBC) of the current system near the Cabo Frio region with regularly eddy formation and the possible surface intensification due to Agulhas eddies interactions[10] form a complex and potentially dangerous system for offshore operations. A good observing system regarding the mass field is important due to the dominant baroclinic nature of this system.

To our knowledge, Ref. 11 is one of the most complete works with mooring measurements in the region and describes in detail the BC downstream changes within Santos Basin. The authors analyze and compare data from four deep water moorings and quasi-synoptic hydrographic sections with other similar work for the region (e.g., Refs. 12–14). Their results suggests an intensification of BC related to the Santos Bifurcation and the consequent end of the BC-IWBC shearing layer, with a southward flow varying from 0.31 ± 0.12 m/s near 23°S up to 0.5 ± 0.05 m/s when the flow reaches 25.5°S. Data from another mooring line located at 27°S shows a southward flow down to 900 m (maximum sensor depth), confirming the intensification and the thickening (from 350 m north of Santos Basin down to 900 m at its southern limit) of BC.

3. OBSERVATIONS AND DATABASE

Offshore operations in Santos Basin are concentrated in deep waters, frequently at water columns of up to 2000 m. The understanding of the vertical structure of the currents system and the mesoscale activity in the upper layers is key to activities such as rigging and blowout contingency planning.

The main goals of the observation module at Projeto Azul are to provide long time series of quasi-synoptic hydrographic data on the vertical structure of the Santos Basin ocean dynamics and to generate relevant surface current data ($z \leq 15$ m) for currents model evaluation and turbulence studies. Data collection methods chosen were gliders[15] (used for the first time in Brazilian waters), floats,[16] and SVP drifters.[17]

All those methods provide near real-time data and allow for a maximized spatial coverage without requiring extensive usage of vessels, minimizing operational risks and interruptions in the time series due to bad weather. Near real-time availability and spatial coverage also makes the dataset well suited for assimilation in operational forecasts. Vertical profilers such as gliders and floats have great impact on the assimilation scheme as presented by Pan et al.[18] and Cummings and Smedstad.[19]

3.1 GLIDERS

Gliders are autonomous underwater vehicles (AUVs) capable—within some constraints—to follow routes set by the user while diving to 1000 m depth. The glider model used in Projeto Azul is the SeaGlider,[15] equipped with a Seabird Free Flow CT-Sail, an Aanderaa oxygen optode, and a WetLabs Eco Triplet sensor

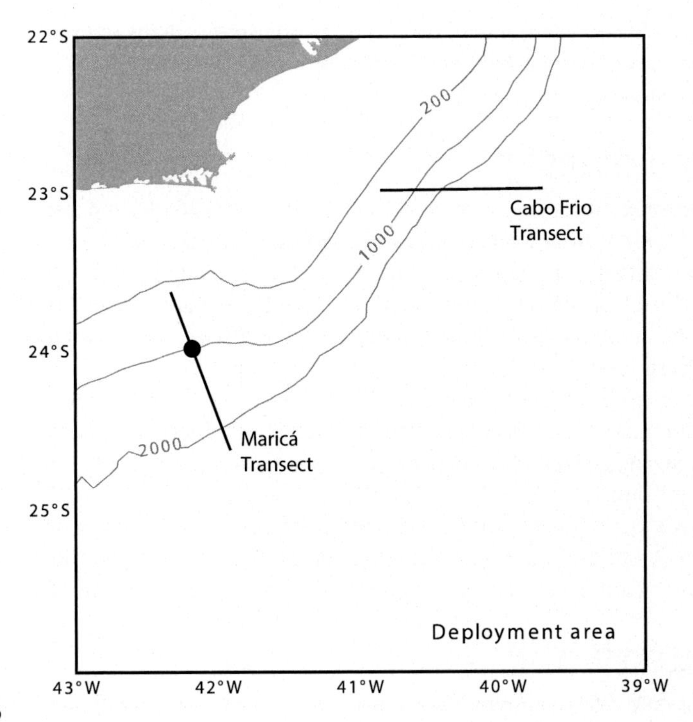

FIGURE 2

Planned glider transects (black lines) and standard deployment location (black dot) for floats and drifters.

for optical scattering, chlorophyll-a, and C-DOM. The gliders first missions were to investigate the region of the BC meandering and formation of the CFE. To that purpose, two hydrographic sections were proposed: the *Cabo Frio* transect, aligned at the 23°S parallel just before the BC overshooting the shelf break, and the *Maricá* transect, located further south in the expected region for the meandering of BC and shedding of the CFE (Figure 2). The transects should be covered simultaneously by two gliders, providing hydrographic sections of the BC-IWBC system before and after the Cabo Frio, allowing for studies on the CFE origin and behavior. Each transect is approximately 100-km wide, departing from the 200-m isobaths and crossing the shelf break. As the glider's horizontal velocity is near 0.25 m/s, the whole section should be covered in around five days, permitting quasi-synoptic assumptions.

3.2 FLOATS

The deployment of profiling floats was planned to increase the spatial coverage of the CTD profiles (as the gliders mission are concentrated in the northern part of Santos Basin), to collect data down to 2000 m, and to have estimates of subsurface currents at selected parking depths. In order to increase the number of CTD profiles in the Santos Basin, the standard dive cycle period for the floats in Projeto Azul was set to three days, with parking depths varying from 500 to 1500 m according to the feature of interest.

3.3 SVP DRIFTERS

Due to its water-following characteristics, satellite-tracked ocean drifters are useful tools for tracking surface oil spills and the associated dynamics.[20,21] During the first year, SVP drifters were deployed monthly in clusters of three (two regular and one SVP-B, with a barometric pressure sensor). This deployment strategy objective is to increase the statistics on Lagrangian dispersion in the area, in continuation of the work of Berti et al.[22]

The increased number of active SVP drifters in the area aims to aid the rapid interpretation of surface dynamics and the evaluation of model nowcasts. As a later outcome, this data should benefit studies of surface circulation such as, Ref. 23 as the statistics for binning may suffice for seasonal analysis.

Standard deployment coordinates for drifters and floats were 24°00′S and 42°10′W (Figure 2), in the intersection of the *Maricá* transect and the 1000-m isobaths. This location varied according to the location of CFE and BC meandering.

3.4 COLLECTED DATA

To present date, 20 oceanographic campaigns were completed with the deployment of 60 SVP drifters and 12 floats. The gliders performed 12 months of uninterrupted hydrographic sections at *Maricá* transect and two months at the *Cabo Frio* section. Gliders accomplished 6087 dives, and floats performed 1062 profiles. Drifter trajectories and the location of each glider and float profiles are presented in Figure 3.

3.5 DATA PROCESSING AND DISSEMINATION

All data collected is passed through a first-level automated quality control and stored in a database. SVP drifters positions are filtered using the forward–backward technique.[24] At present, a simple statistical quality control (flagging data points outside 2 standard deviations of the distribution mean) is applied to gliders and floats data, while a robust algorithm (following the proposed on Ref. 25) is under development. All data is stored in a database hosted in a cloud environment in order to minimize risks of data loss by hardware failure.

As a program targeted not only to the scientific community but also to the technical staff involved in offshore operations, Projeto Azul's Website (http://projetoazul.eco.br) was created to present the information in a clean and pleasant way that could be easily absorbed by a wider audience, and not only by specialists. On the Website, visitors can explore an interactive map displaying real-time monitoring of the equipment, browse information and beautiful imagery captured by the field team, understand the dynamic of the region through periodic analysis made by oceanographers, and freely download all collected data (Figure 4).

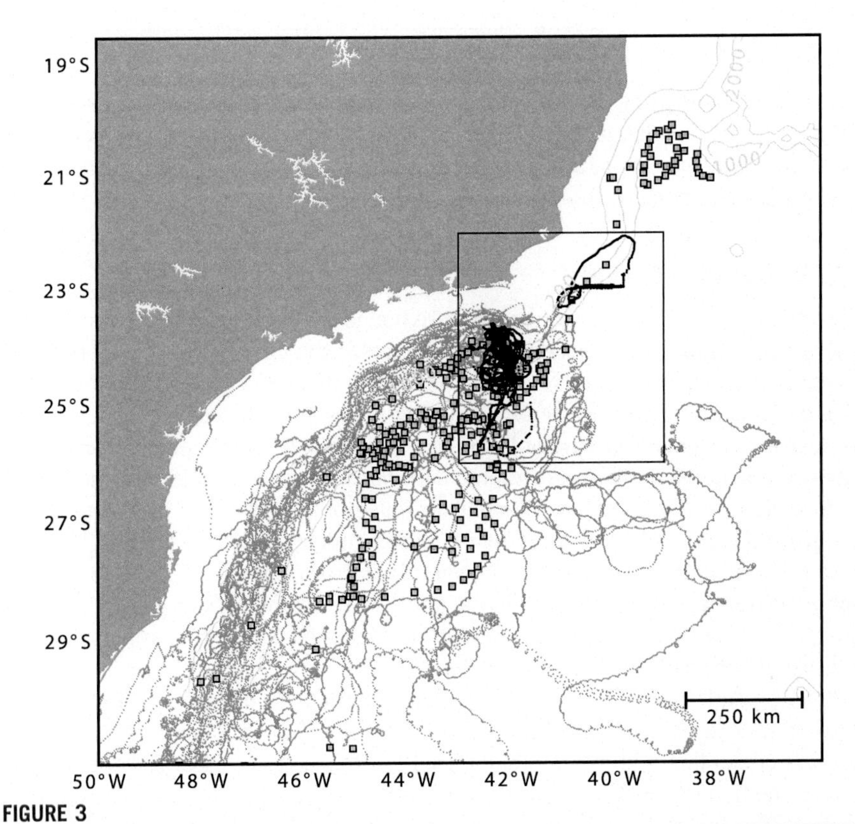

FIGURE 3

Drifter trajectories (gray lines), floats CTD profiles (gray squares), and glider dive positions (black dots) from March 2013 to October 2014. The black square shows the deployment area presented in Figure 2.

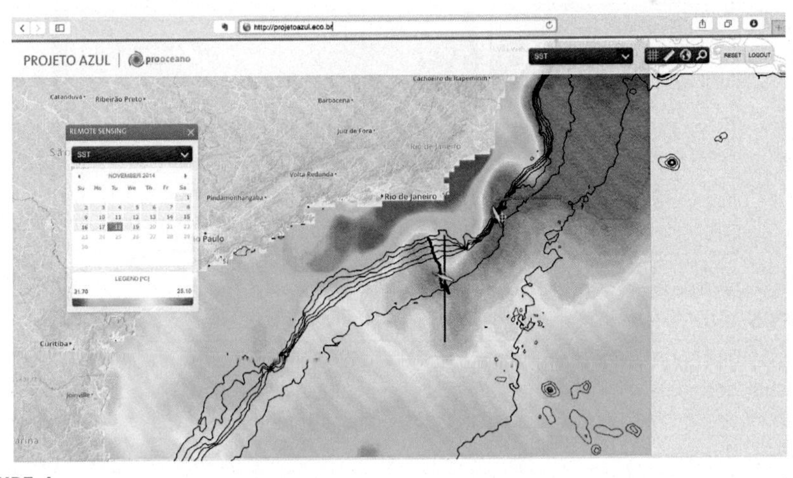

FIGURE 4

Projeto Azul Website (http://projetoazul.eco.br).

4. RESULTS

The dataset obtained by the project is quite extensive and allows many possibilities for analysis and results. In this chapter, two brief examples of such possibilities are provided, focusing on peculiar dynamic aspects of Santos Basin, namely, the BC-IWBC system and Cabo Frio Eddy.

4.1 BC-IWBC SYSTEM

The dynamics of the Brazil Current western boundary system can be effectively followed using gliders, profilers, drifters, and remote sensing data. From gliders' CTD profiles, it is possible to derive relative velocities using the geostrophic method.[26] Moreover, satellite measurements of surface slope are used to calculate surface geostrophic currents. Combining both data, the absolute geostrophic velocities are obtained, using the surface geostrophic currents as the "level of known motion."[10] To that purpose, the Absolute Dynamical Topography (produced by Ssalto/Duacs and distributed by AVISO at http://www.aviso.altimetry.fr/duacs/) is used. Figure 5 shows an example of absolute geostrophic velocities obtained by this technique where positive values are entered in the paper (directed to NE). Because the glider takes, on average, five days to complete its programmed path, a geostrophic velocity section can be obtained on a weekly basis.

It is possible to verify the current shear between the BC and IWBC varying between 300 and 500 m. The mean value of this interface is found at 400 m depth, and it is coincident with the value obtained by Rocha et al.[11] The BC core can be seen in the first 100 m depth, with strong currents reaching 0.68 m/s. The currents on the left of the profile are a consequence of the cross-profile shear on the surface velocities resulting from a cyclonic meander of the BC.

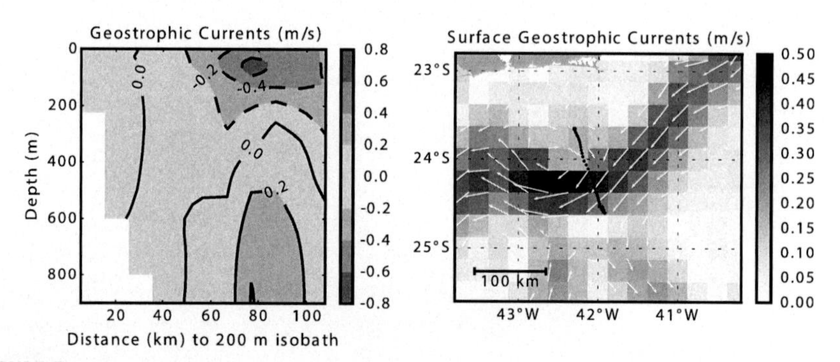

FIGURE 5

Absolute geostrophic velocities profile obtained from glider dives. Solid contour lines represent northeastward flow and dashed lines, southwestward (left). AVISO geostrophic currents used as "level of known motion" for the geostrophic method (color bar and vectors) and glider's trajectory (black dots) (right).

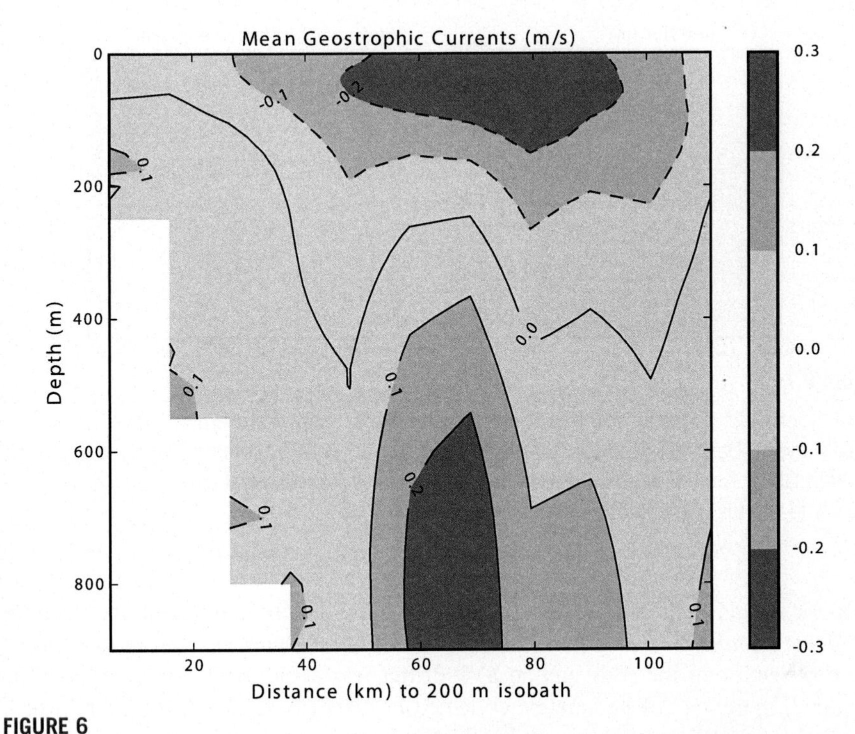

FIGURE 6

Absolute geostrophic velocities mean profile obtained from 30 sections from gliders in 2013. Solid contour lines represent northeastward flow and dashed lines, southwestward flow.

The continuous monitoring allowed for a long-term study of the features present on the region dynamics, generating enough data to produce annual and seasonal studies. During the year of 2013, the gliders from Projeto Azul performed 30 missions on the Maricá transect, considered straight enough to calculate the geostrophic velocity profile across them. Based on these data, a mean profile of geostrophic velocity was calculated (Figure 6). Once more, the shear interface between the BC and the IWBC is clearly present.

This mean profile unveils an average level of 350 m depth to the interface between the currents in the outer-shelf region. The maximum velocity of the BC part (negative values) is 0.3 m/s, whereas the maximum at the IWBC is slightly lower, 0.25 m/s.

4.2 CABO FRIO EDDY

An example of the CFE investigation occurred in November 2014. Evidences of the CFE presence were identified in remote sensing products such as the Mean Sea Level Anomaly (MSLA) produced by Ssalto/Duacs (distributed by AVISO at

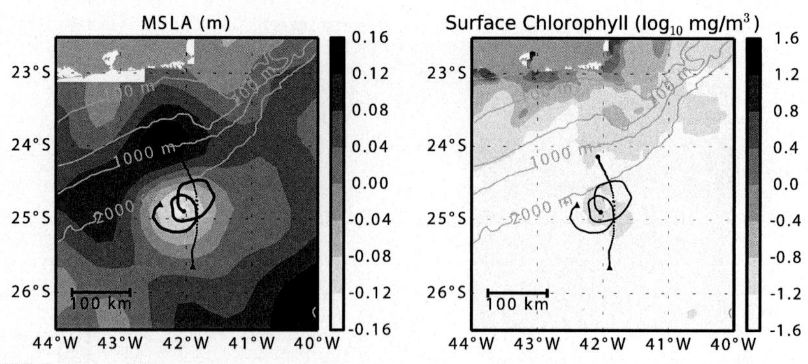

FIGURE 7

MSLA (left) and surface chlorophyll (right) maps from satellite at November 10, 2014, during a CFE event. Dashed lines represents the trajectory of an SVP drifter deployed in the eddy center. Glider route is represented by the dotted line from the black circle (November 4, 2014) to the black triangle (November 13, 2014).

http://www.aviso.altimetry.fr/duacs/) and the surface chlorophyll map produced by *Collecte Localisation Satellites*.[27] Glider routes were changed in order to perform a cross-section of the eddy and an SVP drifter was deployed in the expected CFE center (Figure 7). All parameters collected in the CFE vertical cross-section performed by the glider are presented in Figure 8.

As shown by the satellite surface maps, the glider passed near the core of CFE, and its presence is quite marked by the rise on the isotherms and isohalines at the center of the salinity profile. The signature of the CFE core was found in all other sensors, as greater values are displayed in the middle of the section.

4.3 SUBSURFACE CURRENTS

Figure 9 shows the trajectories of seven profilers released in the first year of Project Azul. Among them, it is possible to see that some were trapped by the CFE.

Profilers parking data were used to estimate deep currents in the vertical structure of the eddy, using the profilers parking data by the method described in Ref. 28, where velocities are assessed along with errors due to vertical shear of the horizontal flow. As an example, the velocity of one profiler from Projeto Azul that drifted with CFE was calculated. It has been trapped in the eddy for three months, diving with parking depth of 600 m, and its mean deep velocity is 0.11 ± 0.02 m/s. Because the error is one order of magnitude smaller, it is possible to infer that the deep currents are the main forcing of the profiler's drift and that the CFE is still present at this depth.

It was also possible to obtain the mean deep currents at different latitude sections for each parking depth available. In Projeto Azul, three different levels were used, 300, 600, and 1500 m depth. Table 1 summarizes the results.

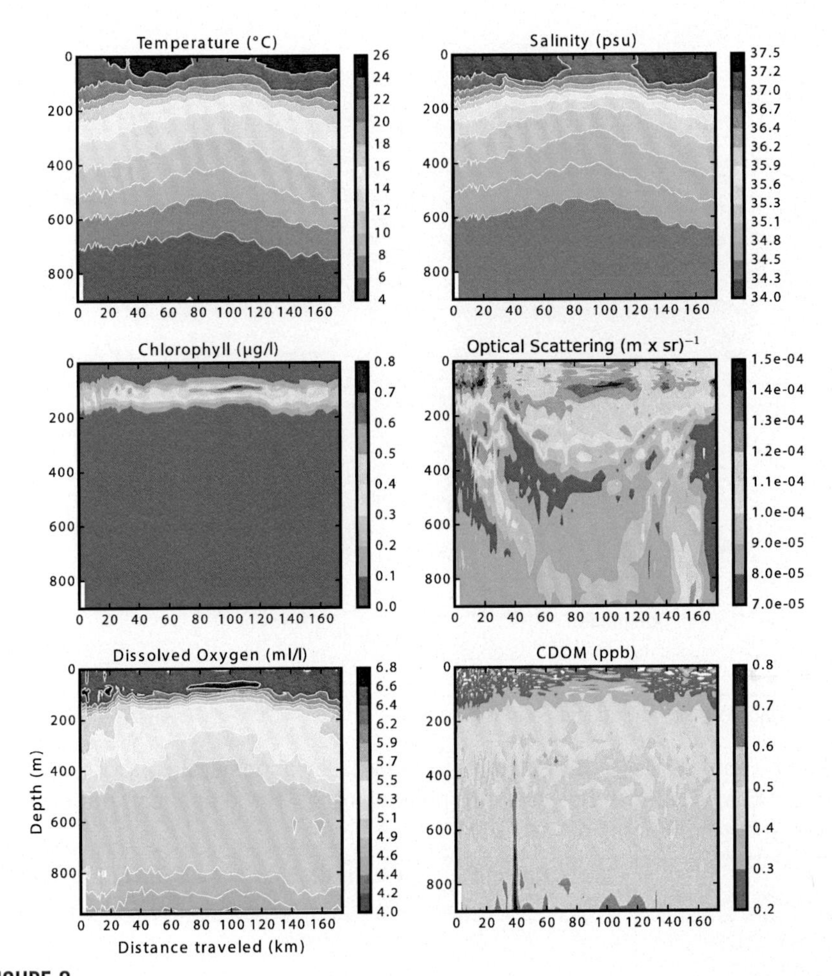

FIGURE 8

From top-left to bottom-right, temperature, salinity, chlorophyll, optical scattering, dissolved oxygen, and CDOM vertical cross-section of CFE performed by a Projeto Azul glider from November 4 to 13, 2014. Distance traveled refers to the route described in Figure 7.

By interpolating the values in depth, it is possible to compare the moorings measurements from Ref. 11 with the estimates obtained from the profilers in the same levels (highlighted values in Table 1). Table 2 gives the interpolated values of the moorings at the profiler's parking depths.

Because the moorings do not reach the 1500 m depth, the last point of each mooring was extrapolated to this level. The results show a good accordance between both datasets; eight of 11 values are consistent, considering their standard deviation. This result indicates that this method can provide a good estimate about the deep currents and the vertical structure of CFE.

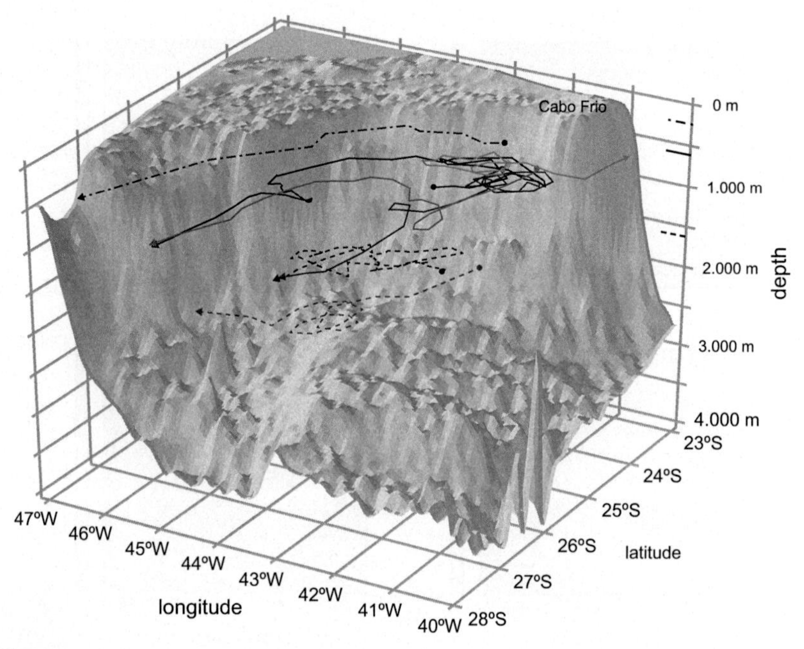

FIGURE 9

Floats trajectory at their specific parking depth.

Table 1 Profilers Deep Currents and Standard Deviations at 300, 600, and 1500 m (values in bold are interpolations to allow comparison if the mooring measurements of Ref. 11)

Latitude Section	300 m Current (m/s)	600 m Current (m/s)	1500 m Current (m/s)
20°S to 21°S	–	0.09 ± 0.02	–
21°S to 22°S	–	0.09 ± 0.02	–
22°S to 23°S	–	0.25 ± 0.01	–
23°S to 24°S	**0.05 ± 0.01**	**0.16 ± 0.02**	–
24°S to 25°S	**0.07 ± 0.00**	**0.12 ± 0.02**	**0.04 ± 0.02**
25°S to 26°S	**0.10 ± 0.00**	**0.11 ± 0.02**	**0.03 ± 0.02**
26°S to 27°S	0.17 ± 0.00	0.10 ± 0.02	0.04 ± 0.02
27°S to 28°S	**0.23 ± 0.00**	**0.11 ± 0.02**	**0.06 ± 0.02**
28°S to 29°S	0.13 ± 0.00	0.11 ± 0.02	0.12 ± 0.02
29°S to 30°S	0.08 ± 0.00	–	0.11 ± 0.02
30°S to 31°S	0.09 ± 0.00	–	0.05 ± 0.02
31°S to 32°S	0.11 ± 0.00	–	0.11 ± 0.02
32°S to 33°S	0.14 ± 0.00	–	0.11 ± 0.02
33°S to 34°S	0.13 ± 0.00	–	0.10 ± 0.02
34°S to 35°S	0.25 ± 0.00	–	–
35°S to 36°S	0.17 ± 0.00	–	–

Table 2 Interpolated Velocity Intensities from the Moorings Extracted from Rocha et al.[11]

Latitude Section	300 m Current (m/s)	600 m Current (m/s)	1500 m Current (m/s)
22.7°S	0.07 ± 0.06	0.18 ± 0.02	0.21 ± 0.03
24.2°S	0.06 ± 0.10	0.08 ± 0.03	0.10 ± 0.03
25.5°S	0.22 ± 0.04	0.07 ± 0.03	0.09 ± 0.03
27.9°S	0.26 ± 0.05	0.09 ± 0.02	0.02 ± 0.02

5. HYDRODYNAMIC MODELING AND DATA ASSIMILATION

5.1 CHALLENGES IN REGIONAL OPERATIONAL FORECAST

Data assimilation (DA) is one of the three main components of operational oceanography, along with observation system and numerical modeling. DA has the main objective of producing estimates of the ocean state and flow, which are simultaneously consistent with observed data and numerical model dynamics. These estimates can be used to produce long-term hindcast analysis and to serve as the initial condition to a short-range forecast.

On the scope of Projeto Azul, the data assimilation system is designed to allow accurate representation and forecasting of the challenging mesoscale oceanographic features of the region, by combining the data collected by the project observing system, including gliders, drifters, profilers, and remote sensing data with the numerical ocean model set up for the region.

Forward ocean modeling, if well configured, will be able to represent mesoscale features in a climatological perspective but may be unsatisfactory for operational purposes. As described in Section 2, the area of interest is forced by a highly unstable baroclinic current system, with a vertical structure of currents flowing in opposite directions and several meanders and eddy detachment throughout its path. This represents a challenge to ocean forecasting because its space–time variability adds a high degree of complexity in local circulation and contributes to a lower forecast skill of short and medium period. Sequential assimilation of observations is crucial to have a proper representation of such features in real time. Moreover, it is not sufficient to have a sequential assimilation system in an oceanic model if the observation data are sparse or nonexistent in depth. For instance, two global ocean circulation models with a robust DA system assimilating satellite SST and SSH data present difficulties in representing, for instance, the Cabo Frio eddy position (Figure 10).

5.2 PROJETO AZUL OCEAN MODELING AND DA SYSTEM

Projeto Azul modeling and DA system was designed under Dr Andrew Moore's (from University of California at Santa Cruz) orientation and refers to the extensive

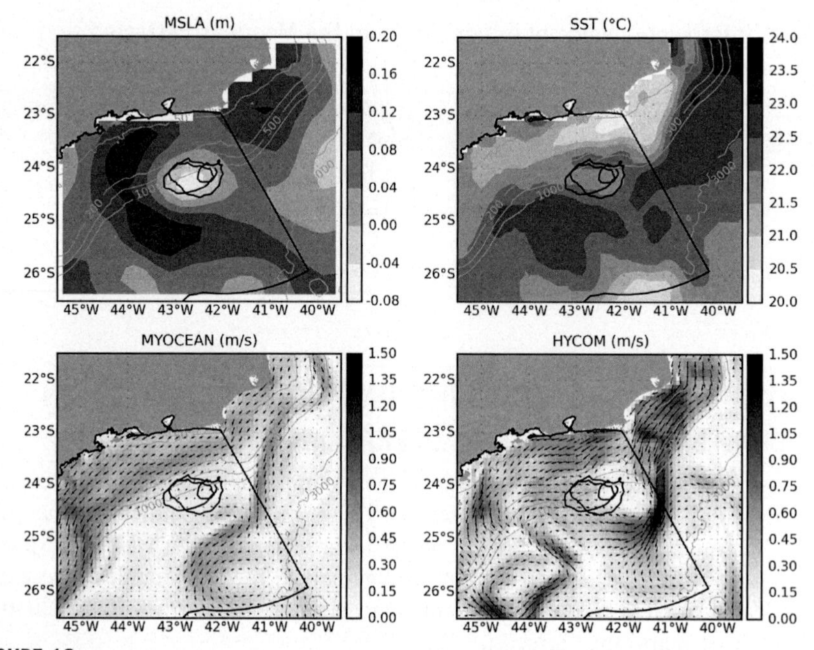

FIGURE 10

From upper left corner, clockwise, SSH and SST satellite data for September 24, 2014, and drifters trajectories showing clear evidences of a cyclonic eddy (Cabo Frio Eddy), while two global circulation models forecast current fields (HYCOM and MyOcean) are not able to represent it at the right position. White line represents drifter trajectory.

CNES/CLS (SSH and SST), HYCOM Consortium (OGCM), MyOcean (OGCM), and Projeto Azul observing system (Drifter).

work of Moore, Arango, Broquet, Powell, Weaver, and Zavala-Garay (described in Refs. 29–31). It uses the Regional Ocean Modeling System (ROMS), a community ocean general circulation model that includes drivers for strong (S4DVAR, IS4DVAR) and weak (W4DVAR) constraint variational data assimilation.[32,33] ROMS is a primitive equation ocean model that uses a curvilinear orthogonal coordinate system in the horizontal and terrain-following coordinates in the vertical. It is a hydrostatic model, has a wide range of user-controlled options for the numeric and physical parameterization, and options for prescribing open boundary conditions, and the source code is freely available. A full description of ROMS can be found in Refs. 34, 35. With the 4DVAR-developed capability, ROMS is used in conjunction with available observations to identify a best estimate of the ocean circulation based on a set of hypotheses about errors in the model and observation variables.[29] ROMS 4D-Var has been successfully implemented in diverse areas around the world.[36–40]

A ROMS grid was configured within 30°S–15°S and 50°W–24°W with 1/12° of horizontal resolution and 40 vertical levels (ROMS-SBB). ROMS-SBB is forced at the surface by atmospheric variables obtained from the National Centers for

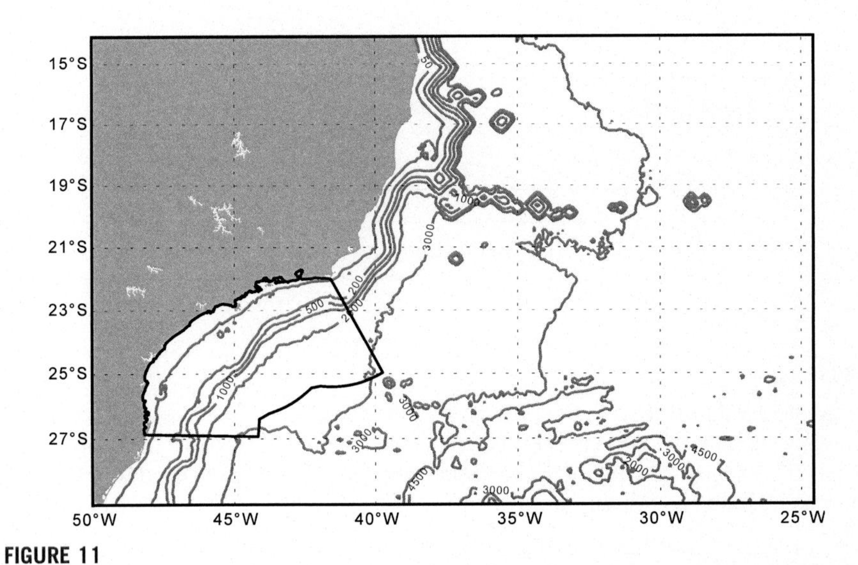

FIGURE 11

Model domain with bathymetry (gray contoured) and focused interest region (black polygon).

Environmental Prediction (NCEP) Global Forecast System (GFS) global forecasts and by project specific Weather Research and Forecast (WRF) model runs. Open boundary conditions are provided from MyOcean Project Analysis and Forecast results, which offers daily fields of the required variables (ocean currents, temperature, salinity, and SSH) in 1/12° spatial resolution.

The region of interest for monitoring and forecasting is focused on Santos Basin (Figure 11).

The primal formulation of incremental strong constraint (IS4DVAR) was chosen for the project's DA system. ROMS-SBB is run sequentially using data assimilation windows (cycles) that span the analysis period of seven days, creating the initial conditions for a forecast run of eight days. Prior estimates of the initial conditions (x^b), boundary conditions (b^b), and surface forcing (f^b) are required for data assimilation. For each cycle, the prior initial condition is the best circulation estimate of the previous cycle, the prior surface forcing is the NCEP-GFS atmospheric data, and the boundary conditions are provided by MyOcean outputs (Figure 12).

In addition, for the incremental strong constraint 4DVAR, prior error covariance matrices are required for the initial and boundary conditions and surface forcing. The error covariance matrix is expressed as a block diagonal, factorized by a diagonal matrix of standard deviations, univariate correlation matrix, and a multivariate dynamical balance operator. The univariate correlation matrix is modeled as the solution of a pseudo-heat diffusion equation, where the de-correlation length scale is approximately the natural de-correlation length scale of the related variable and controls the magnitude of the extrapolation of sparse observation data. The standard deviations obtained from a long, nonassimilative ROMS-SBB run account for the

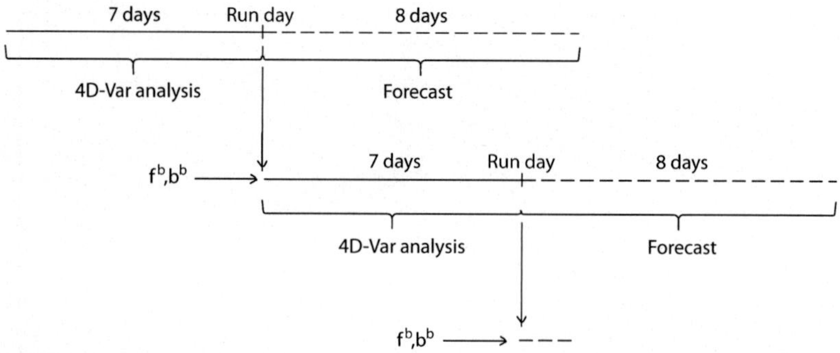

7 days Run day 8 days
4D-Var analysis Forecast
f^b, b^b
7 days Run day 8 days
4D-Var analysis Forecast
f^b, b^b

FIGURE 12

Schematic illustration of 4D-Var analysis cycles and forecast runs. The prior initial conditions (x^b) are taken from the best circulation estimate of the previous cycle, the prior surface forcing (f^b) is the NCEP-GFS atmospheric data, and the prior boundary conditions (b^b) are provided by MyOcean results.

main source of the background error, linked to model bias and to position error of circulation features in the model. Combined with the dynamical balance operator, it produces a spatial extrapolation of observational data related to ocean dynamics and natural length scale. Therefore, modeling of prior error covariance is crucial, and several adjustments are made to achieve best results.

The observations to be used in the assimilation cycles include SST from POES AVHRR, one-day gridded composite that covers the whole domain of the model with a 0.1° resolution; SSH from AVISO one-day gridded composite of the MDT, covering the whole domain with 0.3° resolution; surface and subsurface hydrographic measurements of temperature and salinity collected with gliders; and ARGO profiling floats and drifters from Projeto Azul.

The elements of the observation error covariance are derived from several sources of errors such as instrument error, interpolation errors, and representativeness error, which is a measure of the uncertainty of the ability of a single observation to describe the circulation in a single model grid (thus, it is grid size dependent). The observation errors used on ROMS-SBB are ~2 cm for SSH, ~0.4 °C for SST, ~0.1 °C for hydrographic temperature, and 0.01 for hydrographic salinity.

Because data assimilation can be viewed as the identification of a control vector (initial condition, boundary condition, and surface forcing) that minimizes the cost or penalty function (J) and thus maximizes the conditional probability, the performance of this system was assessed by monitoring J during a one-year multiple cycles experiment, assimilating SST (POES AVHRR), SSH (AVISO), and hydrographic profiles from UK Metoffice Hadley Centre observations datasets.[41] The left panel of Figure 13 shows that the posterior cost function is always smaller than the prior, indicating that the minimization algorithm is behaving correctly.

Also, for each hydrographic station assimilated (1546 temperature profiles), a comparison was made between the observation and the corresponding profiles of

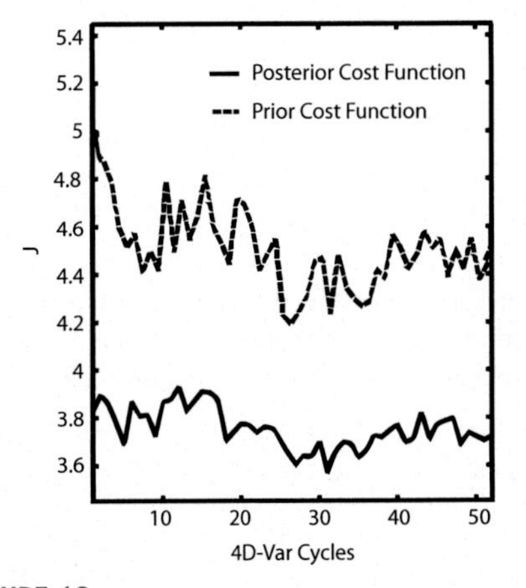

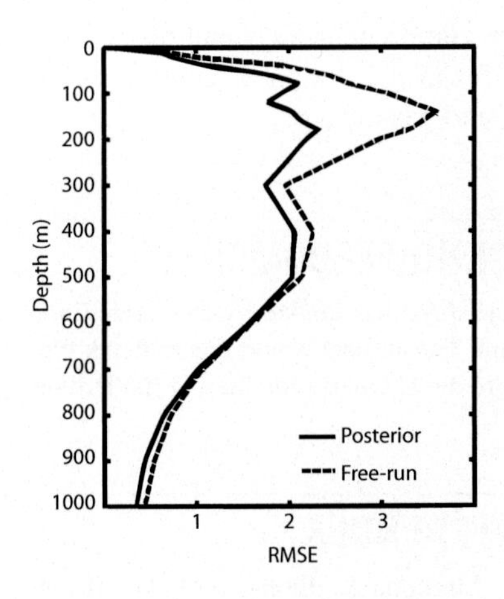

FIGURE 13

Left panel: Time series of the prior and posterior cost function for each assimilation cycle, spanning the one-year experiment (52 cycles). Right panel: Vertical profile of the RMSE ($n = 1546$) between model and assimilated temperature profiles.

temperature from the posterior estimate and from a run without data assimilation (right panel of Figure 13). The analysis presented a significant reduction of the error over the whole water column but remarkably at the thermocline region.

The results obtained so far at Projeto Azul hydrodynamic modeling and data assimilation system clearly show an improvement in the skill of ocean nowcast and short-range forecast for the South Brazilian Bight when the assimilation of the extensive in situ dataset collected systematically by the project observing system takes place. It is the first time that a 4D-Var DA system was successfully implemented in a regional ocean modeling system in Brazil, showing a great potential to improve forecast skill of SBB ocean dynamics.

6. FINAL REMARKS AND FUTURE STEPS

During its first half, Projeto Azul is successfully progressing toward its main objectives, with an observation program capable of extensively sampling the oceanographic features of the region of interest and making them publicly available. The following steps are directed at the southern part of Santos Basin, where the Santos Bifurcation has great influence on the thickening of BC and into completing an operational forecast system that assimilates local data to provide reliable forecasts in feasible time.

Projeto Azul data has a great potential for contributing to different topics of the Santos Basin ocean dynamics ranging from the improvement of Lagrangian dispersion models to the understanding of the vertical structure of the BC-IWBC system.

The comprehensive ocean observing and modeling efforts in Projeto Azul will be useful in the design of contingency plans for any oil spill incidents in this active oil and gas area.

ACKNOWLEDGMENTS

Projeto Azul is funded by BG-Brasil in the scope of ANP Research and Development program. The authors would like to thank Flávia Adissi for the corporate support and Luiz Alexandre de A. Guerra for sharing the Python routines for "level of known motion" calculation.

REFERENCES

1. Stramma L, England M. On the water masses and mean circulation of the South Atlantic Ocean. *J Geophys Res* 1999;**104**(C9):20863–83. http://dx.doi.org/10.1029/1999JC900139.
2. Campos EJD, Gonçalves JE, Ikeda Y. Water mass characteristics and geostrophic circulation in the South Brazil Bight: Summer of 1991. *J Geophys Res* 1995;**100**(C9):18537–50. http://dx.doi.org/10.1029/95JC01724.
3. Silveira ICA. On the baroclinic structure of the Brazil Current–Intermediate Western Boundary Current system at 22°–23°S. *Geophys Res Lett* 2004;**31**:L14308. http://dx.doi.org/10.1029/2004GL020036.
4. Boebel O, Schmid C, Zenk W. Flow and recirculation of Antarctic Intermediate Water across the Rio Grande Rise. *J Geophys Res* 1997;**102**(C9):20967–86. http://dx.doi.org/10.1029/97JC00977.
5. Boebel O, Davis RE, Ollitrault M, Peterson RG, Richardson PL, Schmid C, et al. The intermediate depth circulation of the western South Atlantic. *Geophys Res Lett* 1999;**26**:3329–32. http://dx.doi.org/10.1029/1999GL002355.
6. Legeais J-F, Ollitrault M, Arhan M. Lagrangian observations in the Intermediate Western Boundary Current of the South Atlantic. *Deep-Sea Res II* 2013;**85**:109–26. http://dx.doi.org/10.1016/j.dsr2.2012.07.028.
7. Soutelino RG, Gangopadhyay A, Silveira ICA. The roles of vertical shear and topography on the eddy formation near the site of origin of the Brazil Current. *Cont Shelf Res* 2013;**70**:46–60. http://dx.doi.org/10.1016/j.csr.2013.10.001.
8. Campos EJD, Ikeda Y, Castro BM, Gaeta SA, Lorenzzetti JA, Stevenson MR. Experiment studies circulation in the western South Atlantic. *Eos Trans* 1996;**77**:253–9. http://dx.doi.org/10.1029/96EO00177.
9. Silveira ICA, Lima JAM, Schmidt ACK, Belo WC, Sartori A, Francisco CPF, et al. Is the meander growth in the Brazil Current system off Southeast Brazil due to baroclinic instability? *Dynam Atmos Ocean* 2008;**45**:187–207. http://dx.doi.org/10.1016/j.dynatmoce.2008.01.002.
10. Guerra LA, de A. *Vórtices das Agulhas Colidem com a Corrente do Brasil?* [PhD. thesis]. Universidade Federal do Rio de Janeiro; 2012.
11. Rocha CB, Silveira ICA, Castro BM, Lima JAM. Vertical structure, energetics, and dynamics of the Brazil Current System at 22°S–28°S. *J Geophys Res* 2014;**119**(C9):52–69. http://dx.doi.org/10.1002/2013JC009143.

12. Evans DL, Signorini SS. Vertical structure of the Brazil Current. *Nature* 1985;**315**: 48–50. http://dx.doi.org/10.1038/315048a0.
13. Müller TJ, Ikeda Y, Zangenberg N, Nonato LV. Direct measurements of western boundary currents off Brazil between 20°S and 28°S. *J Geophys Res* 1998;**103**(C9):5429–37. http://dx.doi.org/10.1029/97JC03529.
14. Silveira ICA, Schmidt ACK, Campos EJD, Godoi SS, Ikeda Y. A Corrente do Brasil ao Largo da Costa Leste Brasileira. *Rev Bras De Oceanogr* 2000;**48**:171–83.
15. Eriksen CC, Osse TJ, Light RD, Wen T, Lehman TW, Sabin PL, et al. Seaglider: a long-range autonomous underwater vehicle for oceanographic research. *IEEE J Ocean Eng* 2001;**26**:424–36. http://dx.doi.org/10.1109/48.972073.
16. Davis RE, Sherman JT, Dufour J. Profiling ALACEs and other advances in autonomous subsurface floats. *J Atmos Ocean Technol* 2001;**18**:982–93. http://dx.doi.org/10.1175/1520-0426(2001)018<0982:PAAOAI>2.0.CO;2.
17. Sybrandy AL, Niiler PP, Martin C, Scuba W, Charpentier E, Meldrum DT. *Global drifter programme barometer drifter design reference, DBCP Report, 4*. 2009.
18. Pan C, Zheng L, Weisberg RH, Liu Y, Lembke C. Comparisons of different ensemble schemes for glider data assimilation on West Florida Shelf. *Ocean Model* 2014;**81**: 13–24. http://dx.doi.org/10.1016/j.ocemod.2014.06.005.
19. Cummings JA, Smedstad OM. Ocean data impacts in global HYCOM. *J Atmos Ocean Technol* 2014;**31**:1771–91. http://dx.doi.org/10.1175/JTECH-D-14-00011.1.
20. Liu Y, Weisberg RH, Vignudelli S, Mitchum GT. Evaluation of altimetry-derived surface current products using Lagrangian drifter trajectories in the eastern Gulf of Mexico. *J Geophys Res* 2014;**119**(C9):2827–42. http://dx.doi.org/10.1002/2013JC009710.
21. Liu Y, Weisberg RH, Hu C, Kovach C, Riethmüller R. Evolution of the Loop Current system during the Deepwater Horizon oil spill event as observed with drifters and satellites. In: Liu Y, et al., editors. *Monitoring and modeling the Deepwater Horizon oil spill: a record- breaking enterprise. Geophys. monogr. ser*, vol. 195. Washington, D.C: AGU; 2011. p. 91–101. http://dx.doi.org/10.1029/2011GM001127.
22. Berti S, dos Santos FA, Lacorata G, Vulpiani A. Lagrangian drifter dispersion in the Southwestern Atlantic ocean. *J Phys Oceanogr* 2011;**41**:1659–72. http://dx.doi.org/10.1175/2011JPO4541.1.
23. Oliveira LR, Piola AR, Mata MM, Soares ID. Brazil Current surface circulation and energetics observed from drifting buoys. *J Geophys Res* 2009;**114**(C9):C10006. http://dx.doi.org/10.1029/2008JC004900.
24. Hansen DV, Poulain PM. Quality control and interpolations of WOCE-TOGA drifter data. *J Atmos Ocean Technol* 1996;**13**:900–9. http://dx.doi.org/10.1175/1520-0426(1996)013<0900:QCAIOW>2.0.CO;2.
25. IOOS. *Manual for real-time quality control of in-situ temperature and salinity data*. 1st ed. 2014.
26. Gill AE. *Atmosphere-ocean dynamics*, vol. 30. Academic Press; 1982.
27. Pottier C, Garçon V, Larnicol G, Sudre J, Schaeffer P, Le Traon PY. Merging SeaWiFS and MODIS/Aqua ocean color data in North and Equatorial Atlantic using weighted averaging and objective analysis. *IEEE Trans Geosci Remote Sens* 2006;**44**(11): 3436–51. http://dx.doi.org/10.1109/TGRS.2006.878441.
28. Yoshinari H, Maximenko NA, Hacker PW. YoMaHa'05. velocity data assessed from trajectories of Argo floats at parking level and at the sea surface. *IPRC Tech Note* 2006;**4**:20pp.

29. Moore AM, Arango HG, Broquet G, Powell BS, Weaver AT, Zavala-Garay J. The Regional Ocean Modeling System (ROMS) 4-dimensional variational data assimilation systems: Part I—System overview and formulation. *Prog Oceanogr* 2011;**91**(1):34–49. http://dx.doi.org/10.1016/j.pocean.2011.05.004.
30. Moore AM, Arango HG, Broquet G, Edwards C, Veneziani M, Powell B, et al. The regional ocean modeling system (ROMS) 4-dimensional variational data assimilation systems: Part II— performance and application to the California Current System. *Prog Oceanogr* 2011;**91**(1):50–73. http://dx.doi.org/10.1016/j.pocean.2011.05.003.
31. Moore AM, Arango HG, Broquet G, Edwards C, Veneziani M, Powell B, et al. The Regional Ocean Modeling System (ROMS) 4-dimensional variational data assimilation systems: Part III—Observation impact and observation sensitivity in the California Current System. *Prog Oceanogr* 2011;**91**(1):74–94. http://dx.doi.org/10.1016/j.pocean.2011.05.005.
32. Moore AM, Arango HG, Di Lorenzo E, Cornuelle BD, Miller AJ, Neilson DJ. A comprehensive ocean prediction and analysis system based on the tangent linear and adjoint of a regional ocean model. *Ocean Model* 2004;**7**(1):227–58. http://dx.doi.org/10.1016/j.ocemod.2003.11.001.
33. Di Lorenzo E, Moore AM, Arango HG, Cornuelle BD, Miller AJ, Powell B, et al. Weak and strong constraint data assimilation in the inverse Regional Ocean Modeling System (ROMS): development and application for a baroclinic coastal upwelling system. *Ocean Model* 2007;**16**(3):160–87. http://dx.doi.org/10.1016/j.ocemod.2006.08.002.
34. Shchepetkin AF, McWilliams JC. The regional oceanic modeling system (ROMS): a split-explicit, free-surface, topography-following-coordinate oceanic model. *Ocean Model* 2005;**9**(4):347–404. http://dx.doi.org/10.1016/j.ocemod.2004.08.002.
35. Haidvogel DB, Arango HG, Hedstrom K, Beckmann A, Malanotte-Rizzoli P, Shchepetkin AF. Model evaluation experiments in the North Atlantic Basin: simulations in nonlinear terrain-following coordinates. *Dynam Atmos Ocean* 2000;**32**(3):239–81. http://dx.doi.org/10.1016/S0377-0265(00)00049-X.
36. Janeković I, Powell BS, Matthews D, McManus MA, Sevadjian J. 4D-Var data assimilation in a nested, coastal ocean model: a Hawaiian case study. *J Geophys Res* 2013; **118**(10):5022–35. http://dx.doi.org/10.1002/jgrc.20389.
37. Broquet G, Edwards CA, Moore AM, Powell BS, Veneziani M, Doyle JD. Application of 4D-Variational data assimilation to the California Current System. *Dynam Atmos Ocean* 2009;**48**(1):69–92. http://dx.doi.org/10.1016/j.dynatmoce.2009.03.001.
38. Powell BS, Arango HG, Moore AM, Di Lorenzo E, Milliff RF, Foley D. 4DVAR data assimilation in the intra-Americas sea with the Regional Ocean Modeling System (ROMS). *Ocean Model* 2008;**25**(3):173–88. http://dx.doi.org/10.1016/j.dynatmoce.2009.04.001.
39. Zavala-Garay J, Wilkin JL, Arango HG. Predictability of mesoscale variability in the East Australian current given strong-constraint data assimilation. *J Phys Oceanogr* 2012;**42**:1402–20. http://dx.doi.org/10.1175/JPO-D-11-0168.1.
40. Chen K, He R, Powell BS, Gawarkiewicz GG, Moore AM, Arango HG. Data assimilative modeling investigation of Gulf Stream Warm Core Ring interaction with continental shelf and slope circulation. *J Geophys Res* 2014;**119**(9):5968–91. http://dx.doi.org/10.1002/2014JC009898.
41. Ingleby B, Huddleston M. Quality control of ocean temperature and salinity profiles—historical and real-time data. *J Mar Syst* 2007;**65**:158–75. http://dx.doi.org/10.1016/j.jmarsys.2005.11.019.

CHAPTER

8

Zooplankton Data from High-Frequency Coastal Transects: Enriching the Contributions of Ocean Observing Systems to Ecosystem-Based Management in the Northern California Current[1]

Eric P. Bjorkstedt[1,*], William T. Peterson[2]

NOAA Fisheries, Southwest Fisheries Science Center, Santa Cruz, CA, USA[1];

NOAA Fisheries, Northwest Fisheries Science Center, Seattle, WA, USA[2]

**Corresponding author: E-mail: eric.bjorkstedt@noaa.gov*

CHAPTER OUTLINE

[1]The opinions expressed herein are those of the authors and do not represent official policy of NOAA or the National Marine Fisheries Service.

1. INTRODUCTION

Collection of ocean data has grown tremendously in recent years, spurred in no small part by mandates to manage marine ecosystems holistically and the need to understand and forecast the consequences of anthropogenic climate change.[1,2] A broad suite of remote sensing technologies, whether satellite-borne or land-based, can be used to collect information on sea surface temperature, salinity, currents, winds, sea surface height, and ocean color (phytoplankton, CDOM, sediment) across broad swaths of the ocean. In situ sensors on moorings or at shore stations complement remote sensing by collecting high temporal resolution data near the coastline and throughout the water column—places where remote sensing instruments cannot see. Mobile observation platforms, whether ship-based or aboard autonomous underwater vehicles (AUVs), provide quasi-synoptic snapshots of a dynamic ocean.

In combination, existing data streams from these several approaches provide an increasingly rich view of the ocean's state and dynamics. Time series of physical observations are essential inputs for state-of-the-art data-assimilative models that, in turn, provide nowcasts of the current state of the ocean as well as short-term (e.g., 72-h) forecasts of ocean conditions.[3,4] Building on these successes, modelers are working to integrate remotely sensed biological proxies (e.g., chlorophyll concentration) into coupled biophysical models of the plankton ecosystem.[5] Such modeling efforts represent a significant step toward understanding how changes in physical forcing might impact ecosystems.

Notwithstanding the growing success and clear value of existing ocean observing system (OOS), physical observations and direct or proxy measures of nutrient or phytoplankton concentrations are one or more steps removed from higher trophic levels of marine ecosystems and species of more direct value to society. Even as ecosystem considerations become more prevalent in the management of living marine resources, there are few strong examples where observations at the base of the physics-to-fisheries system are rigorously and quantitatively integrated into managers' understanding of the dynamics of fisheries stocks or protected species.[6,7]

Data on zooplankton are essential for bridging this gap and, as we review below, can be supplied to resource managers in timely fashion by ship-based, sea-going elements of an OOS. Direct observations of zooplankton are a rich source of information on ecosystem state and of how pelagic ecosystems respond to climate forcing.[8] This is true for several reasons. First, because most zooplankton have short life cycles (weeks to months), their population dynamics are tightly coupled to physical variability and environmental change. Second, given their role in food webs as consumers of primary production and prey for fish and higher trophic levels, changes in zooplankton species composition or community structure can serve as a robust index of climate-driven changes in ecosystem structure. Third, many zooplankton species are so-called "indicator species", the presence of which in a zooplankton sample is nearly as accurate as physical data in telling us the source of water that has been sampled (and, thus, the influence of large-scale transport processes).[9] Finally, zooplankton are very abundant and readily quantified using relatively simple methods.

Despite the high information content of zooplankton observations, such data are remarkably sparse in space and time when considered in the broad context of OOS as a whole. At present, some of the longest times series of zooplankton data are based on plankton net samples collected over the course of infrequent (e.g., three or four times per year), quasi-synoptic surveys (e.g., CalCOFI off southern California,[10] IMECOCAL off Baja California, Mexico,[11] and Line P off Vancouver Island, Canada[12]) or from relatively frequent (i.e., monthly) occupation of a station deemed representative of a large region of interest (e.g., the Hawai'i Ocean Time-series (hahana.soest.hawaii.edu/hot) and the Bermuda Atlantic Time-series Study (bats.bios.edu)). In this context, we argue that high-frequency coastal transects (HFCTs) have an essential role in a comprehensive OOS, and that this role will likely continue well into the future even as advanced technologies for autonomous sampling continue to develop and be implemented in the field. We base this argument on our experience with sampling along such lines in the northern California Current (NCC) and bringing these data into coast-wide syntheses of the California Current Ecosystem (CCE)[13] and the California Current Integrated Ecosystem Assessment (CCIEA; www.noaa.gov/iea/regions/california-current-region/index.html). However, we expect that many of the themes we touch upon next will apply to similar efforts in other coastal regions and provide strong motivation for establishing such lines in a coast-wide network.

2. HIGH-FREQUENCY COASTAL TRANSECTS

High-frequency coastal transects (HFCTs) are ship-based surveys of short transects that span the continental shelf and extend to waters over the upper to mid-slope. Survey cruises are executed rapidly (<12 h) at intervals of two to four weeks throughout the year. HFCTs rely on research vessels of modest size (12–30 m) and running cost ($3000–5000 per day), sailing out of local home ports. Flexibility in vessel scheduling is essential, as it allows HFCTs to be sampled during windows of favorable (or at least safely workable) weather and sea conditions.

In the NCC, opportunities to conduct effective surveys of any sort can be quite limited during the winter storm season and when strong upwelling-favorable winds affect the region during the spring and summer months; yet, we have had good success in maintaining year-round sampling along two HFCTs in this region. We describe these two HFCTs—the Newport Hydrographic Line and the Trinidad Head Line—as examples of what we mean by these sampling protocols and as a basis for highlighting what has been learned from such sampling in the NCC. These HFCTs are separated by about three degrees of latitude (approximately one-fifth of the coastline between Cape Flattery, Washington, and Point Conception, California), and they bracket the northern extent of a transitional zone between the relatively simple upwelling system along Oregon's relatively linear coast and regions south of Cape Blanco where interactions between stronger winds, complex orography, and bottom topography result in stronger mesoscale activity.

2.1 THE NEWPORT HYDROGRAPHIC LINE

The Newport Hydrographic Line (NHL) extends west from Newport, Oregon (44.6°N; Figure 1). This line was first sampled by physical oceanographers from Oregon State University from 1961 to 1973 with support from the Office of Naval Research, directed toward study of the (then) poorly understood hydrography of the northern California Current. From 1969 to 1973, Oregon Sea Grant supported early work on the zooplankton and ichthyoplankton in continental shelf waters along the NHL.[14–16] After 1973, apart from a handful of cruises that sampled zooplankton during the summers of 1976 to 1978, 1983, and 1990 to 1992,[17–19] the NHL was not systematically sampled again until 1996. Since 1996, the inner 40 km of the Newport Line (continental shelf and slope) has been sampled on a fortnightly basis. The offshore extent of the NHL, which extends into oceanic waters 140 km from shore, was sampled quarterly from 1998 to 2005 and two or three times per year since. Sampling along the inner 40 km of the NHL has been supported largely by two vessels, the 12-m *R/V Sacajawea* and the 19-m *R/V Elakha*.

2.2 THE TRINIDAD HEAD LINE

The Trinidad Head Line (THL) extends due west from Trinidad Head (41.05°N; Figure 1) in northern California and is anchored at the coast by a shore station maintained by CeNCOOS and Humboldt State University at Trinidad Wharf. The THL is situated near the midpoint of what was then an extensive latitudinal gap in year-round ocean observing efforts between the NHL in the north, and to the south, the northern lines of the CalCOFI grid, especially CalCOFI Line 67 off Monterey Bay. The THL lies approximately 53 km south of where the Klamath River—a key watershed in management of West Coast salmon fisheries—drains into the Pacific Ocean (approximately 41.54°N). Initial sampling along the THL began in 2006, with consistent monthly (and occasionally biweekly) sampling established in early 2008. Sampling along the THL is supported by Humboldt State University's 27-m *R/V Coral Sea*.

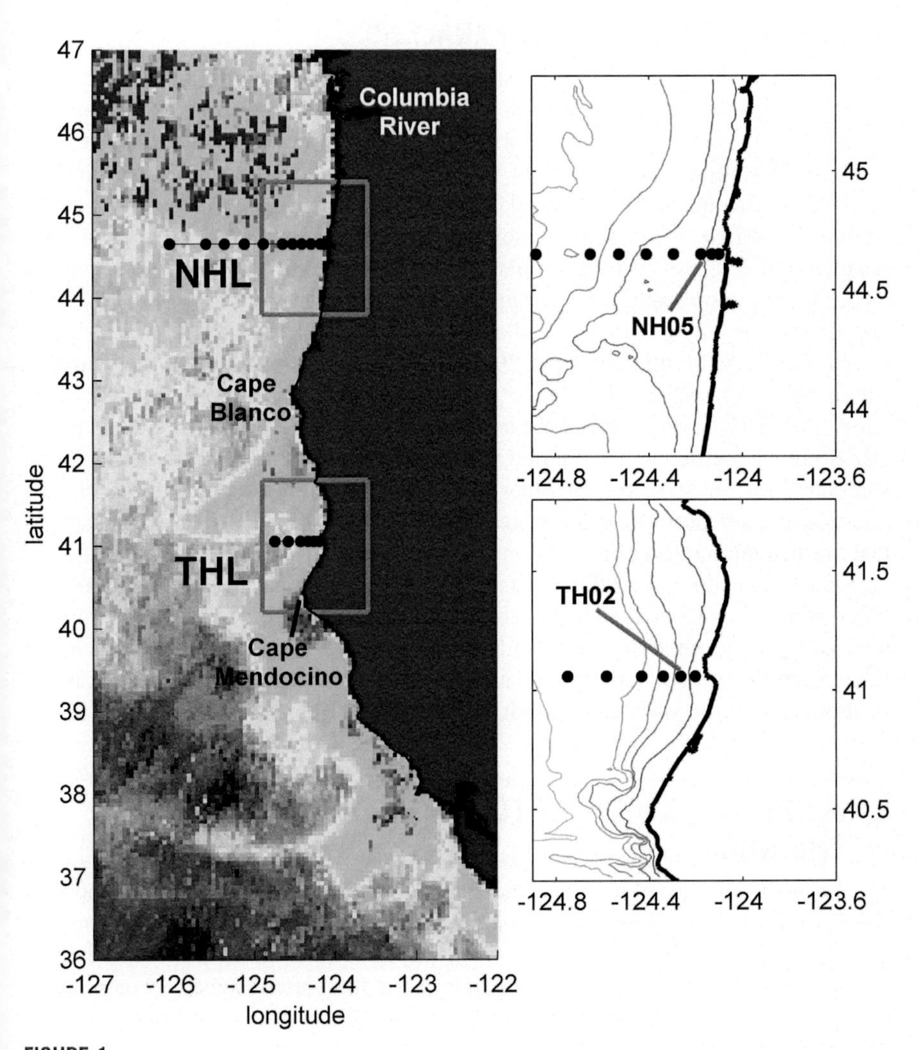

FIGURE 1

Newport Hydrographic Line (NHL) and Trinidad Head Line (THL) overlain on sea surface temperature (eight-day composite centered on 1 June, 2013, downloaded from coastwatch. pfeg.noaa.gov/erddap). Gray boxes indicate location of detailed charts of coastal stations and bathymetry. Contours indicate, in order of dark gray to light gray, 50-, 100-, 200-, 500-, 1000-, and 2000-m isobaths. Observations from highlighted stations are discussed in greater detail in the text.

2.3 SAMPLE COLLECTION AND ANALYSIS

Along both lines, zooplankton are sampled at each station with a 0.5-m ring net fitted with 202-μm mesh and a TSK flowmeter. The net is lowered to a maximum depth of 100 m at deeper stations or within a few meters of the sea floor at shallower shelf stations, then retrieved vertically to the sea surface. These samples are analyzed to quantify the abundance of copepod species, many of which are further identified to stage and sex, euphausiid eggs and larvae, and other zooplankton (see Ref. 20 for further details on methods of analysis). Along the NHL, larger zooplankton (e.g., euphausiids, pteropods, ichthyoplankton, etc.) are sampled at night with 0.5-m diameter Bongo nets fitted with 333-μm mesh towed obliquely through the upper 20 m of the water column. During the quarterly (more offshore) cruises, the Bongo net is fished obliquely to a depth of 100 m. Along the THL, larger zooplankton are sampled with oblique tows of 0.60-m diameter Bongo nets fitted with General Oceanics flowmeters and 505-μm mesh on one side and 335-μm mesh on the other, fished to a maximum depth of 100 m; this allows a greater portion of the water column to be sampled at on-shelf stations that are occupied during daylight on the THL.

Standard in situ observational data are collected during surveys along both lines. CTD casts are conducted at each station to obtain depth profiles of water temperature, salinity, chlorophyll fluorescence, and dissolved oxygen. Bottle and surface water samples are collected for assay of nutrient and chlorophyll concentrations (and phytoplankton species composition along the NHL).

3. WHAT CAN ZOOPLANKTON DATA TELL US ABOUT THE NCC?

The answer to this question lies in understanding what drives variability in copepod communities and recognizing that copepod community structure strongly indexes the ecosystem response to variability in physical forcing. As reviewed below, copepod community data effectively integrates the (often lagged or nonlinear) response of coastal ecosystems to physical forcing acting across a range of spatial and temporal scales.

Why focus on copepods? In part, because copepods are highly abundant and are one of the major links between phytoplankton and fishes in marine food webs. Just as importantly, many of the numerous species of copepod that frequently appear in our samples can be classified according to their affinity for colder versus warmer water and offshore versus neritic habitats.[20,21] The consistency of these affinities underpins the utility of copepod community structure as an indicator of ecosystem state and how the ecosystem is responding to large-scale forcing. Indeed, several individual species are so strongly linked to water masses with particular characteristics that they serve as effective indicators of seasonal changes in circulation patterns, the onset of the upwelling season, oceanographic regime shifts, and potential responses to climate change.[20–23]

Based on analysis of data collected along the NHL and, in particular, the time series of observations of zooplankton at the mid-shelf station NH05 (situated at mid-shelf, approximately 9 km from shore in about 60 m of water, Figure 1), we have documented strong variability in copepod communities off the Oregon coast. Climatologically, cold-water taxa dominate the coastal zooplankton community during the summer upwelling season (typically May through September) when equatorward flow along the coast is fed from northern sources. Conversely, a diverse suite of warm-water taxa is dominant during winter when poleward flows (e.g., the Davidson current) occur. Qualitative differences in the energy content of cold- and warm-water copepods magnify the value of understanding variability in coastal copepod communities as a basis for understanding broader ecosystem dynamics. Specifically, two of the cold-water species, *Calanus marshallae* and *Pseudocalanus mimus*, are rich in lipids relative to warm-water taxa.[24] These differences mean that which species are present has important implications for the bioenergetic content of the food chain and subsequent transfer of energy to higher trophic levels.

3.1 TEMPORAL VARIABILITY: CLIMATE VARIABILITY AND CHANGE

Variability in the annual climatological pattern in copepod species composition at both NHL and THL is linked to variability in large-scale forcing of the North Pacific. In particular, the copepod community off Oregon and northern California is especially responsive to changes in the Pacific Decadal Oscillation (PDO), which indexes changes in the strength of equatorward flow carrying subarctic water into the California Current.[25–27] When the PDO is in negative phase, cold water species are more abundant (and have greater biomass), but when the PDO is in positive phase, warm water species tend to dominate. These patterns appear to be driven by variations in large-scale transport associated with the PDO such that when a greater proportion of the water entering the NCC is from the coastal Gulf of Alaska and the subarctic side of the North Pacific Current, lipid-rich copepods dominate; whereas when the PDO is in positive phase, a greater proportion of water entering the NCC is from the subtropical branch of the North Pacific Current and lipid-poor taxa are more common.[26,27] Similar variability has been documented in response to El Niño (warm) and La Niña (cool) conditions.[28] The response to strong El Niño events is especially profound as El Niño disrupts transitions related to seasonal upwelling and equatorward flow. During such events the abundance of warm-water species is typically greater than normal and the abundance of cold-water species is greatly reduced, regardless of season.[28]

This close coupling between copepod community composition and large-scale forcing associated with the PDO and ENSO presages the ability of HFCTs to detect ecosystem responses to ongoing and future climate change. Trends and shifts in physical conditions will be readily observed by several elements of OOS, but they may not be immediately informative with respect to ecosystem state. In contrast, the arrival of new taxa from southern waters, or the failure of northern species to return as expected, would be a clear indicator of change with respect to how zooplankton communities (and, by proxy, ecosystem productivity) are shifting north or are otherwise being altered.[29,30] Moreover, HFCTs can reveal zooplankton responses to changes in upwelling intensity or ocean stratification driven by climate change and whether changes elsewhere in the North Pacific affect zooplankton communities that are transported into coastal waters of the CCE.[31,32]

3.2 TEMPORAL VARIABILITY: THE ANNUAL PRODUCTION CYCLE

One of the strengths of HFCTs is the ability to sample throughout the year, taking advantage of sometimes narrow windows of favorable conditions. This has allowed us to develop a broader perspective on when and how production events occur during different parts of the year, and to develop metrics that capture important variability in the annual cycle.

3.2.1 Ecosystem Preconditioning in Winter

Observations along the NHL have allowed us to resolve aspects of variability in the annual production cycle that would be difficult at best to capture with any rigor in quarterly or annual surveys. Several recent studies have elucidated statistical relationships between integrated measures of productivity and ocean conditions during the winter months preceding the onset of the upwelling season.[33–38] Analogous relationships emerge from analyses of recruitment variability in winter-spawning rockfishes (*Sebastes* spp.),[39–41] the young of which serve as important prey for seabirds and larger fishes later in the spring.[42,43] These relationships suggest that enhanced primary and secondary production during winter "preconditions" the ecosystem to respond robustly to upwelling in the spring, which implies that winter forcing has a disproportionate influence on overall annual productivity.

Off Oregon, significant wintertime productivity events depend on two factors. The first is the emergence of *Neocalanus* spp. and *Calanus marshallae* from diapause in early January, which rapidly increases copepod biomass in surface waters.[44] During this time, both *C. marshallae* and *Calanus pacificus* begin to produce eggs using stored lipids and by feeding on ciliate prey. The second factor is the occurrence of conditions that favor a winter phytoplankton bloom.[45] In the NCC, such blooms can occur anytime between late-January and early March, when an extended period of calm winds and clear skies allows phytoplankton to bloom in response to increased light, stratification (reduced mixing), and sufficiently high nutrient concentrations (ca.10 μM nitrate) that have accumulated over the winter months. This "clear sky" event does not occur every year off Oregon; however, when these conditions do occur, the resulting bloom can support higher egg production by copepods and euphausiids and greater growth and survival of their larvae and juveniles.[36,37] The net effect is that winter blooms enhance the ecosystem's potential to respond to the onset of sustained upwelling in spring, and it is thought to be the basis for the statistical relationships described previously. Off northern California, we have observed distinct winter bloom events, with clear skies coinciding with light upwelling-favorable winds out of the north in January, followed by additional storm activity. These observations contrast with conditions off central and southern California, where late-winter blooms appear to be associated with the onset of the upwelling season, which often (but not always) occurs earlier at such latitudes.[46,47]

3.2.2 Ecosystem Transitions in Spring

Because of the importance of upwelling for enrichment of coastal ecosystems, several indices have been proposed to distill seasonal variability in upwelling and its consequences in the California Current to simpler interannual metrics. Some studies[33,47] have used cumulative daily values of the Bakun upwelling index[48] to estimate date of spring and fall transition, and length of the upwelling season, whereas others used cumulative wind stress from local winds[49] or have focused on changes in sea level at the coast.[50]

These physical indicators, however, are, at best, indirect indicators of ecosystem state, particularly because in many years, when the upwelling season is first initiated, winds often are weak and alternate between southerly and northerly before settling in a persistent pattern of northerly winds. Thus, these indicators and transition metrics do not capture the initiation of biologically significant productivity and, indeed, may be somewhat decoupled from ecosystem responses depending on recent history of the system. For example, during the El Niño event of 1998, upwelling was quite strong, but the waters that upwelled were warm and nutrient poor, having been drawn from above a depressed pycnocline, and productivity was not enhanced (discussed in Ref. 28). In 2005, upwelling was delayed (spring transition date estimated 24 May, nearly six weeks after the climatological mean) but a significant amount of upwelling did not begin until an additional six weeks had passed (12 July). This caused a complete collapse of the normal upwelling-fueled food web and high mortality of juvenile salmon that went to sea that spring and summer, with the end result that returns of salmon two–three years later were so poor that the salmon fishery along the West Coast of the United States was closed for two years.[51] This event also led to recruitment failures in rockfish stocks and poor reproductive success in many seabirds.[52,53]

Direct observation of upwelling and its consequences can improve the utility of indices based on "date of spring transition" for resource managers. This has been done using NHL CTD data to define the "physical spring transition" as the date on which water colder than 8°C is first observed near the sea floor in mid-shelf waters. This definition effectively integrates physical processes leading to the presence of cold, nutrient-rich water that will upwell at the coast with the onset of strong northerly winds and, thus, indicates conditions favorable to high plankton production rates. In years where shelf waters remain warm, as can occur during strong El Niño events, it is possible that this threshold might never be achieved and no transition to upwelling defined. We define an analogous "biological spring transition" as the date when the copepod community has transitioned from the winter

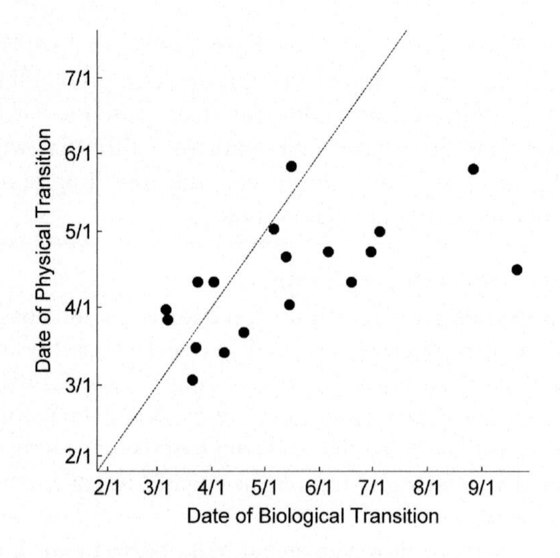

FIGURE 2

Comparison of date of biological transition based on appearance of cold-water copepod taxa along the Newport Hydrographic Line and the date of spring transition based on physical observations using the definition developed by Logerwell et al.[33] Dashed line is 1:1 relationship to illustrate progressive delay of biological transition when physical transition occurs later in a given year.

warm-water community to a summer cold-water community. Corresponding indices of "fall transitions" marking the end of the productive upwelling season are based on displacement of cold bottom water from the shelf and the disappearance of cold-water copepod communities.

Comparisons to physical indices of spring transitions, such as that proposed by Logerwell et al.[33] illustrate why transition indices based on copepod species abundance data are more useful information for fisheries management. When upwelling begins early in the year (up to day 100 or mid-April), cold-water copepods arrive at approximately the same time, and physical and biological definitions of the spring transition are in general agreement (Figure 2). However, if the physical signs of upwelling appear later in the year, biological responses are disproportionally delayed (Figure 2). This disparity underlies the value of direct biological sampling, as physical indices alone are not effective predictors of ecological conditions, especially in unusual years such as 1998 and 2005.

3.2.3 Upwelling and Productivity

Results from several process studies and research programs, including the NHL time series, have given us a relatively good understanding of production associated with the upwelling season off Oregon. The "spring transition" (typically in April or May) is marked by the onset of sustained upwelling that fuels a burst of primary and secondary production, and strong upwelling events occur at five to 10 day intervals during the peak of the upwelling season (July to August), resulting in dense phytoplankton stocks. These blooms support massive egg production by the euphausiid *Euphausia pacifica*[36] and maximum egg production rates in *Calanus marshallae*, *Pseudocalanus mimus*, and *Acartia longiremis* (Refs. 54,55 and Peterson *unpublished data*).

3.2.4 Fall Blooms

Year-round sampling along the NHL has allowed us to characterize the occurrence of fall blooms as well. These blooms usually occur in October, in response to stratification and clear skies after the first major storms of the season mix the water column and raise nutrient concentrations near the surface. Following this bloom, southwesterly storms, deep mixing, and reduced insolation prevent any substantial productivity through the late autumn and early winter. The broader implications of such blooms for ecosystem productivity in the CCE are not well understood.

3.3 SPATIAL VARIABILITY: ALONGSHORE COHERENCE AND DECOUPLING

The insights reviewed above are grounded in nearly 20 years of continuous observations along the NHL (primarily at the single station NH05), augmented by historical data from the 1960s and 1970s. However, looking more broadly at the CCE, questions remain regarding how well observations off Newport capture variability in the broader CCE, and even when correlations exist, what are the mechanistic links that underlie the observed relationships? Or, considering part of the motivation for initiating the THL, can we develop similar insights to ocean influences on survival of Klamath River fall-run Chinook salmon (*Oncorhynchus tshawytscha*), one of the linchpin stocks for salmon management along the US West Coast? While we are not yet to the point of addressing these broader questions in a deeply informative way, the time series of copepod community data along the THL has matured sufficiently to allow us to explore how copepod communities vary over time and space over three degrees of latitude within the NCC.

To illustrate the potential insights to be gained from analysis of data from the two HFCTs, Figure 3 shows an ordination based on nonmetric multidimensional scaling (NMDS) of log-transformed copepod abundance data collected at mid-shelf stations (TH02 and NH05) along each transect from 2008 to early 2014. The two stations are both about 9 km offshore; the depth at TH02 is 75 m and at NH05, 60 m. To better balance the data set, NH05 data were thinned to include only those samples that occurred within a few days of a corresponding sample at TH02 (see Refs. 26,52 for description of ordination methods). This ordination indicates substantial separation between the copepod communities observed at NH05 and TH02, structured by variability in the abundance of species with cold-water, neritic affinities versus those more commonly found in warm-water, oceanic habitats (NMDS1) and the abundance of cold-water, oceanic (subarctic) species relative to warm-water, neritic species (NMDS2) (Figure 3). Differences between the two copepod communities

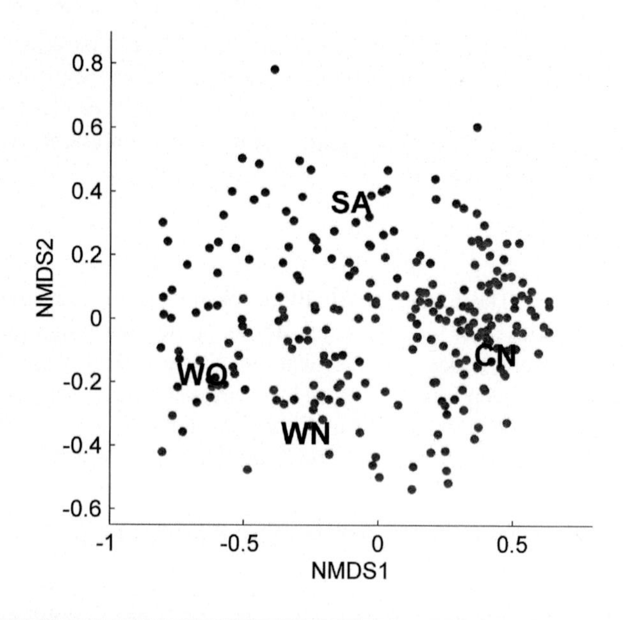

FIGURE 3

Results from NMDS ordination of copepod community data from station TH02 along the Trinidad Head Line (blue) and NH05 along the Newport Hydrographic Line (red), showing the position of individual observations and the mean position of copepod taxa with cold-neritic (CN), subarctic (SA), warm-neritic (WN), and warm-oceanic (WO) affinities (following Ref. 21).

persist throughout the seasonal cycle, with the assemblage off Oregon tending to include more nearshore species, and the assemblage off northern California having a stronger oceanic component.

This preliminary comparison highlights the value of networked HFCTs, as the lipid-rich copepods that commonly dominate off Oregon during cool, productive conditions are not as common off northern California, even though we expect to see similar shifts in ecosystem productivity through the course of the annual cycle. Moreover, this comparison also suggests that the two regions experience climate variability in different ways: following the decay of the 2009–2010 El Niño event, the abundance of cold-water neritic species recovered slowly off Oregon, whereas the copepod community off northern California was dominated by more offshore, subarctic species. Ongoing work is focused on understanding how differences in local environmental conditions, local bathymetry (e.g., retention), and circulation patterns might contribute to these differences, motivated in part by the potential for insight to drivers of alongshore variability in marine survival of salmon stocks.

3.4 SPATIAL VARIABILITY: CROSS-SHELF ZONATION AND HYDROGRAPHIC STRUCTURE

In our review thus far, we have focused on indices derived from data collected at a single mid-shelf station along each of the two transects, however, HFCTs provide these data in the context of nearly synoptic observations of hydrography and plankton ecosystem characteristics across the shelf and upper slope. For example, during the upwelling season off Oregon, we observe the following set of assemblages structured by hydrography and circulation[54,56–59]:

- A unique assemblage of zooplankton in the nearshore zone (the inner 5 km) composed of the larvae of benthic invertebrates (barnacles, bivalves, and several crabs) and copepods (*Acartia hudsonica* and *Centropages abominalis*)
- A high-biomass, low-diversity assemblage dominated by *Pseudocalanus mimus*, *Calanus marshallae*, and *Acartia longirmis* and the eggs and larvae of euphausiids in mid- to outer-shelf waters
- A low-biomass, high-diversity, oceanic assemblage of subtropical and transition zone species offshore of the shelf break

Moreover, cross-shelf sampling yields data on the several euphausiid taxa, including *Thysanoessa spinifera*, which can be highly abundant just inshore of the shelf break, especially in years with a winter "clear sky" bloom,[36] and *Euphausia pacifica*, the dominant grazer and producer at the shelf break and offshore. The presence of other euphausiids (e.g., *Nyctiphanes simplex* and *Stylocheiron* spp.) or other taxa (e.g., *Emeritia* larvae) can corroborate information on transport inferred from copepod community data.[60,61] Sampling along HFCTs has also supported analysis of variability in larval fish community structure and abundance.[62,63]

Integration of cross-shelf data into indices and management has not been well developed as it has for mid-shelf stations that capture conditions that strongly influence important fisheries resources (e.g., salmon and rockfish). Nevertheless, through these observations, HFCTs enrich OOS with information on conditions affecting several parts of the ecosystem, and how cross-shelf zones respond to variation in local- and basin-scale forcing, both directly and in connection to other zones. For example, comparisons along the NHL have documented lower variability in copepod communities in waters over the slope relative to those observed in the more dynamic shelf waters.[64] HFCTs allow us to observe other processes as well, such as the consequences of ongoing acidification of ocean waters and upwelling of deep corrosive waters through analysis of pteropod shell dissolution along cross-shelf gradients of ocean pH,[65] and to develop inferences regarding transport of coastal zooplankton and invertebrate larvae.[66,67] In our ongoing work, we are expanding the data available for offshore stations of the THL and integrating these data to enrich elements of the CCIEA focused on the NCC.

4. ZOOPLANKTON-BASED ECOSYSTEM INDICATORS

From a more applied, informational perspective of OOS, zooplankton data are the basis for several prominent ecosystem indicators included in the CCIEA. Data from the NHL contribute to several indicators for the NCC, including the following:

- Northern Copepod Biomass Anomaly, which is considered as a proxy for the amount of wax esters and fatty acids available to higher trophic levels, and especially to several pelagic fish species for which these energy sources are critical to overwinter growth and survival[20]
- Southern Copepod Biomass Anomaly, an indicator of poleward, onshore transport that displaces productive, lipid-rich copepod communities[20]
- Copepod Species Richness, which captures transitions between relatively species-poor, but productive cold-water communities and the more diverse, less productive warm-water communities[23]
- Copepod Community Index, which is derived from nonmetric multidimensional scaling (NMDS) and highlights variability between cold water, neritic communities, and other less productive ecosystems[26,68]

Of these, Northern Copepod Biomass Anomaly and Copepod Species Richness have proven to be among the most informative of the 40 or so indices considered for the NCC.

Hydrographic and zooplankton data collected from the NHL and THL are analyzed rapidly after each cruise, so that the zooplankton (and transition) indices are updated in timely fashion for resource managers focused on forecasting the status of harvested stocks or evaluating ecosystem state. One prominent example of how data from an HFCT directly informs management can be found in the use of observations from the NHL to generate forecasts for salmon returns to coastal watersheds in Oregon and to the Columbia River.[7,69] This information is obviously useful in the management of fisheries, as managers and fishers peer into the future as part of their regulatory processes and strategic planning. This information is also valuable for evaluating watershed and land-use management activities and hatchery practices that affect production and survival of juvenile salmon in freshwater habitats. Specifically, zooplankton-based indices help to account for variability in marine growth and survival in evaluating whether changes in freshwater habitats have a beneficial or detrimental effect on salmon populations.

In particular, the northern copepod index during the year that juvenile salmon go to sea appears to be a relatively good predictor of salmon returns one to two years later (e.g., Figure 4). These relationships support the hypothesis that salmon survival is enhanced by the presence of a lipid-rich cold-water copepod community in coastal waters. The link between copepods and salmon is almost certainly through the food web, as juvenile Coho and Chinook salmon prey on euphausiids and small fishes rather than directly on copepods. We suspect that cold-water copepods serve as a proxy for the abundance of cold-water coastal fishes such as herring, smelt, and sand lance, as well as being lipid-rich energy sources directly available to these

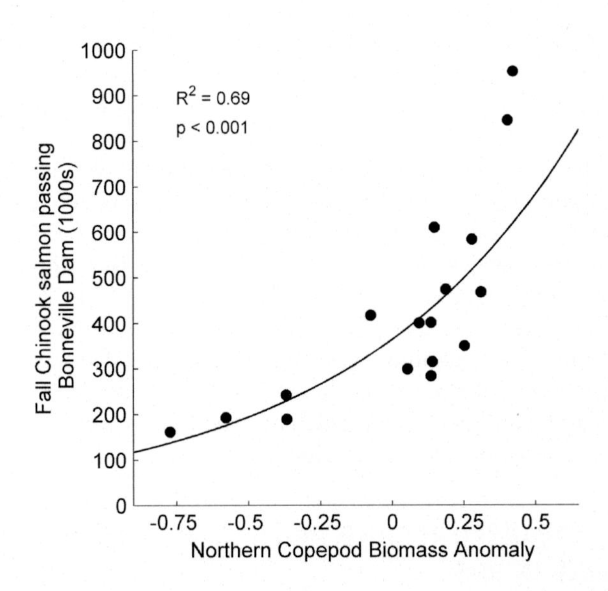

FIGURE 4

Relationship between northern copepod biomass anomaly[21] and returns of fall-run Chinook salmon past Bonneville Dam two years later, illustrating predictive power of zooplankton indices for salmon forecasts. Counts of fall-run Chinook include both lower-river "tules" (believed to be mostly fish that have spent two years at sea) and upriver "brights" (believed to be mostly fish that have spent three years at sea).

planktivorous fishes and to larval fishes that also comprise a substantial portion of juvenile salmon's diets.

5. DISCUSSION

The blend of scientific insights and ecosystem indicators reviewed in this chapter are the result of old-fashioned approaches to ocean observing—going out to sea on a regular and frequent basis to sample the system as holistically as practicable using relatively simple methods. Perhaps the most unique and powerful lesson to be taken from this work is the clear demonstration that zooplankton community data offer greater explanatory power and potential for direct societal benefit (through better informed management) than do physical observations or indices several steps removed from species with high commercial or conservation value. More specifically, the power of zooplankton data emerges largely from knowing what species are present, with the impact of variability in community composition magnified by ecologically significant differences in energetic content among groups with different biogeographical affinities. By demonstrating differences in food chain structure related to climate forcing (especially the PDO), our work lays out a strong hypothesis that links basin-scale climate variability to the local dynamics of sardine, anchovy, and salmon. To develop these indices further, we are developing lipid content time series for key taxa along both lines to examine how lipid content differs seasonally and alongshore.

Moreover, by adopting a strategy that contrasts sharply with infrequent, large-scale hydrographic surveys, HFCTs have allowed us to characterize seasonal variability in production, even during winter periods that present logistical challenges for sampling and had been widely believed to be unproductive. Through this work, we have shown that Oregon waters can be productive from January through October, not just during the April–September upwelling season. Similar observations are emerging from sampling off northern California. Thus, HFCTs can help to resolve variability in the occurrence and timing of production events, as well as their magnitude and duration, all of which are highly informative for understanding ecosystem state and variability in higher trophic levels.

5.1 HFCTs, ADVANCED TECHNOLOGIES, AND BIOLOGICAL SAMPLING IN OOS

Extracting zooplankton community data from zooplankton samples is labor intensive and requires substantial taxonomic expertise, yet this effort is critical to the value of these time series. Yet, had we only a time series of aggregate density measures based on counts or biomass of undifferentiated zooplankton, or volume of entire samples, we would lack the ability to develop insights into the differences in the copepods seen with season, cross-shelf, or alongshore and how these differences correlate with the recruitment dynamics of fishery stocks and other species of interest.

At present, automated sampling technologies and autonomous sampling platforms show great promise, but they have not developed sufficiently to fully supplant at-sea sampling followed by careful microscope work. Indeed, the full extent and richness of the data described here—and the indicators derived from these data—cannot yet be collected in any other way. Optical particle counters can return data on fine-scale patchiness and size distributions in plankton distributions, and biomass can be estimated from the size and number of particles observed in a tow, yet such data have relatively low information content compared to a data set on abundance and biomass resolved to the species level. Optical instruments yield sharp images of individual zooplankters in situ, yet the ability to process voluminous imagery data efficiently, and in particular to discriminate similar species, presents ongoing challenges.[70,71] Genetic techniques hold great promise as approaches for identifying species present in a sample and their relative abundance.[72] Coupled with optical (or other) measurements, genetic analysis of individual zooplankton has the potential to support automated identification of zooplankton by species and developmental stage and, thus, estimation of species-specific biomass. We can imagine automated assays that augment taxonomic data with estimates of energy available to higher trophic levels; however, such technologies still lie well into the future. Acoustic techniques yield information on the distribution and (relative) abundance of selected classes of "echo-targets," yet still require "net-truthing" of what is being quantified and to collect information on diversity.[73,74] Clearly, OOS will benefit greatly from successful research and development of automated technologies and their deployment on autonomous sampling platforms, but much work remains to be done before ship-based surveys are supplanted if detailed data on zooplankton and highly informative zooplankton-based ecosystem indicators are to continue to be available to managers.

5.2 CHALLENGES IN IMPLEMENTING HFCTs AS ELEMENTS OF OOS

Several challenges arise in implementing or maintaining HFCTs as part of an OOS. Like any other element of an OOS, there is an up-front investment in personnel and equipment to stand up a new HFCT, and many of these expenses (ship time, personnel) are ongoing. Another ongoing challenge is maintaining excellent access to locally based coastal research vessels of small-to-moderate size. To be effective, HFCTs must operate with sufficient frequency to remain "priority customers" for vessels, as rare, short cruises will not suffice to warrant a dedicated UNOLS-class vessel, yet competition with other users for local vessels can make it difficult to maintain flexible scheduling in response to variable weather conditions. Under most conditions, larger oceanographic research ships cannot carry out this work cost-effectively, whether because they hail from distant homeports or they are overkill as a platform for this relatively simple observational work. Fortunately, as our experience demonstrates, the smaller vessels ideal for HFCTs also can serve as platforms capable of supporting concurrent (leveraged) research, ancillary observations, technology development, and training.

A second challenge is developing and supporting staff with substantial taxonomic expertise and coordinating this expertise and methodology across HFCTs. Efficient extraction of information from plankton samples requires the dedicated effort and careful attention of para- and master taxonomists.[70,75] Training of taxonomists is a serious investment of time and resources—particularly for challenging taxa such as copepods—yet it remains essential for maintaining ongoing time series and to support the development of reliable automated methods.

A third challenge arises from variability in sampling protocols. Comparisons among disparate sampling programs often requires distillation of each data set to anomalies,[8,76] yet it can be difficult to establish that such measures are indeed comparable. Early in the course of establishing the THL, we implemented the vertical ring-net sampling protocol used on the NHL—the same methods endorsed and used by the U.S.GLOBEC program—to ensure comparability of the two data sets. In contrast, our Bongo sampling protocols have remained slightly different, based on differences in sampling goals.

A fourth challenge in establishing new HFCTs is that zooplankton data may be of limited use until a sufficient time series has been established. In contrast to remote sensing technologies that more or less immediately provide a product useful to OOS missions (e.g., a new HF radar site enhancing search-and-rescue and spill-response capabilities), developing ecosystem information from a new

HFCT may require several years of observation. Insights and indices from the NHL time series are based on nearly 20 years of observation, and they clearly demonstrate the value of long observational time series essential for understanding climate–ecosystem dynamics.[77] Observations along the THL have been integrated in several annual synthesis reports on the California Current since 2008, and they provided useful comparisons to observations to the north (especially the NHL) and to the south (see, e.g., Refs. 78,79). Now, as it approaches seven years in length, the THL time series can support rigorous assessment of variability in zooplankton communities of the NCC, but several more years will be required before comparisons to the dynamics of local salmon stocks (e.g., Klamath River Chinook salmon) can be developed with any rigor. We are now actively engaged in developing indices based on observations from the THL for integration in the CCIEA, but this effort was not considered warranted for the first six years of the THL.

5.3 OPPORTUNITIES AND THE ROLE OF HFCTs IN OOS

If these challenges continue to be met, HFCTs will continue to enhance the information available to OOS, as clearly demonstrated by the ecosystem indicators incorporated in IEAs and in the information available to fisheries managers, and to support efforts on several fronts directed at advancing our understanding of marine ecosystem dynamics and informing management. Based on past data, and indeed, going forward, HFCTs have a role in outlining the range of information that developers of autonomous systems should strive to capture and are well situated to collaborate in this development. Moreover, the data returned by HFCTs are of great value to the modeling community as it advances the frontier of dynamic models beyond realistic representation and forecasting of the physical system to models that achieve greater realism in ecosystem dynamics. Resolving zooplankton (and in some cases phytoplankton[80]) to species provides opportunities for developing and evaluating models that seek to account for importantly differing biological groups within trophic levels as part of the dynamics of the system (e.g., Refs. 81,82).

HFCTs represent one solution to the tradeoff that faces all OOS—synopticity, spatial extent, temporal resolution, and parameter coverage. Thus, even as HFCTs fill a critical gap in ocean observing, rigorous analysis of variability in zooplankton communities and its causes depends on environmental data from other elements of OOS. Indeed, our own ongoing work to understand the dynamics of copepod communities and populations draws on environmental time series and circulation models that assimilate a diverse suite of ocean data.

Our work in the NCC points to the value of networked HFCTs for understanding ecosystem dynamics, particularly in a complex system affected by strong environmental gradients such as the CCE. It is possible to demonstrate significant statistical correlations between indices based on observations along the NHL and other measures of productivity in the CCE. Yet, without observations from the THL (and ideally other points along the coast), it is difficult to evaluate similarities and differences in the mechanisms that translate changes in physical forcing to the ecosystem. Moreover, distributed HFCTs can elucidate local responses to large-scale phenomena, as illustrated by differences in recovery of copepod communities during the decay of the 2009–2010 El Niño. This sort of information may help to understand spatial variability in marine survival of salmon and recruitment to other fishery stocks. More generally, data emerging from networked HFCTs will support efforts to extend our empirical, mechanistic understanding of the interplay between local and larger scale processes in ecosystem responses to climate forcing. Previously developed proposals for OOS have called for establishment of HFCTs at several points along the US West Coast, from the Washington coast south to at least Morro Bay, California (to abut the northern extent of the core CalCOFI region), to resolve temporal variability and spatial structure in zooplankton communities, and to understand how local and regional forcing structure productivity in this highly dynamic ecosystem. Placement of future HFCTs should take into account several factors, both logistical (e.g., locations of ports, vessels, and supporting institutions) and ecological (e.g., placement relative to headlands that structure circulation patterns and zooplankton communities,[83,84] and environmental or biogeographic gradients).[85] Insights emerging from the existing pair of HFCTs, including their contributions to understanding larger scale regional variability in the CCE,[13,53,79] argue strongly for implementing such programs as part of a more comprehensive, biologically informative OOS.

Last, we would emphasize that the value of HFCTs is not entirely in the time series and indices produced, but stems also from the simple opportunity to be at sea observing the environment directly. Such experiences are clearly important for engaging students and training young scientists in the field, but they remain valuable to more seasoned researchers as well. Just as fishers have a sense of the ocean born of experience and close interactions with the environment, it is important that we also strive to maintain this contact, pay attention to noticing something new or unusual that we might otherwise not think to sample, and be open to opportunities for developing new lines of inquiry.

ACKNOWLEDGMENTS

Sampling along the NHL has been supported primarily by NOAA through the U.S. GLOBEC Northeast Pacific Program and the NMFS Stock Assessment Improvement Program (SAIP), with additional support from ONR (National Ocean Partnership Program), NASA, the NMFS-Fisheries and the Environment (FATE) Program. Sampling along the THL has been supported by NOAA's SWFSC, in collaboration with Humboldt State University, with additional support from NMFS's SAIP and FATE programs. Through fair weather and foul, a handful of individuals have carried out much of the sampling at sea along the NHL, including Jay Peterson, Jennifer Fisher, Leah Feinberg, Tracy Shaw, Rian Hooff, Jennifer Menkel, and Karen Hunter. Likewise, sampling along the THL has been led by Roxanne Robertson, Phil White, Kathryn Crane, and Ashok Sadrozinski, assisted by student volunteers and assistants too numerous to list, yet greatly appreciated. We also acknowledge the support and assistance of the captains and crews of the *R/V Sacajawea, R/V Elakha*, and *R/V Coral Sea*, without whom none of this work would be possible.

REFERENCES

1. Collie JS, Botsford LW, Hastings A, Kaplan IC, Largier JL, Livingston PA, et al. Ecosystem models for fisheries management: finding the sweet spot. *Fish Fish* 2014. http://dx.doi.org/10.1111/faf.12093.
2. King JR, Agostini VN, Harvey CJ, McFarlane GA, Foreman MGG, Overland JE, et al. Climate forcing and the California Current Ecosystem. *ICES J Mar Sci* 2011;**68**: 1199–216.
3. Chao Y, Li Z, Farrara J, McWilliams JC, Bellingham J, Capet X, et al. Development, implementation and evaluation of a data-assimilative ocean forecasting system off the central California coast. *Deep-Sea Res Part II Top Stud Oceanogr* 2009;**56**(3): 100–26.
4. Moore AM, Arango HG, Broquet G, Edwards C, Veneziani M, Powell B, et al. The Regional Ocean Modeling System (ROMS) 4-dimensional variational data assimilation systems: part II–performance and application to the California Current System. *Prog Oceanogr* 2011;**91**(1):50–73.
5. Song H, Edwards CA, Moore AM, Fiechter J. Incremental four-dimensional variational data assimilation of positive-definite oceanic variables using a logarithm transformation. *Ocean Model* 2012;**54**:1–17.
6. Beaugrand G, Reid PC. Long-term changes in phytoplankton, zooplankton and salmon related to climate. *Global Change Biol* 2003;**9**(6):801–17.
7. Peterson WT, Fisher JL, Peterson JO, Morgan CA, Burke BJ, Fresh KL. Applied fisheries oceanography: ecosystem indicators of ocean conditions inform fisheries management in the California Current. *Oceanography* 2014;**27**(4):80–9.
8. Mackas DL, Beaugrand G. Comparisons of zooplankton time series. *ICES J Mar Sci* 2010;**79**:286–304.
9. Fager EW, McGowan JA. Zooplankton species groups in the North Pacific. *Science* 1963; **140**:453–60.
10. McClatchie S. *Regional fisheries oceanography of the California Current System: the CalCOFI program*. Springer; 2014. 235 pp.
11. Baumgartner T, Durazo R, Lavaniegos B, Gaxiola G, Gomez G, Garcia J. Ten years of change from IMECOCAL observations in the southern region of the California Current Ecosystem. *GLOBEC Int Newsl* 2008;**14**:43–54.
12. Pena MA, Varela DE. Seasonal and interannual variability in phytoplankton and nutrient dynamics along line P in the NE subarctic Pacific. *Prog Oceanogr* 2007; **75**(2):200–22.
13. Bjorkstedt EP, Goericke R, McClatchie S, Weber E, Watson W, Lo N, et al. State of the California Current 2011-2012: ecosystems respond to local forcing as La Nina wavers and wanes. *CalCOFI Rep* 2012;**53**:41–76.
14. Peterson WT, Miller CB. Year-to-year variations in the planktology of the Oregon upwelling zone. *Fish Bull* 1975;**73**:642–53.
15. Peterson WT, Miller CB. The seasonal cycle of zooplankton abundance and species composition along the central Oregon coast. *Fish Bull* 1977;**75**:717–24.

16. Richardson S, Pearcy WG. Coastal and oceanic fish larvae in an area of upwelling off Yaquina Bay, Oregon. *Fish Bull* 1977;**65**:125–45.
17. Peterson WT. *Life history and ecology of the copepd* Calanus marshallae *Frost in the Oregon upwelling zone* [Ph.D. dissertation]. Oregon State University; 1980. 200 p.
18. Miller CA, Batchelder H, Brodeur R, Pearcy W. Response of the zooplankton and ichthyoplankton off Oregon to the El Niño event of 1983. In: Wooster WS, Fluharty DL, editors. *El Niño north: Niño effects in the eastern subarctic Pacific Ocean*. Washington Sea Grant Program; 1985. p. 185–7.
19. Fessenden LM. *Calanoid copepod diet in an upwelling system: phagotrophic protists vs. phytoplankton* [Ph.D. dissertation]. Oregon State University; 1996. 136 p.
20. Hooff RC, Peterson WT. Recent increases in copepod biodiversity as an indicator of changes in ocean and climate conditions in the northern California current ecosystem. *Limnol Oceanogr* 2006;**51**:2042–51.
21. Peterson WT, Keister JE. Interannual variability in copepod community composition at a coastal station in the northern California Current: a multivariate approach. *Deep-Sea Res* 2003;**50**:2499–517.
22. Mackas DL, Peterson WT, Ohman MD, Lavaniegos BE. Zooplankton anomalies in the California Current System before and during the warm ocean conditions of (2005). *Geophys Res Lett* 2006;**33**(22).
23. Peterson W. Copepod species richness as an indicator of long term changes in the coastal ecosystem of the northern California Current. *CalCOFI Rep* 2009;**50**:73–81.
24. Lee R, Hagen W, Kattner G. Lipid storage in marine zooplankton. *Mar Ecol Prog Ser* 2006;**307**:273–306.
25. Mantua NJ, Hare SR, Zhang Y, Wallace JM, Francis RC. A Pacific interdecadal climate oscillation with impacts on salmon production. *Bull Am Met Soc* 1997;**78**(6):1069–79.
26. Bi H, Peterson W, Strub P. Transport and coastal zooplankton communities in the northern California Current System. *Geophys Res Lett* 2011;**38**(12):L12607. http://dx.doi.org/10.1029/2011GL047927.
27. Keister JE, DiLorenzo E, Morgan CA, Combes V, Peterson WT. Copepod species composition is linked to ocean transport in the northern California Current. *Glob Change Biol* 2011;**17**(7):2498–511. http://dx.doi.org/10.1111/j.1365-2486.(2010).02383.x.
28. Peterson WT, Keister JA, Feinberg LR. The effects of the 1997–99 El Niño/La Niña event on hydrography and zooplankton off the central Oregon coast. *Prog Oceanogr* 2002;**54**:381–98.
29. Climate impacts on U.S. living marine resources: national marine fisheries service concerns, activities and needs. In: Osgood KE, editor. *U.S. Dep. Commerce, NOAA Tech. Memo*. NMFS F/SPO-89; 2008. 118 p.
30. Doney SC, Ruckelshaus M, Duffy JE, Barry JP, Chan F, English CA, et al. Climate change impacts on marine ecosystems. *Ann Rev Mar Sci* 2012;**4**:11–37.
31. Rykaczewski RR, Dunne JP. Enhanced nutrient supply to the California Current Ecosystem with global warming and increased stratification in an earth system model. *Geophys Res Lett* 2010;**37**(21). http://dx.doi.org/10.1029/2010GL045019.
32. Sydeman WJ, Garcia-Reyes M, Schoeman DS, Rykaczewski RR, Thompson SA, Black BA, et al. Climate change and wind intensification in coastal upwelling ecosystems. *Science* 2014;**345**:77–80.
33. Logerwell EA, Mantua N, Lawson PW, Francis RC, Agostini VN. Tracking environmental processes in the coastal zone for understanding and predicting Oregon coho (*Oncorhynchus kisutch*) marine survival. *Fish Oceanogr* 2003;**12**:554–68.

34. Schroeder ID, Sydeman WJ, Sarkar N, Thompson SA, Bograd SJ, Schwing FB. Winter pre-conditioning of seabird phenology in the California Current. *Mar Ecol Prog Ser* 2009;**393**:211–23.
35. Black BA, Schroeder ID, Sydeman WJ, Bograd SJ, Lawson PW. Wintertime ocean conditions synchronize rockfish growth and seabird reproduction in the central California Current Ecosystem. *Can J Fish Aquatic Sci* 2010;**67**(7):1149–58.
36. Feinberg LR, Peterson WT. Variability in duration and intensity of euphausiid spawning off central Oregon, 1996-(2001). *Prog Oceanogr* 2003;**57**:363–79.
37. Feinberg LR, Peterson WT, Shaw CT. The timing and location of spawning for the euphausiid *Thysanoessa spinifera* off the Oregon coast, USA. *Deep-Sea Res. II, Krill symposium special issue* 2010;**57**:572–83.
38. Black BA, Schroeder ID, Sydeman WJ, Bograd SJ, Wells BK, Schwing FB. Winter and summer upwelling modes and their biological importance in the California Current Ecosystem. *Glob Change Biol* 2011;**17**(8):2536–45.
39. Schroeder ID, Black BA, Sydeman WJ, Bograd SJ, Hazen EL, Santora JA, et al. The North Pacific High and wintertime pre-conditioning of California current productivity. *Geophys Res Lett* 2013;**40**(3):541–6.
40. Laidig TE, Chess JR, Howard DF. Relationship between abundance of juvenile rockfishes (*Sebastes* spp.) and environmental variables documented off northern California and potential mechanisms for the covariation. *Fish Bull* 2007;**105**:39–48.
41. Ralston S, Sakuma KM, Field JC. Interannual variation in pelagic juvenile rockfish (*Sebastes* spp.) abundance–going with the flow. *Fish Oceanogr* 2013;**22**:288–308.
42. Mills KL, Laidig T, Ralston S, Sydeman WJ. Diets of top predators indicate pelagic juvenile rockfish (*Sebastes* spp.) abundance in the California Current System. *Fish Oceanogr* 2007;**16**:273–83.
43. Field JC, MacCall AD, Bradley RW, Sydeman WJ. Estimating the impacts of fishing on dependent predators: a case study in the California Current. *Ecol Appl* 2010;**20**:2223–36.
44. Liu H, Peterson W. Seasonal and interannual variations in the abundance and biomass of Neocalanus plumchrus/flemingerii in the slope waters off Oregon. *Fish Oceanogr* 2010;**19**:354–69.
45. Parsons TR, Giovando LF, LeBrasseur RJ. The advent of the Spring bloom in the eastern subarctic Pacific ocean. *J Fish Res Bd Can* 1966;**23**:539–46.
46. Henson SA, Thomas AC. Interannual variability in timing of bloom initiation in the California Current System. *J Geophys Res Oceans (1978–2012)* 2007;**112**(C8). http://dx.doi.org/10.1029/2006JC003960.
47. Bograd SJ, Schroeder I, Sarkar N, Qiu X, Sydeman WJ, Schwing FB. Phenology of coastal upwelling in the California Current. *Geophys Res Lett* 2009;**36**. http://dx.doi.org/10.1029/2008GL035933.
48. Bakun A. *Coastal upwelling indices: west coast of North America 1946–1971. NOAA technical report NMFS SSRF-671*. 1973. 103 pp.
49. Pierce SD, Barth JA, Thomas RE, Fleischer GW. Anomalously warm July 2005 in the northern California Current: historical context and the significance of cumulative wind stress. *Geophys Res Lett* 2006;**33**. http://dx.doi.org/10.1029/2006GL027149.
50. Kosro PM, Peterson WT, Hickey BM, Shearman RK, Pierce SD. Physical versus biological spring transition: 2005. *Geophys Res Lett* 2006;**33**. http://dx.doi.org/10.1029/2006GL027072.
51. Lindley ST, Grimes CB, Mohr MS, Peterson W, Stein J, Anderson JT, et al. What caused the Sacramento River fall Chinook stock collapse? In: *U.S. Dep. Commerce, NOAA Tech. Memo*. NOAA-TM-NMFS-SWFSC-447; 2009. p. 121.

52. Sydeman WJ, Bradley RW, Warzybok P, Abraham CL, Jahncke J, Hyrenbach KD, et al. Planktivorous auklet *Ptychoramphus aleuticus* responses to ocean climate, 2005: unusual atmospheric blocking? *Geophys Res Lett* 2006;**33**. http://dx.doi.org/10.1029/2006GL026736.
53. Peterson WT, Emmett R, Goericke R, Venrick E, Mantyla A, Bograd SJ, et al. The state of the California Current, 2005–2006: warm in the north, cool in the south. *CalCOFI Rep* 2006;**47**:30–74.
54. Gómez-Gutiérrez J, Peterson WT. Egg production rates of eight copepod species during the summer of 1997 off Newport, Oregon, USA. *J Plankton Res* 1998;**21**:637–57.
55. Peterson WT, Gómez-Gutiérrez J, Morgan C. Cross-shelf variation in calanoid copepod production during summer 1996 along the Oregon coast, USA. *Mar Biol* 2002;**141**:353–65.
56. Keister JE, Peterson WT. Zonal and seasonal variations in zooplankton community structure off the central Oregon coast, 1998-(2000). *Prog Oceanogr* 2003;**57**:341–61.
57. Morgan CA, Peterson WT, Emmett RL. Onshore-offshore variations in copepod community structure off the Oregon coast during the summer upwelling season. *Mar Eco Prog Ser* 2003;**249**:223–36.
58. Lamb J, Peterson W. Ecological zonation of zooplankton in the COAST study region off central Oregon in June and August 2001 with consideration of retention mechanisms. *J Geophys Res* 2005;**110**. http://dx.doi.org/10.1029/2004JC00250.
59. Peterson WT, Morgan CA, Fisher JP, Casillas E. Ocean distribution and habitat associations of yearling coho and Chinook salmon in the northern California Current. *Fish Oceanogr* 2010;**19**:508–25.
60. Sorte CJ, Peterson WT, Morgan CA, Emmett RL. Larval dynamics of the sand crab, *Emerita analoga*, off the central Oregon coast during a strong El Niño period. *J Plankton Res* 2001;**23**:939–44.
61. Keister JE, Johnson TB, Morgan CA, Peterson WT. Biological indicators of the timing and direction of warm-water advection during the 1997/98 El Niño off the central Oregon coast, USA. *Mar Ecol Prog Ser* 2005;**295**:43–8.
62. Brodeur RD, Peterson WT, Auth TD, Soulen HL, Parnel MM, Emerson AA. Abundance and diversity of coastal fish larvae as indicators of recent changes in ocean and climate conditions in the Oregon upwelling zone. *Mar Ecol Prog Ser* 2008;**366**:187–202.
63. Auth TD, Brodeur RD, Soulen HL, Ciannelli L, Peterson WT. The response of fish larvae to decadal changes in environmental forcing factors off the Oregon coast. *Fish Oceanogr* 2011;**20**:314–28.
64. Bi H, Peterson WT, Peterson JO, Fisher J. A comparative analysis of coastal and shelf-slope zooplankton communities in the northern California Current System: synchronized response to large scale forcing? *Limnol Oceanogr* 2012;**57**:467–1478.
65. Bednaršek N, Feely RA, Reum JCP, Alin S, Hales B, Peterson W. Impact of ocean acidification on *Limacina helicina* shell dissolution in the California Current Ecosystem. *Proc Roy Soc Lond B* 2014;**28**:20140123.
66. Peterson WT, Miller CB, Hutchinson A. Zonation and maintenance of copepod populations in the Oregon upwelling zone. *Deep-Sea Res* 1979;**26A**:467–94.
67. Fisher J, Peterson W, Morgan S. Does larval advection and survival explain latitudinal differences in recruitment across upwelling regimes? *Mar Ecol Prog Ser* 2014;**503**:123–37.
68. Batchelder HP, Botsford L, Daly K, Davis C, Ji R, Ohman M, et al. Climate impacts on animal populations and communities in coastal marine systems: forecasting change through mechanistic understanding of population dynamics. *Oceanography* 2013;**26**(4):34–51.

69. Burke BJ, Peterson WT, Beckman BR, Morgan C, Daly EA, Litz M. Multivariate models of Adult Pacific Salmon returns. *PLoS One* 2013;**8**(1):e54134. http://dx.doi.org/10.1371/journal.pone.0054134.
70. Benfield MC, Grosjean P, Culverhouse PF, Irigoien X, Sieracki ME, Lopez-Urrutia A, et al. RAPID: research on automated plankton identification. *Oceanography* 2007;**20**: 172–87.
71. Cowen RK, Guigand CM. In situ ichthyoplankton imaging system (ISIIS): system design and preliminary results. *Limnol Oceanogr Method* 2008;**6**:126–32.
72. Ryan JP, Harvey JBJ, Zhang Y, Woodson CB. Distribution of invertebrate larvae and phytoplankton in a coastal upwelling retention zone and peripheral front. *J Exp Mar Biol Ecol* 2014;**459**:51–60.
73. Sutor M, Cowles TJ, Peterson WT, Lamb J. Comparison of acoustic and net sampling systems to determine patterns in zooplankton distribution. *J Geophys Res* 2005;**110**. http://dx.doi.org/10.1029/2004JC00250.
74. Santora JA, Sydeman WJ, Schroeder ID, Reiss CS, Wells BK, Field JC, et al. Krill space: a comparative assessment of mesoscale structuring in polar and temperate marine ecosystems. *ICES J Mar Sci* 2012;**69**:1317–27.
75. Duffy JE, Amaral-Zettler LA, Fautin DG, Paulay G, Rynearson TA, Sosik HM, et al. Envisioning a marine biodiversity observation network. *Bioscience* 2013;**63**:350–61.
76. Mackas DJ, Peterson WT, Zamon J. Interannual-to-decadal anomalies of zooplankton communities along the continental margin of British Columbia and Oregon. *Deep-Sea Res* 2004;**51**:875–96.
77. Ducklow HW, Doney SC, Steinberg DK. Contributions of long-term research and time-series observations to marine ecology and biogeochemistry. *Ann Rev Mar Sci* 2009;**1**:279–302.
78. Bjorkstedt EP, Goericke R, McClatchie S, Weber E, Watson W, Lo N, et al. State of the California Current 2010–2011: Regionally variable responses to a strong (but fleeting?) La Nina. *CalCOFI Rep* 2011;**52**:36–68.
79. Wells BK, Schroeder ID, Santora JA, Hazen EL, Bograd SJ, Bjorkstedt EP, et al. State of the California Current 2012–13: no such thing as an "average" year. *CalCOFI Rep* 2013; **54**:37–71.
80. Du X, Peterson WT. Seasonal cycle of phytoplankton community composition in the coastal upwelling system off central Oregon in 2009. *Estuaries Coasts* 2013;**37**: 299–311.
81. Goebel NL, Edwards CA, Zehr JP, Follows MJ. An emergent community ecosystem model applied to the California Current System. *J Mar Syst* 2010;**83**:221–41.
82. Goebel NL, Edwards CA, Zehr JP, Follows MJ, Morgan SG. Modeled phytoplankton diversity and productivity in the California Current System. *Ecol Model* 2013;**264**: 37–47.
83. Peterson WT, Keister JE. The effect of a large cape on distribution patterns of coastal and oceanic copepods off Oregon and northern California during the 1998–1999 El Niño–La Niña. *Prog Oceanogr* 2002;**53**:389–411.
84. Keister JE, Cowles TJ, Peterson WT, Morgan CA. Do upwelling filaments result in predictable biological distributions in coastal upwelling ecosystems? *Prog Oceanogr* 2009;**83**:303–13.
85. Checkley Jr DM, Barth JA. Patterns and processes in the California Current System. *Prog Oceanogr* 2009;**83**:49–64.

CHAPTER

The IMOS Ocean Radar Facility, ACORN

9

Lucy R. Wyatt

ACORN, College of Science, Technology and Engineering, Centre for Tropical Water and Aquatic Ecosystem Research, James Cook University, Townsville, QLD, Australia; School of Mathematics and Statistics, University of Sheffield, Sheffield, UK

E-mail: lucy.wyatt@jcu.edu.au

CHAPTER OUTLINE

1. INTRODUCTION

High Frequency (HF) radar, as a technology for coastal ocean observation, has been around for many years, but only in the last decade has it really taken off as an essential component of coastal ocean observing systems. In most cases, the use of radar was originally driven by scientific groups and goals resulting in local systems addressing specific scientific questions. In the United States, this began to change when California legislated in 2002 to provide resources to support coastal and beach protection; the U.S. Coast Guard realized the advantage of using the data for search and rescue operations; Homeland Security became a strong driving force for infrastructure funding. The United States now has HF radar systems operating around most of its coastline.[1] Networks are being developed elsewhere in the world including Spain,[2] Korea and Taiwan,[3] and Australia, where the Australian Coastal Ocean Radar Network (ACORN)[4] is part of the Australian Integrated Marine

Observing System (IMOS).[5,6] ACORN is not (yet) providing the sort of comprehensive coverage seen in many parts of the United States; rather, it is a collection of individual deployments of paired radars to provide data in regions of particular interest to the IMOS science community. There continue to be many such deployments in different parts of the world, often with aspirations to become part of a more substantial network. There is now a global effort, the Global High Frequency Network Component of the Group on Earth Observations (GEO),[7] to support further development of HF radar networks and to promote the standardization of data products and QA/QC methodologies. ACORN is participating in this effort. This chapter will focus mainly on ACORN experiences and results.

As usage grows, the requirements on radio frequency bandwidth increase to accommodate this growth. This has been the subject of international negotiations to find solutions for all such systems to coexist. Fixed frequency bands have been allocated to oceanographic radars, and methods to share this limited resource are being developed. Radars also have to share bandwidth with other radio users, and interference from these other users, as well as radio noise from other sources, e.g., thunderstorms, is often a problem limiting the range of oceanographic measurements. This variability in coverage can be ameliorated by combining the radar data with models through assimilation.[8] Coexisting with interference through the use of sophisticated signal processing techniques is an area of active research,[9,10] much of which has not yet made its way into operational applications but has the potential to significantly improve the performance of these systems.

In nearly all these cases, the emphasis has been primarily on surface current measurements. Surface wave measurement has been under development for many years (e.g., Refs. 11–22), but it is more vulnerable to interference as well as to antenna pattern phase offsets that can arise as a result of environmental factors or faults somewhere in the radar system. These measurement also require higher signal-to-noise, so they are always more range-limited than currents.

Most of the deployments around the world use either the US SeaSonde[23] or the German WERA[24] radars, and there are a few, including ACORN, that use both (e.g., Refs. 25,26). Both these radars use frequency modulation (FM) where the range of a target (in this case a patch of sea surface) is determined by the frequency difference between the transmit and receive signals. SeaSonde is a compact radar with a receive antenna (Rx) system comprising a monopole and two cross-loop antennas all mounted on the same pole. The transmit antenna (Tx) is normally on a second pole. At high-HF frequencies, the Tx antenna can be combined with the Rx. These antenna configurations are possible because the SeaSonde uses frequency-modulated, interrupted, continuous wave (FMICW) modulation, where the *I* means that the Tx and Rx are not on at the same time. Sometimes, two Tx antennas are used to improve directionality. Direction-finding with the Multiple Signal Classification algorithm, MUSIC,[27] is used to determine the direction of arrival of the surface current signals. This method involves an eigenanalysis at each Doppler frequency of the covariance matrix of the complex sample-averaged, spectrally analyzed signals from the Rx antennas. Wave measurements can be made in small regions close to the radar sites with the assumption of spatial homogeneity over annular regions at fixed ranges from these sites. WERA normally uses four Tx antennas in a rectangle to increase the directionality of the signal in the main-beam direction, thus limiting the amount of energy that is transmitted landward and, most importantly, toward the Rx array. A similar configuration Rx can be used with direction finding, in a similar manner to SeaSonde. More often, a phased array of 8 to 16 antennas and digital beam-forming is used to isolate signals from particular regions on the sea surface. This approach also makes it easier to make wave measurements over a wider area than SeaSonde, although these can be limited by antenna side lobe problems when short antenna arrays are used. Because WERA uses FMCW modulation, the Rx has to be very carefully aligned with the Tx and power levels kept low to minimize mutual interference. A number of other radar systems have been developed, some of which (e.g., Pisces[28] and OSMAR[29]) are available commercially but have not yet made the same impact. Pisces is also a phased-array system but uses FMICW modulation, so higher power can be used than is the case for WERA, which has made it possible to measure waves at much longer ranges.[28]

2. ACORN

ACORN is currently operating 12 radars arranged in six pairs located as shown in Figure 1, providing surface current data in near real time. Radar parameters and the scientific goals for these radars are summarized in Table 1. ACORN uses two

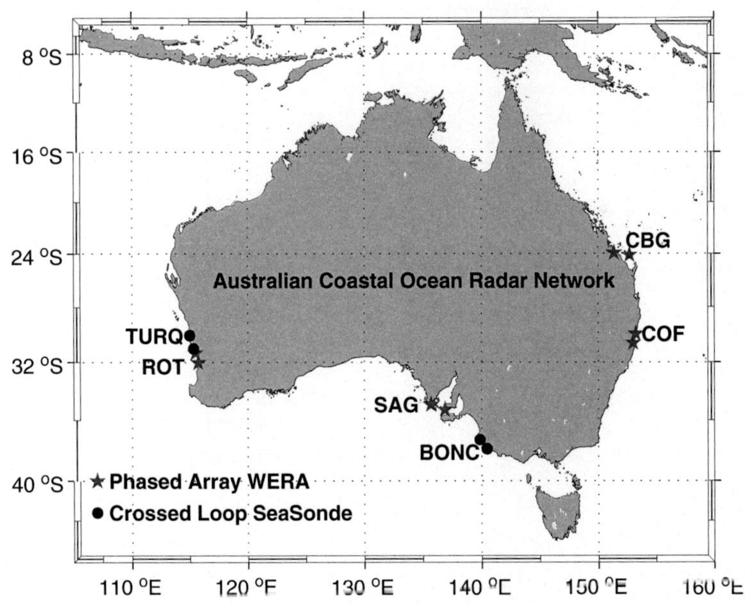

FIGURE 1

Map of Australia showing locations of the ACORN radars.

Table 1 ACORN Radar Configuration

Site	Radar	Stations	Installation Date	Frequency MHz	Range resolution km	Main Science goals
CBG	WERA	LEI/TAN	2007	8.34	4.5	Synoptic current patterns
COF	WERA	RRK/NNB	2012	13.92 initially Now 13.5	1.5	EAC variability and waves
BONC	SeaSonde	NOCR/BFCV	2010	5.12	3	Cross-shelf exchange
SAG	WERA	CSP/CWI	2009	8.512	4.5	Cross-shelf exchange, wind direction and waves
ROT	WERA	FRE/GUI	2010	8.512	4.5	Leeuwin current, its eddies, wind and waves
TURQ	SeaSonde	SBRD/CRVT	2009–2012	5.21	3	Leeuwin current and eddies
		LANC/GHED	2013	5.21	3	

radar technologies, the phased array WERA and the direction-finding SeaSonde. Although Australia has a long history in HF radar oceanographic research,[30–32] this had been quite small in scale, and most Australian oceanographic scientists had not been exposed to the potential benefits of radar in general, or to the relative merits of these two technologies. Thus, to maximize the long-term benefits of the initial installations, in terms of future decision-making on particular technologies for particular applications, examples of both technologies were included in the network. Where there was a specific requirement for wave or wind mapping, WERA was selected because such mapping is not possible with SeaSonde.

ACORN is currently based at James Cook University, Townsville, QLD, but it is moving in late 2014 to the University of Western Australia, Perth, WA. A team of three technicians operate, maintain, and manage the data for these radars. They are supported by caretakers at most of the sites, who can respond quickly to problems, and by the IMOS eMII Facility to which the data is transmitted and then archived and made available as maps and as netcdf files via the IMOS portal.[33]

ACORN is asked to achieve >90% data availability at each site. This is defined as the percentage of time for which a file containing radial currents (see Section 3) is delivered to eMII and appears on the IMOS server. The results obtained for the years 2011–2013 are shown in Table 2. Sites where the goal of 90% has not been achieved are shown in bold italics. The mean or median figures are greater than the requirement in all three of these years. There was a slight improvement in 2012 as a result of significant changes in the way sites were being monitored and faults responded to.

Table 2 Data Availability for Years 2011–2013

Site	% Data 2011	% Data 2012	% Data 2013
TAN	98.1	95.9	94.1
LEI	***85.7***	95.3	***87.6***
CSP	***89.8***	90.1	***88.6***
CWI	***87.6***	99.5	99.4
FRE	99.0	96.4	97.6
GUI	93.8	96.9	90.2
SBRD/LANC	***89.3***	***89.0***	90.6
CRVT/GHED	93.9	***88.9***	99.0
NOCR	95.9	95.2	95.0
BFCV	96.5	95.6	96.4
RRK	–	98.9	96.1
NNB	–	98.6	98.5
Mean	**93.0**	**95.0**	**94.4**
Median	**93.9**	**95.7**	**95.6**

Figures in bold italics are below the target of 90%.

The relocation of the TURQ radars at the end of 2012 seems to have been justified by the increase in data availability in 2013. Overall, though, there was a slight fall in availability in 2013 that may be reflecting the age of some of these systems. A more substantial replacement plan is forming part of the ongoing ACORN strategy. Oceanographic data availability over space and time is of more importance to users and is also a useful metric to identify changes in radar performance and to make decisions about future planning, and this information is also being collected and evaluated.[34]

3. CURRENT MEASUREMENTS, ACCURACY, AND APPLICATIONS

3.1 HOW THE MEASUREMENTS ARE MADE

Currents are measured from the Doppler frequency of the backscattered signal from ocean waves with half the radio wavelength traveling toward and away from the radar.[8] These waves are sometimes referred to as Bragg-matched. The Doppler frequency contains information about the linear phase speed of this wave component, plus the component of surface current in the direction of the radar from the scattering patch, usually referred to as radial current. The phase speed of the waves toward and away from the radar are equal with opposite directions, and hence, this part of the signal is symmetric about zero Doppler frequency. A surface current imposes a positive (if the current is toward the radar) or negative shift to this symmetry and, hence, can be easily separated from the phase speed. Phased-array radars, like WERA, make this measurement from the Doppler spectrum of the signal at the scattering patch. Direction-finding radars, like SeaSonde, use the Doppler spectrum for signals at one range that includes Doppler shifts from all azimuths. Particular shifts are mapped to particular azimuths using MUSIC or other direction-finding methods.

In order to get vector currents, it is necessary to have two or more radars looking at the same patch of ocean from different directions, hence, the pairing of radars in ACORN. The WERA radial measurements are on a prescribed rectangular grid, and the radials from each radar, at each point on that grid, can be combined easily knowing their amplitude and the direction of the grid point to the two radar sites.[35] However, where the difference in direction is small (less than 30° is a good rule of thumb), this geometrical combining method has large errors, usually referred to as geometric dilution of precision (GDOP), and these locations have to be flagged. SeaSonde radials are on polar grids emanating from the radar sites, and these are combined to produce vectors on a grid by imposing an averaging circle at each grid point and least-squares fitting all radials that lie in that circle (provided that at least one comes from the other radar) to the east-west (U) and north-south (V) components of the vector current.[36] GDOP errors still have to be identified and flagged. In both cases, a minimum signal-to-noise of the Bragg peaks is imposed.

3.2 ACCURACY

Figure 2 shows examples of surface current maps from the SeaSonde (a) and WERA (b) systems in West Australia. Note that the scales and ranges of the maps are different, and the time periods are not exactly aligned. The black and red rectangles outline the same areas in both maps. There is some overlap to the south of the SeaSonde coverage and the north of the WERA coverage, so there is an opportunity to compare current measurements from the two systems. In the smaller red rectangle, it is clear that the measurements are different at this time; although, there are some similarities elsewhere within the larger black rectangle. At other times, better agreement has been found. The southern SeaSonde radar at LANC is only a few km from the northern WERA radar at GUI, and for these, the radial currents sufficiently far from the sites should be nearly equal and can, therefore, be directly compared to try to understand the source of any differences. A preliminary radial current analysis[37] has shown that differences are smaller toward the center of the coverage of the two radars with increasing differences toward the outer edges in range and azimuth. This analysis is summarized in the Taylor diagram[38] shown in Figure 3. The correlation coefficient and rms difference between the two radials (which have been averaged spatially) and the individual standard deviations (the last three normalized to the standard deviation in the WERA radials) can be plotted as a single point on the diagram; the closer to the point denoted as A (correlation coefficient = 1, rms = 0, equal variances) on the plot, the better the agreement. The central points are identified by the black oval.

The presence of larger differences away from this central region has motivated a review of the SeaSonde and WERA quality control procedures adopted by ACORN. This work is still underway but includes calculating (where necessary) and making use of the standard deviation in the vector current estimates, ensuring the GDOP is fully accounted for, and flagging cases where the number of radials included in the averaging procedures used during the vector calculations (in time with WERA and in time and space with SeaSonde) are below selected thresholds. The outcome will be validated, and the hoped for improvement quantified, using measurements from other instruments. ADCPs are deployed within some of the ACORN radar coverage regions and are being used for such validations[37,39] as have drifter tracks.[39]

3.3 APPLICATIONS

IMOS was developed to provide data for oceanographic science, and to date, this has been where the major uptake of HF radar data has occurred (some examples are[39–42]). The data are now being used for large operational modeling projects, e.g., Bluelink, for validation purposes and potentially for assimilation. Interest is also developing for applications in search and rescue, in the prawn and scallop fisheries in the Great Barrier Reef and tuna fisheries in South Australia, and even for long-distance swimming between Perth and Rottnest Island. There is great potential in Australia, and there is also a big need, particularly in the vulnerable Great Barrier Reef, to use HF radar for port and harbor management. The ACORN radar coverage in Queensland is off the port of Gladstone, and Figure 4 shows that the radar covers the main shipping lanes from this port.

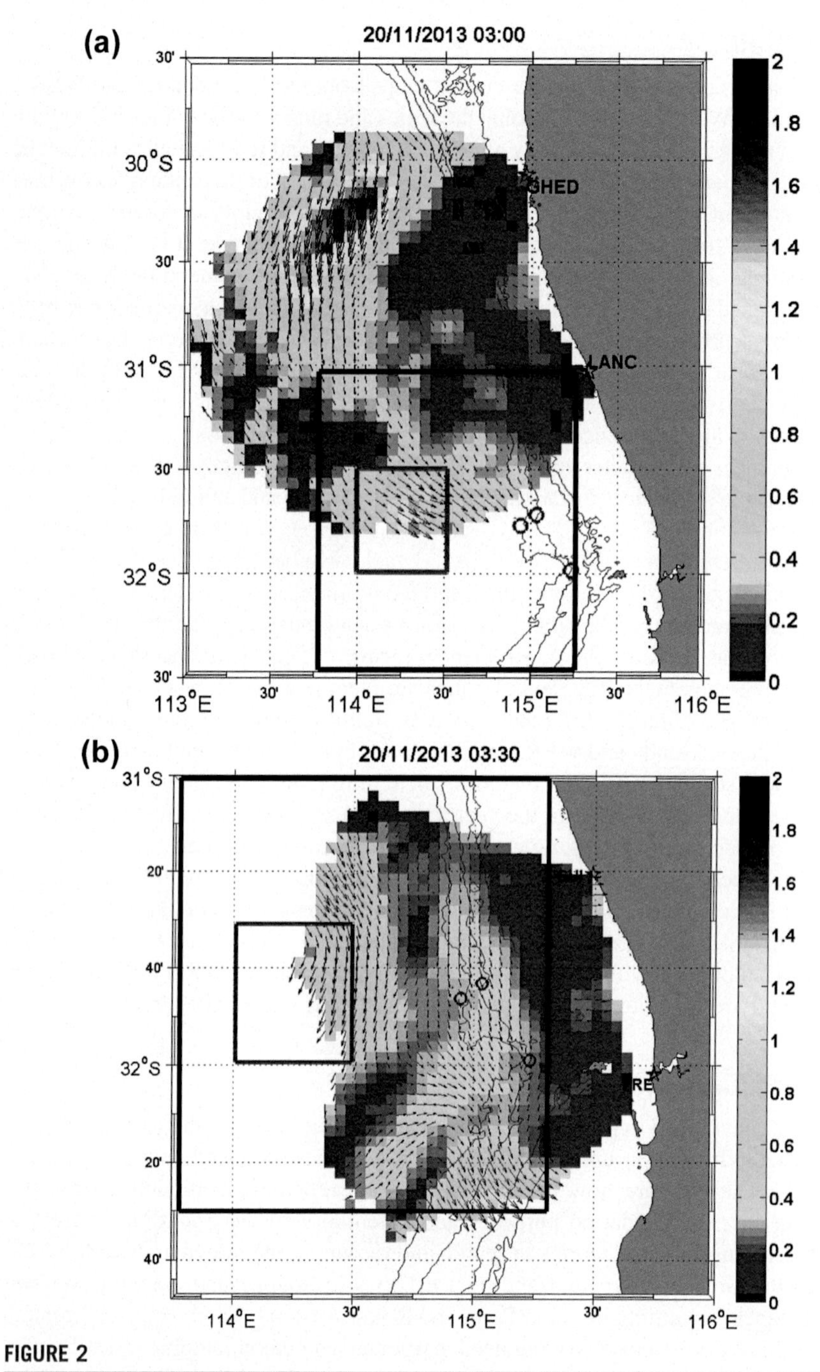

FIGURE 2

(a) SeaSonde currents at TURQ WA, (b) WERA currents at ROT WA.

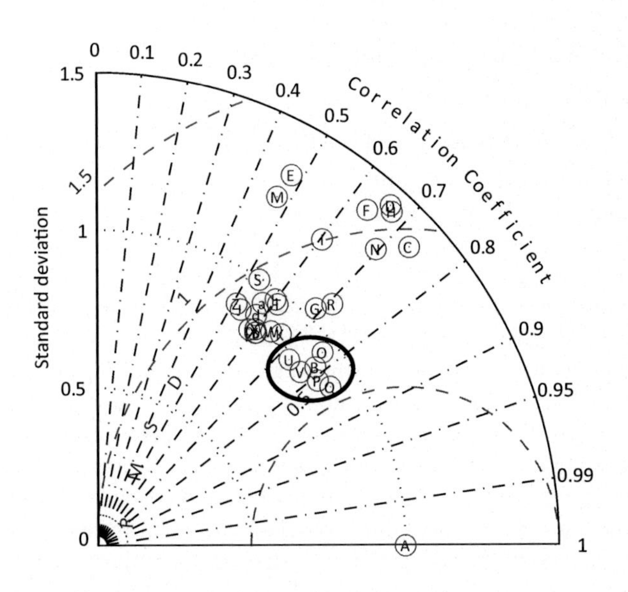

FIGURE 3

Taylor plot of the SeaSonde–WERA radial current comparison.

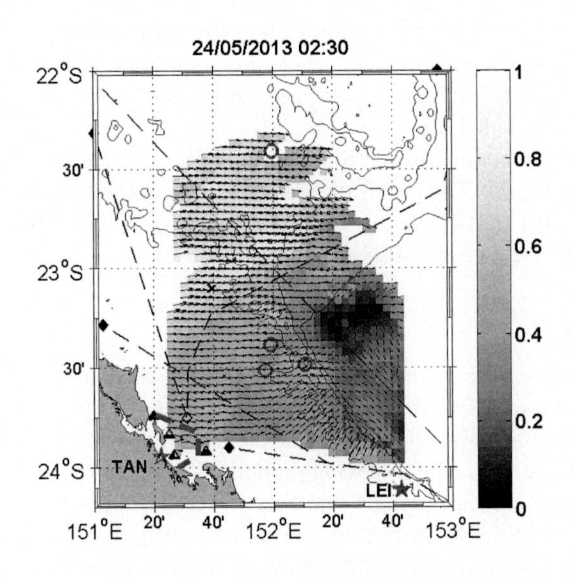

FIGURE 4

A sample ACORN radar current map in the southern Great Barrier Reef east of the port of Gladstone. Dashed lines mark the main shipping channels; the thick dashed line marks the edge of the compulsory pilotage area for the harbor; triangles are way markers at the main entrance to the harbor; diamonds are ship reporting points; circles are the locations of IMOS moorings.

4. WAVE AND WIND MEASUREMENTS, ACCURACY, AND APPLICATIONS

4.1 HOW THE MEASUREMENTS ARE MADE

In addition to the so-called first-order part of the signal that provides current measurements, the power (Doppler) spectrum of the received signal also contains so-called second-order contributions that arise from ocean waves with half the radio wavelength but with phase speeds that depend on second-order wave–wave (hydrodynamic) interactions and on multiple electromagnetic scattering. A model for these two processes was derived by Barrick[43,44] and takes the form of a set of nonlinear integral equations, the inversion of which can provide the full ocean wave directional spectrum.[12,17–20] Empirical methods have also been used to derive relationships between specific wave parameters (e.g., significant wave height) and buoy data (e.g., Refs. 15,16,21). ACORN has implemented both these approaches. Wave measurements using the Gurgel et al.[21] empirical methods are included in the near real-time radial current files on the IMOS server. Wave measurements (both empirical and the full spectrum and parameters derived therefrom) using the Seaview Sensing wave package (Ref. 45 based on Refs. 16,17,20,46) should be appearing on the IMOS server later this year. Examples of the latter approach are included here.

The first-order signal also provides wave information. The direction and directional spreading of the Bragg-matched waves can be inferred from the relative amplitude of the signals of the wave traveling toward and away from the radar.[46] This direction is often identified as wind direction. Although there have been a number of attempts to estimate wind speed, this is still at the research stage, and no examples are included here.

Obtaining wave measurements from the radar data is very computer intensive in terms of time and space. The provision of such resources through the Research Data Infrastructure Project (RDSI) is now making it possible to process the archived radar data and provide the data in IMOS netcdf format.

4.2 ACCURACY

The ACORN radars were not optimally configured for wave measurements because their primary goal was to provide current measurements. Wave measurements have a much higher signal-to-noise requirement, so the range that can be achieved is always less than that for currents. The following limitations have been identified. The radars are too far apart (or equivalently, the available power is too low) to provide sufficient signal-to-noise in low seas (less than ~1-m wave height). High noise levels are experienced at some sites, particularly in West Australia, but all sites are affected at some times of the day, resulting in limited range and availability. Local mains interference is also a problem with the WERA radars at some sites. ACORN has limited bandwidth for data transmission, which does not allow for real-time wave measurement. Spatial averaging to reduce data quantity may be a solution for real-time operations. Wave measurements still tend to be noisier with more significant outliers than is the case for current measurements. New approaches to minimize this problem are being investigated.

Figure 5 shows an example of wave measurements from the SAG radar system. This is a region exposed to the Southern Ocean and often dominated with swell as is the case in this figure. The parameters shown in Figure 5(a) and (b) are derived from the directional spectra that have been measured at each point where there is an arrow. The wind directions (Figure 5(c)) are opposed to those of the swell on this occasion, and the directional spectrum (Figure 5(d)) shows no evidence of a wind-driven component within the measurable range of up to 0.2 Hz. The directional spreading of the short waves (Figure 5(c)) is lower close to Kangaroo Island (to the east of the coverage), perhaps indicating fetch-limited wave development at this location. The accuracy of the wave measurements in comparison with wave buoy, ADCP, and wind measurements is currently being assessed.

4.3 APPLICATIONS

Although the ACORN wave data are not yet in the IMOS archive, they are already being used in model comparisons to both assess the mode performance and that of the radar data. The Australian Bureau of Meteorology in South Australia is particularly interested in the wind direction data because this has the potential to improve bush fire modeling on land to the north of the radar coverage, an area that is very susceptible to serious bush fire events. The agencies of the governments of NSW and QLD responsible for wave monitoring for flooding, erosion, and climate change among other applications have expressed interest in this technology with a view to complementing their wave buoy networks. There is also interest from the marine renewable industry who are beginning to test devices in Australian coastal waters.

5. PROSPECTS FOR FURTHER DEVELOPMENT

The most recent addition to the ACORN network was the Coffs Harbor radar that was installed in 2012. However, all the radar systems were included in the original IMOS plan; it took longer to implement the plans in some locations due primarily to local planning permissions and licensing issues. The success of the systems has now raised interest in new deployments. There is interest in positioning a radar system to monitor the separation zone of the East Australian Current in NSW south of Coffs Harbor. The initial feasibility studies for search and rescue applications in WA have shown that additional coverage provided by additional radar systems would be an advantage. The potential for an HF radar network along the Pilbara and Kimberley coasts to provide data for the offshore oil and gas sector in this region is being explored. Cyclones and strong swell events are of concern in this area as is the potential damage to a largely unspoiled coastline from any spillage. The potential use of radar to support existing and proposed new ports along the coast of the Great

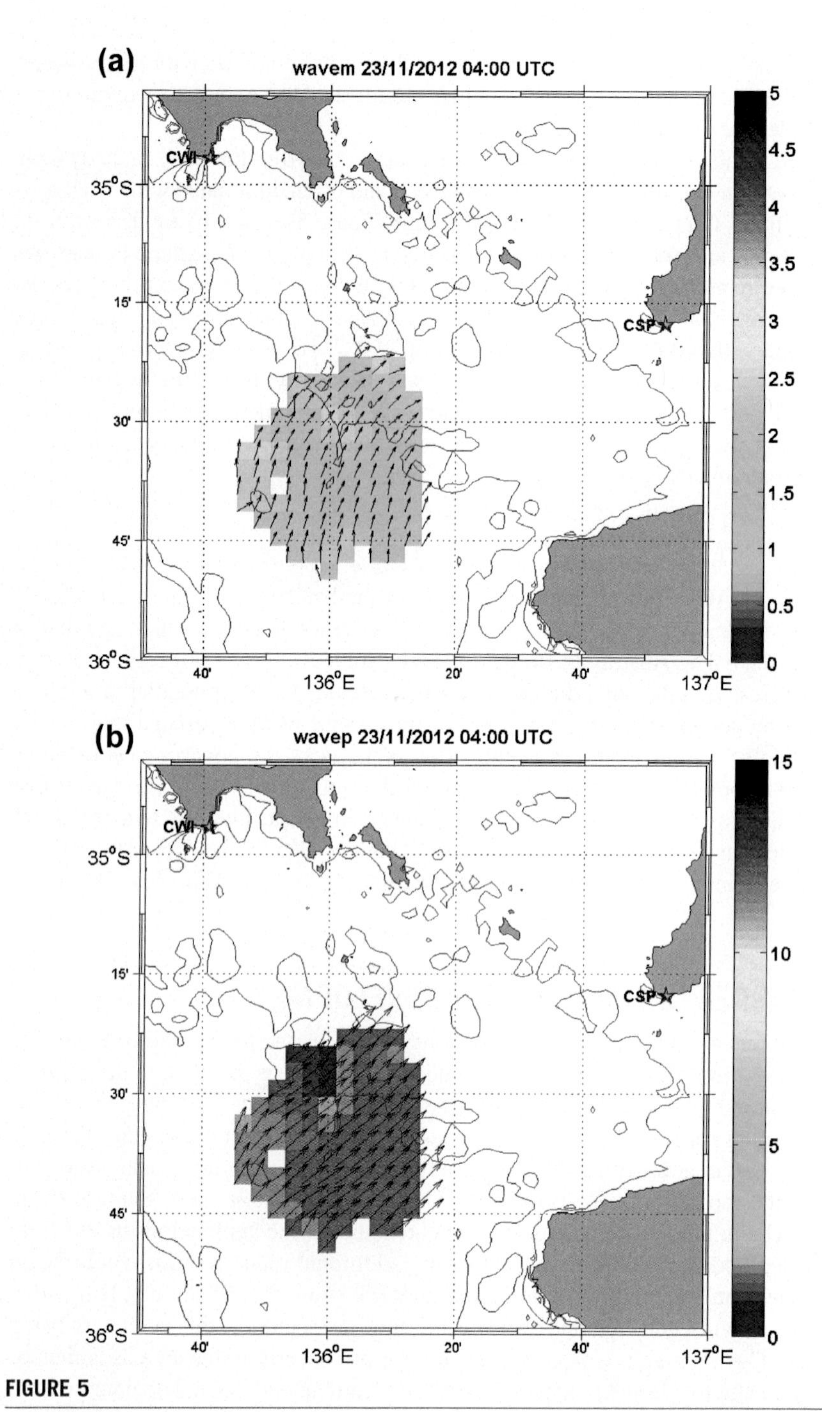

FIGURE 5

(a) Significant wave height (color-coded) and mean direction (arrows), (b) peak period (color-coded) and direction (arrows).

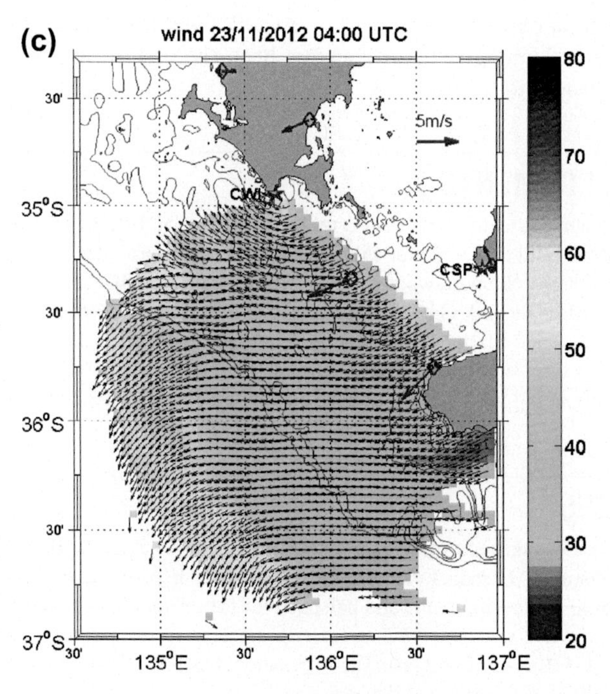

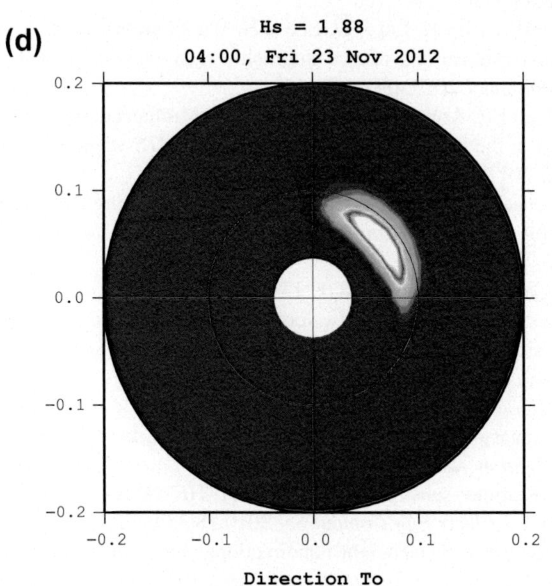

FIGURE 5 Continued

(c) shortwave or wind direction (arrows) and spread (color-coded) compared with BoM measurements (red arrows scaled as shown), (d) directional spectrum (scaled to the peak with colored levels at <10% (dark blue) to >90% (white) of the peak), measured with the SAG radar system.

Barrier Reef has already been referred to. ACORN and IMOS would benefit from a longer financial planning horizon than has been the case in recent years to bring some of these ideas to fruition.

ACKNOWLEDGMENTS

The ACORN technical team, Arnstein Prytz, Dan Atwater, and Paul Lethaby, together with Guillaume Galibert and colleagues at eMII ensure that IMOS provides the best data possible from the radar systems. Support, when needed, from the radar companies CODAR Ocean Sensors and Helzel MessTechnik GmbH are also gratefully acknowledged. ACORN is a facility of Australia's Integrated Marine Observing System (IMOS). IMOS is a national collaborative research infrastructure, supported by the Australian Government. It is led by the University of Tasmania in partnership with the Australian marine and climate science community.

REFERENCES

1. http://www.ioos.noaa.gov/hfradar/welcome.html; [last accessed 08.09.14].
2. Álvarez-Fanjul E, Losada I, Tintoré J, Menéndez J, Espino M, Parrilla G, et al. The ESEOO project: developments and perspectives for operational oceanography at Spain. In: *Proc. ISOPE-2007: the 17th international offshore ocean and polar engineering conference*, vol. 3. Lisbon (Portugal): The International Society of Offshore Ocean and Polar Engineers (ISOPE); 2007. p. 1708–15.
3. Fujii S, Heron ML, Kim K, Lai J-W, Lee S-H, Wu Xiangbai, et al. An overview of developments and applications of oceanographic radar networks in Asia and Oceania countries. *Ocean Sci J* 2013;**48**(1):69–97.
4. Heron ML, Wyatt LR, Atwater DP, Prytz A. The Australian coastal ocean radar network: lessons learned in the establishment phase. In: *IEEE/MTS oceans 2012*. Yeosu (Korea): IEEE Explore; 2012.
5. Roughan M, Schaeffer A, Suthers IM. *Coastal ocean observing systems. Sustained ocean observing along the coast of southeastern Australia: NSW-IMOS 2007–2014*. 1st ed. Elsevier; 2015. p. 76–98.
6. http://imos.org.au; [last accessed 08.09.14].
7. http://www.ioos.noaa.gov/globalhfr/welcome.html; [last accessed 08.09.14].
8. Paduan JD, Washburn L. High-frequency radar observations of ocean surface currents. *Annu Rev Mar Sci* 2013;**5**:115–36.
9. Fabrizio GA, Abramovich YI, Anderson SJ, Gray DA, Turley MDE. Adaptive cancellation of nonstationary interference in HF antenna arrays. In: *IEE Proceedings – radar, sonar and navigation*, vol. 145; 1998. p. 19–24.
10. Anderson S. Remote sensing applications of HF skywave radar: the Australian experience. *Turk J Electr Eng Comput Sci* 2010;**18**:339–72.
11. Barrick DE. The ocean waveheight nondirectional spectrum from inversion of the HF sea-echo Doppler spectrum. *Remote Sens Environ* 1977;**6**:201–27.
12. Lipa BJ. Derivation of directional ocean-wave spectra by inversion of second order radar echoes. *Radio Sci* 1977;**12**:425–34.

13. Lipa BJ, Barrick DE. Extraction of sea state from HF radar sea echo: remote sensing applications of HF skywave radar: the Australian experience mathematical theory and modelling. *Radio Sci* 1986;**21**:81–100.
14. Lipa BJ, Barrick DE. Methods for the extraction of long-period ocean wave parameters from narrow beam HF radar sea echo. *Radio Sci* 1980;**15**:843–53.
15. Maresca JW, Georges TM. Measuring rms wave height and the scalar ocean wave spectrum with HF skywave radar. *J Geophys Res C* 1980;**85**:2759–71.
16. Wyatt LR. Significant waveheight measurement with HF radar. *Int J Remote Sens* 1988; **9**:1087–95.
17. Wyatt LR. A relaxation method for integral inversion applied to HF radar measurement of the ocean wave directional spectrum. *Int J Remote Sens* 1990;**11**:1481–94.
18. Hisaki Y. Nonlinear inversion of the integral equation to estimate ocean wave spectra from HF radar. *Radio Sci* 1996;**31**:25–39.
19. Hashimoto N, Tokuda M. A Bayesian approach for estimating directional spectra with HF radar. *Coastal Eng J* 1999;**41**:137–49.
20. Green JJ, Wyatt LR. Row-action inversion of the Barrick–Weber equations. *J Atmos Oceanic Technol* 2006;**23**:501–10.
21. Gurgel K-W, Essen H-H, Schlick T. An empirical method to derive ocean waves from second-order Bragg scattering: prospects and limitations. *IEEE J Oceanic Eng* 2006; **31**:804–11.
22. Wyatt LR, Green JJ, Middleditch A. HF radar data quality requirements for wave measurement. *Coastal Eng* 2011;**58**:327–36.
23. Barrick D. 30 years of CMTC and CODAR. In: *Proc. IEEE OES ninth working Conf. Curr. Meas. Technol., Charleston, South Carolina, USA*; 2008.
24. Gurgel K-W, Antonischki G, Essen H-H, Schlick T. Wellen radar (WERA): a new ground-wave HF radar for ocean remote sensing. *Coastal Eng* 1999;**37**:219–34.
25. Merz CR, Weisberg RH, Liu Y. Evolution of the USF/CMS CODAR and WERA HF radar network. In: *Proc. of MTS/IEEE oceans'12*; 2012. p. 1–5. http://dx.doi.org/10.1109/OCEANS.2012.6404947.
26. Liu Y, Weisberg RH, Merz CR. Assessment of CODAR SeaSonde and WERA HF radars in mapping surface currents on the West Florida Shelf. *J Atmos Oceanic Technol* 2014; **31**:1363–82. http://dx.doi.org/10.1175/JTECH-D-13-00107.1.
27. Barrick DE, Lipa BJ. Evolution of bearing determination in HF current mapping radars. *Oceanography* 1997;**10**:72–5.
28. Wyatt LR, Green JJ, Middleditch A, Moorhead MD, Howarth J, Holt M, et al. Operational wave, current and wind measurements with the pisces HF radar. *IEEE J Oceanic Eng* 2006;**31**:819–34.
29. Wu X, Yang S, Cheng F, Wu S, Yang JI, Wen B, et al. Ocean surface currents detection at the eastern China sea by HF surface wave radar. *Chin J Geophys* 2003;**46**:340–6.
30. Dexter PE, Heron ML, Ward JF. Remote sensing of the sea-air interface using HF radars. *Aust Meteorol Mag* 1982;**30**:31–42.
31. Heron SF, Heron ML. A comparison of algorithms for extracting significant wave height from HF radar ocean backscatter spectra. *J Atmos Oceanic Technol* 1998;**15**:1157–63.
32. Mao Y, Heron ML. The influence of fetch on the response of surface currents to wind studied by HF ocean surface radar. *J Phys Oceanogr* 2008;**38**:1107–21.
33. http://imos.aodn.org.au/imos123/; [last accessed 08.09.14].

34. Wyatt LR, Atwater D, Mantovanelli A, Prytz A, Rehder S. The Australian coastal ocean radar network data availability and quality. In: *Radar symposium (IRS), 2013 14th international*, vol. 1. Dresden (German): IEEE; 2013. p. 405–10. 19–21.
35. Chapman RD, Shay LK, Graber HC, Edson JB, Karachintsev A. On the accuracy of HF radar surface current measurements: intercomparisons with ship-based sensors. *J Geophys Res* 1997;**102**:18737–48.
36. Lipa BJ, E Barrick D. Least-squares methods for the extraction of surface currents from CODAR crossed-loop data: application at ARSLOE. *IEEE J Oceanic Eng* 1983;**8**: 226–53.
37. Wyatt LR, Atwater D, Mantovanelli A, Prytz A, Rehder S. A comparison between SeaSonde and WERA HF radar current measurements. In: *Proc. of oceans 2013, Bergen*; 2013.
38. Taylor KE. Summarising multiple aspects of model performance in a single diagram. *J Geophys Res* 2001;**106**:7183–92.
39. Mantovanelli A, Heron ML, Heron SF, Steinberg CR. Relative dispersion of surface drifters in a barrier reef region. *J Geophys Res C* 2012;**117**. http://dx.doi.org/10.1029/2012JC008106.
40. Penton J, Pattiaratchi C. The effects of wind forcing on surface currents on the continental shelf surrounding Rrottnest Island. *Int J Environ Ecol Geol Min Eng* 2013;**7**:13–8.
41. Schaeffer A, Roughan M, Wood JE. Observed bottom boundary layer transport and uplift on the continental shelf adjacent to a western boundary current. *J Geophys Res Oceans* 2014;**119**. http://dx.doi.org/10.1002/2013JC009735.
42. Mao Y, Luick JL. Circulation in the southern Great Barrier Reef studied through an integration of multiple remote sensing and in situ measurements. *J Geophys Res Oceans* 2014;**119**. http://dx.doi.org/10.1002/2013JC009397.
43. Barrick DE. First order theory and analysis of MF/HF/VHF scatter from the sea. *IEEE Trans Antennas Propag* 1972;**20**:2–10.
44. Barrick DE, Weber BL. On the nonlinear theory for gravity waves on the ocean's surface. Part II: interpretation and applications. *J Phys Oceanogr* 1977;**7**:11–21.
45. www.seaviewsensing.com; [last accessed 08.09.14].
46. Wyatt LR. Shortwave direction and spreading measured with HF radar. *J Atmos Oceanic Technol* 2012;**29**:286–99.

CHAPTER

10

How High-Resolution Wave Observations and HF Radar–Derived Surface Currents are Critical to Decision-Making for Maritime Operations

Julie Thomas[1,*], Lisa Hazard[1], Robert E. Jensen[2], Mark Otero[1], Eric Terrill[1], Carolyn Keen[1], Jack Harlan[3], Todd Fake[4]

Scripps Institution of Oceanography, La Jolla, CA, USA[1]; US Army Corps of Engineers, Environmental Research and Development Center, Washington, DC, USA[2]; NOAA/U.S. Integrated Ocean Observing System, Silver Spring, MD, USA[3]; University of Connecticut, Storrs, CT, USA[4]

**Corresponding author: E-mail: jot@cdip.ucsd.edu, jothomas@ucsd.edu*

CHAPTER OUTLINE

1. INTRODUCTION

It is critical that the ocean observation infrastructure is in place to assure the safety of our maritime community, to promote the economic health of maritime transportation, and to protect our environment. Observations such as waves and surface currents are essential: Wave measurements are used for ship operations; wave forecasts are used for ship planning. Wave information is particularly essential to the maritime shipping community as vessel length and draft are increasing and under-keel clearance is becoming a critical issue. Surface currents are used operationally for search and rescue, oil spill tracking, port and harbor operations, and recreational boating. Both waves and surface currents are integrated into NOAA operational products such as the Physical Oceanographic Real Time System (PORTS®) and the National Weather Service (NWS) Advanced Weather Interactive Processing System (AWIPS). Weather forecasters use the surface current data to improve their situational awareness for daily marine forecasts. A key element is the ability to decipher whether the currents are wave-following or wave-opposing because the local wave height may be reduced or enhanced, respectively. Additional maritime uses include identifying spatial extents and trajectories of surface-following marine larvae populations to assist with marine protected area (MPA) evaluation, assisting in tracking coastal plumes, and discharges for water quality management.

Two programs, the Coastal Data Information Program (CDIP) and the Coastal Observing Research and Development Center (CORDC), based at the Scripps Institution of Oceanography (SIO) in La Jolla, California, participate in the near real-time data feed of operational information for maritime operations. CDIP, primarily funded by the U.S. Army Corps of Engineers (USACE), focuses on ocean wave measurements and is considered one of the USACE's contributions to the Integrated Ocean Observing System (IOOS). Many of the wave buoy stations that CDIP manages are cost shared with the California Department of Parks and Recreation, IOOS, U.S. Navy, U.S. Geological Survey, and industry partners. CORDC provides data acquisition and near real-time processing of the national High Frequency Radar network (HFRNet) that has been established to measure surface currents throughout the US coastal ocean waters. CORDC developed and has operated data management for integration, distribution, and visualization of HFRNet surface currents for over 10 years, as described in Terrill et al.[1] Central repository nodes have been deployed and are maintained on the East Coast (Rutgers University), West Coast (SIO), and at the National Data Buoy Center (NDBC) to demonstrate an end-to-end distributed data system that links multiple regions to a central repository of data. Operations and maintenance of HFRNet are funded primarily by IOOS and involve numerous participating organizations, including but not limited to Bodega Marine Laboratory; California Polytechnic State University; Humboldt State University; Naval Post Graduate School; Oregon State University; Rutgers University; San Francisco State University; Scripps Institution of Oceanography; Texas A&M; Autonomous University of Baja California (UABC); University of California, Santa Barbara; University of Hawaii; University of Maine, University of Miami; University of South Florida; and University of Southern Mississippi.

2. WAVE AND SURFACE CURRENT MEASUREMENT PROGRAM OVERVIEW AND SUPPORTING INFORMATION

2.1 WAVE PROGRAM OVERVIEW

CDIP was founded in 1975, and it has provided high-resolution, reliable wave measurements since inception (http://cdip.ucsd.edu). By 2014, CDIP disseminated the data for 60 coastal wave buoys, many of them at entries to ports and harbors, supporting near shore navigation. CDIP is providing publically available, long-term, sustained observations, focusing on the quality control of the real-time dissemination. Through a communication path, the data are transmitted every 30 min from the buoy to the Department of Defense gateway in Honolulu, back to Scripps for analysis and quality control, and then disseminated to NDBC where they are assigned an identification number and posted on their Website. The data are further transmitted to the National Weather Service (NWS) for marine broadcast. All historic data are available in network common data form (netCDF) or text formats, with the appropriate metadata. The data are all accessible via web services from the CDIP Website or from the federal archive at the National Oceanographic Data Center (NODC).

Over the 39 years that CDIP has been in existence, there have been many requests from federal, state, local, industry, and academia users for high-resolution wave data. In the beginning years, CDIP deployed mostly Paros pressure sensors that were either mounted at the end of piers or cabled to within 1500 m of shore. At certain cabled sites, these would be grouped by four sensors per location, in a 6-m square configuration, providing directional capability. In the 1980s, CDIP started deploying the Dutch-manufactured Datawell Waverider buoys. The early buoys were nondirectional and high-frequency transmission only. A shore-based data acquisition system was needed to capture the data. Placement of the buoys was restricted by high-frequency transmission line-of-sight, and shore station reliability was dependent on power and communications. Early in 1990, Datawell provided a directional buoy and, later, communications. Once there were no longer the restrictions of the shore station data acquisition system, which often failed during the peak of the storms causing data loss, the data collection per station increased. Today, particularly with the flash card on-board storage addition, the amount of xyz time series and spectral file collected is 98–100% per station. These reliability metrics, plus all raw, spectral, and parameter data, including numerous statistical products, are available from the CDIP site.

2.1.1 Directional Wave Measurements

Although waves are a fundamental oceanographic variable and measurement systems exist, the total number of in situ real-time wave observations is relatively small and very unevenly distributed. Even for the United States, the most well-sampled geographical region in the world, the total number of in situ real-time wave observations for the nation's approximate 27,358 km (17,000 mile)-long coastline is only about 200 nationwide, and only about one-half report some measure of wave direction (A National Operational Observation Wave Plan: NOAA and USACE).[2]

Global and regional wave observational requirements are dependent on the application and include the following: (1) assimilation into wave forecast models, (2) validation of wave forecast models, (3) ocean wave climate and its variability on seasonal to decadal time scales, and (4) role of waves in ocean–atmosphere coupling. Additionally, wave observations are also required for short-range forecasting and nowcasting, as well as for warning of extreme waves associated with extra-tropical and tropical storms. In situ wave observations are also needed for calibration/validation of satellite wave sensors. The key observations needed are the following: (1) significant wave height, (2) dominant wave direction, (3) wave period, (4) one-dimensional frequency spectral wave energy density, and (5) two-dimensional frequency–direction estimators (e.g., directional moments). Also important and desirable are observations of individual wave components (sea and swell).

Accuracy levels of directional wave measurements required by various user groups vary considerably. However, if the most stringent requirements are followed, then the requirements of the diverse user groups and applications will be met. The most stringent requirements come from the wave physics groups followed by operational forecast offices and, ultimately, numerical wave modelers. Tolerance requirements suggested by these groups are on the order of centimeters in amplitudes, tenths of seconds in periods (inverse frequencies), and directional estimates on the order of 2 to 5°. The latter includes the higher directional moments of spread, skewness, and kurtosis, which can only be successfully estimated from high-quality spectra over the entire frequency range of surface gravity waves. If this requirement is met for any directional wave measurement, ground truth would be established, and analyses of these data sets would no longer require a priori assumptions for the type of device, hull design, mooring system, and transfer functions used to approximate surface gravity waves. Increasing the number of directional wave measurements with high-resolution directional capabilities will directly lead to improvements of modeling technologies and will translate into better wave forecasts for the user community.

2.1.2 Wave Quality Control

The overriding objective of the global evaluation is to ensure consistent wave measurements to a level of accuracy that will serve the requirements of the broadest range of wave information users. Swail et al. (2010)[3] identified this concern:

> *Continuous testing and evaluation of operational and pre-operational measurement systems are an essential component of a global wave observing system, equal in importance to the deployment of new assets.*

Historically, interplatform tests have been pursued. However, with the global variations in hulls, sensors, and processing systems, evolution of sensors, changes in buoy designs, and new platform systems, a fresh look is required. Investigations have shown that large systematic differences are seen between different observing networks, including a systematic 10% difference in significant wave height measurements between the US and Canadian networks. In October 2008, a wave measurement technology workshop was held in New York (www.jcomm.info/wavebuoys) with broad participation from the scientific community, wave sensor manufacturers, and wave data users, following a March 2007 Wave Sensor Technologies Workshop (www.act-us.info). The overwhelming community consensus resulting from those workshops was the following:

- The success of a wave measurement network is dependent in large part on reliable and effective instrumentation (e.g., sensors and platforms).
- A thorough and comprehensive understanding of the performance of existing technologies under real-world conditions is currently lacking.
- An independent performance testing of wave instruments is required.

The workshop also confirmed the following basic principles for wave measurements, where participants agreed that the basic foundation for all technology evaluations is to build community consensus on a performance standard and protocol framework:

1. Multiple locations are required to appropriately evaluate the performance of wave measurement systems given the wide spectrum of wave regimes that are of interest.
2. An agreed upon wave reference standard (e.g., instrument of known performance characteristics, Datawell Directional Waverider MK Series) should be deployed next to existing wave measurement systems for extended periods (e.g., 6–12 months, including a storm season) to conduct "in-place" evaluations of wave measurement systems.

All integral wave properties—height, period and direction—are derived from the motion of the platform. This depends both on the capability of the sensor being used and the influence of the platform. Because of this complexity, the measurement of waves is dependent on the capabilities of the specific system being used and is, therefore, unlike the in situ measurement of other oceanographic variables, such as ocean temperature, which tend to be independent of the sensor used (excepting for measurement accuracy).

In order to serve the full range of users, a wave observation network should accurately resolve the details of the directional spectral wave field as well as provide the standard integrated parameters. It is strongly recommended that all directional wave measuring devices should reliably estimate the so-called "First-5"standard parameters, Alliance for Coastal Technologies.[4] Technically, First-5 refers to five defining variables at a particular wave frequency (or period). The first variable is the wave energy, which is related to the wave height, and the other four are the first four coefficients of the Fourier series that defines the directional distribution of that energy. At each frequency band, not only is the mean wave direction defined but also the spread (second moment), skewness (third moment), and kurtosis (the fourth moment). The skewness resolves how the directional distribution is concentrated (to the left or right of the mean), and the kurtosis defines the peakedness of the distribution. Obtaining these three additional parameters (spread, skewness, and kurtosis) for each frequency band yields an improved representation of the wave field. High-quality First-5 observations can also be used to resolve two component wave systems at the same frequency, if they are at least 60° apart. Although there are more than five Fourier coefficients, the First-5 variables provide the minimum level of accuracy required for a directional wave observing system, as they cover both the basic information, which is the significant wave height (Hs), peak wave period (Tp), and the mean wave direction at the peak wave period (θm), along with sufficient detail of the component wave systems to be used for the widest range of activities.

An international effort that addresses the quality of wave measurements is the WMO-IOC Joint Technical Commission for Oceanography and Marine Meteorology (JCOMM) wave sensor comparison project (World Meteorological Organization, Intergovernmental Oceanographic Commission, United Nations Environment Programme, and International Council for Science).[5] A description and published data are available at http://www.jcomm.info/pp-wet.

Another effort addressing the quality control tests and flagging required for wave data analysis is available at the IOOS-funded Quality Assurance of Real Time Ocean Data (QARTOD). A revised and updated Manual for the Real-Time Quality Control of In Situ Surface Wave Data was published in June 2013 (http://www.ioos.noaa.gov/qartod/waves/qartod_waves_manual.pdf). The manual documents a series of test procedures for data quality control. Its goal is to provide the foundation for high-quality marine observations to ensure credibility and value to the operators and data users. These procedures are written as a high-level narrative from which a computer programmer can develop code to execute specific data flags (data quality indicator) within an automated software program. Both the JCOMM/Wave Eval Tool and IOOS/QARTOD efforts should be commended for supporting critical work within the observational arena.

2.2 HF RADAR PROGRAM OVERVIEW

Local, state, regional, and federal discussions directed toward the IOOS emphasized a desire for the installation, development, and operation of a network of surface current mapping systems for use by a broad range of end users. This network not only brings together and synthesizes physical data, but it also builds relationships throughout the oceanography community. The HF-Radar Network started as a prototype in 2004 at SIO with data collected from local radars as well as systems installed by Rutgers University and the University of California at Santa Barbara. The network has since grown to become an operational system with contribution from 31 organizations collecting data from 130 radars. To date, over nine million radial files have been collected contributing to 10 TB of radial and near real-time total vector products. Central to the operational success of a large-scale network is an efficient data management, storage, access, and delivery system.

The surface current mapping network is characterized by a tiered structure that extends from the individual field installations of HF radar equipment (a site), a regional operations center that maintains multiple installations (an aggregator), and centralized locations that aggregate data from multiple regions (a node); see Terrill et al.[1] The receive antenna from a CODAR Ocean Sensor 5 MHz SeaSonde system is shown in Figure 1 as an example coastal installation.

Data aggregators are currently deployed at 10 institutions: Scripps Institution of Oceanography; Rutgers University; University of Southern Mississippi; Monterey Bay Aquarium Research Institute; University of Miami, Rosenstiel School of

FIGURE 1

Example Codar Ocean Sensors SeaSonde (5 MHz receive antenna) installed at Scripps Institution of Oceanography, La Jolla, California.

Marine and Atmospheric Science; University of Maine; Oregon State University, College of Oceanic and Atmospheric Sciences; University of California, Santa Barbara; San Francisco State University; and the California Polytechnic State University (Cal Poly). The architecture of the HF-Radar Network lends itself well to a distributed real-time network and serves as a model for networking sensors on a national level. This joint University–IOOS partnership is focused on defining and meeting the expressed needs for a national network of surface current mapping data systems (National Surface Current Plan).[6]

HF radar–derived surface current data are made available through online visualizations, an application programming interface (API) (which can be incorporated into any web view), and a web service—the Thematic Real-Time Environmental Distributed Data Services (THREDDS) Data Server (TDS) at both SIO http://hfrnet.ucsd.edu/thredds/catalog.html and NDBC http://sdf.ndbc.noaa.gov:8080/thredds/catalog.html.

2.2.1 HF Radar–Derived Surface Currents

High-frequency radar (HF radar) systems measure radio waves scattered off the surface of the ocean. HF radar has proven to be an effective method for coastal sea surface current mapping for a number of reasons. First, the targets required to produce coherent sea echo using HF are surface gravity waves, typically of 3 to 30 m wavelength, which are well understood and nearly always present in the open ocean. Second, vertically polarized HF waves can propagate over conductive seawater via coupling to the mean spherical sea surface producing measurement ranges beyond line-of-sight, out to approximately 200 km offshore. Third, Doppler sea echo at HF, under most wave conditions, has a well-defined signal from wave–current interactions that is easily distinguishable from wave–wave processes. This allows for robust extraction of current velocities. It is primarily these three features, along with the spatial resolutions, that are possible due to the frequency modulation discussed next, which place the HF band in a unique status for coastal current monitoring; see Barrick.[7–9]

HF radar land-based installations are located near the coastline and include one to an array of antennas depending on the type and frequency of the system. A radio signal is broadcast across the ocean's surface, and the receive antenna(s) listen for the signal scattered by the ocean's waves. Any deviation in Doppler shift from the theoretical wave speed is attributed to the surface current velocity. Because the radar measures these velocities in directions radial to the receive antenna, the surface ocean current measurements are called radial velocities. Data from neighboring antennas are aggregated through HFRNet, processed and displayed to the user as surface currents maps—showing velocity (speed and direction) in near real-time.

2.2.2 HF Radar–Derived Surface Current Quality Control

Once radial data arrives at an HFRNet Node, it is available for integration with other radial velocity measurements from neighboring sites through surface current mapping. HFRNet's primary operational product is the generation of near real-time velocities (RTV) that are ocean surface currents mapped from radial component measurements. There are three general steps in producing RTVs:

1. Radial Velocity QC
2. Surface current mapping
3. Resolved surface current QC

2.2.2.1 Radial Velocity QC

Questionable radial velocity measurements are removed prior to mapping surface currents in order to reduce error. Two criteria must be met in order for a radial measurement to be used in deriving RTVs. The radial velocity must be (1) below the maximum radial magnitude threshold and (2) located over water. The maximum radial magnitude threshold represents the maximum reasonable radial magnitude for the given domain.

2.2.2.2 Surface Current Mapping

Surface currents are mapped onto regional grids based on equidistant cylindrical projections with resolutions of 500 m, 1, 2, and 6 km. Regional grids have been developed for the West Coast and the Gulf/East Coast of the Continental United States, the Gulf of Alaska, the North Slope of Alaska, the Hawaiian Islands, Puerto Rico, and the US Virgin Islands. In order to reduce the solution space, grid points over land and near the coast (within 0.5 km) are removed.

Surface currents are derived using a least squares fit to radial velocities within a predefined distance from each grid point, as described by Lipa and Barrick.[10] Radials must come from at least two different sites, and there must be at least three radials available in order to produce a velocity estimate for a given grid point. The search radius around each grid point is approximately 30% greater than the grid resolution.

The contribution of each site's radials to solutions for a given resolution is currently determined solely by the site's operating frequency. Sites operating near 25 MHz and higher contribute to solutions at 1 km resolution, 12 MHz and higher sites contribute to solutions at 2 km resolution, and all sites contribute to solutions at 6 km resolution.

2.2.2.3 Resolved Surface Current QC

Surface currents derived from integrated radial velocity measurements must not exceed the following thresholds:

1. Maximum total speed threshold
2. Maximum geometric dilution of precision (GDOP) threshold

The maximum total speed thresholds are similar to those used for radial velocities. The maximum total speed threshold is 1 m/s for the West Coast of the United States and 3 m/s for the East/Gulf Coast domain. GDOP is a scalar representing the contribution of the radial (bearing) geometry to uncertainty in velocity at a given grid point. Higher GDOP values indicate larger covariances associated with the least squares fit used in obtaining the solution. The GDOP maximum threshold is 10 for all domains. However, near real-time applications apply a more conservative maximum threshold of 1.25 to RTV solutions. There are ongoing research and development efforts for furthering HF radar–derived surface current quality control.

3. CASE STUDIES

3.1 CASE STUDIES OF EXAMPLES WHERE HIGH-RESOLUTION WAVE OBSERVATIONS ARE ESSENTIAL

3.1.1 Entrance to Ports and Harbors

Maritime trade has long defined our region's identity, culture, and economy. International trade in the Pacific depends on direct access to world markets.

Wave measurements are used for the ship operations, whereas wave forecasts are used for ship planning. This information is particularly essential as vessel length and draft are increasing and under-keel clearance (UKC) is becoming a critical issue. In support of safe and efficient operations, CDIP and SCCOOS have developed customized products. Two of the cases are described below.

3.1.1.1 The Los Angeles and Long Beach Harbors

Combined, the Ports of Long Beach and Los Angeles handle more than 40 percent of the nation's imports. Just within the Port of Long Beach, more than $150 billion worth of goods move through each year. The Port of Long Beach serves more than 140 shipping lines with connections to 217 seaports worldwide. The issue is how to keep the ports commercially viable with the increasing draft on the trans-Pacific and Panamax cargo vessels.

The Port's Safety Committee Plan for 2013 defines UKC as "the minimum clearance available between the deepest point on the vessel and the bottom in still water." UKC is not only a concern in the harbor, but also for the approaches to the port complex, specifically before the federal channels and areas off to the side of the channels when escape routes are used, if required. In the Port of Long Beach, Jacobsen Pilots have noted that because of their design, ultra-large crude carriers are being impacted by 12 to 14 period energetic swells. In a 365-m (1197-ft) vessel, a 12–14 s swell approaching from the stern causes a 1° pitch that results in an increase of draft by 3.2 m (10.5 ft). During large swell conditions, knowing when to change course before entering the Federal Channel with a least depth that exceeds the UKC is challenging.

Currently the channel at Long Beach is dredged to 19.812 m (65 ft). The oil on the supertankers is lightered offshore, and then transferred into port on the smaller vessels. However, the UKC on the vessels that do transit directly into port is monitored closely. One of the products that assists is the information that SCCOOS/CDIP transmit to Jacobsen Pilots, alerting of certain exceedance wave conditions. If the waves are from the west, the significant wave height (Hs) is greater

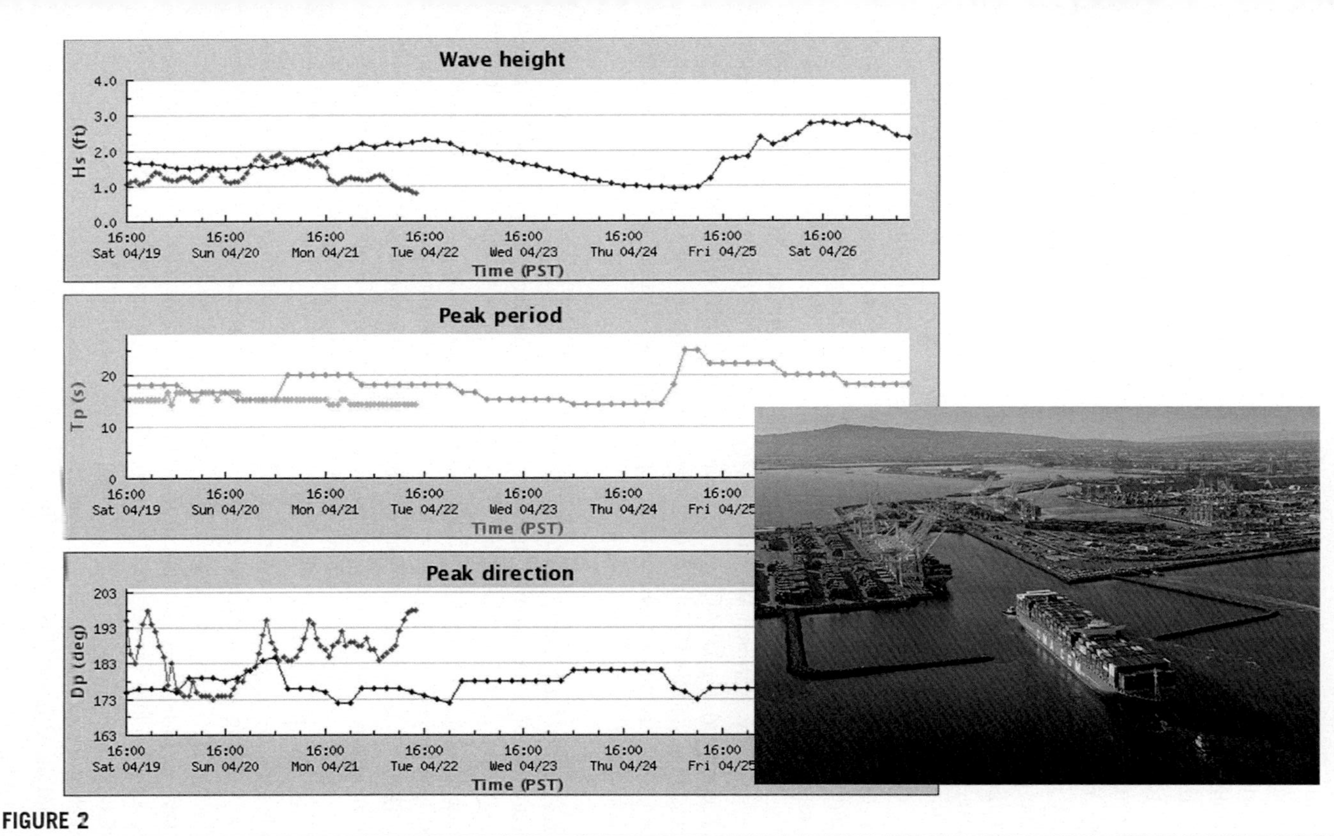

FIGURE 2

Port of Long Beach. Comparison of San Pedro Buoy real-time observations versus WW3 8-day model predictions, April 2014.

than 1 m, and the wave period is greater than 12 s, a bulletin is sent to the pilots and to the Port of Long Beach cruise ships.

In order to keep our ports commercially viable, we must have up-to-date hydrographic surveys, knowledge of the ship specifications entering the port complex, and high-resolution wave observations and models that indicate real-time and forecast conditions.

Figure 2 represents the comparison of the 8-day global Pacific model WaveWatch3 forecast and the observations from the CDIP San Pedro Buoy that is moored close to the entrance channel. The user can then denote how accurate the WaveWatch3 model is portraying the wave conditions and determine if the model is over- or under-predicting. This is critical knowledge for planning transits to and from the ports.

As there are additional requests to accommodate the large supertankers, a comprehensive wave model is needed to ensure safe navigation during transit of large ships impacted by sizeable swell conditions. Since high resolution and accuracy of the model are critical, the challenge is how to integrate the best available sources of wave observations and wave models. The observations are key for boundary conditions and validation of the models. The models provide the needed spatial and temporal coverage for the nowcast and forecast conditions. In addition to the Port, the National Oceanic and Atmospheric Administration (NOAA–IOOS, Office of Coast Survey, National Weather Service, and the National Centers for Environment Prediction), the USACE, U.S. Coast Guard, California's Oil Spill Prevention and Response, and SIO/CDIP have all been discussing the best options for assuring that commerce remains viable and access to the Port continues in an environmentally safe manner.

3.1.1.2 Mouth of the Columbia River

The federal navigation channel in the Lower Columbia River is 177 km (110 miles) long and now 13.1 m (43 ft) deep. The channel supports over 40 million tons of cargo each year, valued at $16 billion. Over 40,000 local jobs are dependent on this trade. The UKC is of concern to commercial maritime traffic on the Columbia River also. This area is the number one bulk exporter in the United States, including wheat and corn. The Pacific Northwest exports around 10 million metric tons of wheat annually around the globe, with primary regular export customers being Japan, the Philippines, Korea, and Taiwan.

Energetic wave conditions are one of the greatest challenges for this area. Since fall 2009, through partnerships with the USACE in Portland and the Columbia River Bar Pilots, CDIP has maintained three buoys at this area. Two of the buoys are deployed, and one buoy is configured and housed in a nearby warehouse, ready to deploy as a backup as needed. The two offshore buoys are deployed at the south entrance to the Columbia River (Clatsop Spit), and the other is 43 km (27 miles) due west of the Columbia River. Thus, the local seas are captured by the near shore buoy, and the offshore Pacific swells are captured by the offshore Astoria Canyon buoy. The largest wave measured by Astoria Canyon was 15.95 m, 12 s period in

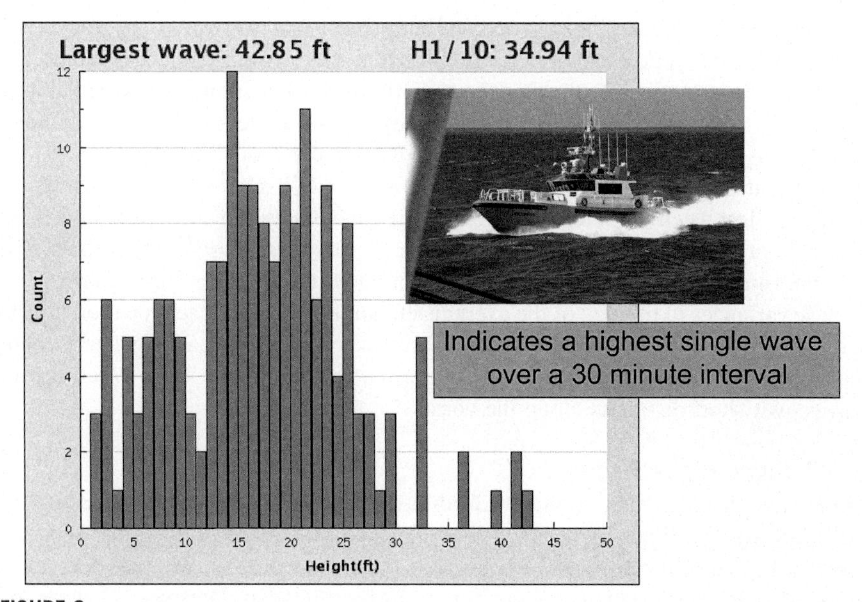

FIGURE 3

Histogram of wave height distribution over 30-min time spans from the CDIP buoy at the mouth of the Columbia River, Pacific Northwest, United States.

March 2012. The largest wave at Clatsop Spit occurred in October 2010, and it measured 14.3 m at 13 s period. Figure 3 depicts one of the products developed at the request of the Bar Pilots. The product shows the distribution of the spectral wave over a 30-min period. This information is useful during pilot transfers as it gives the pilots an indication of how many "peaks" will occur during the 30-min period. The pilots can then have an idea about the timing when transferring to the commercial vessel.

Since 2005, the Bar Pilots have contracted with the Australian company OMC International for developing a UKC system. Wave energy and directional spectral data are ingested real time by OMC for the analysis of the system. At this point, the system is still experimental and undergoing validation.

The issue of UKC necessitates a diverse field of experts in hydrographic surveying, vessel pilotage, port administration, tides and currents, coastal zone management, geodesy, marine transportation, and academia. CDIP has a key role of providing high-resolution wave data in order to solve this difficult issue of UKC in an ever changing environment. The wave buoys are proving essential to this process.

3.1.2 Tracking and Validation of Hurricane Models

Providing real-time observations as validation for wave forecast models is essential. During Hurricane Sandy, October 2012, CDIP had 14 buoys on the East Coast, spanning from the Caribbean to New Hampshire, that were used by the NWS Offices and emergency planners to monitor real-time conditions and validate the wave forecasts, noting if the model predictions were accurate. These 14 buoys had 100% reliability during Hurricane Sandy. Wave models versus observations continue to be accessible on the CDIP and Regional IOOS sites.

A plot that shows the wave height observed by CDIP's buoys on the East Coast during the last week of October 2012, as Hurricane Sandy moved south to north, is available at http://cdip.ucsd.edu/themes/topics/hurricanes?d2=p12. The chart also lists the single largest wave recorded by each buoy during that same period.

The variances in the size of the average and largest waves at each location underscore the importance of a robust wave observation network in coastal waters, where the impacts of coastal land and bathymetric features can cause large variations in waves over short distances along the coast.

3.1.3 Commercial Fishing

Commercial fishing ranks among the deadliest professions in America, with a fatality rate typically five times higher than that of police officers or firefighters. Between 2000 and 2010, more than 545 commercial fishermen died on the job, according to U.S. Coast Guard statistics (Commercial Fishing Incident Database, Centers for Disease Control and Prevention). In July 2005, in the offshore waters of Humboldt Bay, California, one of the long-term, well-equipped, 14-m (46-ft) fishing vessels disappeared. This vessel was piloted by one of the well-respected veteran captains, along with two crew members. The offshore conditions were a sloppy 1-m seas, well within the realm of the fishermen's capability. As seen by the data from the CDIP Mendocino Buoy, the waves were approaching from both the south and the north, so wave convergence was occurring. Operating with full loads can be deadly, particularly in convergence zones with bidirectional waves approaching. The fishing industry depends upon accurate wave data and models for their operations and planning.

At the August 2014 California Ocean Observing Marine Symposium, held at California Polytechnic State University, one of the long-time commercial fisherman, Peter Hansen, stated that "the integration of data he can now receive digitally greatly reduces his carbon footprint and increases catch efficiency. The important thing is staying alive—many people have been lost due to inability to access data. That has changed drastically—CDIP and NOAA buoys and weather forecasting data are aggregated and transmitted via satellites are literally a life saver. There has been a major decrease in deaths." Obviously, data providers with technical expertise should take very seriously the role of providing as high-resolution observations possible.

3.1.4 Military Support

There are two projects where high-resolution data are critical to military operations on a daily basis. The following three projects demonstrate the value of high-resolution wave data to models.

3.1.4.1 Kings Bay, Georgia

Since July 1995, funded by the USACE, CDIP has provided real-time wave data and NOAA tides to the Naval Submarine Base Kings Bay. These data are one component of the Navy's Carderock-developed Environmental Monitoring and Operator Guidance System (EMOGS), a vessel guidance system provided for the submarines entering and departing the base. The Ohio-class submarines have a draft of 35 ft. St Mary's entrance channel to the base along the Georgia/Florida border has a dredged channel approximately 46 to 52 ft in depth. The EMOGS operationally assists deep draft Navy ships transiting shallow channels. It predicts the UKC in advance of the ship transit. As with the commercial traffic mentioned previously, it is not only the draft of the vessel but the wave height and tides that affect the clearance. A customized report from CDIP is downloaded every morning at the base, which has the necessary information for inclusion in the EMOGS system. This report indicates whether or not the necessary UKC for vessel passage is available.

3.1.4.2 Pt Mugu, California

As requested by NAVAIR at Pt Mugu Naval Base, California, both SCCOOS and CDIP are providing real-time data in support of Navy operational testing and training. Through SCCOOS, a data portal was developed (http://www.sccoos.org/data/harbors/navair/fullscreen.php) to provide detailed information for waves and surface currents in their operational region. A user can click any point on the map for the updated currents and click on the green dots for the wave spectra. The yellow dots represent the CDIP wave buoys that are deployed in the area. The Navy is using these data for a variety of reasons where the coastal wind and wave field information is crucial.

3.2 CASE STUDIES OF EXAMPLES WHERE SURFACE CURRENT OBSERVATIONS ARE ESSENTIAL

3.2.1 U.S. Coast Guard Search and Rescue Operations

Beginning in 2000, the U.S. Coast Guard (USCG) Research and Development Center began a multiyear investigation into the utility of near real-time HF radar–derived surface current measurements for search and rescue (SAR). This assessment showed improved performance using radar-derived currents when compared against available NOAA tidal current predictions. Additionally, a key element using the HF radar currents was the development of the short-term predictive system (STPS), a forecasting model that uses statistical information for surface current prediction. Following these evaluation studies, available in situ surface current velocities were used to evaluate and define appropriate parameters for integration in the USCG Search and Rescue Optimal Planning System (SAROPS), as the inclusion of HF radar currents significantly reduced the search area for USCG search and operators, as detailed in Ullman et al.[11] and Roarty et al.[12]

The University of Connecticut developed the STPS and now operates the model. STPS runs automatically but has human technical support and troubleshooting on

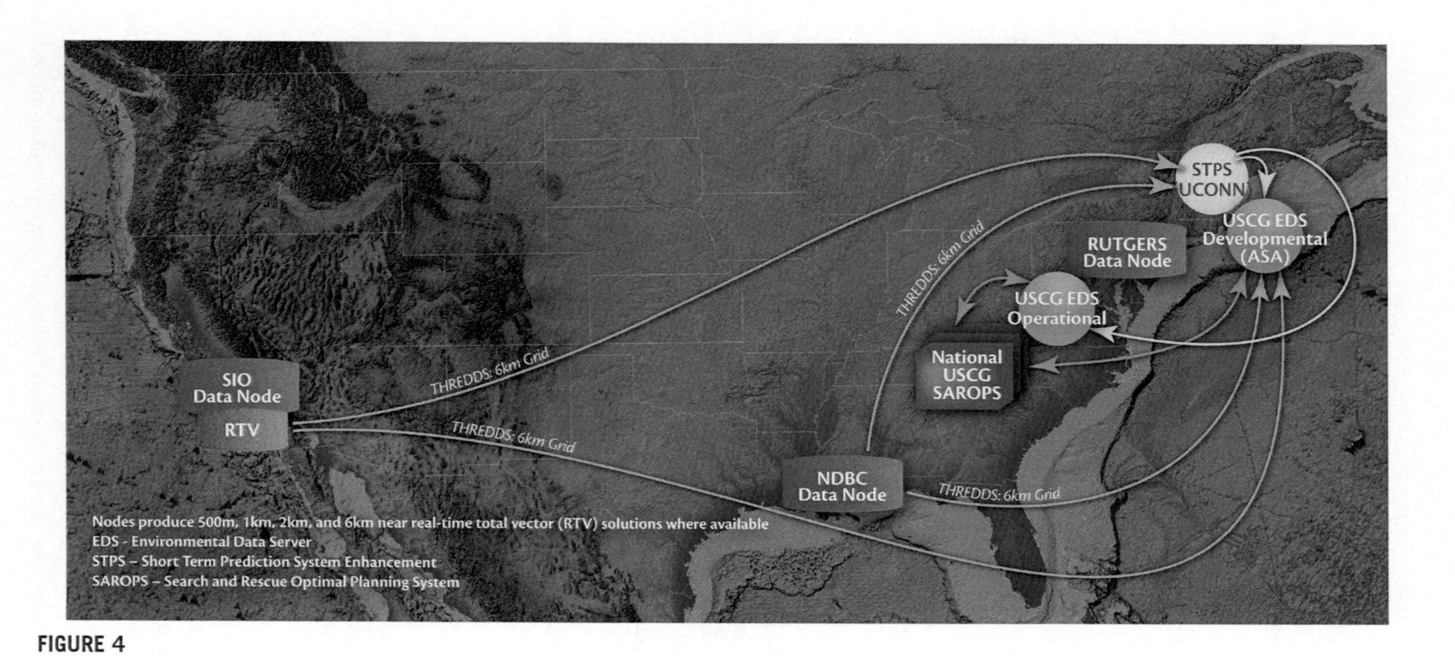

FIGURE 4

HFRNet data flow for ingestion into the search and rescue optimal planning system (SAROPS) tool.

call. Every hour, it creates a 24-h forecast of current field evolution that is consistent with the most recent data and the statistics of the observed current variability. The STPS forecast is created by exploiting the periodicity of the tides and the fact that weather systems move slowly. The tidal part of the current can be predicted using traditional methods, and then, the less-regular, weather-forced part can be isolated and extrapolated. Adding these parts together then results in a complete forecast of the currents. The approach is applicable to any coastal area regardless of coastal geometry and bathymetry, and it does not make any assumptions about the underlying circulation dynamics. The primary advantage of this approach is that the algorithm can be readily applied with limited effort and expense in any area with an operational surface current data set.

Current velocities from HFRNet and the STPS forecasts are now included in the USCG SAROPS as seen in Figure 4. Data is made available in an easily digestible format through web services that were previously mentioned in the HF Radar program overview.

3.2.2 Oil Spill Response

Once a spill has occurred, tracking its movement, especially in dark or foggy conditions, is the first challenge in response and mitigation efforts. HF radar has the ability to track ocean surface currents during the night, in fog, or when conditions do not allow for direct observation of the spill. When the Cosco Busan collided with the Bay Bridge in San Francisco in 2007, spilling over 53,000 gallons of fuel oil, dense fog hampered containment efforts. Once news of the spill was released, HF radar–derived surface currents were able to show the path the spill had taken and provide a track of where the oil would travel next. As HF radar capabilities are further integrated into spill response, immediate response will be unimpeded by lack of visual data. Additionally, predictive tools will increase response time and allow for better protection of critical habitats and resources both on land and at sea. The data are used in the NOAA Office of Response and Restoration (OR & R) General NOAA Operational Modeling Environment (GNOME) for spill trajectories.

Within California, investigators participated in several exercises including Safe Seas 2006, a NOAA-led multiagency simulated spill off the San Francisco coast, and the National Preparedness for Response Exercise Program simulation involving over 200 participants off the coast of San Diego and Santa Barbara in 2008 and 2009, respectively. These simulations allow the many state and federal regulatory agencies involved in oil spill response to practice working together in the event of an actual spill and to demonstrate the value of real-time surface current maps and forecasts in response management and decision-making. Because repeated demonstrations have highlighted the accuracy and importance of surface current data to oil spill response, HF radar data are now being integrated with NOAA spill response models, and they will enable spill responders to predict the pathway of a spill, allowing pinpoint targeting for containment and cleanup. David Panzer, an oceanographer with the Minerals Management Service (now Bureau of Ocean Energy Management (BOEM)), writes that surface currents "greatly enhance our ability to calculate oil spill trajectories" (David Panzer, Minerals Management Service Letter of Support, October 2007). Immediately following the Deepwater Horizon incident in April 2010, personnel from the University of Southern Mississippi were able to re-deploy three HF radars along the Gulf of Mexico and staff at the Coastal Observing Research and Development Center (CORDC), Scripps Institution of Oceanography were able to bring those online through the national HFRNet, providing the first surface current data near the site of the spill. Both SAROPS and oil spill response applications are referenced in Harlan et al. (2010).[13]

3.2.3 Coastal Recreation

HF radar technology is also useful for coastal recreational activities. A prime example of such an activity is sailing, and HF radar–derived surface currents can assist with course planning. Ray Huff and John Ugoretz, co-captains of the 34-ft chartered yacht, the Getaway, plotted their course using wind forecasts and surface current web-based data products in the sixty-first Newport to Ensenada Yacht Race on 25 to 26 April, 2008, and they won. Co-captain John Ugoretz writes, "I used both wind forecast and ocean current information to help plan my route for the race. I reviewed the information for about one week prior to the race, right up until the morning of the start. Perhaps the most important factor in our strategy was a decision on where to be at night, when the winds are the lightest. Using the plan derived from SCCOOS information, we were able to average 2.5 knots of boat speed all night long. This may seem slow, but I've had years where we drifted backwards at night with no wind and a counter current. This year the boat never stopped, with steady progress towards Ensenada throughout the night."

Because the Getaway was chartered, the race team had little time to prepare before the race. Participants are not allowed to access outside information during the race, so race routes are planned using forecasting tools. Ugoretz, the official team navigator and tactician, used SCCOOS-provided wind forecasts and surface current maps to plot the team's winning route. Ugoretz commented that his proposed route nearly matched that of a competitor's that had been developed using complex sailing models. The Getaway finished 1.5 h ahead of the number two boat in their category on corrected time.

The sailors used both the SCCOOS 48-h wind forecasts and the near real-time surface current visualizations, provided as interactive online web visualizations. A 25-h average product of the surface currents depicts flow conditions that have tidal influences removed. This product is useful for sailing and recreation as it gives the user an indication of dominant flow patterns that may affect their operations. In addition to helping sailors plot their course, mariners, scuba divers, and recreational boaters can also use web-based data products to check on conditions in coastal areas.

Scuba diving is another example of a coastal recreational activity that benefits from accessible ocean observations. Many factors play into a safe open ocean dive that could be compromised if the winds are too high, the waves are too large, the water is too cold, or the surface currents are too strong. Recently, the president of the Sole Searchers Dive Club in Pasadena, California, reached out to SCCOOS to brief the club on what observations are available to help plan diving operations. The Sole Searchers consider safety to be paramount and near real-time ocean observations provide the backdrop for safe diving excursions. Having access to wind, wave, water temperature, and surface currents are critical for making "go/no go" decisions on whether or not to dive that day. Participants were not only interested in surface currents for diver safety, but they also appreciated the integration into search and rescue operations (should there be a lost diver) and oil spill response and recovery as water quality is a concern. Knowledge of and access to ocean observations, including both waves and surface currents, allows users to make safer decisions on the water.

4. SUMMARY

As noted earlier, there are a large number of users who require both wave and surface current information covering a broad range of complexity. Simple measures of wave height and period, to separations of the sea and swell components, to full two-dimensional spectral wave measurements for vessel response and shoreline erosion studies have value depending on the operational scenario. Similarly, general surface current flow to integration into products such as the short-term prediction system (STPS) and oil spill forecasts to future use in maritime domain awareness are useful to a number of different organizations and agencies. Wave and surface current measurements help enable safe navigation, allow for economic growth, and can provide insight into environmental conditions in our climate changing world. It is important to establish a consistent, common framework for these measurements that is scalable and accessible. The success and continued expansion of these networks for the distribution of waves and coastal surface currents will be possible through the dedication and partnerships of multiple institutions, federal and nonfederal agencies, local and state governments, and private companies. It is important to combine these efforts, and focus on interoperable products that are useful and freely available to everyone. The U.S. IOOS is dedicated to maintaining the US network and is reaching out to global partners as these data are used in universal applications for the health and safety of human and marine populations.

REFERENCES

1. Terrill E, Otero M, Hazard L, Conlee D, Harlan J, Kohut J, et al. *Data management and real-time distribution in the HF-radar national network*. OCEANS; 2006.
2. *A national operational wave observation plan, NOAA and USACE*. 2009. http://www.ioos.noaa.gov/library/wave_plan_final_03122009.pdf.
3. Swail VR, Komen G, Ryabinin V, Holt M, Taylor PK, Bidlot J. Waves in the global ocean observation system. In: Koblinsky CJ, Smith NR, editors. *Observing the oceans*

in the 21st Century: a strategy for global ocean observations. Melbourne, Australia: Bureau of Meteorology; 2010. p. 149–76.

4. Alliance for Coastal Technologies. Observations of directional waves, wave sensor technologies. In: *Dr W.A. O'Reilly Keynote Presentation*. St Petersburg, FL: Alliance for Coastal Technologies (ACT); 2007. March, http://aquaticcommons.org/3108/1/ACT_WR07-03_Wave_Sensor.pdf.
5. World Meteorological Organization, Intergovernmental Oceanographic Commission, United Nations Environment Programme, and International Council for Science. *Implementation plan for the global observing system for climate in support of the UNFCCC*. 2004. WMO/TD No. 1219, http://www.wmo.int/pages/prog/gcos/Publications/gcos-92_GIP_ES.pdf.
6. *A Plan to meet the nation's needs for surface current mapping*. 2013. http://www.ioos.noaa.gov/library/national_surface_current_plan.pdf.
7. Barrick DE. A review of scattering from surfaces with different roughness scales. *Radio Sci* 1968;**3**:865–8.
8. Barrick DE. First-order theory and analysis of MF/HF/VHF scatter from the sea. *IEEE Trans Antennas Propag* 1972;**AP-20**(1):2–10. http://dx.doi.org/10.1109/TAP.1972.1140123.
9. Barrick DE. *FM/CW radar signals and digital processing*. 1973. NOAA Technical Report ERL 283-WPL 26, July 1973.
10. Lipa B, Barrick DE. Least-squares methods for the extraction of surface currents from CODAR crossed-loop data: application at Arsloe. *IEEE J Ocean Eng* October 1983; **OE-8**(4).
11. Ullman D, O'Donnell J, Edwards C, Fake T, Morschauser D, Sprague M, et al. *Use of coastal ocean dynamics application radar (CODAR) technology in U.S. Coast guard search and rescue planning*. 2003. US Coast Guard Report No. CG-D-09–03.
12. Roarty H, Glenn S, Kohut J, Gong D, Handel E, Rivera E, et al. Operation and application of a regional high-frequency radar network in the Mid-Atlantic Bight. *Mar Technol Soc J* November/December 2010;**44**(6):133–45.
13. Harlan J, Terrill E, Hazard L, Keen C, Barrick D, Whelan C, et al. The integrated ocean observing system high-frequency radar network: status and local, regional, and national applications. *Mar Technol Soc J* November/December 2010;**44**(6):122–32.

CHAPTER

11

Observing Frontal Instabilities of the Florida Current Using High Frequency Radar

Matthew R. Archer*, Lynn K. Shay, Benjamin Jaimes, Jorge Martinez-Pedraja

Department of Ocean Sciences, Rosenstiel School of Marine and Atmospheric Science, University of Miami, FL, USA

**Corresponding author: E-mail: matthew.robert.archer@gmail.com*

CHAPTER OUTLINE

1. INTRODUCTION

The Florida Current (FC) is a western boundary current regime characterized by large velocities, strong horizontal shears, and relative vorticities that can approach $10f$, where f is the local Coriolis parameter.[1] This current is constrained in close proximity to the coastline by the narrow Straits of Florida (SOF) channel, which guides it from the Gulf of Mexico to the South Atlantic Bight. It is one of the most studied ocean currents in the world because of its importance in the meridional transport of heat from the tropics to the poles, its proximity to the United States coastline, and the experimental convenience of the channel-like bathymetry within the SOF. The mean structure, volume transport, and low-frequency variability of the FC have been examined in detail by numerous studies (e.g., Refs. 2–7). Instabilities along the frontal region of the FC have been observed and modeled.[1,8–14] However, there are still many gaps in our understanding. Questions are now directed at the dynamics of the smaller scale shear-zone instabilities, especially on the eastern front of the FC, which has not yet been addressed because of lack of observations.

Research aimed at understanding the coastal ocean circulation in the SOF is extremely valuable to policymakers who must balance societal interests and environmental concerns. Improving our capability to measure and predict the ocean currents will benefit a broad spectrum of societally relevant applications: search and rescue (SAR), maritime security, navigation, fisheries management, commercial shipping interests, and oil spill mitigation. However, near coastal ocean processes are difficult to study because of the variability in the forcing mechanisms that occur over a wide range of temporal and spatial scales. In the coastal SOF, the primary forcing is from the meandering FC and the wind, but the ocean response is complicated by the geometry of the coastline and underlying bathymetry. The effect of stratification (e.g., buoyant riverine output) and tidal currents also influence the circulation pattern inshore.

Previous research in the FC has been largely lead by in situ measurements, such as hydrographic sections or moorings, which have provided reasonable coverage in space or time, but not both. Traditional in situ instruments and satellites do not as easily observe flow features with smaller horizontal scales (the submesoscale) that evolve more quickly in time (on the order of hours). HF radar can provide two-dimensional maps of coastal ocean surface currents in near real time, with the ability to sample at intervals as little as a few minutes and a spatial resolution of less than a kilometer. This allows us to resolve small-scale flow features at the surface within the SOF.

This chapter documents new observations of frontal instabilities of the FC using HF radar. Two case studies demonstrate the power of HF radar for coastal ocean observing. In the first case, a study of a submesoscale frontal eddy in the cyclonic shear-zone of the FC is presented. The emphasis is on the ability of HF radar to provide new insight into spatial variability of these features, using the two-dimensional velocity field and its derivatives to investigate their kinematics. Understanding the flow field provides insight into particle dispersion, which if known could help in SAR operations and pollution mitigation. These eddies also contribute to cross-shelf exchange of mass and nutrients, which has implications for biological productivity along the Florida Keys and South Florida coastlines.[15]

In the second case study, a near-inertial signal in the eastern anticyclonic flank of the FC is presented for the first time. This is an example of HF radar's unique ability to measure transient events that are difficult to capture with ship and in situ point measurements, or to resolve using satellite imagery. These features could have implications for mixing and cross-shelf exchange on the eastern side of the channel.

A background review of the scales of variability of the FC, together with a discussion of open questions, is provided in Section 2. Instrumentation is described in Section 3. The two case studies are presented in Sections 4 and 5 of cyclonic and anticyclonic shear-zone instability, respectively. Finally, the results are summarized and the direction for future work is discussed.

2. BACKGROUND: THE FLORIDA CURRENT

2.1 INSTABILITY OF THE FLORIDA CURRENT

The FC transports warmer water of equatorial origin northward through cooler local waters. This results in two frontal regions on either side of the jet core, which are narrow zones of enhanced horizontal gradients of velocity and temperature. These high-shear frontal regions are subject to instabilities that can result in mixing of the water masses. Such instabilities are an important link between littoral and offshore waters as they can act to redistribute heat, salt, momentum, and nutrients. For example, cyclonic eddies have been shown to play a significant role in larval recruitment along the Florida Keys reef tract, where upwelling in the eddy core produces favorable conditions during the early stages of development.[15,16]

A summary of the time and space scales of the instabilities in the SOF, based on observations published in the literature, is presented in Figure 1. It reveals a broad spectrum of scales, ranging from slower, larger meanders and Tortugas eddies to smaller, rapidly evolving features such as submesoscale vortices and a super-tidal oscillation. These features are not independent but are strongly influenced by one another and the dynamics of the FC.

2.2 MEANDERING

Meanders are characterized by a lateral wave-like movement of the FC axis. These waves are asymmetric, with the crests (shoreward displacement) and troughs

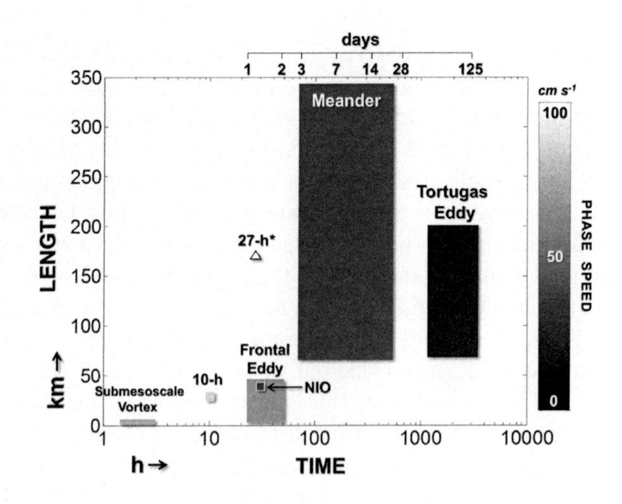

FIGURE 1

The scales of time, length, and phase speed of the instabilities that have been observed in the Straits of Florida. *The 27-h signal phase speed observed by Peters et al.[1] is 170 cm/s.

Refs 1,5,8–11,13–15,19–21,23–27.

(offshore displacement) leading on the eastern side of the channel.[5,17] These unstable barotropic modes can be forced by small perturbations in regions of high horizontal shear.[8] Their wavelengths range from 70 km to a few hundred kilometers, with periods of several days to weeks. In the meridional SOF, the most energetic meandering signals are centered at 5 and 12 days, with phase speeds (wavelengths) of 40 cm/s (170 km) and 25 cm/s (340 km).[5] Cross-channel amplitudes decrease from $\mathcal{O}$ (100 km) in the western entrance to $\mathcal{O}$ (10 km) offshore of Miami because of the narrowing channel and shoaling topography.[18]

2.3 CYCLONIC EDDIES

Associated with the meander troughs are cyclonic eddies that are advected downstream in between the Florida coastline and the FC. Quasi-stationary mesoscale "Tortugas" eddies (100–200 km diameter) are observed in the southern SOF off the Dry Tortugas.[9,19] Their generation mechanism is connected to an extreme southward orientated Loop Current (LC) as it enters the straits, and their subsequent advection is forced by approaching upstream LC frontal eddies. Smaller scale (10–50 km) frontal eddies (also termed spin-off eddies, or edge-eddies) have been measured throughout the year over different bathymetric features along the SOF.[8,14,20,21] They are not directly related to wind forcing, although strong wind events perturb the high-shear regime that may lead to FC instabilities and eddy generation.[8] The lifespan of frontal eddies is estimated to be between one to three weeks, with an average one week occurrence.[8] The passage of these eddies distorts

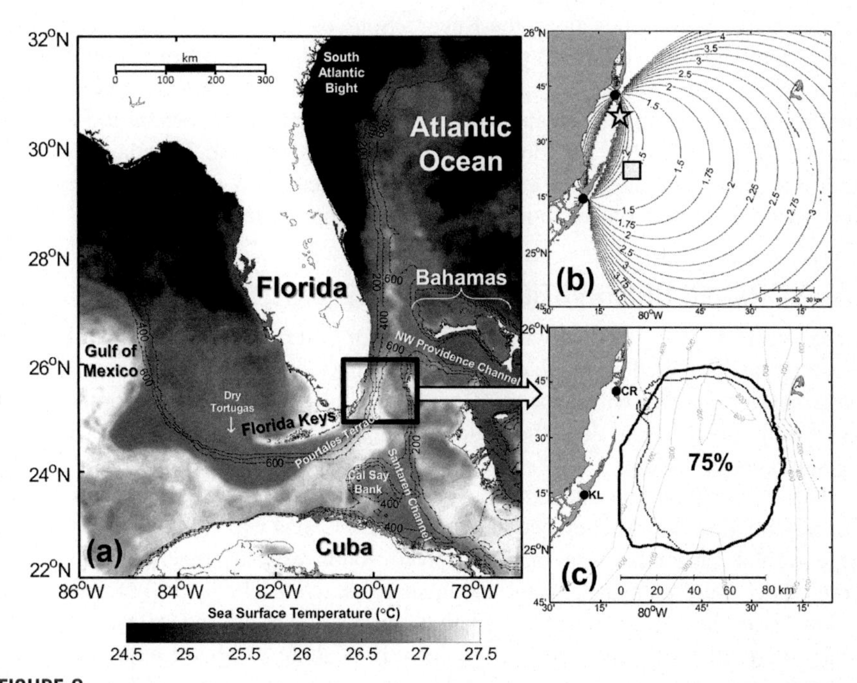

FIGURE 2

(a) 1-day mean SST from MODIS at 1-km resolution for May 7, 2006 (http://mur.jpl.nasa.gov/). Depth contours in meters. (b) Geometric Dilution of Precision (GDOP; see Section 3 for a definition). Star: Fowey Rocks C-MAN station. Square: Acoustic Doppler Current Profiler used in comparison by Parks et al.[14] (c) The 75% radar data coverage contour during the cyclonic case study (dashed line) and the anticyclonic case study (solid line). Black circles pinpoint radar site locations at Crandon Park (CR) and Key Largo (KL).

the thermal structure over the shelf break, which is visible in satellite sea surface temperature (SST) imagery (Figure 2(a)). Uplifted isotherms associated with the cyclonic circulation create a cool surface band of water near the center of the vortex. This produces a strong SST signature characterized by a warm tongue-like extrusion forced by the southward-oriented currents, with a cold upwelling region between the FC and the extruding filament that has an isopycnal uplift of approximately 10 m per day in the upper 200 m.[8] In this sense, they are more akin to roll-vortices produced by wavelike rolling up of the shear-zone than isolated rings observed in the Gulf Stream after it detaches from the coastline.[22]

2.4 INTERNAL WAVES

A complex internal wave (IW) field is created by the strongly sheared current velocity regime and narrow channel with steep topographic gradients.[28] IWs contribute to the generation and distribution of turbulent mixing and mass transfer to coral reef communities along the Keys.[29] Sources of IW energy are the interaction of the barotropic (surface) tide with the along-shelf topography and FC fluctuations.[13,26] The IW field is particularly energetic in the spring and summer months (March–October) because of the uplift of the pycnoclines on the western side of the SOF.[30] Propagation of frontal eddies along the Florida Keys reef tract has been associated with enhanced high frequency IW energy and a peak of diurnal-band spectral power in near shore barotropic currents.[16,26]

2.5 HF RADAR OBSERVATIONS OF FC INSTABILITIES

Since the mid-1990s the deployment of HF radar systems along the South Florida coastline has provided a unique opportunity to investigate the spatial variability of ocean surface currents within the SOF (e.g., Refs. 1,10,11,14). Several frontal eddies have been observed and described, along with new velocity signals. The findings are summarized below.

Two cyclonic frontal eddies were mapped by Haus et al.[21] seaward of Hawk Channel off Key Largo. They occurred when the FC axis was further offshore, with large frontal shears because of wind-forced southwestward inner-shelf currents. Both eddies were elongated in the along-shore direction (19 by 15 km and 47 by 25 km) with fast downstream translation speeds (53 cm/s and 80 cm/s). Parks et al.[14] described a cyclonic submesoscale eddy offshore of Miami with a diameter of approximately 15 km and moving at 45 cm/s. A moored acoustic Doppler current profiler (ADCP) revealed it had a barotropic structure, characterized by a westward u-component through the entire water column over a 48-h period.

A near-inertial oscillation translating eastward along the 150-m isobath off Key West was observed by Shay et al.[10] The signal moved along the inshore edge of the FC at 30 cm/s, exhibiting a dipole-like structure of the current vectors in space. The signal was embedded in the near-inertial passband and absent in the subinertial band. They suggest it was forced by an abrupt change in wind stress, consistent with analytical model results of Kundu.[31] The FC jet trapped the higher frequency near-inertial motions because of negative vorticity, and it amplified these motions in the positive vorticity (cross-shelf gradient $\pm 2f$) regime.

Shay et al.[11] utilized very high frequency (VHF) radar with a horizontal resolution of 250 m. Several submesoscale vortices were observed with diameters of 2 to 3 km over the shelf break at Ft Lauderdale. The translation speed of these features was approximately 30 cm/s, consistent with that of frontal eddies and near-inertial motions; although these vortices are an order of magnitude smaller. They concluded that the vortex they investigated was linked to FC intrusions over the shelf break because it occurred during a period of weak wind conditions.

Two dominant modes of narrowband frequency embedded in the subinertial FC flow were investigated by Peters et al.[1]: a 10-h (super-tidal) signal with amplitude near 50 cm/s and an equally strong 27-h signal (close to the local inertial period). Both signals appeared barotropic in shallow water (50 m depth), but farther offshore

(160 m depth), the 10-h signal exhibited baroclinicity with a phase reversal at depth. The 10-h signal "leaned against the shear" (phase trend in the east–west direction), which is consistent with unstable, growing waves that draw energy from the mean flow. Neither signal could be associated with mesoscale meandering or the near-inertial broadband oscillation observed by Shay et al.[10]

2.6 WIND FORCING

The wind field in the SOF is dominated by the easterly trade wind regime, with a significant northeasterly component during the wintertime.[32] From October to March, cold frontal passages provide forcing with a period range of 4 to 12 days. In summer, the disturbances are due to tropical and subtropical depressions with periods from 15 to 30 days.[33] Strong fluctuations in along-channel wind stress produce a high correlation with measured large-scale transport variations, which vary by time scale depending on the season.[33] Observed frontal instabilities may be a result of wind stress forcing on the shear region of the FC; however, wind does not appear to be a controlling forcing mechanism because these instabilities are prevalent across all seasons.[8]

2.7 OPEN SCIENTIFIC QUESTIONS

In a numerical study of the Gulf Stream along the South Atlantic Bight, Xue and Bane[34] investigated frontal instabilities on either side of the jet core. They note clockwise rotation on the offshore side of the meander crests. Fiechter and Mooers[12] modeled FC instabilities, and though their focus was on cyclonic eddies (like all studies in the SOF), they noted the presence of frontal instabilities in the eastern shear-zone.

Observations from moored current meters through the SOF have hinted at anticyclonic shear-zone instabilities.[2,19] When the FC is in an offshore meander over the Pourtales Terrace, it interacts with the Cal Say Bank.[2] Lee et al.,[19] using SST imagery, observed an offshore FC meander that was partially diverted clockwise around the Cal Say Bank and into the Santaren Channel, which set up a cyclonic rotation. These results suggest that eddies could be formed upstream because of instability of the meandering FC impinging on the steep shelf break of the Cal Say Bank. This interaction may form either anticyclonic (directly) or cyclonic (indirectly through the Santaren Channel) circulations. Another mechanism could be wind stress perturbations on the laterally sheared jet, which because of its unstable nature can encourage fast-growing modes.[8]

Anticyclonic shear-zone instabilities in the SOF have not received attention in the research literature, thus very little is known about their kinematics or dynamics. In this study, HF radar will be used to investigate the spatial and temporal characteristics of an anticyclonic instability, to begin to elucidate the kinematics of these features.

3. INSTRUMENTATION AND EXPERIMENTAL DESIGN

3.1 HF RADAR—PRINCIPLES OF OPERATION

The basic physics of backscattering of electromagnetic waves from the sea surface was identified by Crombie,[35] who observed that the sea echo spectra showed a slight Doppler shift from the transmitted signal. The Doppler shift is the change in frequency (and wavelength) emitted or reflected by an object because of motion. Surface waves with one-half of the incident wavelength produce an enhanced backscatter phenomenon known as resonant Bragg scattering. Bragg scattering results in two distinct peaks in the Doppler spectrum, shifted from the transmit frequency by an amount proportional to the deep water phase speed of the Bragg waves based on linear wave theory. The Doppler frequency shift can be calculated as follows:

$$f_B = \pm\sqrt{\frac{gf_R}{\pi c}} \tag{1}$$

where g is acceleration because of gravity, f_R is the radar transmit frequency, and c is the speed of light. The presence of an underlying surface current will further shift the Bragg peaks by an amount Δf:

$$\Delta f = \frac{2V_r f_R}{c} \tag{2}$$

where V_r is the radial current along the look direction of the radar. By measuring Δf, V_r can be calculated. At least two radar sites are required to resolve vector current velocities from radial measurements.

The effective centroid depth of the measurement depends on the depth of influence of the Bragg waves, shown by Stewart and Joy[36] to be $d = \lambda/8\pi$ (λ is the transmit wavelength). Stable estimates require scattering from hundreds of wave crests plus ensemble averaging of the spectral returns, which sets the time–space resolution limits of the instruments.

3.2 EXPERIMENTAL DEPLOYMENT

Two WERA (Wellen Radar) systems were deployed from June 2004 to July 2011 at Crandon Park on Key Biscayne (CR, 25°42.84′N, 80°9.06′W) and north Key Largo (KL, 25°14.46′N, 80°18.48′W) in South Florida (Figure 2). The WERA systems were chosen for operational flexibility; the user sets the parameters that determine the desired capabilities for the experiment (see Table 1). These WERAs operate in beamforming mode, where a narrow beam is electronically steered over the illuminated ocean, attaining more accurate data returns.[37] Each site operates at 16.045 MHz, obtaining data on a radial grid with resolution of 1.2 km range and 7.5° azimuth, every 20 min, out to approximately 80 km.

Conversion from a radial grid (range and azimuth) to a 1.2 km resolution Cartesian grid is performed on the Doppler spectra. For a given Cartesian grid point, the four closest spectra are identified, two in range and two in azimuth. These four spectra, weighted by distance, are interpolated onto the Cartesian grid point (Klaus-Werner

Table 1 Parameters and Capabilities of the WERA System Operating at 16.045 MHz

Parameters	Value	Capabilities	Value
Operating frequency (MHz)	16.045	Average range (km)	80
Transmit wavelength (m)	18.7	Range cell resolution (km)	1.2
Bragg wavelength (m)	9.35	Measurement depth (m)	0.75
Bragg deep water phase speed (cm/s)	38	Sampling interval (min)	4.5
Bragg frequency shift (Hz)	0.408	Azimuth resolution (°)	7.5
Chirp length (#)	1024		
Chirp duration (s)	0.26		
Modulation bandwidth (KHz)	125		
Transmit elements (square array) (#)	4		
Receive elements (#)	16		
Transmitter peak power (W)	30		

Gurgel, personal communication). Radial velocities are calculated from interpolated spectra every 20 min using manufacturer supplied software. Vectors are calculated every 10 min, using the alternating 20 min radial measurements, with an unweighted least squares method.[38] Velocity field derivatives are calculated using centered differencing.[39] Vector current accuracy is a function of the angle of intersection between the radials from each site, termed the geometric dilution of precision (GDOP), and it can be thought of as a multiplier of the measurement noise.[40] The GDOP ranges from 1 to 2.5 in the radar domain (Figure 2(b)). For this analysis, computed current vectors were filtered with a 7-point Hanning window in time at each grid point. Data points over 3 standard deviations (STD) from a 5 day running mean, and grid points that exceeded a threshold STD of 40 cm/s, or with less than 25% data coverage, were removed from the analysis (see Figure 2(c) for spatial coverage). Comparison with an acoustic Doppler current profiler (location in Figure 2(b)) at the 14 m bin depth revealed root mean square differences between 10 and 30 cm/s.[14]

4. CYCLONIC SHEAR-ZONE INSTABILITY

4.1 OBSERVED SURFACE CURRENT FIELD

A cyclonic frontal eddy was observed translating downstream, inshore of the FC, from 18 to 21 January, 2005 (Figure 3). The eddy was almost stationary in the southernmost part of the domain for ~48 h. Then on 20 January at approximately 12:00 (all times in UTC), it began propagating northward along the 200 m isobath, over a 36-h period. During the passage, downstream current velocities in the FC approached 200 cm/s, and the southward tangential flow of the eddy reached 80 cm/s. The eddy was nestled in the trough of an FC meander, which translated with the feature.

The length scale of the eddy was approximately 20 km. This value is estimated based on the tangential velocities not contaminated by the strong mean flow. As the

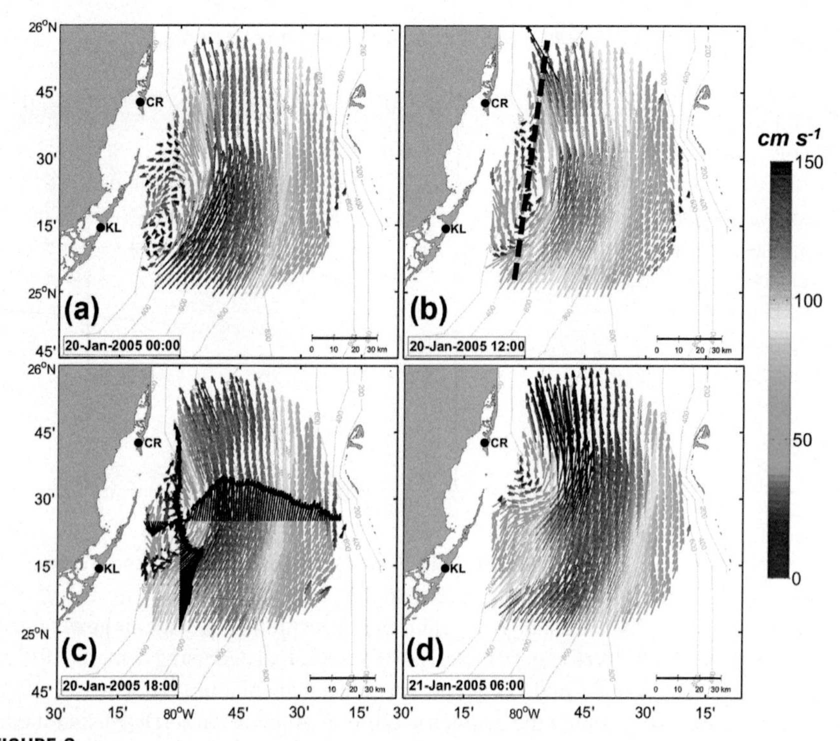

FIGURE 3

Four snapshots of surface current vectors (hourly averaged) at (a) 20 Jan 00:00 UTC, (b) 20 Jan 12:00 UTC, (c) 20 Jan 18:00 UTC and (d) 21 Jan 06:00 UTC depict a submesoscale frontal eddy, which propagates northward along the inshore edge of the Florida Current. Dashed line in (b) depicts transect used for Hovmöller (Figure 4), and the black velocity vectors in (c) highlight the radial horizontal profile of the eddy and the FC.

eddy translated downstream, it moved inshore, and thus, its shoreward side gradually exited the HF radar footprint. The length scale is near the first baroclinic mode Rossby radius of deformation, which has been measured to be between 15 and 30 km in the SOF.[1,11] The eddy is defined as submesoscale, based on its $\mathcal{O}(1)$ Rossby number (see later discussion). The signal propagated north at ~46 km per day (53 cm/s), measured using the slope (Δ *latitude*/Δ *time*) of the *u*-component (Figure 4).

Associated with the passage of this frontal eddy was a strong SST front along the western wall of the FC (Figure 5). The 1 day mean cross-frontal SST gradient was 0.9°C/km on 20 January, compared to 0.04°C/km along the anticyclonic front east of the FC core, and 0.02°C/km at the cyclonic front at a latitude outside the feature (Figure 5(d)). A map of the 1 day mean surface current velocity field superimposed on the SST image reveals the warm FC meander, and within its trough the cold frontal eddy. The translating feature has been smeared by the 1 day average, but there is a correlation between the vectors and SST gradients. Warmer water surrounds a

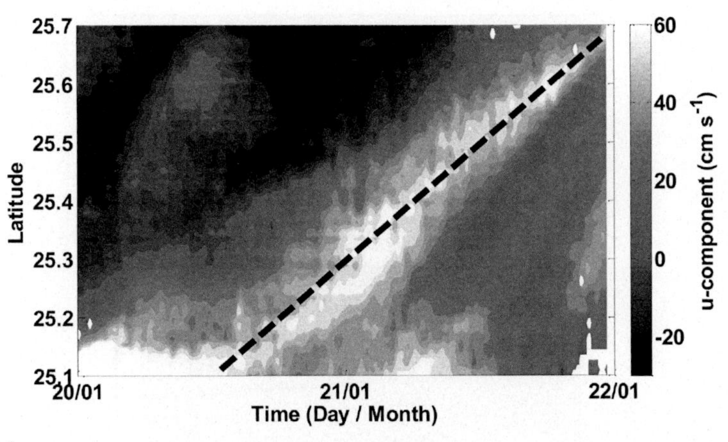

FIGURE 4

Hovmöller diagram of the *u*-component of velocity (cm/s) as a function of latitude (*y*-axis) and time in day/month (*x*-axis). Black dashed line indicates the slope (= speed) of the signal propagation: 66 km/34 h = 46 km per day, or 53 cm/s. The transect follows the path of the eddy, shown in Figure 3(b).

core of cooler water, presumably upwelled because of divergence at the surface since horizontal temperature advection is unlikely considering the temperature of the ambient surface water. The cross-frontal structure at the latitude of the eddy's core is shown in Figure 5(b)–(d). The flow is cyclonic and divergent to the west of the jet core, and maximum values correspond to the FC front where the gradient in the velocity is greatest.

Pressure charts (not shown) indicate the progression of a cold front that passed the SOF on 15 January. A southward wind approached 10 m/s on 17 January with wind stress (surface frictional velocity) over 50 cm/s,[14] which forced a southward countercurrent. As the cold front moved through, the wind weakened and shifted to the southwest on 20 January. At this time, the eddy began propagating north.

4.2 FLOW FIELD KINEMATICS

The dispersion of passive tracers, such as phytoplankton or oil, is primarily controlled by ocean currents and wind. Understanding transport of passive tracers on the ocean surface has practical application, most notably for SAR operations and oil spill mitigation. Disregarding the effects of wind and unresolved small-scale processes, passive tracer (or particle) dispersion is a function of the velocity gradient tensor, the components of which can be calculated with the HF radar dataset. It is of interest to compare how the flow field kinematics change between an eddy event (e.g., a frontal eddy) and normal background conditions.

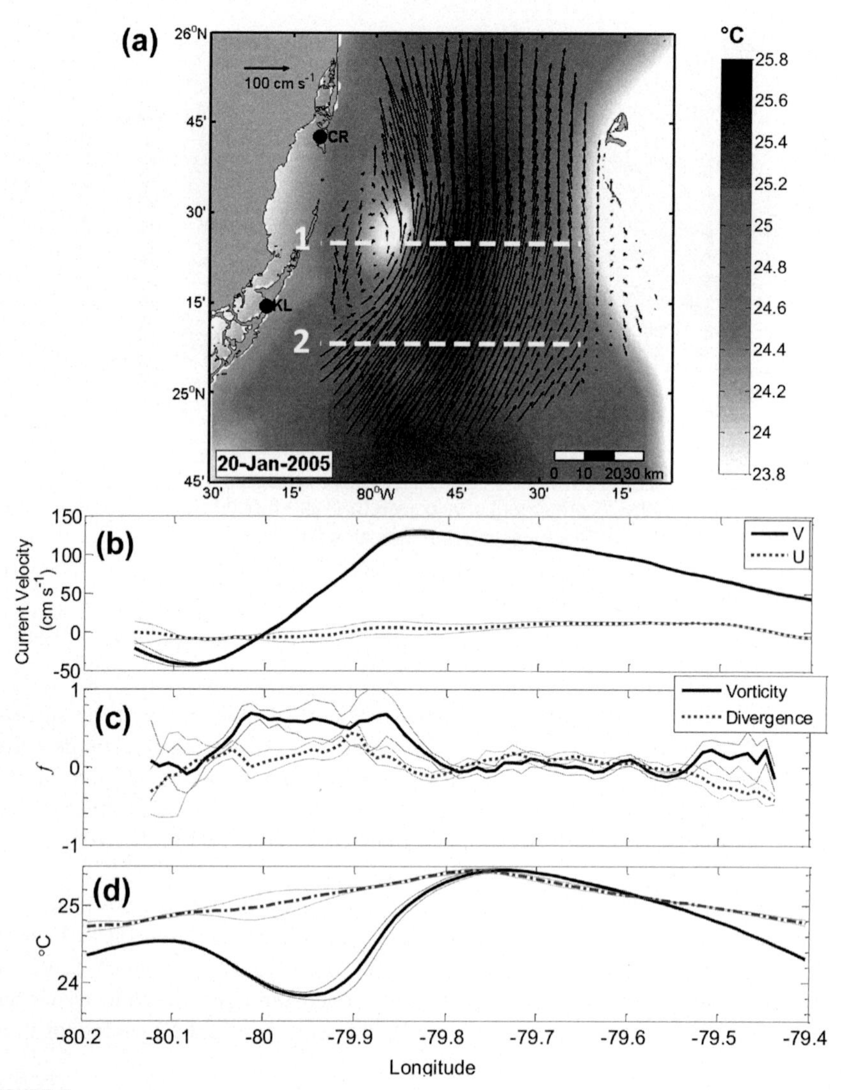

FIGURE 5

(a) 1 day mean SST (°C) from MODIS at 1 km resolution (http://mur.jpl.nasa.gov/). Superimposed on this is the 1 day mean HF radar–derived surface current field, which reveals upwelling in the core of the cyclonic submesoscale eddy. White dotted lines 1 and 2 denote the latitude of the cross-sections plotted below. (b) Cross-section of *u*- and *v*-component of velocity along line 1, (c) vorticity and divergence normalized by local Coriolis frequency along line 1, and (d) SST along line 1 (solid line) and line 2 (dot-dashed line). Thin dotted lines indicate the standard deviation over 25.34–25.47°N (line 1) and 25.09–25.20°N (line 2).

4.3 VELOCITY GRADIENT TENSOR

A two-dimensional surface velocity field, $\boldsymbol{u}(\boldsymbol{x},t) = (u(x,y,t),v(x,y,t))$, can be expanded into a Taylor's series near a reference point x_0.[41] Discarding higher order terms gives the following:

$$\boldsymbol{u}(\boldsymbol{x},t) = \boldsymbol{u}_0 + \boldsymbol{a}\,(x(t) - x_0(t)) \tag{3}$$

where $\boldsymbol{u}_0 = \boldsymbol{u}(\boldsymbol{x}_0)$ is the mean velocity and $\boldsymbol{a}$ is the second-order velocity gradient tensor:

$$\boldsymbol{a}_{ij} \equiv \nabla u = \begin{bmatrix} \dfrac{\partial u}{\partial x} & \dfrac{\partial u}{\partial y} \\ \dfrac{\partial v}{\partial x} & \dfrac{\partial v}{\partial y} \end{bmatrix} \tag{4}$$

The following elemental components may be defined:[42]

$$\text{vorticity} \quad \frac{\partial v}{\partial x} - \frac{\partial u}{\partial y} \equiv \zeta \tag{5}$$

$$\text{divergence} \quad \frac{\partial u}{\partial x} + \frac{\partial v}{\partial y} \equiv d \tag{6}$$

$$\text{normal strain} \quad \frac{\partial u}{\partial x} - \frac{\partial v}{\partial y} \equiv s_n \tag{7}$$

$$\text{shear strain} \quad \frac{\partial v}{\partial x} + \frac{\partial u}{\partial y} \equiv s_s \tag{8}$$

A purely rotational flow (vorticity) does not separate particles. Particle separation is controlled by the combined effect of divergence and nondivergent strain.[43] An eddy core is an "elliptic" regime where vorticity dominates over strain, and particle trapping and transport occurs. The core is surrounded by a hyperbolic regime where strain dominates over vorticity,[44] and filamentation and mixing occurs leading to dispersion.

4.4 LAGRANGIAN AND EULERIAN DIAGNOSTICS OF THE FLOW FIELD

There are numerous techniques to quantify dispersion and resolve coherent flow features in a horizontal velocity field (e.g., Refs 45–47). Some methods are based on Lagrangian, time-dependent information (e.g., Lyapunov exponents), whereas others require only a Eulerian snapshot of the velocity field (e.g., Okubo-Weiss). In general, Lagrangian techniques are the preferred approach because they integrate in time, allowing resolution of coherent structures in the flow field; whereas Eulerian methods will resolve an instantaneous flow field that cannot distinguish between coherent and transient features. Advantages to the Eulerian approach when using real data are the ease of calculation and retained spatial coverage. Several studies have shown the utility of the Eulerian strain and divergence field for measuring particle dispersion.[43,47,48]

Here we apply the Eulerian method, based on a consideration of the oceanography in the SOF; the strong FC advects flow patterns quickly through the domain. Techniques based on integration time that attempt to capture coherent features of the flow would suffer from either lack of data (because seeded particles quickly exit the region) or would have to drastically reduce the length of integration to retain spatial information, so would converge to near-instantaneous values.

An *instantaneous* rate of separation (IROS) is the Eulerian metric that determines how an infinitesimally small particle will be moved by an instantaneous velocity field, and it is equal to the finite-time Lyapunov exponent (FTLE) at time $t = 0$.[43] It can be calculated from the sum of divergence (d) and total strain (s_n and s_s). The FTLE picks out features that dominate over longer time periods, whereas IROS acts as a guide to how the particles react in the moment.[43] High values of IROS indicate regions of elevated particle dispersion.

4.5 EULERIAN VELOCITY FIELD DURING AN EDDY EVENT

The components of the velocity gradient tensor exhibited large magnitude changes during the eddy passage (Figure 6). In the absence of an eddy event, the vorticity structure is dominated by the FC shear; there is uniform positive vorticity to the west of the jet axis, switching to negative vorticity on the eastern side, with magnitudes close to f (Figure 6(c)). However, when the submesoscale frontal eddy moved through the domain, the vorticity showed strong nonuniform fluctuations in time and space (of both positive and negative sign) that approached $11f$. For this reason, the dynamics of the frontal eddy are clearly within the submesoscale because the Rossby number (vorticity normalized by f) is very large.[49]

The IROS field during the eddy passage was similarly complex with strong magnitudes that revealed regions with a strong dispersive nature. A comparison of IROS and vorticity reveals co-location of peak values, which indicates that regions of strong vorticity in eddy cores do not necessarily correspond to particle trapping in the FC. This is because the eddy core is not purely rotational, as deformation plays a significant role. During a quiescent period with no eddy activity, the IROS field comprised mostly low values across the domain (Figure 6(d)).

The field of maximum value (divergence and IROS) extracted from each period of interest (the eddy and no eddy cases) reveals the nature of the flow field (Figure 7). For the eddy passage, there is a clear "track" in which maximum values exceed background levels. There was strong divergence associated with the passage of the eddy, which peaked at $4f$ (Figure 7(a)). This is consistent with the pattern of SST discussed previously (Figure 5). IROS exhibits strong values during the event, implying that there is strong particle dispersion because of the presence of the eddy. Regions of strong divergence and IROS translate downstream with the eddy, which

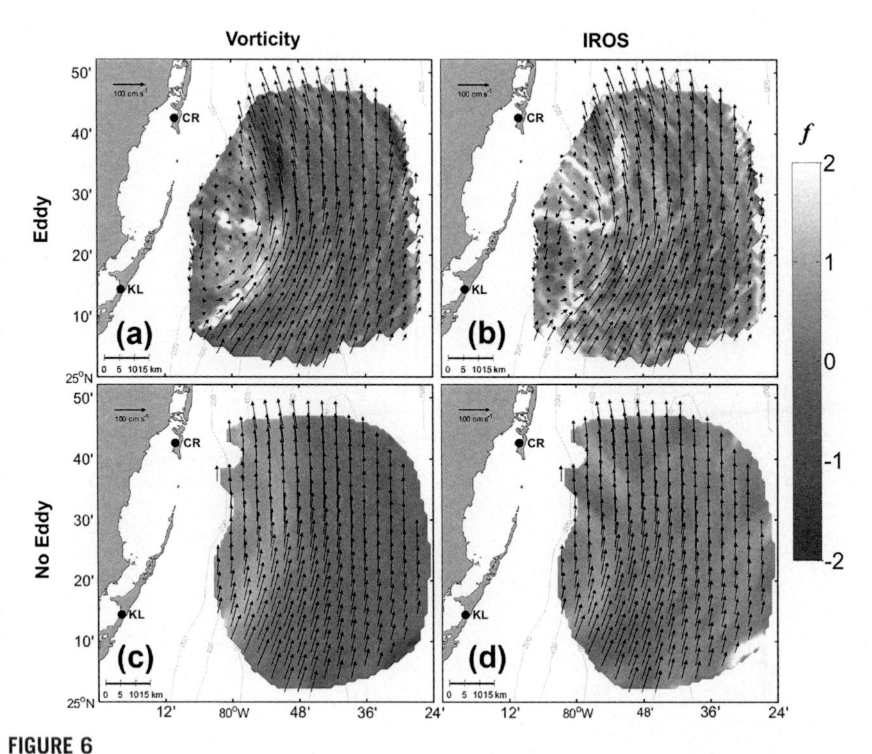

FIGURE 6

A snapshot of current vectors superimposed with fields of vorticity (a, c) and IROS (b, d) during an eddy event (January 20, 2005 at 16:00; a, b) and during a time with no eddy activity (October 4, 2006 at 00:00; c, d). The "no eddy" time period was identified as an example with relatively uniform downstream velocity, to contrast to the eddy event.

produces the track-like pattern in the maximum value field. When there is no eddy activity, the domain comprises small background levels, except along the periphery of the footprint where the GDOP is higher.

Velocity gradients during an eddy event exhibit very strong fluctuations in comparison to the FC flow field with no eddy activity. There was a complicated pattern of vorticity and deformation suggestive of strong particle leakage out of the eddy core, associated with high values of IROS. Divergence was strongly positive and consistent with concurrent MODIS SST imagery of cold water anomaly near the eddy core, associated with upwelling. These results indicate the energetic nature of these frontal eddies. Using the HF radar dataset to study the flux of kinetic energy between the mean and perturbations during both an eddy and no eddy period could shed light on the impact of eddies with respect to the energetics of the FC.

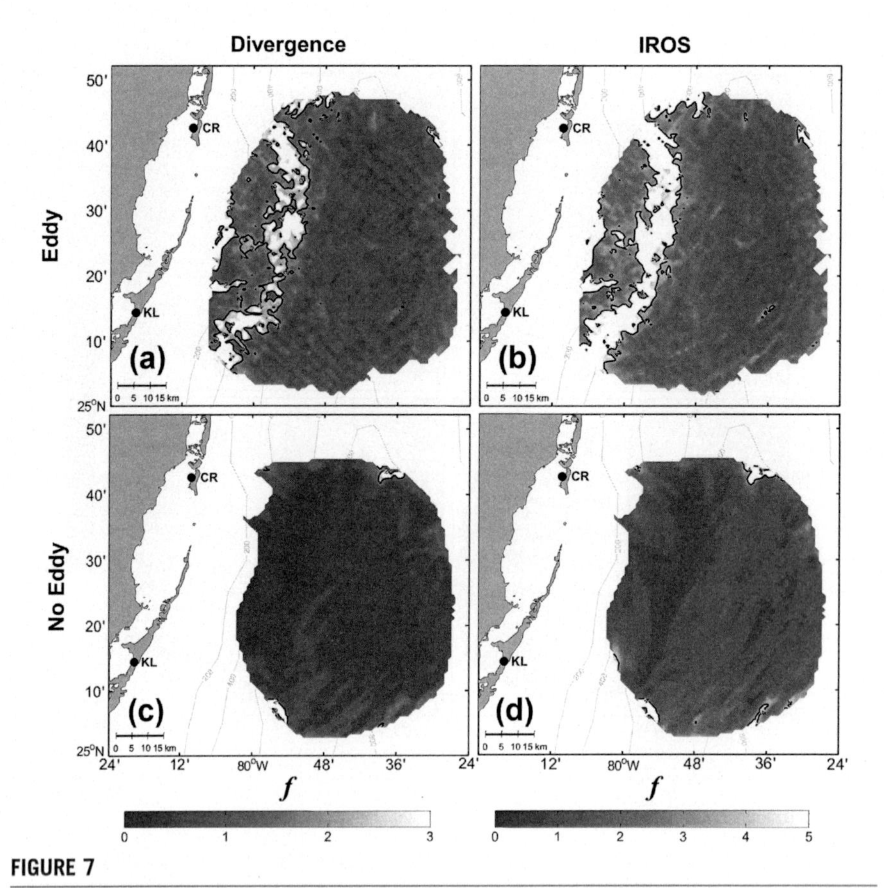

FIGURE 7

Maximum values of the field extracted for the time periods (a, b) 2005: 20 January 00:00 to 21 January 12:00 (eddy event) and (c, d) 2006: 3 October 12:00 to 4 October 16:00 (no eddy). Divergence is plotted with a $2f$ solid line contour (a, c) and IROS is plotted with a $4f$ solid line contour (b, d).

5. ANTICYCLONIC SHEAR-ZONE INSTABILITY

5.1 OBSERVED SURFACE CURRENT FIELD

Four consecutive eddy-like features were observed translating through the radar domain from 15 to 21 October, 2006. However, unlike the near-ubiquitous cyclonic frontal eddies observed along the western SOF, these features moved along the outer eastern flank of the FC. All four features exhibited clockwise rotation at the surface, in a water depth of approximately 650 m. The 19 October feature, which was best resolved in the radar footprint, is shown in Figure 8. This event produced a strong propagating signal in the time—longitude Hövmöller plot of u-component velocity (Figure 9(a)). The phase propagated northward at approximately 80 cm/s. By

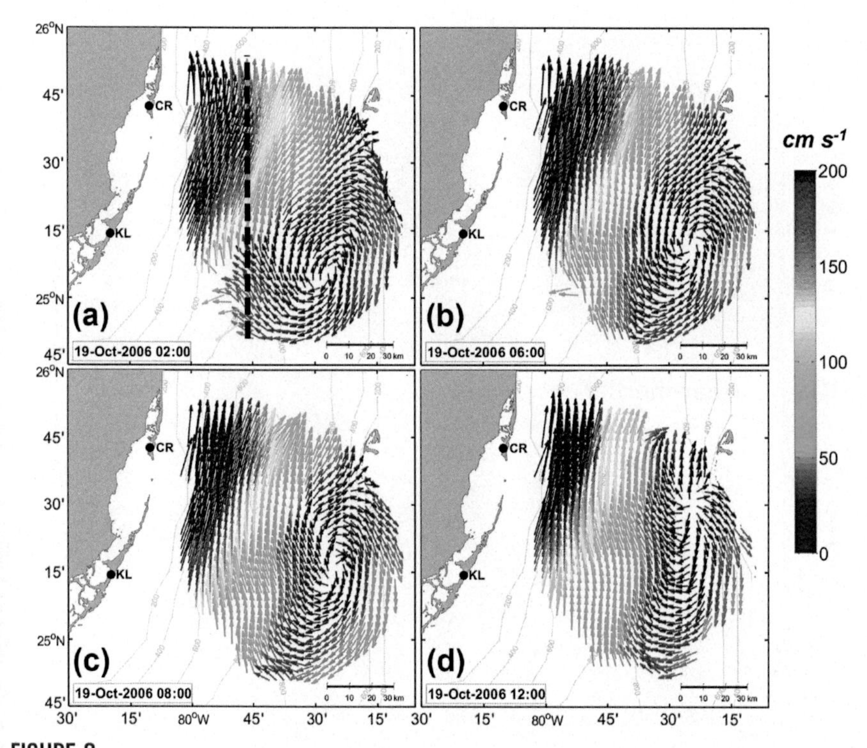

FIGURE 8

Four snapshots of the surface current vectors (hourly averaged) on 19 October 2006 at (a) 02:00 UTC, (b) 06:00 UTC, (C) 08:00 UTC and (d) 12:00 UTC reveal the evolution of a clockwise-rotating eddy observed by WERA HF radar. Dashed line in (a) depicts transect used for Hovmöller (Figure 9).

contrast, observed SOF cyclonic eddy translation speeds range between 6 and 19 cm/s (Tortugas eddies), 46–93 cm/s (frontal eddies), and 17–46 cm/s (submesoscale features).[8–10]

On 14 October, the wind speed increased from 5 to 12 m/s and shifted from a variable northerly wind to a steady easterly throughout the event. An easterly wind could force the currents shoreward (to the west), which was observed on 18 October, and may increase FC magnitude in the surface layer via wind-driven Ekman velocity. There was an observable increase in FC surface velocity at this time, peaking on 19 October (Figure 9(a)). However, because the disturbance was generated upstream of the observational domain, without additional data the contribution from the wind cannot be determined.

5.2 SEPARATING THE SIGNAL FROM THE BACKGROUND FLOW

The signal exhibited a periodicity close to the local inertial period ($2\pi f^{-1}$), which for latitudes from 25° to 25.7° ranges from 27.6 to 28.3 h. To separate from the

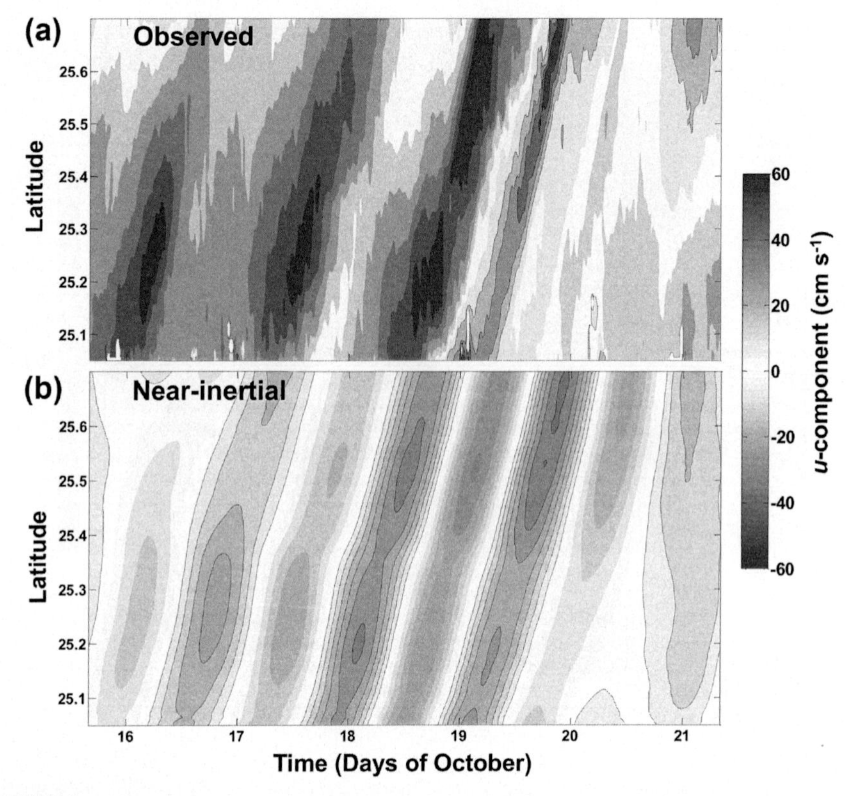

FIGURE 9

Hövmöller plots (*u*-component contours plotted on time vs latitude axes) at longitude 79.8°W (transect plotted in Figure 8(a)) for (a) observed and (b) near-inertial currents. Solid contour lines denote negative values. In (b), note the clear propagation of negative-*u*, which is masked out in the observed currents by the strong northward FC.

background flow (background is herein defined as the current field unassociated with the signal), the time series at each grid point was decomposed into subinertial (>48 h), near-inertial (20–36 h) and high frequency (<20 h) currents. The near-inertial bandwidth was assigned based upon the analysis of Mooers and Brooks,[6] who noted that because of the strongly sheared background flow, the inertial frequency can be shifted by up to 30% of f in the SOF. After conducting sensitivity tests, the Fourier filter proved optimal for the decomposition.[50] The Fourier filter requires a complete time series, which imposed restrictions on the spatial coverage of the dataset. The diminished spatial coverage does not fully cover the features that pass through (Figure 10(b)), although it does capture the rotation along the western periphery. This does not significantly affect the outcome; it can be shown that Reynolds decomposition, which uses all the data, gets the same qualitative result.

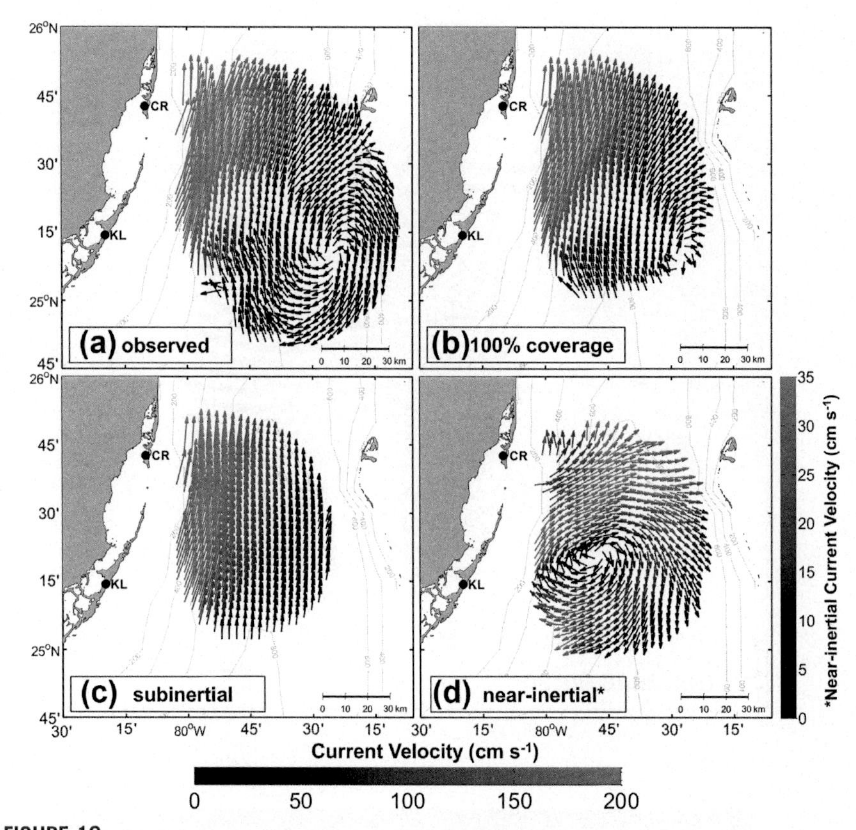

FIGURE 10

Frequency decomposition of the surface velocity field at 19 October 02:00 for (a) observed, (b) observed: region of 100% coverage that can be filtered, (c) subinertial, and (d) near-inertial components (*color bar scale for near-inertial currents is from 0 to 35 cm/s).

Tidal constituents were not removed because of the complication of contamination by the episodic FC meandering over daily time scales. Previous studies have shown that tidal velocities in the Straits are <10 cm/s.[1,51,52] Tidal forcing is continuous and periodic, whereas eddy events are transient and highly intermittent.

5.3 NEAR-INERTIAL OSCILLATION

The signal was embedded in the near-inertial band, as shown by decomposed surface current maps (Figure 10). The subinertial band comprised the FC meandering, whereas the high frequency band (not shown) exhibited neither coherent structure nor significant amplitude. Once isolated from the background flow field, the signal is oscillatory (Figure 9(b)). In the surface vector maps, the near-inertial currents reveal what can be interpreted as the crest and trough of a wave (Figure 11).

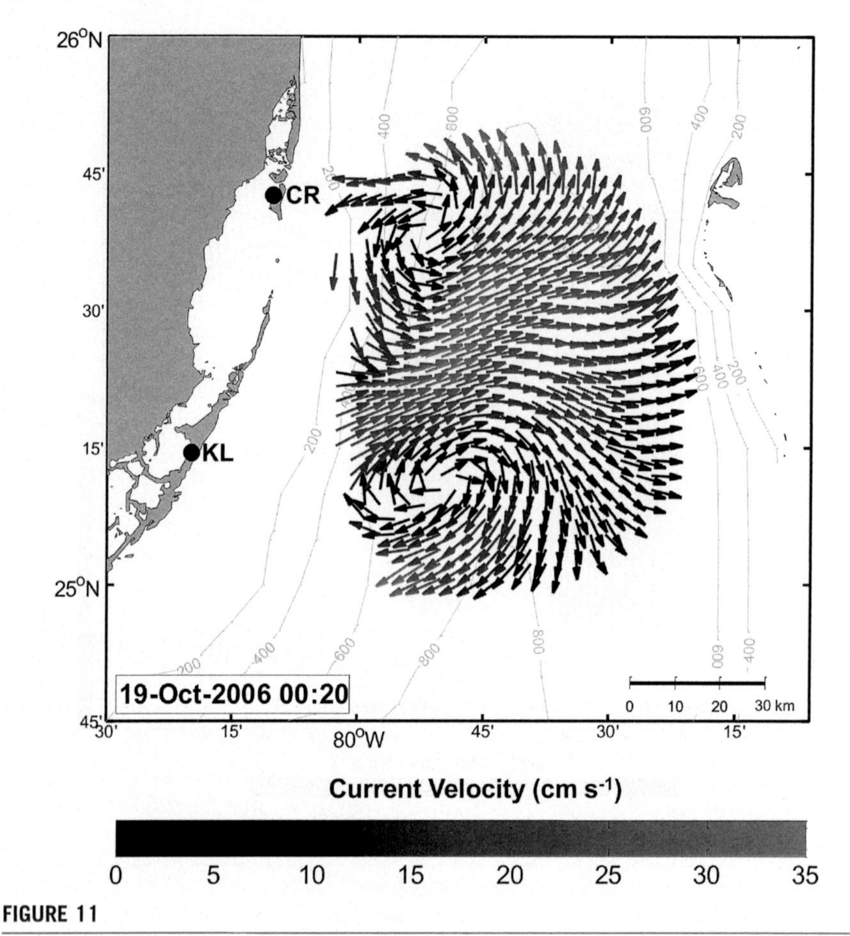

FIGURE 11

Near-inertial current vector map on 19 October 00:20 reveals the crest (convergence) and trough (divergence) of a near-inertial wave.

A clockwise rotation of the vectors in time produces horizontal convergence (crest) and divergence (trough) of the near-inertial currents.

Near-inertial motions can be generated by fluctuations in local wind stress,[53] or "loss of balance/spontaneous adjustment" by western boundary currents, mesoscale eddies, and submesoscale frontogenesis.[54–56] After a transient forcing event, and in the absence of all other forces, horizontal currents move under their own inertia, and on a rotating Earth in the Northern (Southern) Hemisphere will complete clockwise (counter-clockwise) oscillations at the inertial frequency f.[57] However, in the real ocean, these motions are often shifted off f because of other forces. A horizontal-sheared background flow, with relative vorticity ζ_g, can lower the bound of the internal waveband from f to an effective frequency $f_{eff} = f + \zeta_g/2$.[58] Kunze[59] showed that when a near-inertial wave propagates through a horizontal gradient of f_{eff}, its wave vector must evolve to satisfy the dispersion relation, which leads to refraction and partial or total reflection. Horizontal gradients in f_{eff} result in a nonuniform wave field. Waves generated in regions where $f_{eff} < f$ are trapped, as they encounter turning points outside of the negative vorticity trough.[60] Elevated near-inertial kinetic energy on the anticyclonic (negative vorticity) side of a front has been observed in numerous field studies (e.g., Refs. 61–64).

Within the SOF, Shay et al.[10] documented near-inertial motions with horizontal wavelengths of 40 km that were trapped and advected by the FC. Vertical current structure measurements from an ADCP revealed vertical wavelengths between 50 and 100 m, and phase propagation reversals at a critical layer (the depth where the speed of the wave group equals that of the current).[59] Our case differs in that the near-inertial signal was observed in the anticyclonic shear-zone of the FC. The strongly sheared background flow partially masked the near-inertial current field, which is manifested as a succession of clockwise-rotating eddies in the observed surface current maps. The wave trough is not evident in the total surface currents when embedded in a laterally sheared flow regime. Some caution should be invoked, however, because this method filters Eulerian data to look at a translating Lagrangian feature. Further analysis must be conducted to relate this signal to near-inertial wave dynamics.

5.4 IDEALIZED MODEL

To elucidate the geometric effects of a background shear flow on the signal pattern observed in our HF radar domain, a simple analytical model of an asymmetric jet with lateral shear is superimposed with a dipole perturbation. Stream functions for the jet (ψ_J), perturbation (ψ_e), and total flow (ψ_T) are as follows:

$$\psi_J = A \cdot e^{-aL_x} \tag{9}$$

$$\psi_e = B \cdot \sin\left(m\pi^{-1}\right) \cdot \sin(l + \phi) \tag{10}$$

$$\psi_T = \psi_J + \psi_e \tag{11}$$

Where A is the amplitude of the jet core, a is a scaling factor for the lateral shear, and $L_x = L_y = 100$ are the zonal and meridional extent of the domain (size is arbitrary). B is the wave amplitude, $m = \pi/40$ is defined at $x = 10{:}50$ (where 40 is the width of the jet), $l = 2\pi/L_y$ is the meridional wave number, and ϕ the phase. The model has been assigned parameters to resemble the data: specifically, the wavelength/domain ratio and phase of the disturbance. The idealized fields are compared to the observations of the total surface currents, the band-passed near-inertial currents and the low-passed subinertial flow (Figure 12).

The model confirms that for a dipole perturbation embedded in a laterally sheared anticyclonic background flow, only closed clockwise rotation is apparent in the total flow field. The counter-clockwise rotating eddy acts to distort the stream function contours in the region (see the total fields in Figure 12), but there is no

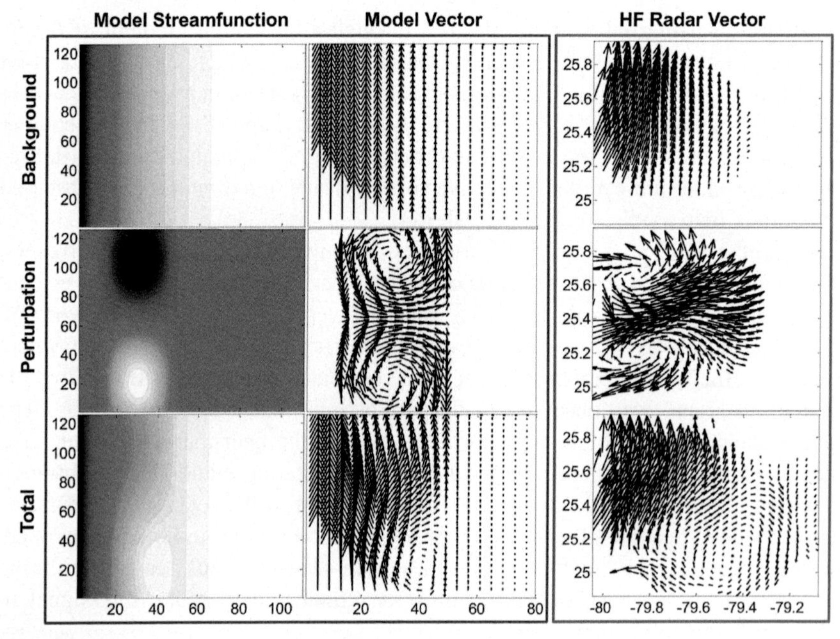

FIGURE 12

(left column) Model stream function, (middle column) model vector plots, and (right column) observed vector plots for (top) laterally sheared jet, (middle) dipole perturbation, and (bottom) total field.

closed circulation. Accounting for complications in the observed currents, such as differences in jet and eddy orientation, the simple model can replicate the basic flow pattern. This model reveals qualitatively how a horizontally sheared mean flow can mask a rotary perturbation signal.

5.5 NEAR-INERTIAL WAVE KINEMATICS

The hypothesis was that these transient clockwise-rotating features are a succession of stationary eddies advected northward by the FC. However, a systematic frequency analysis of the signal reveals these features to be strongly embedded in the near-inertial passband, and absent from the subinertial. The characteristics of the signal in frequency and space resemble a near-inertial oscillation. In this section, the properties of the signal are examined and compared to near-inertial wave theory.

5.6 SUBINERTIAL VELOCITY FIELD

The jet core was located in the western part of the radar domain, along the continental shelf. There was a thin region of cyclonic shear to the west of the axis and a much wider anticyclonic shear region to the east (Figure 10(c)). Within the core, subinertial velocities (v) reached 200 cm/s, decreasing to less than 40 cm/s toward the east, over a distance (x) of 40 km. This equates to a sheared background flow with a normalized vorticity of $f^{-1}\ \partial v/\partial x = -0.6$. Similarly, the cyclonic shear-zone exhibited equal magnitudes in vorticity. This suggests that any near-inertial wave propagating in this field will experience strong horizontal gradients of f_{eff}, which could lead to frequency shifts and trapping in the negative vorticity trough.

5.7 WAVELENGTH

To determine the horizontal wavelength of the near-inertial signal, a series of trial wavenumbers ($2\pi/L$, where L is the wavelength) were fitted to the band-passed data at grid points along lines of constant longitude, using a plane wave model[10]:

$$u(y) = A_1 \cos(ly) + B_1 \sin(ly) + u_r(y) \tag{12}$$

$$v(y) = A_2 \cos(ly) + B_2 \sin(ly) + v_r(y) \tag{13}$$

where (u, v) are the observed near-inertial data, $A_{1,2}$ and $B_{1,2}$ are the velocity amplitudes (Fourier coefficients), l is the meridional wave number (trial wavelengths defined between 1 and 300 km), and (u_r, v_r) is the residual current not explained by the model. A "carrier" wave number is defined, which maximizes the correlation coefficient R between the observed and modeled data[65]:

$$R = \sqrt{\frac{\left(r_u^2 + r_v^2\right)}{2}} \tag{14}$$

where $(r_u,\ r_v) = s_{xy}^2/s_{xx}s_{yy}$, r_u and r_v are the correlation coefficients between observed and modeled velocity for the u- and v-components of velocity, s_{xx} and s_{yy} are the variance matrices of observed and modeled velocities, respectively, and s_{xy} is the covariance matrix. For each longitude, the latitudinal average was removed at each grid point in latitude, and the least squares fit was performed over two inertial periods (IP), the time period when the signal was at its strongest. Note: This approach assumes there is a dominant single carrier wave number for each longitude.

The model reveals an average wavelength of ~110 km (Figure 13). The wavelength is close to the Eady model most unstable mode of $3.9R_d$, where R_d is the Rossby radius of deformation, which in the SOF is ~30 km.[1]

5.8 FREQUENCY

The dominant frequency of the oscillation is calculated with Eqns (12) and (13) by substituting in frequency and temporal variations at each grid point in place of wave number and spatial variations. The near-inertial components at each grid point were fit to a series of trial frequencies between 0.5 and $1.5f$ (intervals of 0.05). The carrier frequency (2π/T, where T is the wave period) is defined as the value that maximizes the correlation between observed and modeled data,

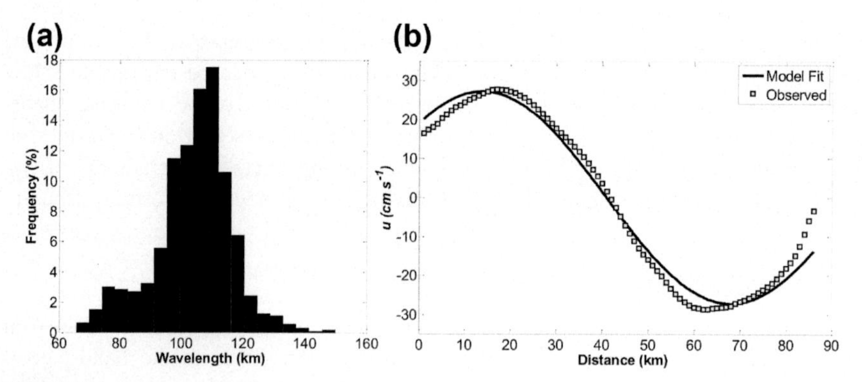

FIGURE 13

(a) Histogram of the modeled wavelengths between longitudes 79.5°W to 79.9°W and over a time period of 1.5 IP (43 hrs) and (b) results of least squares fit of the near-inertial currents at longitude 79.8°W.

over the two IPs. Over the domain, calculated carrier frequencies range between 0.8 and 1.3*f*, although a shifting to lower frequencies dominates, and the average is 0.87*f*, with correlations between the model and data as high as 0.95 (Figure 14). The width of the peaks is because of the broadband character of near-inertial motions.

5.9 EFFECTS OF SUBINERTIAL VORTICITY

The frequency shift of the signal below *f* agrees with theoretical results of a near-inertial wave propagating in a region of negative vorticity.[59] The mean near-inertial

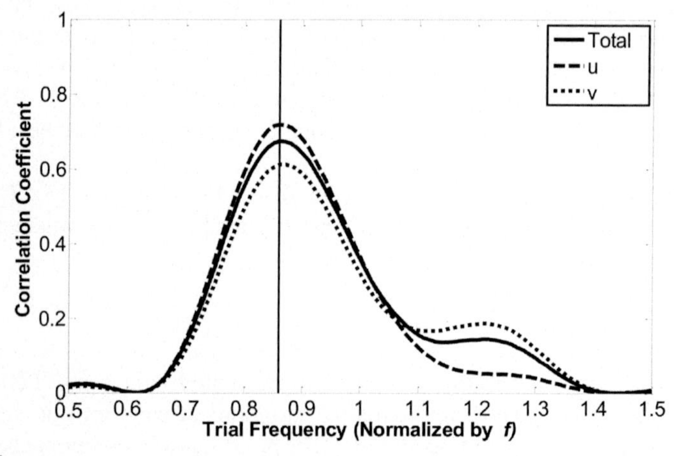

FIGURE 14

Trial frequency versus correlation between the model and data. Black vertical line denotes the carrier frequency, which has the best fit to the data.

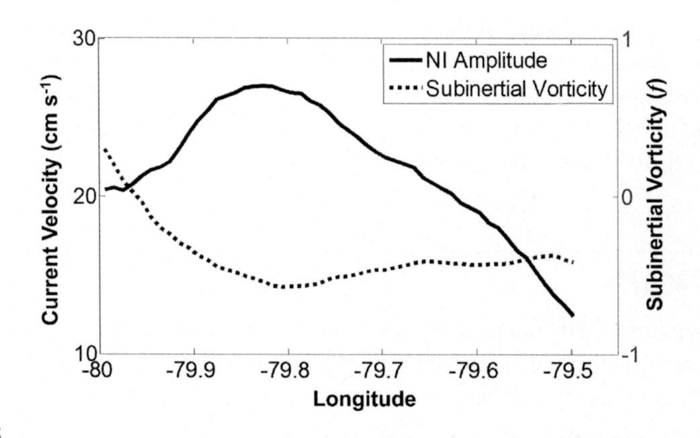

FIGURE 15

Near-inertial wave current amplitude associated with the dominant wavelength (solid) and the subinertial vorticity normalized by f (dashed), as a function of longitude over two inertial periods.

current amplitude distribution exhibits a peak aligned with the subinertial vorticity trough (Figure 15). This suggests trapping of the near-inertial signal generated in the vorticity trough. Trapping occurs because of wave refraction in a spatially nonuniform vorticity field—the wave cannot propagate freely away as it is refracted back and forth between regions of less negative vorticity, and it leads to peaks in near-inertial energy within the trough.[59,60]

5.10 OPEN QUESTIONS

The observed signal is consistent with near-inertial wave propagation in geostrophic shear.[59] The vector rotation at each grid point over most of the HF radar footprint is clockwise, which is consistent with near-inertial oscillations in the northern hemisphere. However, the rotation of vectors within the jet core and the strong cyclonic shear region is counter-clockwise (not shown). HF radar measures the Eulerian frequency $\omega = \omega_0 + \vec{k} \cdot \vec{V}$, where ω_0 is the intrinsic frequency, $\vec{k}$ is the wave vector, and $\vec{V}$ is the subinertial velocity. The Eulerian frequency is constant in a steady flow, but in our domain, it has a spatial gradient because of nonuniform f_{eff} and $\vec{V}$. The term $\vec{k} \cdot \vec{V}$ represents the Doppler shift (advection) by the background flow. In the North Atlantic subtropical zone, Mied et al.[66] found evidence of near-inertial waves strongly influenced by Doppler shifting. Preliminary results here indicate this Doppler shift may be significant enough to dominate the near-inertial oscillation frequency within the core. However, without additional observations, it is difficult to form a solid conclusion of the mechanism of the observed signal. One avenue for further insight could be to develop the idealized model by individually applying the dynamics of a vortex (e.g., a Rankine model) and a near-inertial wave. The lateral shear and magnitude of the jet core can also be adjusted to best fit the observations. Because this study is the initial analysis of a new signal observed in the anticyclonic shear-zone of the FC, more observations of such events, covered by both remote and in situ instrumentation, as well as numerical modeling efforts, are required to fully explain these complex features.

6. SUMMARY

The deployment of HF radar along the South Florida coastline has improved our ability to monitor the ocean surface currents within the SOF. This has been shown by two case studies, which have demonstrated HF radar's ability to (1) examine how the flow field kinematics are significantly altered during the passage of a submesoscale frontal eddy, and (2) document a near-inertial velocity signal along the anticyclonic flank of the FC that has not been studied before.

In the first case study, the passage of a submesoscale cyclonic frontal eddy moving quickly downstream was captured in the HF radar footprint. In contrast to conditions recorded in a period of no eddy activity, during the event the vorticity field revealed a complex structure, with significant contributions from strain and a Rossby number that greatly exceeded unity, implying the flow field was governed by submesoscale dynamics. Indeed, there was strong horizontal current divergence near the core of the eddy, associated with anomalously cold water brought to the surface by upwelling, observed in MODIS SST satellite imagery. IROS, which is a metric of particle dispersion, exhibited high values that translated with the eddy, indicating the potential for strong dispersion of a passive tracer. This has important implications for cross-shelf exchange of water properties between offshore and coastal regions, and it is important information for SAR operations and pollution mitigation.

In the second case study, a transient, coherent signal in the near-inertial passband was identified. It was found that the strongly sheared FC partially masked the structure of the near-inertial oscillation, which was manifested as a succession of clockwise-rotating eddies in the observed surface currents. The wave trough was not evident when embedded in a laterally sheared northward background flow. The dominant frequency was shifted by $\sim$13% below f in the average, which is consistent with a near-inertial wave propagating in a background regime with negative vorticity. The spatial pattern of frequency was highly anisotropic because of the variations in the subinertial current velocity and its associated vorticity. Near-inertial energy peaked in the negative vorticity trough along the FC's eastern flank, indicative of wave trapping in the horizontal. These results suggest the observed signal was governed by near-inertial wave dynamics. However, because this is a preliminary study of these features, further work is required to clarify their mechanisms.

These example cases, in addition to previous modeling and observational studies, reveal the highly intermittent nature of the flow within the SOF, which comprises periods of strong fluctuations on both the cyclonic and anticyclonic shear-zones of the FC. Future work needs to take the big step forward from individual case studies to long-term time series analysis, which can determine the quantitative and statistical details of the time and space scales of these instabilities and whether they exhibit change over time. The ultimate goal is to incorporate this information into improving model forecasts of the current and wave field in the SOF.

ACKNOWLEDGMENTS

The authors gratefully acknowledge support by NOAA IOOS-supported South East Coastal Ocean Observing Regional Association (SECOORA) through grant NA11NOS0120033. We thank Pierre Flament, who suggested the comparison with an idealized model. Arthur Mariano, Rick Lumpkin, Ed Ryan, and Mitch Roffer contributed with productive discussions regarding data analysis. We are thankful for technical support from Jodi Brewster and Claire McCaskill. MODIS SST data was obtained from http://mur.jpl.nasa.gov/.

REFERENCES

1. Peters H, Shay LK, Mariano AJ, Cook TM. Current variability on a narrow shelf with large ambient vorticity. *J Geophys Res Oceans (1978–2012)* 2002;**107**(C8):2-1.
2. Leaman KD, Vertes PS, Atkinson LP, Lee TN, Hamilton P, Waddell E. Transport, potential vorticity, and current/temperature structure across Northwest Providence and Santaren Channels and the Florida Current off Cay Sal Bank. *J Geophys Res Oceans (1978–2012)* 1995;**100**(C5):8561–9.
3. Meinen CS, Baringer MO, Garcia RF. Florida Current transport variability: an analysis of annual and longer-period signals. *Deep Sea Res Part I: Oceanogr Res Pap* 2010; **57**(7):835–46.
4. Schott F, Düing W. Continental shelf waves in the Florida Straits. *J Phys Oceanogr* 1976; **6**(4):451–60.
5. Johns WE, Schott F. Meandering and transport variations of the Florida Current. *J Phys Oceanogr* 1987;**17**(8):1128–47.
6. Mooers CN, Brooks DA. Fluctuations in the Florida Current, summer 1970. *Deep Sea Res* 1977;**24**(5):399–425.
7. Lee TN, Williams E. Wind-forced transport fluctuations of the Florida Current. *J Phys Oceanogr* 1988;**18**(7):937–46.
8. Lee TN, Mayer DA. Low-frequency current variability and spin-off eddies along the shelf of southeast Florida. *J Mar Res* 1977;**35**:193–220.
9. Fratantoni PS, Lee TN, Podesta GP, Muller-Karger F. The influence of loop current perturbations on the formation and evolution of Tortugas eddies in the southern Straits of Florida. *J Geophys Res Oceans (1978–2012)* 1998;**103**(C11):24759–79.
10. Shay LK, Lee TN, Williams EJ, Graber HC, Rooth CG. Effects of low-frequency current variability on near-inertial submesoscale vortices. *J Geophys Res Oceans (1978–2012)* 1998;**103**(C9):18691–714.
11. Shay LK, Cook TM, Haus BK, Martinez J, Peters H, Mariano AJ, et al. VHF radar detects oceanic submesoscale vortex along Florida coast. *Eos, Trans Am Geophys Union* 2000;**81**(19):209–13.

12. Fiechter J, Mooers CN. Simulation of frontal eddies on the East Florida Shelf. *Geophys Res Lett* 2003;**30**(22).
13. Boudra DB, Bleck R, Schott F. A numerical model of instabilities in the Florida Current. *J Mar Res* 1988;**46**(4):715–51.
14. Parks AB, Shay LK, Johns WE, Martinez-Pedraja J, Gurgel KW. HF radar observations of small-scale surface current variability in the Straits of Florida. *J Geophys Res Oceans (1978–2012)* 2009;**114**(C8).
15. Lee TN, Rooth C, Williams E, McGowan M, Szmant AF, Clarke ME. Influence of Florida Current, gyres and wind-driven circulation on transport of larvae and recruitment in the Florida Keys coral reefs. *Cont Shelf Res* 1992;**12**(7):971–1002.
16. Sponaugle S, Lee T, Kourafalou V, Pinkard D. Florida Current frontal eddies and the settlement of coral reef fishes. *Limnol Oceanogr* 2005;**50**(4):1033.
17. Bane JM, Brooks DA. Gulf Stream meanders along the continental margin from the Florida Straits to Cape Hatteras. *Geophys Res Lett* 1979;**6**(4):280–2.
18. Schmitz Jr WJ, Richardson PL. On the sources of the Florida Current. *Deep Sea Res Part A. Oceanogr Res Pap* 1991;**38**:S379–409.
19. Lee TN, Leaman K, Williams E, Berger T, Atkinson L. Florida Current meanders and gyre formation in the southern Straits of Florida. *J Geophys Res Oceans (1978–2012)* 1995;**100**(C5):8607–20.
20. Lee TN. Florida Current spin-off eddies. *Deep Sea Res Oceanogr Abstract* 1975;**22**(11): 753–65. Elsevier.
21. Haus BK, Wang JD, Rivera J, Martinez-Pedraja J, Smith N. Remote radar measurement of shelf currents off Key Largo, Florida, USA. *Estuar Coast Shelf Sci* 2000;**51**(5): 553–69.
22. Lee TN, Atkinson LP, Legeckis R. Observations of a gulf stream frontal eddy on the Georgia continental shelf, April 1977. *Deep Sea Res Part A. Oceanogr Res Pap* 1981; **28**(4):347–78.
23. Brooks DA, Mooers CN. Wind-forced continental shelf waves in the Florida Current. *J Geophys Res* 1977;**82**(18):2569–76.
24. Chew F. The turning process in meandering currents: a case study. *J Phys Oceanogr* 1974;**4**(1):27–57.
25. Brooks IH, Niiler PP. Florida Current at Key West – summer 1972. *J Mar Res* 1975; **33**(1):83–92.
26. Davis KA, Leichter JJ, Hench JL, Monismith SG. Effects of western boundary current dynamics on the internal wave field of the Southeast Florida shelf. *J Geophys Res Oceans (1978–2012)* 2008;**113**(C9).
27. Soloviev AV, Luther ME, Weisberg RH. Energetic baroclinic super-tidal oscillations on the southeast Florida shelf. *Geophys Res Lett* 2003;**30**(9).
28. Winkel DP, Gregg MC, Sanford TB. Patterns of shear and turbulence across the Florida Current. *J Phys Oceanogr* 2002;**32**(11):3269–85.
29. Leichter JJ, Deane GB, Stokes MD. Spatial and temporal variability of internal wave forcing on a coral reef. *J Phys Oceanogr* 2005;**35**(11):1945–62.
30. Leichter JJ, Wing SR, Miller SL, Denny MW. Pulsed delivery of subthermocline water to Conch Reef (Florida Keys) by internal tidal bores. *Limnol Oceanogr* 1996;**41**(7): 1490–501.
31. Kundu PK. Generation of coastal inertial oscillations by time-varying wind. *J Phys Oceanogr* 1984;**14**(12):1901–13.

32. Peng G, Mooers CN, Graber HC. Coastal winds in south Florida. *J Appl Meteorol* 1999; **38**(12):1740–57.
33. Schott F, Lee TN, Zantopp R. Variability of structure and transport of the Florida Current in the period range of days to seasonal. *J Phys Oceanogr* 1988;**18**:1209–30.
34. Xue H, Bane Jr JM. A numerical investigation of the Gulf stream and its meanders in response to cold air outbreaks. *J Phys Oceanogr* 1997;**27**(12):2606–29.
35. Crombie DD. Doppler spectrum of sea echo at 13.56 MHz. *Nature* 1955;**175**:681–2.
36. Stewart RH, Joy JW. HF radio measurements of surface currents. *Deep Sea Res and Oceanogr Abstract* December 1974;**21**(12):1039–49. Elsevier.
37. Gurgel KW, Antonischki G, Essen HH, Schlick T. Wellen Radar (WERA): a new ground-wave HF radar for ocean remote sensing. *Coast Eng* 1999;**37**(3):219–34.
38. Gurgel KW. Shipborne measurement of surface current fields by HF radar. In: *OCEANS'94. Oceans Engineering for Today's Technology and Tomorrow's Preservation. Proceedings*, vol. 3. IEEE; 1994. p. III–23.
39. Haltiner GJ, Williams RT. *Numerical prediction and dynamic meteorology*, vol. 2. New York: Wiley; 1980.
40. Chapman RD, Shay LK, Graber HC, Edson JB, Karachintsev A, Trump CL, et al. On the accuracy of HF radar surface current measurements: intercomparisons with ship-based sensors. *J Geophys Res Oceans (1978–2012)* 1997;**102**(C8):18737–48.
41. Flament P, Armi L. The shear, convergence, and thermohaline structure of a front. *J Phys Oceanogr* 2000;**30**(1):51–66.
42. Saucier WJ. Horizontal deformation in atmospheric motion. *Trans Am Geophys Union* 1953;**34**:709–19.
43. Futch V. *The Lagrangian properties of the flow west of Oahu* [thesis]. Manoa: University of Hawaii; 2009.
44. McWilliams J. The emergence of isolated coherent vortices in turbulent flow. *J Fluid Mech* 1984;**146**:21–43.
45. d'Ovidio F, Isern-Fontanet J, López C, Hernández-García E, García-Ladona E. Comparison between Eulerian diagnostics and finite-size Lyapunov exponents computed from altimetry in the Algerian basin. *Deep Sea Res Part I: Oceanogr Res Pap* 2009;**56**(1):15–31.
46. Beron-Vera FJ, Olascoaga MJ, Goni GJ. Oceanic mesoscale eddies as revealed by Lagrangian coherent structures. *Geophys Res Lett* 2008;**35**(12).
47. Poje AC, Haza AC, Özgökmen TM, Magaldi MG, Garraffo ZD. Resolution dependent relative dispersion statistics in a hierarchy of ocean models. *Ocean Model* 2010;**31**(1):36–50.
48. Haza AC, Özgökmen TM, Griffa A, Molcard A, Poulain PM, Peggion G. Transport properties in small-scale coastal flows: relative dispersion from VHF radar measurements in the Gulf of La Spezia. *Ocean Dyn* 2010;**60**(4):861–82.
49. Thomas LN, Tandon A, Mahadevan A. Submesoscale processes and dynamics. *Ocean Model Eddying Regime* 2008:17–38.
50. Walters RA, Heston C. Removing tidal-period variations from time-series data using low-pass digital filters. *J Phys Oceanogr* 1982;**12**(1):112–5.
51. Kielmann J, Düing W. Tidal and sub-inertial fluctuations in the Florida Current. *J Phys Oceanogr* 1974;**4**(2):227–36.
52. Mayer DA, Leaman KD, Lee TN. Tidal motions in the Florida Current. *J Phys Oceanogr* 1984;**14**(10):1551–9.

53. D'Asaro EA. The energy flux from the wind to near-inertial motions in the surface mixed layer. *J Phys Oceanogr* 1985;**15**(8):1043–59.
54. Ford R. Gravity wave radiation from vortex trains in rotating shallow water. *J Fluid Mech* 1994;**281**:81–118.
55. D'Asaro E, Lee C, Rainville L, Harcourt R, Thomas L. Enhanced turbulence and energy dissipation at ocean fronts. *Science* 2011;**332**(6027):318–22.
56. Alford MH, Shcherbina AY, Gregg MC. Observations of near-inertial internal gravity waves radiating from a frontal jet. *J Phys Oceanogr* 2013;**43**(6):1225–39.
57. Cushman-Roisin B. *Introduction to geophysical fluid dynamics*. Englewood Cliffs, New Jersey: Prentice Hall; 1994.
58. Mooers CN. Several effects of a baroclinic current on the cross-stream propagation of inertial-internal waves. *Geophys Astrophys Fluid Dyn* 1975;**6**(3):245–75.
59. Kunze E. Near-inertial wave propagation in geostrophic shear. *J Phys Oceanogr* 1985; **15**(5):544–65.
60. Lee CM, Eriksen CC. Near-inertial internal wave interactions with mesoscale fronts: observations and models. *J Geophys Res Oceans (1978–2012)* 1997;**102**(C2):3237–53.
61. Kunze E, Sanford TB. Observations of near-inertial waves in a front. *J Phys Oceanogr* 1984;**14**(3):566–81.
62. Granata T, Wiggert J, Dickey T. Trapped near-inertial waves and enhanced chlorophyll distributions. *J Geophys Res Oceans (1978–2012)* 1995;**100**(C10):20793–804.
63. Rainville L, Pinkel R. Observations of energetic high-wavenumber internal waves in the Kuroshio. *J Phys Oceanogr* 2004;**34**(7):1495–505.
64. Nagai T, Tandon A, Yamazaki H, Doubell MJ, Gallager S. Direct observations of microscale turbulence and thermohaline structure in the Kuroshio front. *J Geophys Res Oceans (1978–2012)* 2012;**117**(C8).
65. Jaimes B, Shay LK. Near-inertial wave wake of hurricanes Katrina and Rita over mesoscale oceanic eddies. *J Phys Oceanogr* 2010;**40**(6):1320–37.
66. Mied RP, Lindemann GJ, Trump CL. Inertial wave dynamics in the North Atlantic subtropical zone. *J Geophys Res Oceans (1978–2012)* 1987;**92**(C12):13063–74.

CHAPTER

12

Fine-Scale Tidal and Subtidal Variability of an Upwelling-Influenced Bay as Measured by the Mexican High Frequency Radar Observing System

Xavier Flores-Vidal[1,*], Reginaldo Durazo[2], Rubén Castro[2], Luis F. Navarro[1], Feliciano Dominguez[1], Eduardo Gil[1]

Instituto de Investigaciones Oceanológicas, Universidad Autónoma de Baja California, Ensenada, Baja California, México[1]; Facultad de Ciencias Marinas, Universidad Autónoma de Baja California, Ensenada, Baja California, México[2]

**Corresponding author: E-mail: floresx@uabc.edu.mx*

CHAPTER OUTLINE

1. INTRODUCTION

In 2009, a system of two HF radars was installed at the Bay of Todos Santos (BTS) in northwestern Mexico, about 60 miles south of the international border between Mexico and the United States of America. These two radars were the precursors of the Coastal and Regional Oceanographic Observatory (http://oorco.ens.uabc.mx), currently operated by the Autonomous University of Baja California in Mexico, in collaboration with the University of Hawaii, the Coastal Observing Research and Development Center (CORDC), and Scripps Institution of Oceanography (SIO) in the United States. Since 2012, the Coastal and Regional Oceanographic Observatory, or OORCo by its Spanish acronym, has measured and displayed data in real time from HF radars, weather stations, ocean drifters, and video monitoring. To date, OORCo operates three HF radars, which map hourly sea surface currents inside the bay. Installation of three new HF radars is expected shortly to expand coverage up to 200 km offshore. Additionally, OORCo developed its own low-cost GPRS telemetry drifters[1,2] that are released inside the BTS as often as possible. OORCo uses exclusively open source tools and GNU licenses, making it a low-cost observatory, which in addition measures and reports ocean data in near real time with scientific quality. The location of the BTS was particularly convenient for the development of this low-cost observatory, with the campus of the Autonomous University of Baja California located at the north shore of the bay.

The BTS is a relatively small semi-enclosed body of shallow water located at 31.85°N and −116.75°W on Baja California, Mexico. Its main axis is approximately 16 km long, and it has an average depth of 40 m. Two entrances communicate the bay with the open ocean; the north entrance is wider ($\approx$10 km) and shallower ($\approx$30 m) than the south entrance, which is narrow ($\approx$4 km) and has a deep canyon of about 450-m depth (Figure 1). Numerical and observational studies[3–5] suggest that local circulation depends on the conditions of the adjacent Pacific Ocean (i.e., the California Coastal Current, CCC). The first circulation schemes obtained with Lagrangian drifters[6] proposed a southward current that enters the bay through the northern portion and a counter current streaming northward from the southern

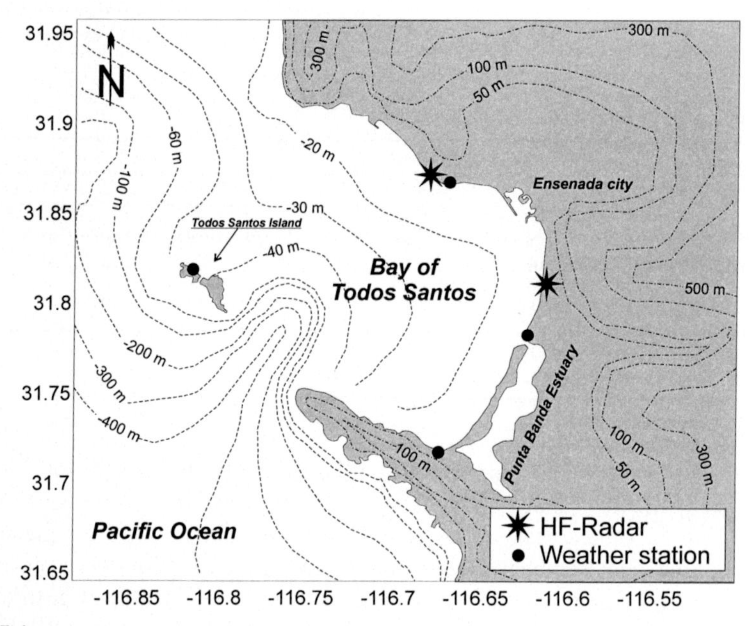

FIGURE 1

Bay of Todos Santos. Dashed lines with negative numbers represent the bathymetry, dash-dot lines with positive numbers represent the topography (both estimated from Etopo1). The instrumental setup is indicated by the stars (HF radars) and by the bold dots (weather stations).

portion of the bay, hence, a convergence zone in the central-eastern zone of the bay. A numerical study[7] proposed wind-forced patterns by means of a two-dimensional numerical model forced with homogeneous wind fields. This study suggested a very similar circulation pattern to that proposed by Alvarez-Sanchez et al.,[6] in which the northwestern portion of the BTS shows a clockwise rotation while the southeastern portion a counterclockwise rotation. To date, neither the scheme proposed by Alvarez-Sanchez et al.[6] nor that by Argote-Espinoza et al.[7] (hereinafter Al–Ar) have been clearly observed. It should be noted that the Al–Ar studies are considered classic, and they are practically accepted as facts by the local scientific community.

Recently, the local circulation of the BTS was studied by numerical simulation using ROMS (Regional Ocean Modeling System) forced with the CCC.[3,4] The CCC conditions were found to modify the mean circulation inside the BTS on time scales of a few weeks. Moorings, SST derived from satellite products, and oceanographic cruises[5,8,9] reported a mean circulation similar to the classical scheme of Al–Ar, and they confirmed dependency of BTS circulation on CCC conditions, in agreement with Mateos et al.[3]

However, none of these studies proposed a different scheme, forcing mechanism, or physical process from those reported in the pioneering studies of Al–Ar. Furthermore, diurnal and semidiurnal variability has not been addressed yet; hence, the spatial and time scales relevant in the interactions between the CCC and the BTS remain to be determined.

In this work, we used modern synoptic data measured by HF radars to examine the importance of the CCC and its interaction with the currents inside the BTS. In addition, we studied the influence of the synoptic wind field on the BTS surface currents.

1.1 METHODOLOGY

The present study used surface current measurements registered along a period of two years, 2010–2011. This lapse was selected for the quality of data and because gaps in the time series were almost negligible. Data were obtained by two HF radars well known as Coastal Ocean Dynamics Application Radars (CODAR) SeaSonde, operating in direction-finding mode at 25 and 25.5 MHz, respectively, and yielding a radar wave length of 12 and 11.76 m, hence an effective depth of 0.54 and 0.52 m, respectively. Both radars operated using transmission bandwidth of 300 kHz (i.e., radial resolution of 500 m) and transmission peak power of 50 W (i.e., range of about 20 km). The integration time to compute radial currents was 1 h, yielding 6 cm/s of accuracy. Total sea surface currents, composed of radial currents measured at each radar site, were computed over a Cartesian grid of 1 km resolution. Figure 2 shows the percent of data return at each grid cell, from which the highest percentage associated with more orthogonal radial currents can be inferred. Hourly time series of two years of data are between 70% and 100% complete at the inner bay.

In order to study the frequency-dependence variability, sea surface currents mapped by the HF radar were band-pass filtered using a Lanczos filter with windows

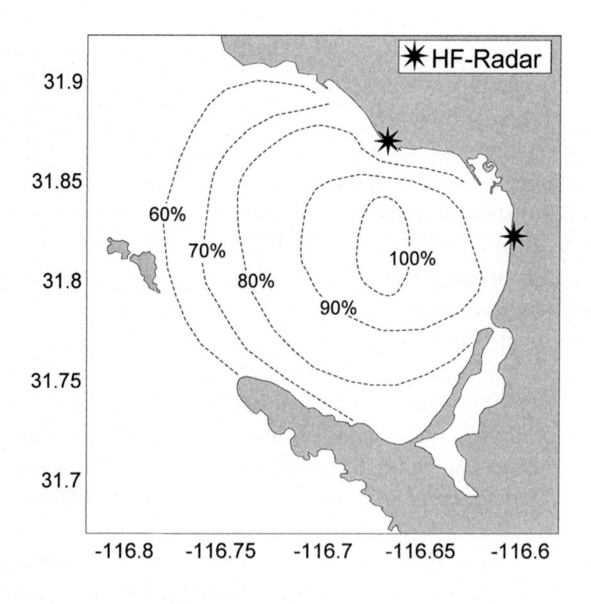

FIGURE 2

Percentage of data return at each grid cell; integrity of the time series can be inferred.

centered at 12 and 24 h, respectively, as well as a low-pass filter with a cutoff period of 72 h. The length of the time series allowed spectral partitions of 10 days, yielding spectral partitions of $N = 240$ points and a Nyquist frequency of 12 cycles per day (cpd). A Blackman-Harris window was applied to reduce side-lobe leakage at each partition. For each case, the 95% confidence interval was estimated.

In addition to the spectral and filtering analysis, we least-square fitted six tidal harmonics constituents (M2, S2, L2, K1, O1, and Q1) to the HFR time series to reproduce the tidal circulation. The divergence—convergence was calculated on the two-dimensional (horizontal) vector field to provide information about the variations in vertical velocity. For an incompressible fluid, the relationship of the three-dimensional divergence to two-dimensional divergence can be represent as follows:

$$\nabla \cdot \vec{U} = \partial u/\partial x + \partial v/\partial y + \partial w/\partial z \approx 0, \tag{1}$$

therefore,

$$DIV_h = -\partial w/\partial z, \tag{2}$$

which means that the horizontal divergence is associated with variations in the vertical velocity.

Another characteristic of the divergence—convergence concept is that $DIV_h = 0$ defines fringes (or borders) where particles flow always tangentially (i.e., cannot cross these areas). Finally, empirical orthogonal functions (EOFs) analysis was applied to separate spatial varying signals into basic modes of variability or empirical modes. For this, we considered the components of the surface currents as

Table 1 Characteristic and Basic Information of the Drifters Deployments

Experiment	Dates	Number of Drifters	Days of Deployment	Buoy-Hours
1	June 28–30, 2012	5	3	156
2	July 5–8, 2012	7	4	450
3	August 28–31, 2012	6	4	498
4	October 9–13, 2012	6	5	278
5	January 31–February 7, 2013	8	8	745
6	April 29–May 4, 2013	7	6	529
7	September 18–26, 2013	5	9	720

independent measurements for each grid node (time series), and then we obtained the empirical modes using singular value decomposition (SVD). Typically, the modes are sorted by their value of global or total explained variance, but sorting modes by their total variability is by no means automatically related to dynamic processes, as the statistical decomposition may fail. Additional information about the local significance of specific modes can be achieved by the local explained variance. The local explained variance is, therefore, the percentage of explained local variability (at a certain grid point), which is captured by the respective mode. The local explained variance or spatial variance has been previously used to study pronounced spatial structures.[10–12]

The HF radar surface currents were complemented by Lagrangian data obtained from several drifters released inside the bay during seven field experiments of about 3 to 9 days length each (Table 1). The drifters were armed with a drogue at 1 m depth and were GPS WAAS-enabled, yielding accuracy of 1 to 3 m on the horizontal plane. Data obtained from drifters were used to evaluate the dispersion and aggregation patterns inside the bay. An Eulerian grid with resolution of 1 km was used to compute all statistics on particle dispersion. Approximately 50 trajectories of 4 to 12 days, each spread over one year, were used to estimate the percent of drifters inside each grid cell, and thus to provide a first view of the recirculation zones inside the bay. A more detailed analysis consisted in the estimation of the aggregation index defined as follows:

$$\langle C_{ij} \rangle = \frac{T^{-1}\sum_{t=0}^{T} C_{ij}(t)}{\overline{c}}, \tag{3}$$

where $\overline{c} = N_o/B$ is the expected number of particles per grid cell, with N_o = number of particles, and B = the number of grid cells. Particle concentration in each grid cell is then $C_{ij}(t)$, where i and j are the grid cell index and t a given time. Hence, the aggregation index is defined as the ratio of $C_{ij}(t)$ and $\overline{c}$ average over the total period T. This index will be >1 when the particles aggregate, and <1 when the particles disperse.[13]

Table 2 Characteristic and Basic Information of the Wind Data. Wind Direction Is Expressed in Oceanographic Convention

Weather Station	Parameters	Accuracy	Dates	Average
North station	Wind speed (m/s)	0.05	February 1–September 30, 2014	4.5
	Wind direction (°N)	0.1		192
East station	Wind speed (m/s)	0.05	May 1–September 30, 2014	4
	Wind direction (°N)	0.1		170
South station	Wind speed (m/s)	0.05	March 1–September 30, 2014	4.8
	Wind direction (°N)	0.1		120
West station	Wind speed (m/s)	0.05	February 1–September 30, 2014	5.1
	Wind direction (°N)	0.1		120

Additionally, we used seven months of wind data measured simultaneously at four locations inside the bay. These data were obtained via low-cost weather stations with enough time resolution and accuracy. Wind data were registered every second and averaged in 10 min bins. Table 2 shows the details of the wind data and the length of the time series.

Although the HFR data analyzed in this work were not obtained simultaneously with the weather stations and the drifters, the long-term circulation (subtidal), mean flow, and diurnal variability (either tidal or sea breeze) agreed between data sets.

2. RESULTS

Figure 3 shows the wind synoptic variability estimated from the four weather stations installed inside the bay. Mean vector (bold arrow), total variability (bold-line ellipse), diurnal variability (dashed ellipses), and synoptic low-frequency variability (dotted ellipses) are plotted at each location. Of the four weather stations, the three at the southern half-portion of the bay (i.e., westernmost, easternmost, and southernmost stations) show the mean vector, total variability, and diurnal variability aligned from northwest to southeast; whereas, the northernmost station is quite different with total variability aligned from northeast to southwest. It should be emphasized that the diurnal variability (sea breeze) is aligned from west to east at the four stations, and it is the most energetic component of the total variability. On the other hand, the synoptic low-frequency variability tends to decrease in importance at the inner southernmost and easternmost stations, while no clear tendency is noticed for the ellipse orientations. Two key features should be noted: first, the differences in behavior of the northernmost station, which has its low-frequency variability as energetic as its diurnal variability, and is normal to the shore and about 90° off compared with the other three stations. The second key feature is the increment in diurnal importance of the inner stations (southernmost and easternmost). These

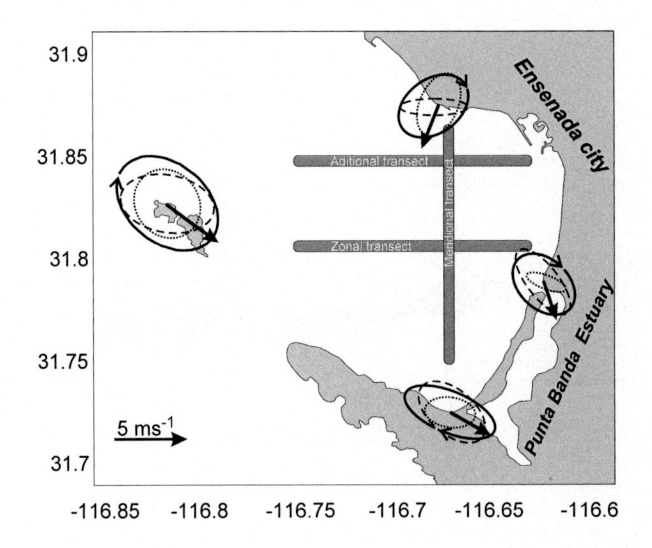

FIGURE 3

Bold-line ellipses are the total wind variability and bold-line arrows are the mean vectors at the four weather stations selected for this work. Additionally, the diurnal variability ellipses (dashed ellipses) and the low-frequency synoptic ellipses (dot ellipses) are plotted. The transects presented in Figures 11–13 are also indicated.

results point out about the importance of the fine-scale variability of the wind field inside the BTS.

The synoptic response of the sea-surface currents inside the bay may be studied from the HFR data. Rotary power spectrum is presented in Figure 4; it is an average of the whole two-dimensional domain (the whole HF radar foot print), and it is meant to depict the most important ocean variability inside the bay. The length of the time series allowed an estimation of the power spectra with enough resolution to distinguish between the M2 and S2 tidal constituents. Clockwise rotation dominated the whole bay in both diurnal and semidiurnal oscillations. The diurnal variability is due to the K1 tidal harmonic constituent and not to inertial motions, which have a period of $T_i = 21.8$ h at this latitude. Likely because the bay is small enough and dynamically disconnected from the open ocean, inertial motions were not detected in the inner bay (at least at the surface). The most energetic signals came from the tide, as is depicted in Figure 5(a)–(c), which illustrate the K1, M2, and S2 amplitudes (cm/s) and phases (degrees from GMT). As shown, the amplitudes for the three tidal constituents increased in the northwestern portion of the bay, where the co-tidal (phase) lines tend to be parallel to the isobaths and where the bay is shallow. Elsewhere inside the bay, the tidal amplitude was negligible, except at the mouth of the Punta Banda estuary, where an important increment of the M2 amplitude was noticed.

Figure 5(d)–(f) show the tidal ellipses. The K1 constituent reproduced ellipses along the isobaths in the northwestern portion and center of the bay, which may

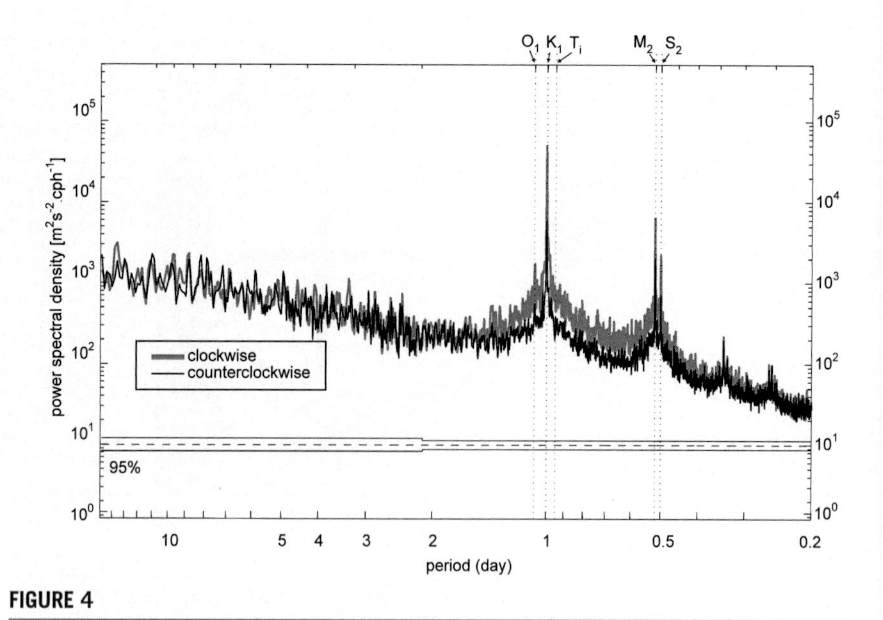

FIGURE 4

Rotational power spectra from the spatially averaged time series. The gray line is the anticyclonic component of the spectrum; the black line is the cyclonic component.

suggest local processes of tidal rectification due to lateral shear caused by the depth-distributed friction force (bathymetry). Co-tidal lines or phases of K1 suggest this tidal component to come from the southwest, in agreement with Godin.[14] Alignment or tendency of the M2 and S2 ellipses are barely noticeable due to their negligible amplitude, except at the northern entrance of the bay and on the entrance-mouth of the Punta Banda estuary, where the M2 amplitude was large enough. At the northern entrance, the M2 ellipsis lacks eccentricity, which suggests that its direction or tendency is not well established. At the southwest of the Punta Banda estuary entrance, the M2 ellipses suddenly become more elongated, which in turn suggests a well-defined tendency from northeast to southwest, as produced by the ebb–flood tidal cycle. The effect of the tide was clearly more evident at the southern portion of the Punta Banda mouth.

Figure 5 indicates that the tidal (K1, M2, and S2) variations are more important in the northwestern portion of the bay, which reveals the importance of fine-scale tidal variability. Particularly, the K1 and M2 tidal constituents were well reproduced by the surface currents data at the northwestern portion of the HF radar footprint. Figure 6 shows the variance explained by the tide (six tidal harmonic constituents M2, S2, L2, K1, O1, and Q1). Two spots of about 50% of tidal-explained variance can be noticed at the northern portion of the bay and at the entrance of the estuary. The latter spot may be attributed to the estuarine flood and ebb cycle, which develops important tidal currents. The former spot (northern spot of 50%

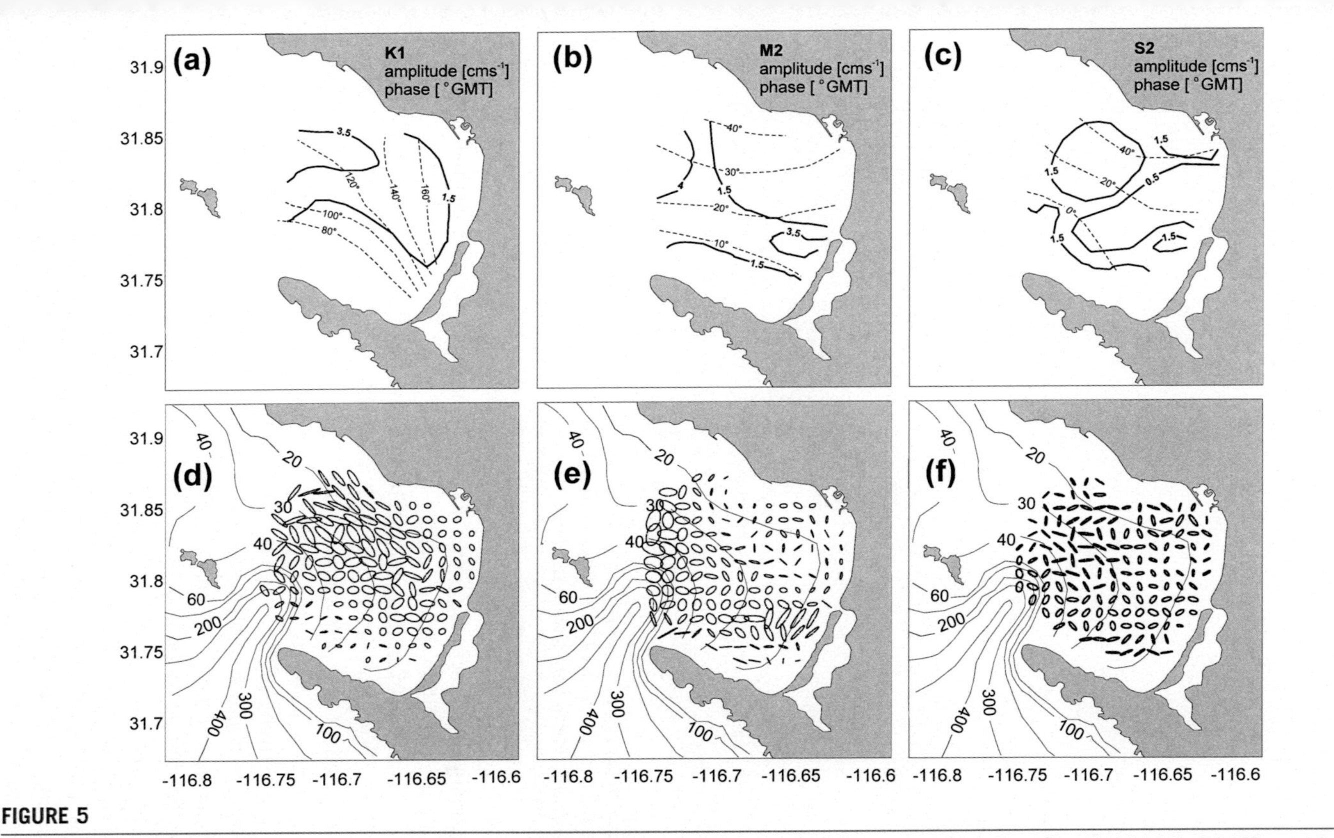

FIGURE 5

Tidal ha-monic analysis obtained from two years of HF radar data. Amplitude (bold lines) and phase (dashed lines) of (a) K1, (b) M2, and (c) S2. Ellipses ɔf tidal variability for the corresponding harmonics, as well as the bathymetry, are shown on panels (d), (e), and (f).

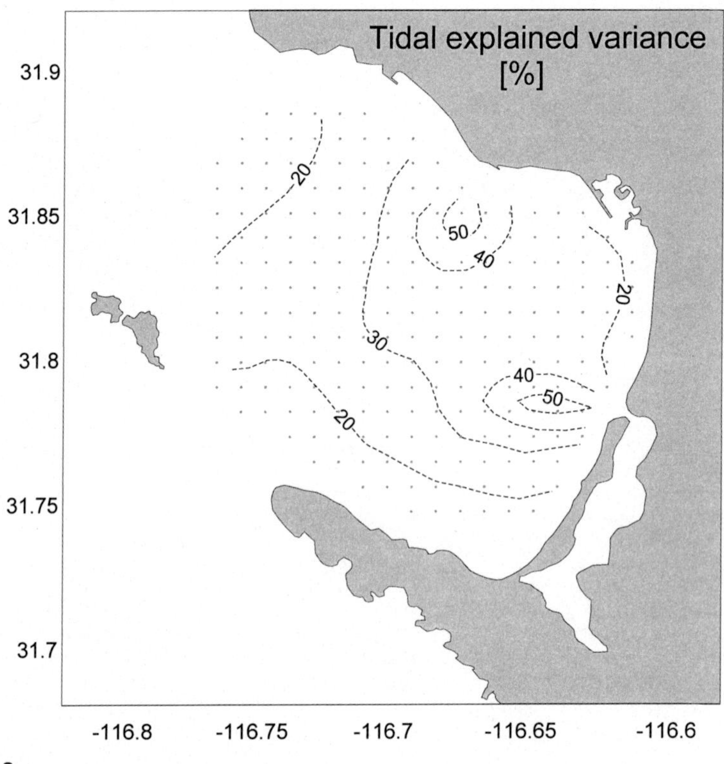

FIGURE 6

Tidal-explained variance obtained from the harmonic analysis. The gray dots represent the grid over which time series were obtained.

tidal-explained variance) requires a detailed revision of the total, diurnal, and semi-diurnal currents within the inner bay.

We will start by examining the mean field, divergence, and variability of the subtidal (Figure 7(a)), diurnal (Figure 7(b)), and semidiurnal (Figure 7(c)) surface currents. In subtidal and semidiurnal currents, the variability ellipses tend to rotate anticyclonically (CW) in the center of the bay and cyclonically (CCW) in the shallower portions of the bay, i.e., in the southern, eastern, and northern shores. However, it should be noted that one fringe or margin of CCW rotation is also observed at the border between the bay and the open ocean. CCW rotation is not present in the diurnal variability, where most of the domain displayed CW rotation, except the area near the Ensenada harbor. The convergence (dark shadow contour) and divergence (light shadow contour), estimated by the vector fields, are in agreement with the rotary ellipses (i.e., convergence spots are related with CW ellipses). The currents pattern in Figure 7(a)–(c) is southward with a curvature toward the east as it approaches the eastern limit of the bay, more evident in the subtidal mean flow

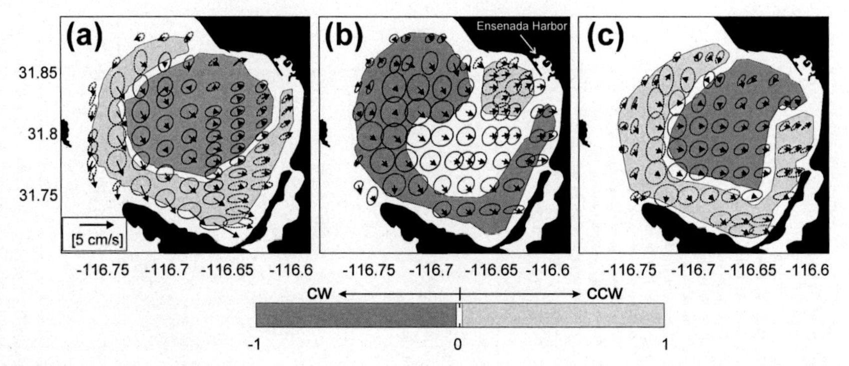

FIGURE 7

Divergence fields (shadow contours), variability ellipses, and mean vector of (a) subtidal, (b) diurnal, and (c) semidiurnal currents. Light-shadow (dark-shadow) contours represent normalized divergence (convergence). Dashed-line (solid-line) ellipses have CCW (CW) rotation.

(Figure 7(a)). This currents pattern could be the effect of a larger dimension current (synoptic or low frequency), which flows almost unidirectional toward the south along the outside edge of the bay with refraction or rectification of the flow as it enters the bay. However, another possibility that can explain the currents pattern is the synoptic wind field (Figure 3). Regardless of its origin, the exact part of the bay where the outside flow turns and enters the bay will certainly cause a CCW curvature of the flow, which may later be corrected by the local forcing of the bay (e.g., bathymetry, tides in shallow waters, and fine-scale wind fields).

To elaborate further on the proposed scheme, we herein present the EOFs analysis applied to the subtidal, diurnal, and semidiurnal surface currents. Figure 8(a)–(c) present the spatial mode 1 and the corresponding spatially or locally explained variance (Figure 8(d)–(f)). In the subtidal mode 1 (Figure 8(a)), the margin that "closes" the bay and divides the inner circulation from the outer southward current is evident. Nevertheless, it is also important to note that the northernmost portion (hereinafter referred to as NMP) near the Ensenada harbor diverges from the rest of the bay by exhibiting quite a different spatial mode (i.e., currents along the shore). The local variance explained at the NMP of the bay is about 60%, whereas in the remaining bay, it is more than 70%. The second mode of this subtidal variability (Figure 9(a)) also shows a clear margin or border between the outer flow and the inner bay, and the local variance explained by that mode in the NMP is approximately the remaining 40%. This means that spatial mode 2 also explains the currents in the NMP as an offshore flow.

Mode 1 of the diurnal band (Figure 8(b)) is almost homogeneous toward the southeast, similar to the mean diurnal flow, except for the NMP, where it appears to have a different direction that extends offshore toward the south. The locally explained variance of mode 1 is higher at NMP (about 85%). Mode 2

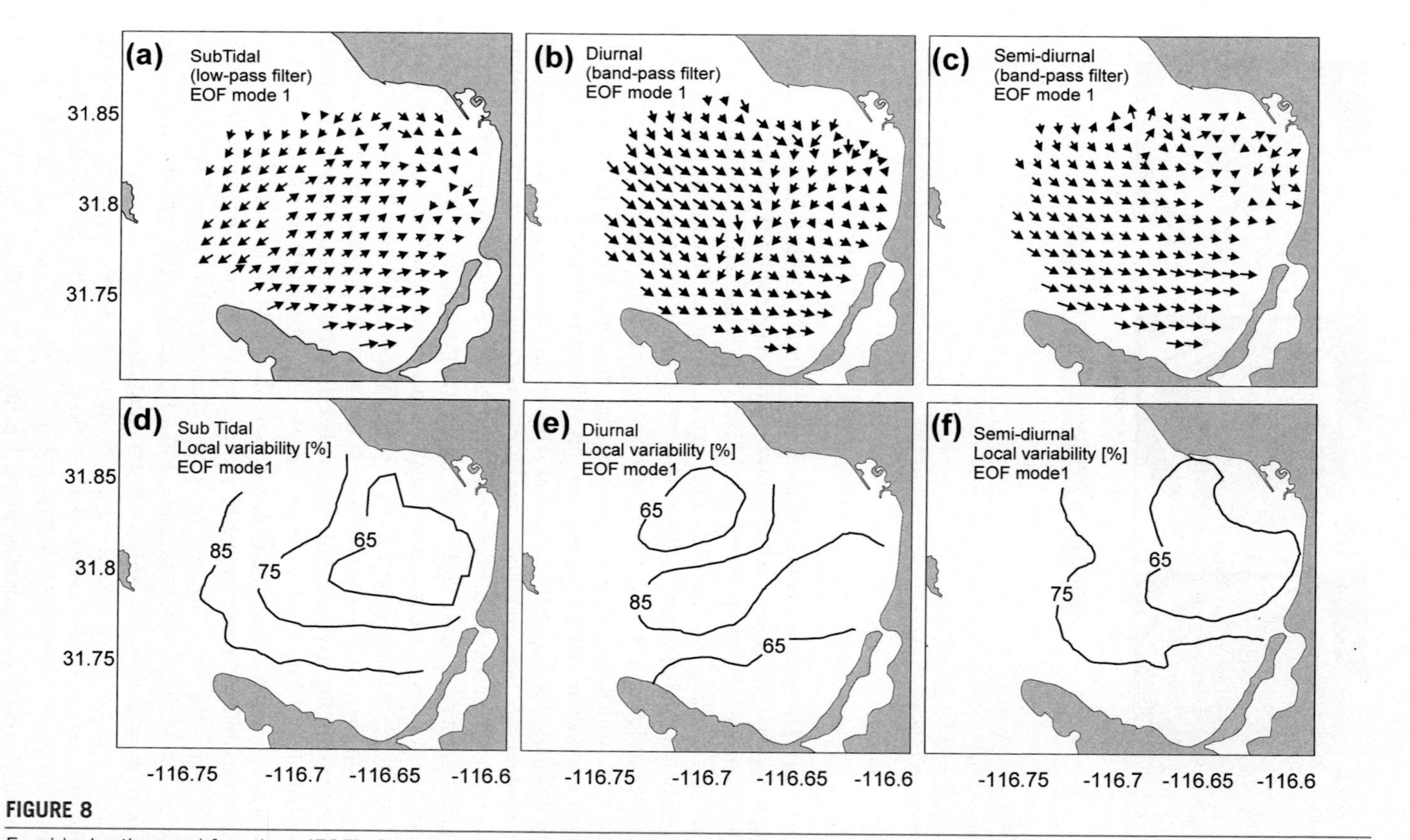

FIGURE 8

Empirical orthogonal functions (EOF). Spatial mode 1 (panels (a), (b), and (c)) and spatially explained variance (panels (d), (e), and (f)) of (a) subtidal, (b) diurnal, and (c) semidiurnal currents.

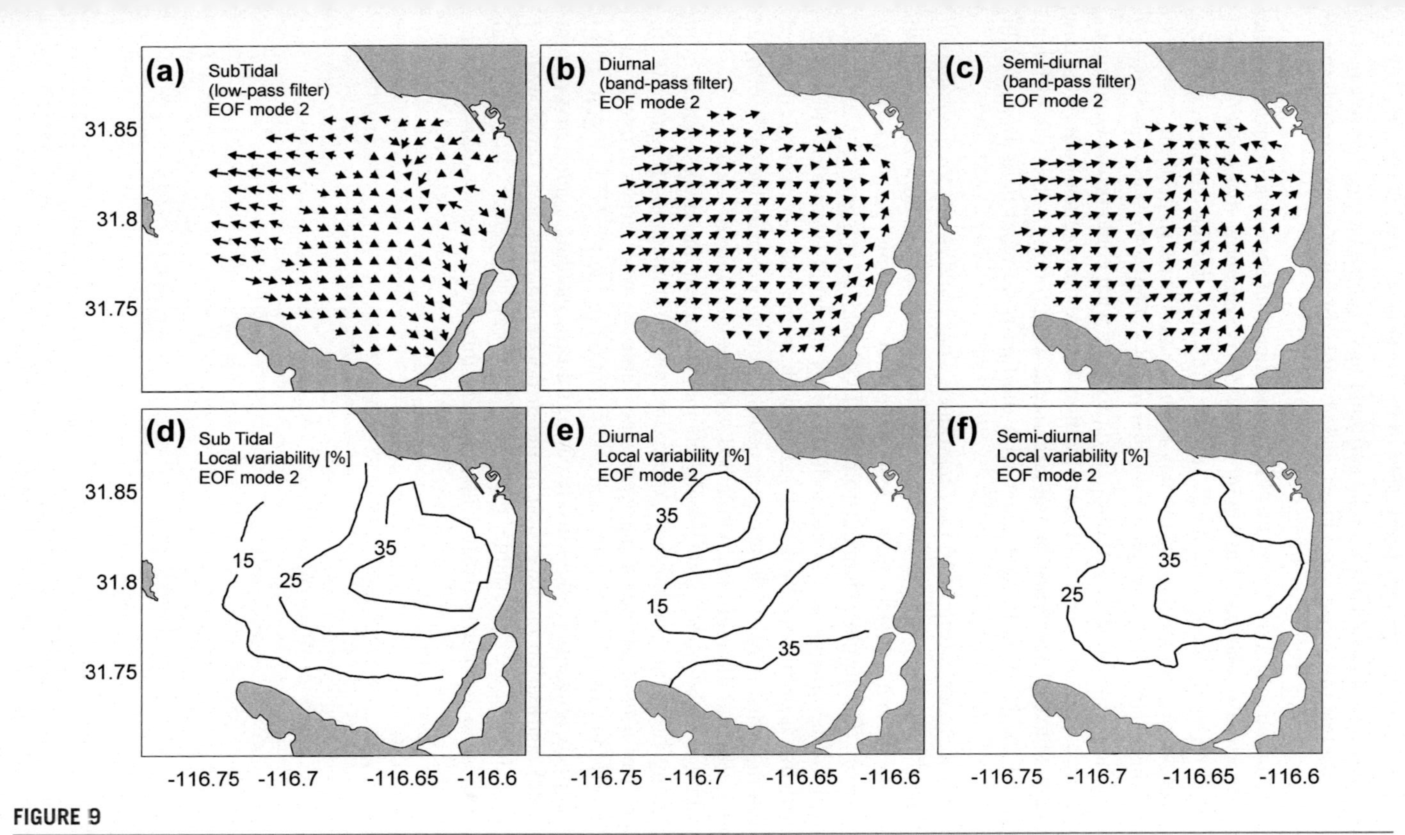

FIGURE 9

Same as Figure 8 but for mode 2.

(Figure 9(b)) explains approximately 15% of the variance in that area, but it explains a greater percentage (about 35%) in the surroundings. Note the alongshore current at the east coast that this mode reproduces, and also that direction of the currents in the northern portion of the bay is similar to K1 variability.

The semidiurnal mode 1 (Figure 8(c)) reproduces the southward current in the southern portion of the bay, but again, the NMP is somehow disconnected from the main flow, while the second mode (Figure 9(c)) also exhibits a fine-scale variability that separates the south of the bay from the northeastern portion.

From this EOF analysis, we can argue that mode 1 (Figure 8) of the three studied bands (subtidal, diurnal, and semidiurnal) reproduces what could be the wind circulation of the bay. In subtidal circulation, the flow toward the southeast out of the bay could be potentially developed by the large-scale wind forcing, while the different behavior of the NMP is in agreement with the fine-scale wind field presented in Figure 3.

Therefore, mode 2 (Figure 9) may reproduce what could be the tidal circulation of the bay because it explains more closely the circulation in the northwest limit (Figure 9(e)) and shows more dependence of shallow parts of the bay (Figure 9(b)). The results obtained from this EOF analysis suggests the existence of a shadow-like area, which is not influenced by the wind (west of NMP) where the circulation is dominated by the tide, along with an area that is directly influenced by the diurnal wind (NMP). These results could potentially explain the CCW ellipses in Figure 7(b) and the high tidal-explained variance in Figure 6.

In addition to the proposed fine-scale scheme, we present the analysis made from the Lagrangian drifter trajectories. Approximately 50 trajectories (Figure 10(a)), spanned over a total period of one year (see Table 1), were gridded into a 1 km by 1 km grid to evaluate the dispersion and retention inside the BTS. Figure 10(b) shows the percentage of drifters inside each grid cell, yielding a good first insight of the total distribution or dispersion of the drifters inside the bay. Three areas of high accumulation are discernible: the NMP of the bay, the center of the bay, and one area close to the estuary (east). Unfortunately, this simple analysis only provides

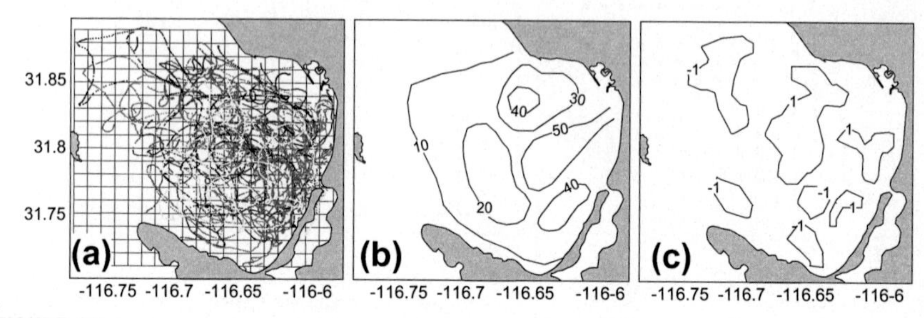

FIGURE 10

(a) Approximately 50 Lagrangian drifter trajectories, (b) percentage of occurrence over a grid of 1 km by 1 km, and (c) index of aggregation–dispersion.

a bulk percent of occurrence but no information about the real accumulation or dispersion of the drifters, which is time dependent. To evaluate the real accumulation or dispersion of the drifters as time evolves, Figure 10(c) shows the results obtained by Eqn (3) (aggregation index), which define more accurately the zones inside the bay where the drifters tended to accumulate over the whole study. This area is precisely linked with the orographic wind shadow-like area proposed in this study, and with the margin that divides the inner bay from the adjacent ocean. It is worth nothing here that not a single drifter left the bay during the present study; moreover, most of them never crossed the fringe or margin shown in Figures 7(a), 8(a), and 9(a). A few drifters were heading very rapidly ($\approx$50 cm/s) toward the northwest (outside of the bay) but turned back as soon as they reached the bay's limit (e.g., northwesternmost trajectories in Figure 10(a)).

However, although the drifters could provide a good insight into the effect of the shadow-like zone and the evidence of a margin between the inner bay and the adjacent ocean, the long-term nature of the data base only allowed a direct comparison between them and the long-term or subtidal circulation. Figure 11 shows one year of subtidal circulation across two transects. Component U (east–west) is plotted for the meridional transect (Figure 11(a)) and component V (north–south) for the zonal transect (Figure 11(b)). For further reference, Figure 3 shows the position of the two transects. As shown, the currents tend to enter the bay in the southern portion, while currents exiting the bay are more common in the northern portion (Figure 11(a)). By continuity, the flow tends to move southward in the west and northward in the east (Figure 11(b)). Although the main core of the southward current spreads almost over the whole transect, some events reach its eastern limit at around $-116.70°$, and they are precisely those that have a counter count flow toward the north at the east of the transect, i.e., a pattern similar to Figure 7(a). This could constitute evidence of the relationship between the intensification of the current at the adjacent ocean and the development of a barrier or dynamic border between the inner bay and the adjacent ocean.

The temporal variations of these features were studied by three-dimensional power spectral analyses. Figure 12 shows the power spectral density of the surface currents as a function of the period at the same transects as in Figure 11. The time series were long enough to define the variability from a few hours up to one month with good confidence. Starting with the diurnal and semidiurnal signals, they were well defined zonally and meridionally all along the transects; only the semidiurnal signal tended to become attenuated at the eastern limit of the bay (Figure 11(b)) and north of the estuary entrance (see Figure 3 for transect location). Longer period signals can be identified at $\approx$18 days in both, south of the meridional transect and west of the zonal transect, and were better defined in the latter. This 18-day variability attenuates as it enters the bay and is apparently restricted at $\approx -116.70°$ (eastern limit) and $\approx 31.77°$ (northern limit). Another identifiable long-term energetic signal is an oscillation of about 28 days. This variability is well defined at the south of the meridional transect between $\approx 31.76°$ and $31.78°$, but it is barely identified in the zonal transect at $\approx -116.68°$. Two schemes are possible: On the one hand,

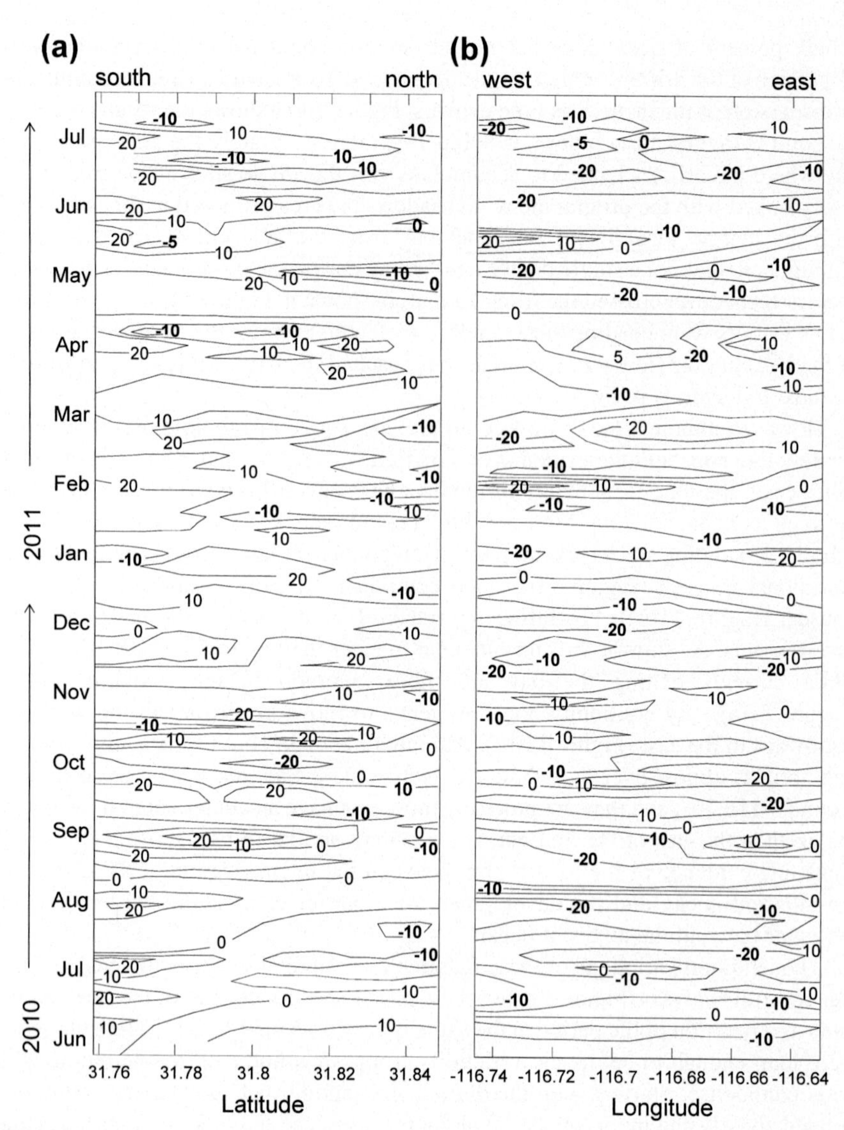

FIGURE 11

Meridional (a) and zonal (b) transects of east–west (a) and north–south (b) speed (cm/s) components of subtidal variability as measured by the HF radars. See Figure 3 for transect position.

the long-term energy (period >10 days) is remote with clear signals that enter the bay and is attenuated (at least at the surface) in the inner bay. On the other hand, the ≈18-day oscillation is just noticed at the submarine canyon region, whereas the ≈28-day oscillation is located in the inner bay at the shallow portion

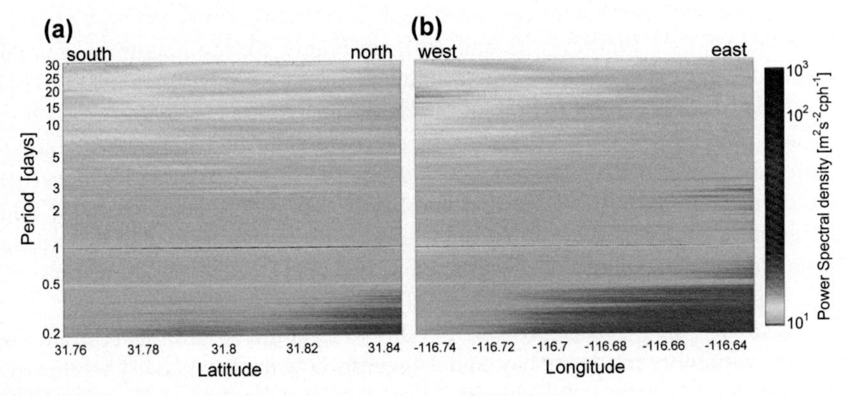

FIGURE 12

Power spectral density obtained over two transects as a function of latitude (a), longitude (b), and period. See Figure 3 for transect position.

of the bay ($\approx$20 depth). The former could be evidence of the influence of larger (mesoscale) features such as the CCC; the latter could be the effect of the shear caused by friction (bathymetry), which transfers dominant frequencies to other harmonics.

Another process that could influence the incursion of the long-term remote variability inside the bay is the local fine-scale synoptic wind field. Figure 13 is the

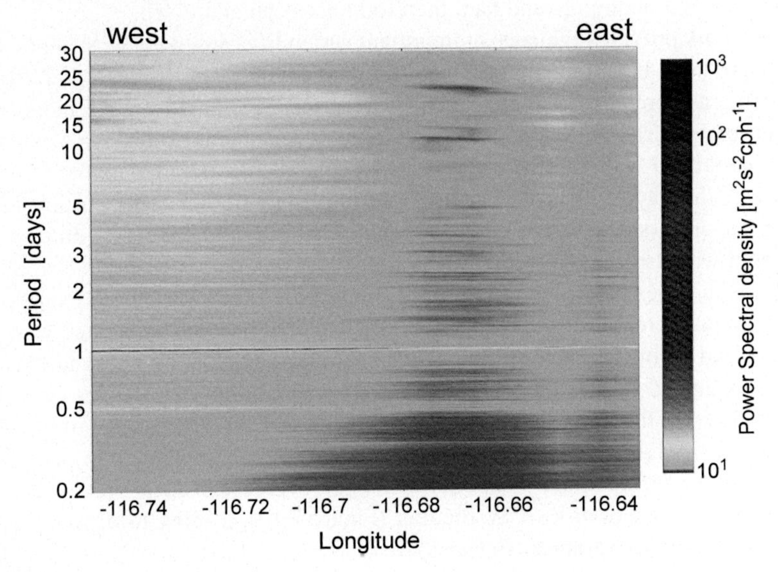

FIGURE 13

Power spectral density obtained over one transect as a function of longitude and period. See Figure 3 for additional transect position.

three-dimensional spectrum of one additional zonal transect near the northernmost part of the bay (see Figures 3, 6, and 7(b)). In Figure 13, the long-term, diurnal, and semidiurnal energy bands show a dramatic attenuation between $\approx -116.68°$ and $-116.66°$. This is right in front of the Ensenada harbor (see Figure 7(b)) where the wind could influence the sea surface by decreasing the tidal and subtidal energy, and probably lowering the sea level, which develops CCW circulation (Figure 7(b)). Next to this area (northward), we find the shadow-like area, a zone where the tidal effect is more evident (see Figure 6) and where the sea surface should show an upset that should be related with a deep thermocline, and convergence, hence the accumulation of particles (see Figure 10(b) and (c)).

In summary, the evidence presented in this work points toward the relevance of fine-scale variability inside the bay, and it suggests how the local and synoptic forcing modulates and defines the complex circulation of this relatively small bay.

3. DISCUSSION AND CONCLUSIONS

Although the BTS is relatively small, its hydrodynamic properties are complex. Previous studies have described its dynamics by isolating different time scales and forcing, such as wind-driven currents,[7] thermohaline and geostrophic circulation,[5,9] influence of the CCC,[3,4] sediment dynamics,[15] dispersion,[6] as well as biological aspects.[16] However, the importance of the synoptic (spatial) variability of the wind forcing has not been previously evaluated. Furthermore, tidal circulation was considered negligible and had, therefore, not been addressed.

This work provides evidence of important variability caused by the synoptic fine-scale wind field, in particular, the development of a wind shadow-like area (WSA) at the northern shore of the bay. This WSA was observed in data from HF radars (surface currents), drifters, and wind records. The development of this WSA is apparently associated with northwest and north winds that impinge on the surface waters of the bay only in its southern portion; thus, the northern portion of the bay is covered (shadowed) by the mountain chain around the bay's north shore. The latter was also noted by Reyes and Parés[17] who reported similar wind variability results as the ones presented in this work (Figure 3). The WSA should then experience a slight increase of local sea level, which should become associated with CW rotation and a convergence process. This was observed as an increase in the variability explained by the tide (Figure 6) and accumulation of particles or drifters (Figure 10). On the other hand, the area to the east of the WSA is the place where diurnal local winds impinge on the surface waters (see Figures 3, 6, 7(b), and 8(b)). There, the sea surface should experience a slight decrease of local sea level, which should be associated with CCW rotation (Figure 7(b)), leading to divergence and less tidal explained variance (Figure 6).

Based on phytoplankton counts, García-Mendoza et al.[16] reported the genesis of a harmful algae bloom on the northernmost part of the inner BTS. This could be explained as recirculation associated with CW circulation near an area of thermocline doming by action of the local wind, causing ventilation of the photic zone and favorable conditions for algae blooms.

The tidal circulation only becomes important in two areas inside the bay, the northern part and the surroundings of the estuary. However, regardless of the energy band (diurnal, semidiurnal, or subtidal), the currents inside the bay tend to have CW rotation, while the outside variability (at the adjacent ocean) apparently has CCW rotation. The proposed scheme should lower the thermocline at the center of the bay, leading to water convergence and a margin of zero divergence in the boundary between the inner bay and the adjacent ocean, which is a physical barrier that divides the water out of the bay from the water inside the bay. It should be noted that this scheme is valid only for surface currents (1-m depth) and might not hold for circulation at other depth levels. Nevertheless, the scheme presented by Mateos et al.,[4] Pérez-Brunius and Calva Chávez,[5] and Kurczyn et al.,[9] which proposed the bay as an upwelling trap influenced by the CCC, is complemented and extended in this work. The upwelling trap needs to be further analyzed, especially in terms of water masses and not only as a surface signal of currents. Finally, the influence of the CCC surface signal was apparently observed only under moderate wind conditions or on the subtidal circulation (long-term variability). Strong wind conditions developed a mean flow toward the inner bay; this flow (Figure 7(a)) brings waters from the adjacent ocean into the bay that could be trapped there if the wind attenuates and allows the CCC to develop the margin (Figure 9(a)) that closes the bay.

ACKNOWLEDGMENTS

This work was supported by grants to X. Flores-Vidal, R. Durazo, and R. Castro, funded by the programs-institutions; PROMEP (ref: 10166), UABC (ref: 376 and 623), and CONACyT-CONAGUA (ref#143803). We sincerely thank professors P. Flament and E. Terril for their support on developing the Mexican High Frequency Radar Network. R. Durazo acknowledges sabbatical scholarships granted by CONACyT and UABC. Finally, we thank the anonymous reviewers who improved this work with helpful suggestions.

REFERENCES

[1] Gómez-Liera JA, Flores-Vidal X, Domínguez Preciado F, García Guerrero EE, Inzunza González E, Navarro Olache LF, et al. Manual de un sistema de telemetría GSM/GPRS acoplado a sensores oceanográficos. In: *Techical report ref: DORIS_RT122013*. Autonomous University of Baja California, Ocean Research Institute UABC-IIO; 2013. p. 101.

[2] Gómez-Liera JA. *Diseño de un dispositivo electrónico autónomo con telemetría* [MSc. thesis]. School of Engineering Architecture and Design. Autonomous University of Baja California; 2015. pp. 88.

[3] Mateos E, Marinone SG, Parés-Sierra A. Towards the numerical simulation of the summer circulation in Todos Santos Bay, Ensenada, B. C., México. In: *Ocean modeling*, vol. 27; 2009. p. 107–12.

[4] Mateos E, Marinone SG, Lavin MF. Numerical Modeling of the coastal circulation off northern Baja California and southern California. In: *Continental shelf research*; 2013. p. 50–66.

[5] Pérez-Brunius P, Calva Chávez MA. *Factores que determinan la variabilidad estacional de la estratificación en la Bahía de Todos Santos, B. C.* 2013. Abstract in Mexican Geophysical Union Meeting, Puerto Vallarta, Jalisco.

[6] Alvarez-Sanchez LG, Hernandez WR, Durazo AR. Patrones de deriva de trazadores lagrangeanos en la Bahía de Todos Santos. *Ciencias Marinas* 1988;**14**(4):135–62. Autonomous University of Baja California.

[7] Argote-Espinoza ML, Gavidia-Medina F, Amador-Buenrostro A. Wind induced circulation in todos santos Bay, B. C. México. In: *Atmosfera*, vol. 4; 1991. p. 101–15.

[8] Miranda LE. *Caracterización de las corrientes en las entradas y alrededores de la Bahía de Todos Santos* [MSc. thesis]. DOF-Centro de Investigación Científica y de Educación Superior de Ensenada; 2012. p. 57. http://biblioteca.cicese.mx/catalogo/tesis/index.php?coleccion=8.

[9] Kurczyn JA, Pérez-Brunius P, García J, Candela J, López M. *Seasonal characteristics of the ocean currents and water masses off Bahía de Todos Santos, Baja California.* 2014. Abstract in Ocean Sciences Meeting, Honolulu 2014.

[10] Janssen F. *Statistical analysis of multi-year hydrographic variability in the North Sea and Baltic Sea. Validation and correction of systematic errors in a regional ocean model* [Ph.D. thesis]. Fachbereich Geowissenschaften, Universität Hamburg; 2002.

[11] Schrum C, Mike SJ, Aleksseva I. ECOSMO, a coupled ecosystem model of North Sea and Baltic Sea: Part II. Spatial-seasonal characteristics in the North Sea as revealed by EOF analysis. In: *Journal of marine systems*, vol. 61; 2006. p. 100–13. http://dx.doi.org/10.1016/j.jmarsys.2006.01.004.

[12] Castro R, Martínez JA. Variabilidad espacial y temporal del campo de viento frente a la península de Baja California. In: Gaxiola-Castro G, Durazo R, editors. *Dinámica del Ecosistema Pelágico frente a Baja California, 1997–2007: Diez Años de Investigaciones Mexicanas de la Corriente de California.* México: Instituto Nacional de Ecología (INE), Centro de Investigación Científica y de Educación Superior (CICESE); 2010, ISBN 978-607-7908-30-2.

[13] Mariani P, MacKenzie BR, Iudicone D, Bozec A. Modelling retention and dispersion mechanisms of bluefin tuna eggs and larvae in the northwest Mediterranean Sea. In: *Progress in oceanography*, vol. 86; 2010. p. 45–58.

[14] Godin G. *Tides*. Centro de Investigación Científica y de Educación Superior de Ensenada; 1988, ISBN 0968210317, 9780968210314. p. 348.

[15] Sánchez A, Carriquiry JD. Sediment transport patterns in todos santos Bay, Baja California, Mexico, inferred from grain-size trends. In: *Sediment transport in aquatic environments*, ISBN 978-953-307-586-0 (A. Manning Ed.).

[16] García-Mendoza E, Rivas D, Olivos-Ortiz A, Almazán-Becerril A, Castañeda-Vega C, Peña-Manjarrez JL. A toxic *Pseudo-nitzschia* bloom in todos santos Bay, northwestern Baja California, Mexico. In: *Harmful algae*, vol. 8; 2009. p. 493–503.

[17] Reyes S, Pares A. Análisis de componentes principales de los vientos superficiales sobre la Bahía de Todos Santos B.C. México. *Geofisica Internacional* 1983;**22**(2):179–203.

CHAPTER

13

Effect of Radio Frequency Interference (RFI) Noise Energy on WERA Performance Using the "Listen Before Talk" Adaptive Noise Procedure on the West Florida Shelf

Clifford R. Merz[1,*], Yonggang Liu[1], Klaus-Werner Gurgel[2], Leif Petersen[3], Robert H. Weisberg[1]

College of Marine Science, University of South Florida, St. Petersburg, FL, USA[1]; Institute of Oceanography, University of Hamburg, Hamburg, Germany[2]; Helzel Messtechnik GmbH, Kaltenkirchen, Germany[3]

**Corresponding author: E-mail: cmerz@usf.edu*

CHAPTER OUTLINE

1. INTRODUCTION

In order to improve our understanding of the workings of the coastal ocean across a variety of West Coast of Florida environmental applications, the University of South Florida's College of Marine Science (USF/CMS) established and hosts a real-time Coastal Ocean Monitoring and Prediction System (COMPS) on the West Florida

Continental Shelf (WFS). COMPS program observing assets consist of an array of offshore buoys for measuring ocean and atmospheric variables, shore-based meteorological and water level stations for coastal inundation applications, shore-based ocean high-frequency (HF) radars (two Wellen Radar [WERA] Phased Array and three Coastal Ocean Dynamics Applications Radar [CODAR] Direction Finding) for near surface current velocity measurements, as well as numerical ocean circulation and wave models[1–4] that may be coupled for hurricane storm surge applications. Such an integrated coastal observing system provides a basis for quantitative predictions as well as cross-comparisons between different observing techniques/instruments.[5–10]

Quality observations require careful attention to system operation and performance metrics. Through careful monitoring of routine maintenance visits, it was observed that the external archival data storage devices attached to each of the two WERA sites were not filling up at the same rate, thus requiring the swapping out of the external storage devices at differing intervals. Examination of the fill rate of the onsite archival data storage devices attached to the WERA systems revealed a site-to-site variation of approximately 20% for identically configured systems. Both WERA sites utilize the "Listen Before Talk" adaptive noise procedure that dynamically adapts the center transmit frequency (Tx) and measurement bandwidth to better react to changes in electromagnetic (EM) propagation (ionospheric and/or radio frequency interference [RFI]) characteristics.[11,12] Ultimately, this results in files of differing sizes being stored and, thus, differing fill rates. Although the two WERA sites are only 68 km (42 statute miles) apart, there must be a significant difference in external RFI noise encountered at each site to account for the difference in storage device fill rates; because from an ionospheric propagation perspective, both WERA sites should behave (nearly) identically. CODAR systems have a set transmit frequency and bandwidth and, therefore, exhibit no significant site-to-site variation; for this reason, they are not included in this investigation.

Based on the WERA site-to-site findings, and driven by curiosity into the actual RFI variations, further examination into external background RFI noise levels ensued, with the intent of documenting this observation and quantifying the overall variations in the offshore coverage extent. It is not the intent of this study to identify the individual RFI noise contributors because the remote sites are located by necessity near airports, military installations, and industrial regions—none of which are likely to change their transmission characteristics. Rather, the resulting contribution of this evaluation lies in the quantification of the Listen Before Talk adaptive procedure in a real, operational situation, while examining the resulting impacts on site operational frequency selection, measurement bandwidth (resolution), S/N, and offshore range extent variations. All of the factors ultimately translate to the influence of background radio noise on ocean observations and the accuracy of the near surface current velocity measurements.[13] For example, it has been ascribed in Ref. 14 that the diurnal variation of HF radar data return is related to the diurnal change in the background radio noise in the ionosphere.

2. BACKGROUND

Using the return signal scattered off the ocean surface, HF radars provide a means for mapping fields of near surface ocean current velocities toward or away from the receive antennas. HF radars operate in the 3–30 MHz frequency band, providing long-range, over-the-horizon capabilities by virtue of ground-wave or sky-wave propagation. Besides echoes from ships or other targets, there is a strong backscatter signal due to the sea surface roughness.[15] The dominant process causing echo returns is Bragg-resonant scattering from ocean waves at half the transmitted electromagnetic wavelength. As these ocean waves travel at a well-defined phase speed superimposed on an underlying current, they generate distinct peaks in the backscatter Doppler spectrum. From these returns, a set of radial velocity components is calculated. The radial velocity components, collected by two independent systems deployed at different locations along the shoreline, are combined to provide simultaneous measurements of a field of near surface velocity vectors, as well as ocean wave variables and wind direction.[16–18]

Two primary types of HF radar systems are available commercially: either direction finding (DF) with a compact antenna system or phased array (PA) beam forming with a linear antenna array.[1] DF HF radars, such as the commercially available Seasonde developed by CODAR Ocean Sensors, Ltd,[19] utilize a compact directional antenna system consisting of three antenna elements (two orthogonally mounted loops and a monopole whip) in a single housing for receiving, and a separate, single omnidirectional antenna for transmitting frequency modulated interrupted continuous wave (FMICW) pulses. The transmissions are omnidirectional, and the receive antenna senses returns along radial directions at fixed angular and radial distances. PA HF radars, such as the commercially available WERA[16] developed by the University of Hamburg and manufactured by Helzel Messtechnik GmbH (Helzel), utilize two separate, simultaneously operated antenna arrays: one for receiving and the other for transmitting frequency-modulated, continuous wave (FMCW) chirps. WERAs utilize a one to four element transmit array, and normally a 12–16 element receive linear array is used to record the backscattered radio wave signals. By virtue of the increased number of antennas, it can operate both in a direction-finding and directional beam-forming mode. WERA systems are in use in many locations around the world, and where in use, the Listen Before Talk Adaptive Noise Procedure is commonly used. Therefore, this topic is of value not only to the WERA user but also as general information to the overall HF Radar community.

The USF/COMPS/Ocean Circulation Group (OCG) presently operates a total of three long-range CODAR Seasonde systems operating at 4.9 MHz and two WERA systems operating between 12.275 and 13.20 MHz along the WFS. One WERA system is located within the Ft De Soto Park at Tierra Verde, Florida. Ft De Soto is the largest park within the Pinellas County Park System, consisting of 1136 acres comprising five interconnected islands (keys). The second WERA system is located at 1200 South Harbor Drive, Venice, Florida, at the U.S. Coast Guard Auxiliary Station number 86, directly across from the Venice city airport. The Venice location

Table 1 USF HF Radar Network Remote Site Location and Array Type Details

Site	Latitude	Longitude	Type	System
Redington Shore (RdSr)	27° 49.952′ N	82° 50.064′ W	DF	CODAR
Venice (Veni)	27° 04.535′ N	82° 27.066′ W	DF	CODAR
Naples (Napl)	26° 09.732′ N	81° 48.630′ W	DF	CODAR
Ft De Soto Park (FDS)	27° 38.150′ N	82° 44.286′ W	PA	WERA
Venice (VEN) co-located with Veni	27° 04.655′ N	82° 27.096′ W	PA	WERA

has had both a CODAR and a WERA system co-located since 2010, and it was the first location in North America to do this. Notch filters placed on the WERA Rx channel input locations are used to remove any in-band CODAR-generated signals. General specifics about these sites can be found in Table 1, with a more detailed discussion of the combined USF HF radar network, overall system layout, and remote site design found in Ref. 1 In a recent assessment against a moored, offshore COMPS buoy-mounted Acoustic Doppler Current Profiler (ADCP), CODAR and WERA current velocity measurements exhibited comparable RMS differences to the in situ ADCP.[7]

Hourly data from each site is pulled via scripting to a central processing station located at the USF/CMS in St Petersburg, Florida, where the data is processed and web served in near real time through the COMPS Website http://comps.marine.usf.edu, the Southeast Coastal Ocean Regional Association (SECOORA) Website http://secoora.org, and the Integrated Ocean Observing System (IOOS) National HF Radar Network Website http://cordc.ucsd.edu/projects/mapping/stats. The combined HF Radar network provides real-time surface current measurements on the WFS with footprints of the network coverage shown in Figure 1.

3. SYSTEM OPERATIONAL CHARACTERISTICS AND PROBLEM DEFINITION

WERA radars apply FMCW modulation for range resolution transmitting a linear frequency chirp covering a bandwidth within the chirp duration T. The bandwidth is directly related to the range resolution. The depth of a range cell in km is provided by the relation $r = c/2000B$: with c being the speed of light (300,000 km/s), B being the bandwidth in kHz, and the factor 2000 accounting for the radar path length to and from the scattering area.[20] For example, the range cell (or resolution) depth for a 100 kHz bandwidth FMCW chirp is 1.5 km, calculated by $300{,}000/2000 \times 100$. The FMCW chirp relationship between bandwidth and range cell depth is illustrated in Figure 2.

Both existing Ft De Soto and Venice WERA systems currently operate using the Listen Before Talk procedure, in which a frequency pre-scan is made across the entire FCC licensed 1-MHz operational band (12.275–13.20 MHz) prior to the commencement of the full acquisition.[11,12] During the pre-scan, regions of lowest RFI external noise are identified with the widest quietest bandwidth determined, and the corresponding mid-span transmit frequency is selected for subsequent use in the following full measurement.

Variations in the local RFI external noise field will result in the possibility of differing bandwidths both spatially and temporally between sites. If the frequency pre-scan detects too much external noise, the measurement bandwidth is reduced, which in turn increases the individual range cell size. This method is implemented in this manner, rather than keeping the number of range cells constant, for instance, to limit the storing of data from far ranges that contain little or no radar echoes. To keep the same approximate maximum offshore range (in our case, 154 km), the software reduces the number of range cell bins for the measurement, so data

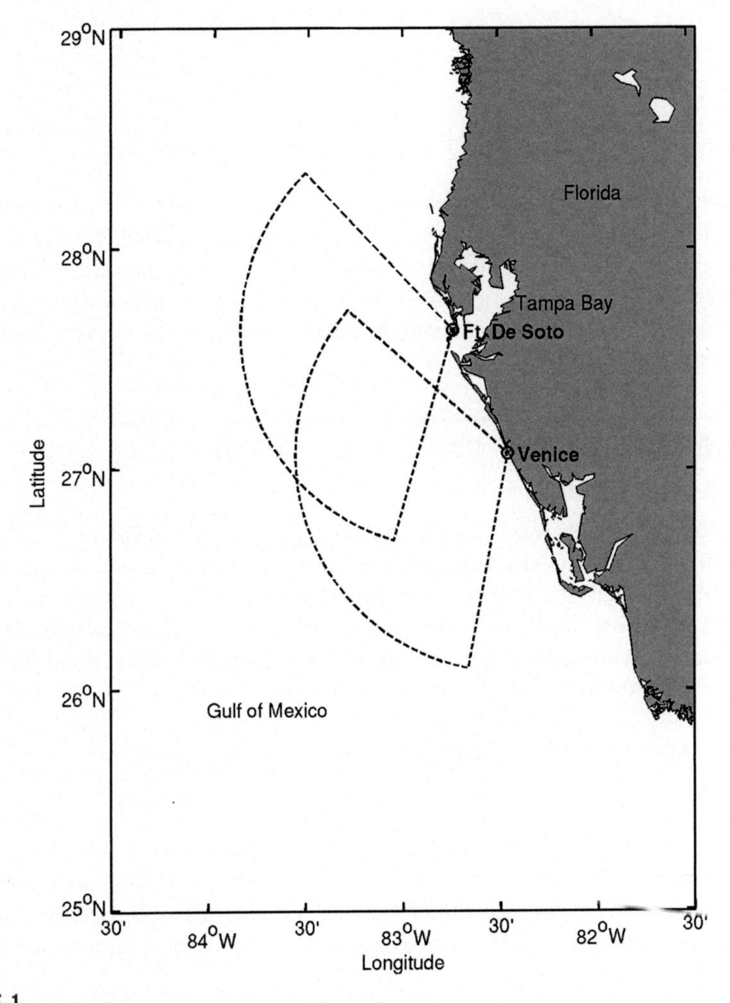

FIGURE 1

USF's WERA PA HF radar system locations and West Florida Shelf coverage areas.

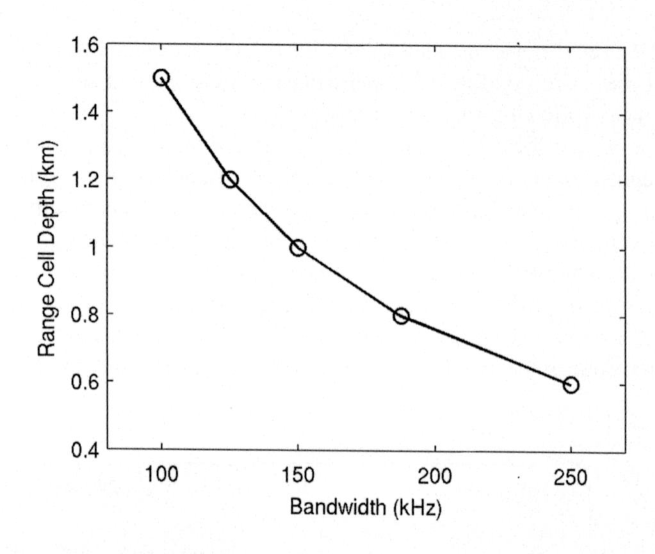

FIGURE 2

Relationship between bandwidth and range cell depth for an FMCW chirp.

for a fewer number of range cells bins will be stored, resulting in smaller processed data (.SORT and .RFI) files. As a result, sites with higher external RFI noise energy levels (Venice) will see their data drives fill up at a slower rate than remote sites with lower external RFI noise energy levels (Ft De Soto). This matches what we saw in our actual individual field site archival disk data fill rates and is further discussed in the next section.

4. QUANTIFYING THE VARIATIONS IN THE LOCAL NOISE FIELD PRESENT

Automated daily scripts were initially created to log into each WERA site and record the remaining archival device storage capacity. Information was then transferred to a spread sheet for analysis, after which it was confirmed that over a 30-day period, the archive data device attached to the "quiet" Ft De Soto WERA system filled up approximately 20% faster than the "noisy" Venice WERA site. Based on this, a WERA site-to-site RFI investigation was initiated.

4.1 RFI PRE-ANALYSIS PREPARATION METHODOLOGY

To put this investigation to practice, several things had to happen:

1. WERA software modifications were needed to automatically record the pre-scan determined widest/quietest bandwidth and corresponding mid-span transmit frequency selected for use in the following full measurement. BASH and CSH scripting programs were written to automatically save these data for future analysis.

2. WERA software modifications were needed to directly compare site-to-site spatial and temporal measurements of Signal Power and RFI Noise. Subsequent discussions with personnel of the University of Hamburg led to the use of bin-to-bin Signal Power and RFI Noise data computed across each measurement for this purpose. However, to make scaled comparisons, modification to various Fortran plot/list programs were made with a fixed and unique reference level of the Analog to Digital Convertor (ADC) full-scale value mapped to zero (0) dB. In addition, a new program called List_Radial_dB was created that included the computed Signal Power and RFI Noise data referenced in dB. BASH shell scripting programs were also created to automatically save the bin-to-bin data for future analysis and S/N calculation.
3. Confirmation that both sites provided comparable transmitted signal power values into the main beam direction. The Ft De Soto site uses a single transmit antenna, whereas, the Venice site uses a standard four-antenna rectangular transmit configuration. Power measurements made at the antenna input location for Ft De Soto and Venice were 32 and 10 W, respectively. The modeled Ft De Soto transmit beam pattern is omnidirectional with an antenna gain of −11.48 dBi. The Venice directional transmit beam pattern has an antenna gain of −6.91 dBi, which is a difference of 4.57 dB (factor of 2.86). Using the omnidirectional antenna as a reference (i.e., factor of 1.0), Venice can be referenced by multiplying the measured input power (P) by the antenna pattern factor to get the equivalent power in the direction main beam. Therefore, $P_{\text{Ft De Soto}} = 32\text{ W} \times 1.0 = 32\text{ W}$, whereas, $P_{\text{Venice}} = 10\text{ W} \times 2.86 = 28.6\text{ W}$. These results are within the error estimate of the power meter used as well as the overall system error of reasonability. Therefore, direct data comparison is acceptable, understanding this is a simple, modeled estimate that does not include the details of the environment.

4.2 INITIAL DATA COLLECTION

Synchronized WERA data acquisition measurements are made for both sites at minutes 0, 20, and 40 of each hour with the frequency pre-scan occurring 1.5 min prior to this. One system sweeps up in frequency and the other sweeps down. The initial Listen Before Talk data collection analysis period occurred from 11/18/13 to 12/20/13. Analysis of these 32.1-day records for both sites is revealed in Figures 3–6. A pre-scan selected transmit frequency versus frequency of occurrence histogram (Figures 3 and 5) and range cell depth versus frequency bar graph (Figures 4 and 6) are shown for Ft De Soto and Venice, respectively.

Examination of Figures 3–6 revealed two very interesting and unexpected findings:

1. A significant difference in the shape and distribution of the transmission frequency histogram (Figures 3 and 5) revealed almost no overlap with only a small region of mid-frequency overlap occurring around 12.580 MHz (seven occurrences out of 4606 measured).

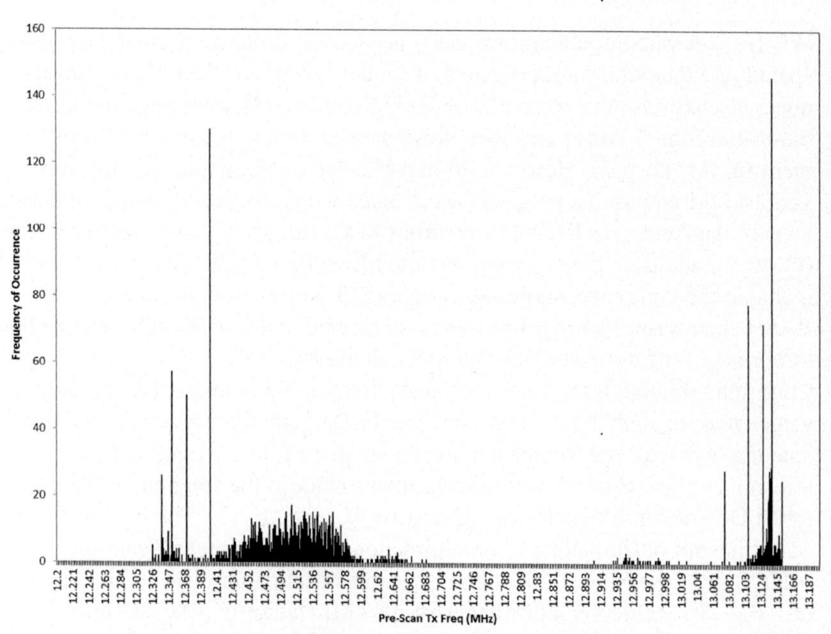

FIGURE 3

Ft De Soto (quiet) WERA pre-scan transmit frequency (Tx) versus frequency of occurrence histogram—11/18/13 to 12/20/13.

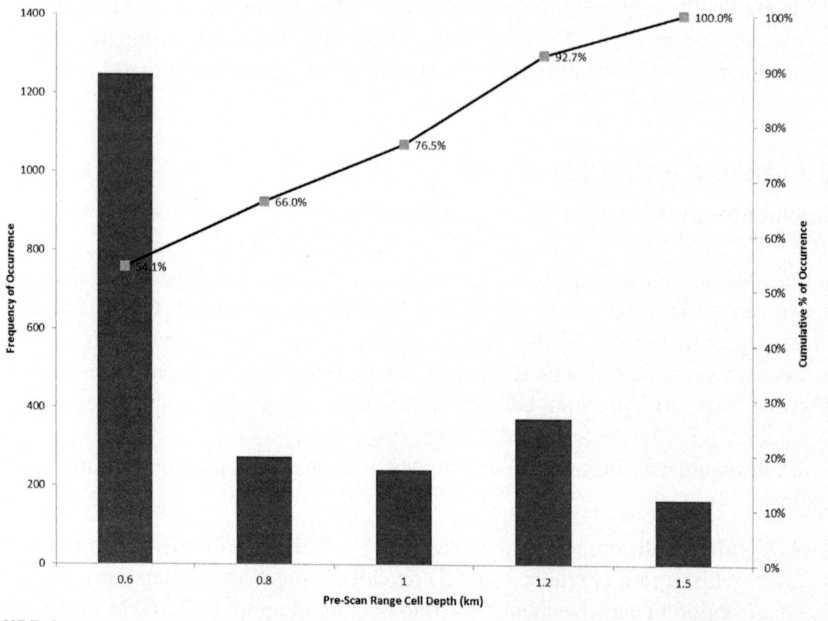

FIGURE 4

Ft De Soto (quiet) WERA range cell depth versus frequency of occurrence bar graph—11/18/13 to 12/20/13. Solid black line is cumulative percentage of occurrence.

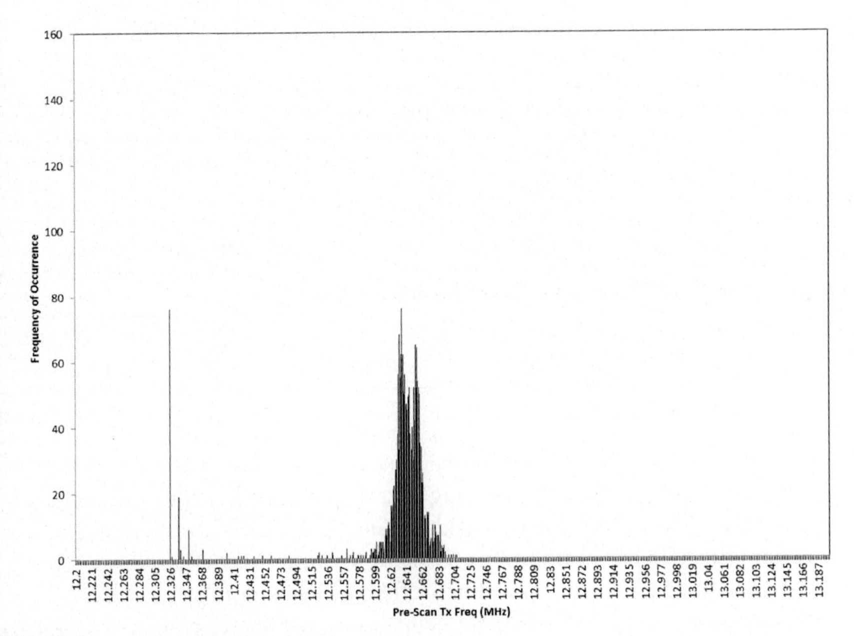

FIGURE 5

Venice (noisy) WERA pre-scan transmit frequency (Tx) versus frequency of occurrence histogram—11/18/13 to 12/20/13.

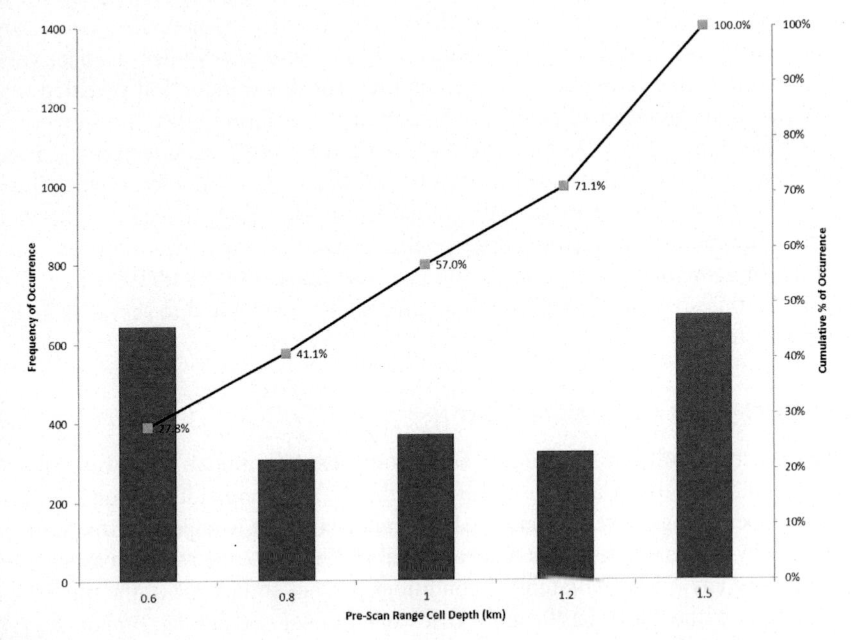

FIGURE 6

Venice (noisy) WERA pre-scan range cell depth versus frequency of occurrence bar graph—11/18/13 to 12/20/13. Solid black line is cumulative percentage of occurrence.

Table 2 Example Modified List_Radial_dB Output Data Header File for the Ft De Soto and Venice Sites

IX	IY	Velocity, m/s	Accuracy, m/s	Std. Deviation, m/s	Power, dB	Noise, dB
Ft De Soto: 18-NOV-2013 20:00:00 UTC 1096 Grid Points						
56	86	0.03395	0.04600	0.26821	−128.246	−140.273
56	87	−0.01772	0.02266	0.13214	−125.470	−140.495
Venice: 18-NOV-2013 20:00:00 UTC 844 Grid Points						
90	122	−0.10299	0.02400	0.13997	−124.267	−138.467
90	123	−0.09966	0.02381	0.13882	−124.310	−138.761

2. A significant difference in the number and distribution of the range cell depth bar graph (Figures 4 and 6) revealing that the quiet Ft De Soto site pre-scan selected a 0.6-km (250 kHz bandwidth) range cell 54.1% of the time and a 1.5-km (100 kHz bandwidth) only 7.3% of the time, versus the noisy Venice site pre-scan that selected a 0.6-km (250 kHz bandwidth) range cell 27.8% of the time and a 1.5-km (100 kHz bandwidth) at a nearly equivalent 28.9% of the time.

A small segment of the modified List_Radial_dB output data header file for Ft De Soto and Venice during this time period is presented in Table 2. As shown in this example header file, there were 1096 (844) grid points actually covered by measurements with sufficient S/N values at Ft De Soto and Venice, respectively. Examination of the individual site pre-scan file for this time period revealed that the Ft De Soto pre-scan selected a 0.6-km range resolution with 256 range cell bins (or 0.6 km × 256 = 154 km), whereas, Venice's pre-scan selected a 1.5-km range resolution with 103 range cell bins (or 1.5 km × 103 = 154 km). Thus, illustrating that because of the increased external RFI noise energy detected at the Venice site, the number of covered grid points (range cell bins) was reduced. This resulted in a smaller corresponding file size for Venice ([1−(844/1096)] or 23% smaller in this specific example), which ultimately contributed to less disk space being consumed.

4.3 EXTENDED DATA COLLECTION

Based on the results of the initial month-long test, additional tests were run to examine and quantify what effects differing the Tx and bandwidth has on the measurement percentage coverage and offshore radial extent. It is important to acknowledge that in comparing test runs done at different times that the results may vary by the differing environmental/ambient conditions present at the time of the run. However, because the "Standard" and "Test" operational conditions cannot be run concurrently, there is little alternative. And, the presence and degree of any

Table 3 Total Six Run Test Duration Summary

Run No.	System Operation	Tx	Bandwidth	Test Duration Dates
1	Standard	Variable	Variable	11/18/13–12/20/13
2	Test	Fixed at 12.580 MHz	Fixed at 100 kHz	2/2/14–3/5/14
3	Test	Fixed at 12.580 MHz	Fixed at 250 kHz	3/5/14–4/8/14
4	Standard	Variable	Variable	4/8/14–5/9/14
5	Test	Variable	Fixed at 100 kHz	5/9/14–6/10/14
6	Test	Variable	Fixed at 250 kHz	6/7/14–7/18/14

measured variation will contribute to the general knowledge base and understanding of our region's local characteristics, as well as assist in the optimized system operational performance settings that result in maximizing the overall offshore radial extent while maintaining high-percentage coverage. This information is important because total velocity vectors are generated from the interpolated overlapping of radials produced from each site. Therefore, improvements in radial extent lead to improvements in the resulting vector velocity field extent.

In total, a continuous temporal data record was acquired over an eight-month period between 11/18/14 and 7/18/14. Table 3 summarizes the specific test duration dates, Tx and bandwidth settings, and system operational conditions of either "Standard Operation" (Listen Before Talk) or "Test Operation" (testing several different configurations) used for this analysis.

The Ft De Soto and Venice List_Radial_dB output time series data is presented in Figures 7 and 8, with the Power, Noise, and S/N computed as an average of all available data points, and the coverage calculated by counting grid points. Figures 9 and 10 present offshore coverage maps created from Figure 7 and Figure 8 data, respectively, and Table 3 test duration dates for Ft De Soto and Venice. Also included for comparison in Figures 9 and 10 are the regions of 80% and 40% coverage along with the number of range cell bins (points) exceeding 80% of the data obtained during the test time period.

Examination of Figures 9 and 10 reveal the following for both WERA sites:

1. The observed offshore range for the quiet Ft De Soto site (Figure 9(a) and (d)) extends further offshore than the noisier Venice site (Figure 10(a) and (d)).
2. The number of range cell bins (points) exceeding 80% of the data obtained during the test periods was consistently higher for the Ft De Soto site versus Venice.
3. Improvement in the data returns occurred when variable Tx frequency was used.
4. Fixing the Tx at the selected "likely" overlap frequency of 12.580 MHz reduced the overall offshore range extent. Varying the bandwidth with the Tx fixed made little change, as shown in slides (b) and (c) from both sites.

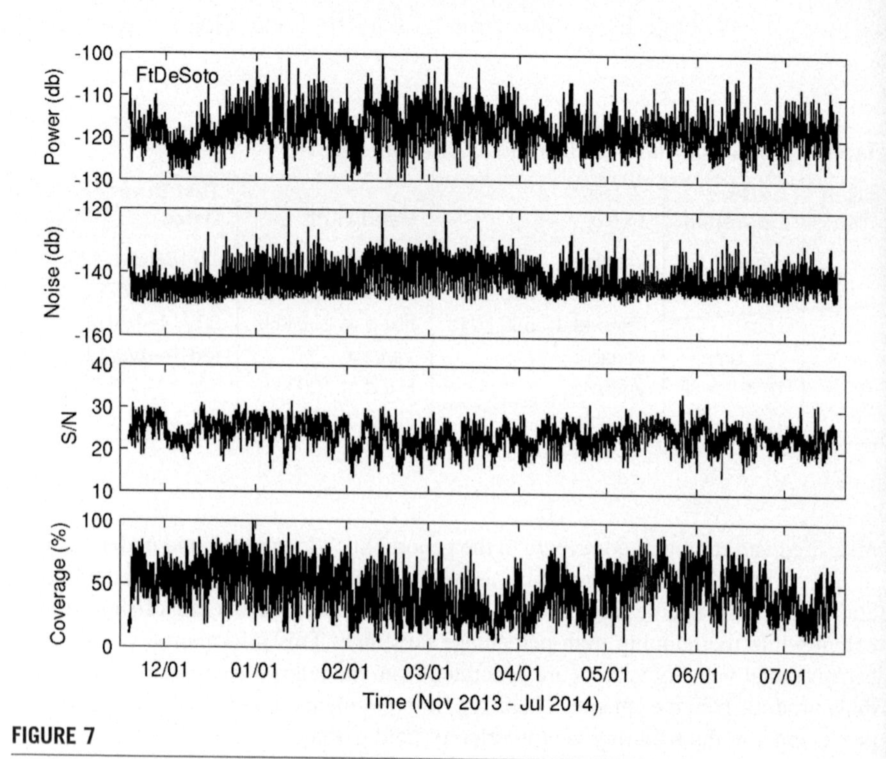

FIGURE 7

Ft De Soto List_Radial_dB output time series data.

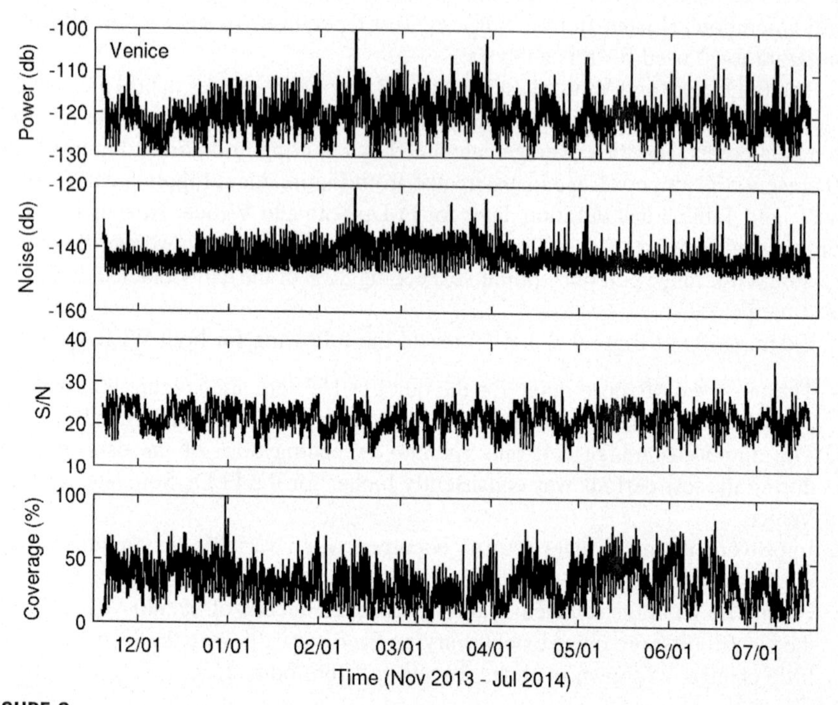

FIGURE 8

Venice List_Radial_dB output time series data.

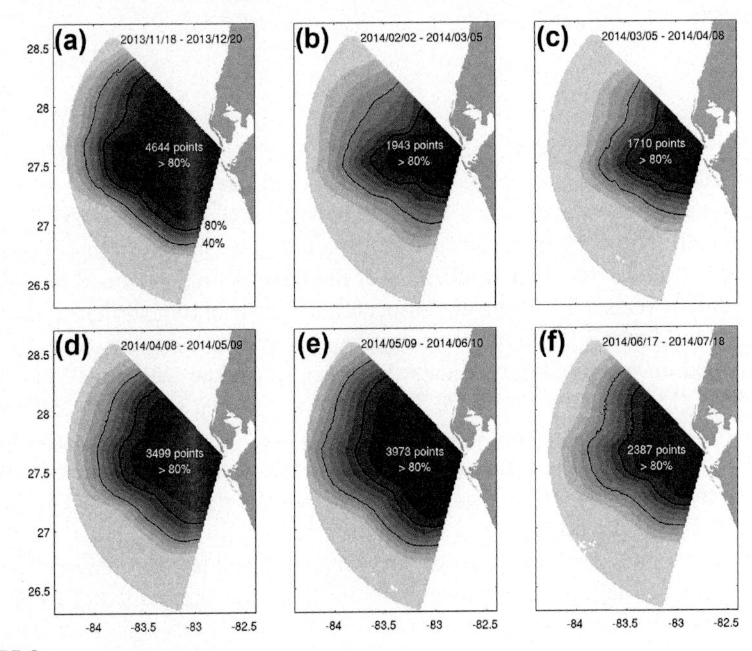

FIGURE 9

Percent coverage maps for Ft De Soto site. Table 3: Runs 1 to 6 correspond to maps (a)–(f), respectively. Regions of 40% and 80% coverage are shown by the darker and lighter line, respectively. Shown also are the number of range cell bins (points) exceeding 80% of the data obtained during the test time period.

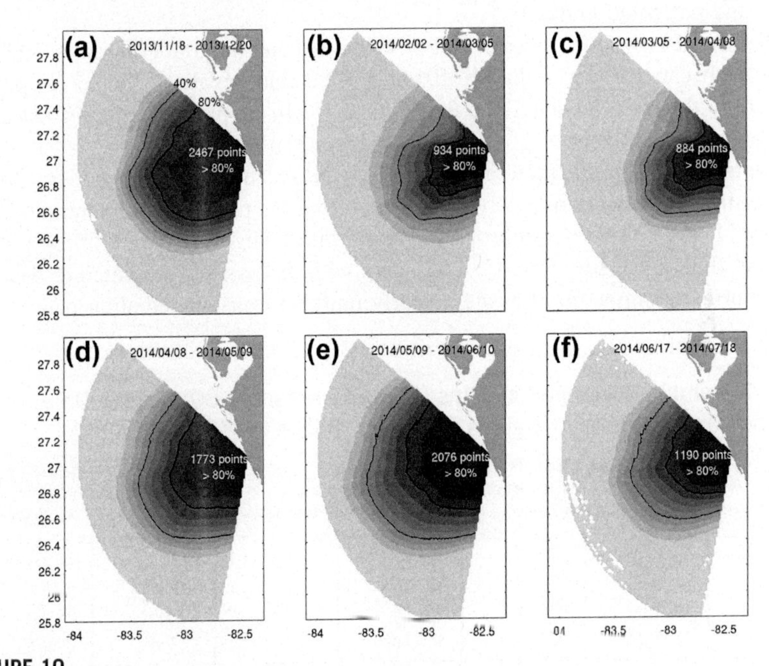

FIGURE 10

Percent coverage maps for Venice site. Table 3: Runs 1 to 6 correspond to maps (a)–(f), respectively. Regions of 40% and 80% coverage are shown by the darker and lighter line, respectively. Shown also are the number of range cell bins (points) exceeding 80% of the data obtained during the test time period.

5. While in the Standard operational mode of Listen Before Talk, some variability is seen between the run 1 slide (a) and run 4 slide (d) data results. Examination of Venice's run 4 Tx and range cell raw data reveals the Tx and range cell frequency of occurrence distribution to be similar to that shown in Figures 5 and 6 (run 1). Examination of Ft De Soto's run 4 Tx and range cell raw data reveals a Tx frequency of occurrence distribution similar to that shown in Figure 3; however, the range cell depth distribution is closer to that observed in Venice's Figure 6 data. Specifically, for the "Standard" 11/18/13 to 12/20/13 test dates, the Ft De Soto pre-scan selected a 0.6-km (250 kHz bandwidth) range cell 54.1% of the time and a 1.5-km (100 kHz bandwidth) 7.3% of the time. Whereas, for the 4/8/14 to 5/9/14 test dates, the Ft De Soto pre-scan selected a 0.6-km range cell 26.3% of the time and a 1.5-km range cell 25.5% of the time.
6. Table 4 summarizes the number of range cell bins (points) exceeding 80% of the data obtained during the test time period presented in Figures 9 and 10 (slides (a)–(f)).

Examination of the Table 4 data reveals that the trends observed and relative percent change between slides (a) through (f) are roughly equivalent for both sites. This illustrates a measure of variability caused by changes in the ambient EM propagation conditions along with the consistency of these changes on overall system performance. Both of which support the validity of this extended eight-month test period and presented findings.

The Signal Power, RFI Noise, and S/N data are presented in Figures 11 and 12. However, because of the reduced offshore extent findings discussed in (d) previously, only the results from Table 3 runs 4 to 6 will be presented for Ft De Soto and Venice, respectively.

The transmit radiation pattern includes both the Tx antenna pattern and the directivity of the Bragg scattering waves (summed from the two peaks). Examination of Figures 11 and 12 shows similar power maps with a slight reduction in the output power of the Venice site, which is consistent with discussions presented in Section 4.1.3. Further examination reveals a nearly uniform transmit radiation pattern for

Table 4 Number of Range Cell Bins (Points) Exceeding 80% of the Data Obtained as Presented in Figures 9 and 10 [Slides (a)–(f)]

Site Name	No. of bins with >80% coverage	No. of bins with >80% coverage	No. of bins with >80% coverage	% Change	% Change	% Change
Ft De Soto	4644 (a)	1943 (b)	1710 (c)	75 (d/a)	42 (b/a)	37 (c/a)
Ft De Soto	3499 (d)	3973 (e)	2387 (f)	–	114 (e/d)	68 (f/d)
Venice	2467 (a)	934 (b)	884 (c)	72 (d/a)	38 (b/a)	36 (c/a)
Venice	1773 (d)	2076 (e)	1190 (f)	–	117 (e/d)	67 (f/d)

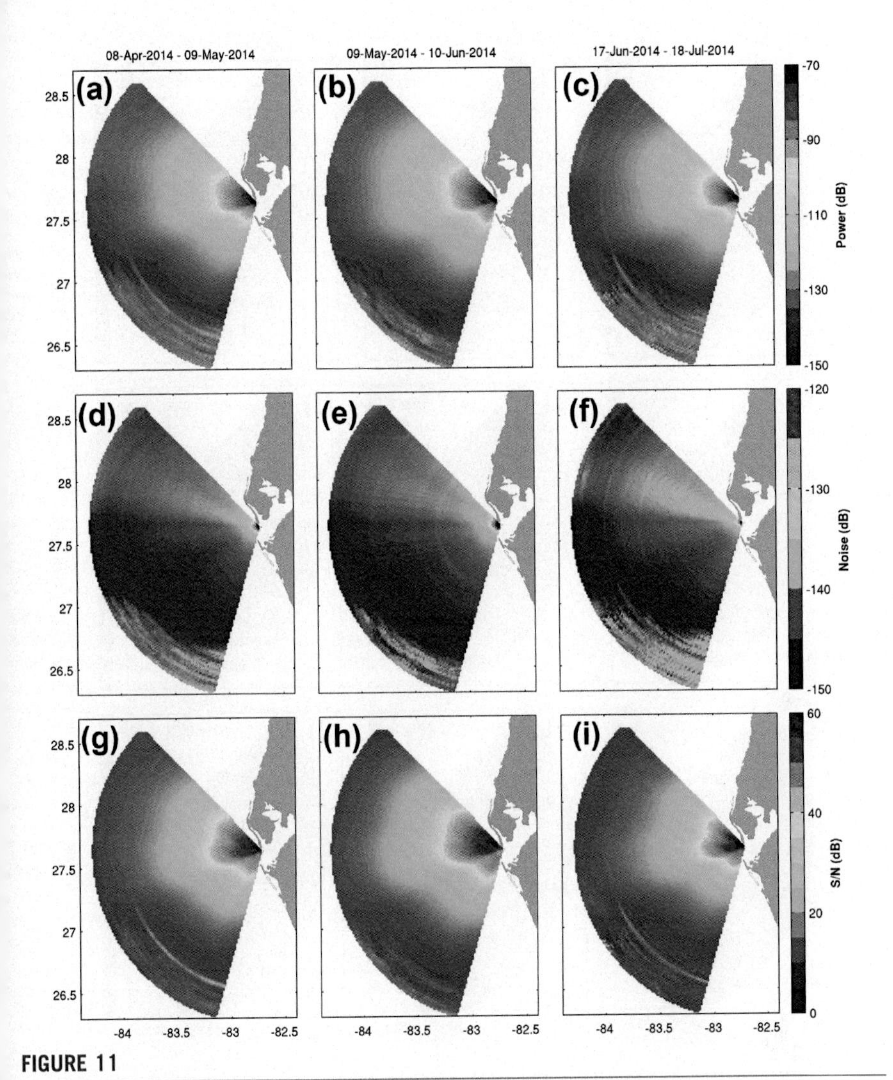

FIGURE 11

Temporal mean maps of Power (a)–(c), Noise (d)–(f), and S/N (g)–(i) for Ft De Soto. Table 3: Runs 4 to 6.

Venice's main beam (Figure 12) with the slightly visible northward shift being the result of an applied phase shift (through cabling) purposely applied to rotate the beam to the right (northward) for increased radial overlap with the Ft De Soto site. The signal attenuation shown in the bottom portion of the Ft De Soto map (Figure 11) is attributable to shoreward park land and Egmont Key (an offshore barrier island) located southwest of the site. Examination of the noise fields between both sites does show differences, especially in the upper and lower quadrant regions

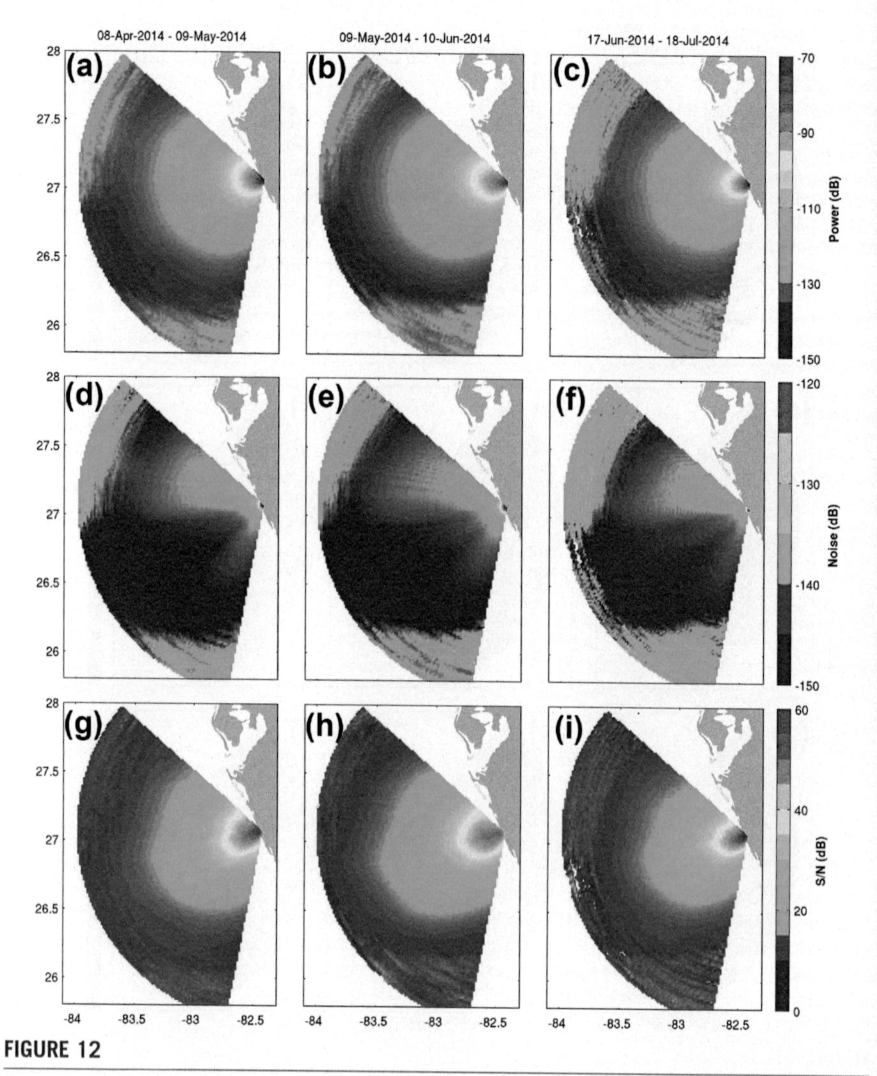

FIGURE 12

Temporal mean maps of Power (a)–(c), Noise (d)–(f), and S/N (g)–(i) for Venice. Table 3: Runs 4 to 6.

across time with the noisy Venice site showing somewhat higher levels than the quieter Ft De Soto site.

5. SUMMARY

This investigation began with an observation based on the uneven data fill rates of attached archival storage devices between two WERA HF radar sites.

Because, from an ionospheric propagation perspective, both WERA sites should behave (nearly) identically, it was hypothesized that there must be a significant difference in external RFI noise encountered at each site. Further investigation confirmed this hypothesis by using offshore bin-to-bin measured Signal Power and RFI Noise data. Besides addressing curiosity, this effort illustrates two items:

1. The importance of having the local system operator significantly involved in both the QA/QC of the data as well as the day-to-day site operation.
2. These two WERA sites were located by necessity, and their operation would be far less productive without the use of the Listen Before Talk adaptive procedure, a variable Tx, and a wide enough operational bandwidth to work within, in this case an FCC-licensed, 1-MHz operational band.

HF radar systems are routinely used for remotely observing coastal and ocean surface currents, and WERA HF radar systems are in use in many locations around the world. The application of the WERA Listen Before Talk adaptive algorithm, along with a wide enough bandwidth to operate within, is shown to increase data coverage and S/N ratio. The results presented herein have direct applicability to the offshore S/N distribution, which ultimately affects integrated coastal and ocean observing system near surface ocean current velocity estimation capability and accuracy.

ACKNOWLEDGMENTS

COMPS operates along the Gulf of Mexico's West Florida coast and was implemented in 1997 as a State of Florida legislative initiative. The USF COMPS program receives partial support from the Southeast Coastal Ocean Observing Regional Association (SECOORA) through the U.S. IOOS Office operated by NOAA (Award No. NA11NOS0120033). WERA HF Radar equipment was originally acquired with USF internal R&D funds. Partial support was also provided by NASA Ocean Surface Topography Science Team (OSTST) (# NNX13AE18G). This is CPR Contribution 38. The authors would like to thank Mr Jeff Donovan of the USF/CMS/OCG group for his effort in the generation of the automated scripting programs required for this analysis. Comments from several anonymous reviewers improved the quality of the manuscript.

REFERENCES

1. Merz CR, Weisberg RH, Liu Y. *Evolution of the USF/CMS CODAR and WERA HF radar network*. Proc. of MTS/IEEE OCEANS'12. 2012. 1–5. http://dx.doi.org/10.1109/OCEANS.2012.6404947.
2. Merz CR. *An overview of the coastal ocean monitoring and prediction system (COMPS)*. Proc. Of MTS/IEEE OCEANS'01, MTS 0-933957-28-9. 2001. http://dx.doi.org/10.1109/OCEANS.2001.968281.

3. Weisberg RH, He R, Liu Y, Virmani JI. West Florida Shelf circulation on synoptic, seasonal, and inter-annual time scales. In: Sturges W, Lugo-Fernandez A, editors. *Circulation in the Gulf of Mexico: observations and models*, vol. 161. Washington (DC): AGU; 2005. p. 325–47.
4. Weisberg RH, Barth A, Alvera-Azcárate A, Zheng L. A coordinated coastal ocean observing and modelling system for the West Florida Shelf. *Harmful Algae* 2009;**8**: 585–98.
5. Liu Y, Weisberg RH, Shay LK. Current patterns on the West Florida Shelf from joint self-organizing map analyses of HF radar and ADCP data. *J Atmos Ocean Technol* 2007;**24**: 702–12. http://dx.doi.org/10.1175/JTECH1999.1.
6. Liu Y, Weisberg RH, Vignudelli S, Roblou L, Merz CR. Comparison of the X-TRACK altimetry estimated currents with moored ADCP and HF radar observations on the West Florida Shelf. *Adv Space Res* 2012;**50**:1085–98. http://dx.doi.org/10.1016/j.asr.2011.09.012.
7. Liu Y, Weisberg RH, Merz CR. Assessment of CODAR Seasonde and WERA HF radars in mapping surface currents on the West Florida Shelf. *J Atmos Ocean Technol* 2014;**31**: 1363–82. http://dx.doi.org/10.1175/JTECH-D-13-00107.1.
8. Gomez R, Helzel T, Petersen L, Kniephoff M, Merz CR, Liu Y, et al. *Real-time quality control of current velocity data on individual grid cells in WERA HF radar.* MTS/IEEE OCEANS'14, Taipei, Taiwan. April 2014. http://dx.doi.org/10.1109/OCEANS-TAIPEI.2014.6964502.
9. Dzvonkovskaya A, Merz CR, Liu Y, Weisberg RH, Helzel T, Petersen L. *Initial surface current measurements on the West Florida Shelf using WERA HF ocean radar with multiple input multiple output (MIMO) synthetic aperture.* Proc. of MTS/IEEE OCEANS'14, Newfoundland, Canada. September 2014. http://dx.doi.org/10.1109/OCEANS.2014.7003235.
10. Dzvonkovskaya A, Helzel T, Petersen L, Merz CR, Liu Y, Weisberg RH. *Initial results of ship detection and tracking using WERA HF ocean radar with MIMO configuration.* Radar Symposium. 2014. p. 1–3. http://dx.doi.org/10.1109/IRS.2014.6869265.
11. Helzel T. *FMCW radar "WERA". Helzel Company internal Document, 10/19/2007.* 2007. Helzel Messtechnik GmbH, Carl-Benz Strasse 9, D-24568 Kaltenkirchen, Germany.
12. Gurgel K-W, Barbin Y, Schlick T. *Radio frequency interference techniques in FMCW modulated HF radars.* Proc. of MTS/IEEE OCEANS'07, Aberdeen. 2007.
13. Pan C, Zhou H, Wen B-Y. Radio frequency interference suppression in high frequency surface wave radar based on range-domain correlation. *J Electromagn Waves Appl* 2013;**27**:448–57. http://dx.doi.org/10.1080/09205071.2013.749773.
14. Liu Y, Weisberg RH, Merz CR, Lichtenwalner S, Kirkpatrick GJ. HF radar performance in a low-energy environment: CODAR seasonde experience on the West Florida Shelf. *J Atmos Ocean Technol* 2010;**27**:1689–710. http://dx.doi.org/10.1175/2010JTECHO720.1.
15. Crombie DD. Doppler spectrum of sea echo at 13.56Mc/s. *Nature* 1955;**175**(4459): 681–2.
16. Gurgel K-W, Antonischki G, Essen H-H, Schlick T. Wellen Radar (WERA), a new ground-wave-based HF radar for ocean remote sensing. *Coast Eng* 1999;**37**:219–34.
17. Huang W, Gill EW, Wu X, Li L. Measurement of sea surface wind direction using bistatic high-frequency radar. *IEEE Trans Geosci Rem Sens* 2012;**50**:4117–22.

18. Shay LK, Cook TM, Peters H, Mariano AJ, An PE, Soloviev A, et al. Very high-frequency radar mapping of surface currents. *IEEE J Ocean Eng* 2002;**27**:155–69. http://dx.doi.org/10.1109/JOE.2002.1002470.
19. Barrick DE, Lipa BJ, Crissman RD. *Mapping surface currents with CODAR*. Sea Technology, Oct. '85. 1985. p. 43–48.
20. Gurgel K-W, Schlick T. *Remarks on signal processing in HF radars using FMCW modulation*. Radar Symposium, Hamburg, Germany. 2009. p. 45–50.

CHAPTER

14

Ocean Remote Sensing Using X-Band Shipborne Nautical Radar—Applications in Eastern Canada

Weimin Huang*, Eric W. Gill

Faculty of Engineering and Applied Science, Memorial University, NL, Canada

**Corresponding author: E-mail: weimin@mun.ca*

CHAPTER OUTLINE

1. INTRODUCTION

Ground-based high-frequency (3–30 MHz) surface wave radar (HFSWR) and X-band nautical radar (around 10 GHz) have been widely used in ocean remote sensing for three decades. HFSWR systems such as CODAR[1] and WERA[2] are able to monitor sea surface currents, winds, waves, and targets "over the horizon" due to the strong interaction between HF radio signals and ocean gravity waves (see recent applications in[3–6]). Unlike HFSWR, which is usually land-based to provide broad coverage at the expense of resolution, compact X-band radar covers the line-of-sight area with a spatial resolution as good as 5 m and can be deployed on ships. This enables X-band radar to be an ideal sensor to fill the nearby blind area (the first one or two ranges) of HFSWR or provide finer measurement for the area of interest.

Since Young et al.[7] first retrieved wave information and current velocity from X-band nautical radar images using a fast Fourier transform analysis, such radars have become widely accepted as cost-effective instruments with high spatial and temporal resolution for sea surface observation. The radar transmitted signals interact strongly with the ocean surface small-scale roughness through Bragg scattering.[8] These short waves are mainly generated by local wind regime.[9] Because the higher energy, long surface waves modulate the small-scale waves, which are predominantly responsible for the scattering at X-band, the backscattered signals contain features from which information regarding the wave spectrum,[7,10–13] and wind field[14–17] parameters and current velocity[18–21] may be determined.

The commercial Wave Monitoring System (WaMoS),[22] developed expressly for extracting oceanographic parameters from X-band radar data, has been extensively deployed on ships and at fixed coastal sites. Successful experimental results have been widely reported. In Canada, the pioneering X-band nautical radar research work started in the early 1990s. The Grand Banks ERS-1 synthetic aperture radar (SAR) wave spectra validation experiment[23] evolved to compare the directional wave spectra obtained from five radar systems and two in situ buoys. Two of the five radars were X-band marine radars operated by the Royal Roads Military College and MacLaren Plansearch Incorporated, and wave analysis results were presented by Buckley et al.[24] Later, the Defence Research and Development Canada (DRDC) Atlantic conducted a number of sea trials off Halifax in 1993, 2004, 2006, and 2008[25] using MacRadar or WaMoS. In order to augment the limited open study on the data from these trials, wave and wind results retrieved from the data collected during the 2008 trial are presented here.

2. WAVE ALGORITHMS

X-band nautical radar backscatter is mainly due to Bragg scattering from centimeter-scale ocean waves[8] and wedge scattering from breaking waves on the long wave crests.[9] The backscatter signal is modulated by the long waves through tilt modulation and shadowing modulation.[10] Thus, the long wave information is mapped to the radar backscatter signals and may subsequently be extracted from the radar gray scale images.

Existing wave information extraction algorithms for nautical radar applications are based on 3-D FFT,[7,10–12] 2-D wavelet,[13] and shadow analysis[26] techniques. The flowchart of the traditional 3-D FFT method, which involves the following key steps, is shown in Figure 1:

1. Apply a 3-D FFT on a temporal sequence of radar subimages to generate 3-D image spectra.
2. Eliminate nonstationary and nonhomogeneous components by high-pass filtering the image spectra.
3. Execute an appropriate algorithm (e.g., Refs. 18–20) to obtain current velocity.
4. Extract the fundamental wave components with a band-pass filter (BPF) constructed based on the current-included dispersion relationship.

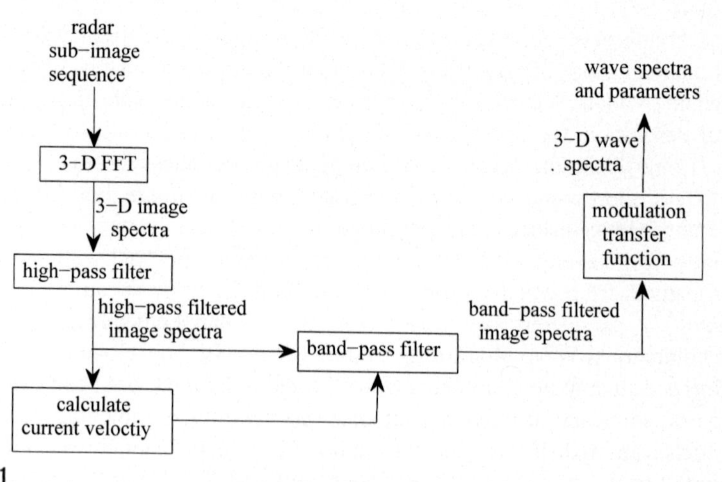

FIGURE 1

Flowchart of the traditional FFT-based wave algorithm.

5. Use a modulation transfer function to convert the filtered image spectra to wave spectra.
6. Derive wave spectra and their associated parameters.

It is clear this method always involves a BPF after the current velocity is obtained. An iterative least squares (LS)-based method that directly uses the classified fundamental and first-order harmonic wave components for wave spectra and parameter retrieval appears by Huang et al.[27] That method enables the elimination of the subsequent band-pass filter that is required to remove the non-wave contributions after the current velocity is obtained. Here, the simplified wave algorithm (without BPF), the details of which the can be found by Huang et al.,[27] will be used and compared with the traditional method (with BPF).

3. WIND ALGORITHMS

The normalized radar cross-section (NRCS) has been shown to be dependent on wind speed[9,28] and direction.[28,29] For horizontally polarized (HH-polarized) radiation at grazing incidence, the NRCS was found to be maximum in the upwind direction and minimum in the crosswind direction,[28,29] changing nonlinearly with wind speed.[14] In 2003, Dankert et al.[15] proposed a method for extracting wind direction from quasi-stationary wind streaks and wind speed from the temporally integrated radar images along with the determined wind direction. This approach can be applied to shipborne radar data with geocoding that is required because extracting wind streaks is difficult due to the platform's horizontal motion.[30] A 180° directional ambiguity exists in the wind direction results. However, this ambiguity can be removed by extracting the movement of wind gusts visible in the radar image sequence.[14] Recently, two methods, which are independent of platform movement, were developed by Lund et al.[16] and Vicen-Bueno et al.[17] The method proposed by Vicen-Bueno et al.[17] is referred to as a backscatter intensity-level-selection (ILS) algorithm, which determines wind direction and speed based on temporally integrated and spatially smoothed radar images and incorporates an empirical third-order polynomial geophysical model function (GMF). The least squares curve-fitting technique that was developed by Lund et al.[16] identifies the upwind peak based on the radar backscatter intensity dependence on the upwind direction and produces wind speed through an empirical third-order polynomial fit. In this chapter, the original curve-fitting-based algorithm and its modified version will be implemented.

3.1 CURVE-FITTING-BASED WIND ALGORITHM

For HH-polarized X-band radars operating at grazing incidence, the radar backscatter intensity varies with antenna look direction and has a single peak in the upwind direction.[28,29] In Lund et al.,[16] the dependence of radar backscatter gray scale intensity on antenna look direction is described by the cosine square function:

$$\sigma_\theta = a_0 + a_1 \cos^2(0.5(\theta - a_2)) \tag{1}$$

where $a_i (i = 1, 2, 3)$ are the regression parameters, and σ_θ represents the averaged radar gray scale intensity over a number of ranges (here from 450 to 1500 m) for an azimuthal direction θ. The regression parameters a_i can be determined by curve fitting the measured range-averaged radar backscatter gray scale intensity and antenna look direction. Figure 2 depicts an example of the measured averaged backscatter intensity variation (dots) with antenna look direction and the best-fit curve (solid). After obtaining the curve-fitted model function of Eqn (1) for each individual radar image, the wind direction can be determined as the direction associated with the upwind backscatter peak. The upwind peak direction is actually given by the regression parameter a_2 of the model in Eqn (1). This corresponds to the peak of the best-fit curve. It has been shown that this method works well even when some sections of the radar field of view are masked.[16]

The relationship between radar backscatter intensity and wind speed is assumed to follow an empirical third-order polynomial, which can be derived using the overall average radar backscatter intensity and the reference wind speed measured by other sensors (e.g., an anemometer).[16] Then, the radar wind speed results can be calculated from the measured overall average intensity value of Eqn (1). The overall average radar backscatter intensity σ_{avg} is defined as follows:[16]

$$\sigma_{avg} = \frac{1}{2\pi} \int_0^{2\pi} a_0 + a_1 \cos^2(0.5(\theta - a_2)) d\theta. \tag{2}$$

Figure 3 shows the anemometer-measured wind speed results and the corresponding radar overall average backscatter intensities σ_{avg}. The best-fit curve (solid) indicated is derived using a least squares method based on a third-order polynomial function:

$$\sigma_{avg} = b_0 + b_1 U_{10} + b_2 U_{10}^2 + b_3 U_{10}^3, \tag{3}$$

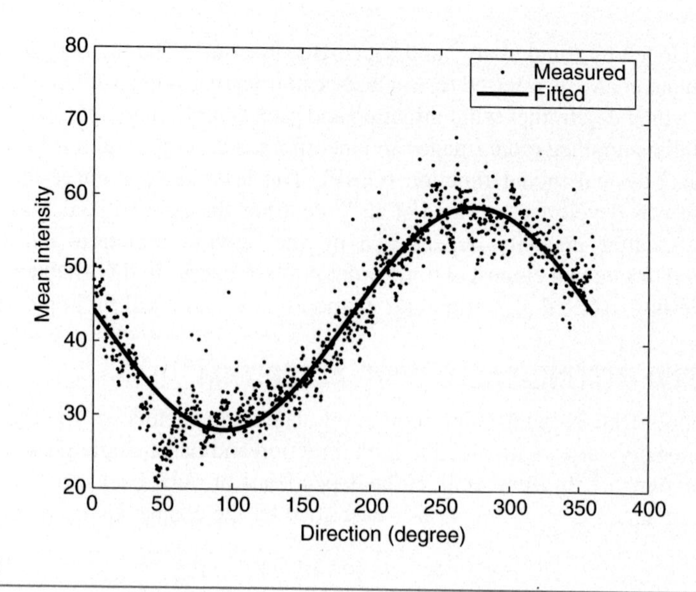

FIGURE 2

Range-averaged radar backscatter gray scale intensity as a function of antenna look direction.

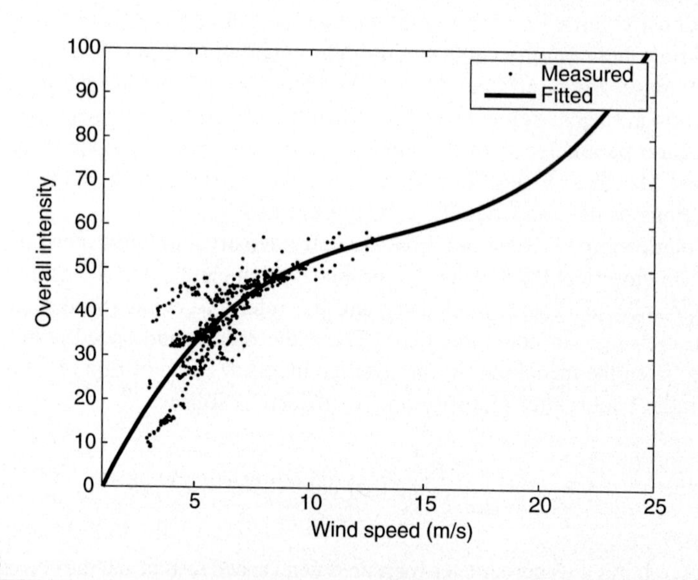

FIGURE 3

A scatter plot showing the relationship between the anemometer wind speed and the corresponding overall average radar backscatter intensity. The solid line represents the best-fit curve based on a third-order polynomial function using rain-free data.

as in Ref. 16 using rain-free data. Here $b_i (i = 0, 1, 2, 3)$ represents coefficients to be determined by curve fitting, and U_{10} is the wind speed at 10 m above the sea surface.

3.2 RAIN EFFECT MITIGATION

It has been found that radar backscatter will be affected by rain through volume scattering and attenuation by raindrops in the intervening atmosphere as well as through changes in sea surface roughness resulting from rain impinging the ocean.[31,32] It is difficult to conclude whether the normalized radar cross-section is increased or decreased by rain because it depends on many factors such as rain rate, raindrop size, radar frequency, and polarization.[31–33] For X-band marine radar, it has been observed that sea surface radar backscatter is generally enhanced by the rain.[16,34–36] As a result, the radar-derived wind speed was found to be overestimated when using radar data collected in the presence of rain. In Ref. 16, Lund et al. presented a method to determine whether or not an image is contaminated by rain by analyzing the intensity histogram of the image. Rain-contaminated images were then discarded from their wind retrieval analysis. Here, a new technique for improving the extraction of wind speed from rain-contaminated data is proposed. The technique includes the following steps:

1. Because rain can strongly impact the number of pixels with zero intensity, the zero-pixel percentage (ZPP, i.e., the ratio, expressed as a percentage, of the number of image pixels with zero intensity to the overall number of pixels) is used to distinguish the rain-free and rain-contaminated images.
2. Two third-order polynomial functions for the relationship between overall average backscatter intensity and wind speed for rain-free and rain-contaminated cases based on least squares fitting with rain-free and rain-contaminated data, respectively, are obtained. Note that these two functions have the same format as Eqn (3), but with different coefficients.
3. Radar images are classified as rain-free and rain-contaminated according to their ZPPs. For the rain-free data, wind speed is retrieved using the rain-free wind speed model obtained in Step 2. For the rain-contaminated data, wind speed is determined using the rain-contaminated wind speed model.

In what follows, this dual-wind-speed-model method is referred to as the two-model algorithm, and that involving the wind speed model derived from rain-free data alone is referred to as the single-model algorithm.

4. EXPERIMENTAL RESULTS

4.1 DATA OVERVIEW

In late November 2008, DRDC conducted a sea trial with the Canadian Navy research ship *CFAV Quest* in the Northwest Atlantic Ocean, approximately

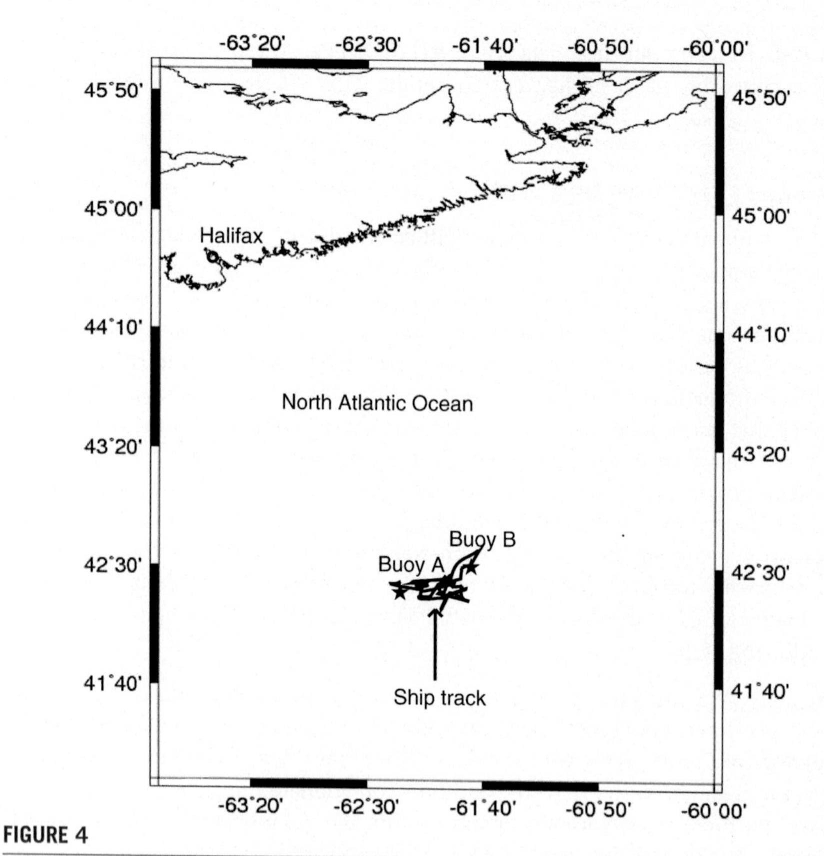

FIGURE 4

Map of the experiment site. Stars represent wave buoys that are close to the GPS track of CFAV Quest.

220 km from the coast of Halifax, Canada.[25] The water depth within the data collection area (near 42°30′N, 62°5′W, see Figure 4) is about 200 m and can be regarded as deep water. A standard ship-borne HH-polarized X-band (9.41 GHz) Decca Bridge-Master II 340 nautical radar and a nonacoustic data acquisition system (NADAS) recording data from two anemometers were deployed on the ship. The nautical radar, which was connected to a Wave Monitoring System II (WaMoS II),[22] covered 360° in azimuth with a beam width of about 2°. The useable range varied from 240 to 2160 m with a radial resolution of 7.5 m. The radar backscatter intensities were digitized and scaled into 8-bit unsigned integers ([0, 255]) by the WaMoS system. Every 32 radar images were sequentially combined into a single file. The time labels were in local standard time coordinates. The radar data collected during the period from 23:43 November 26 to 12:04 November 29 were used in the wave and wind algorithms here. During the sea trial, two nearby TRIAXYS wave buoys, deployed at 42°21′N, 62°15′W (buoy A) and 42°30′N, 61°45′W (buoy B), yielded wave data

averaged every half hour. Note that here only the wave data from the buoy that is closer to the ship are used for comparison with the radar wave results.

4.2 WAVE MEASUREMENT RESULTS

Both the traditional FFT-based wave algorithm and the simplified iterative LS-based method were applied to the radar data. Figure 5 shows the comparison of the radar-derived and buoy-recorded directional wave spectra $E(f, \theta)$ and nondirectional wave frequency spectra $E(f)$. It can be seen that both the simplified algorithm and the traditional method agree well with the buoy measurements.

The peak wave direction and period results derived from the radar data using both the traditional and simplified algorithms are compared with the buoy-recorded result in Figure 6. Note that the radar results are averaged every half

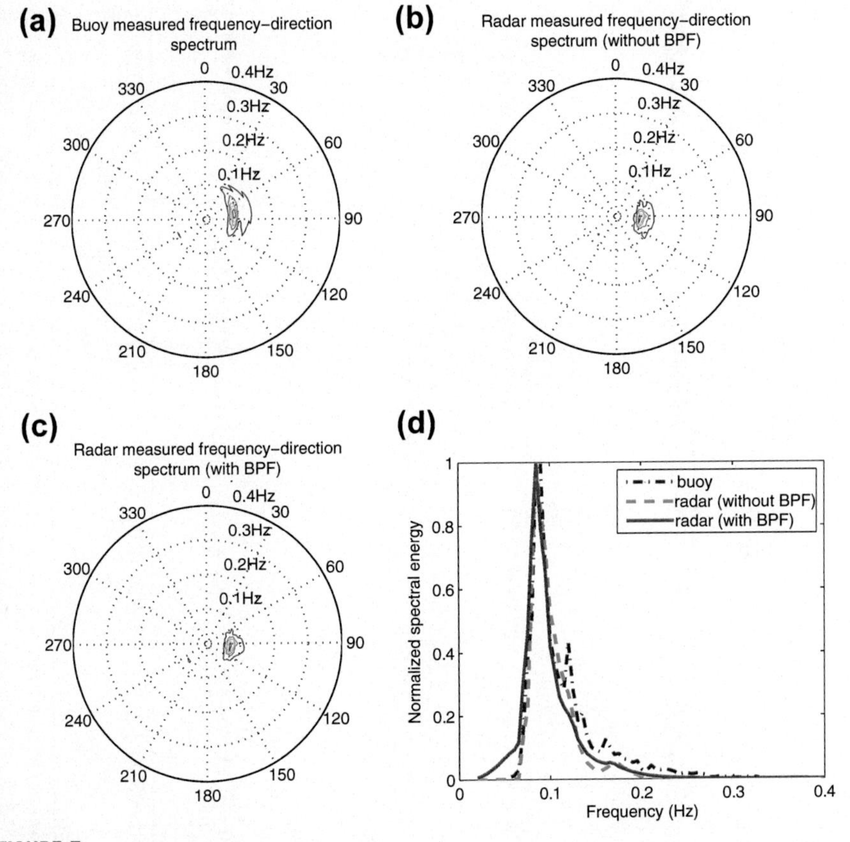

FIGURE 5

Comparison of the radar-derived and buoy-recorded wave spectra: (a) buoy-recorded $E(f, \theta)$; (b) radar-derived $E(f, \theta)$ using the simplified algorithm (without BPF); (c) radar-derived $E(f, \theta)$ using the traditional algorithm (with BPF); (d) 1-D wave frequency spectrum $E(f)$.

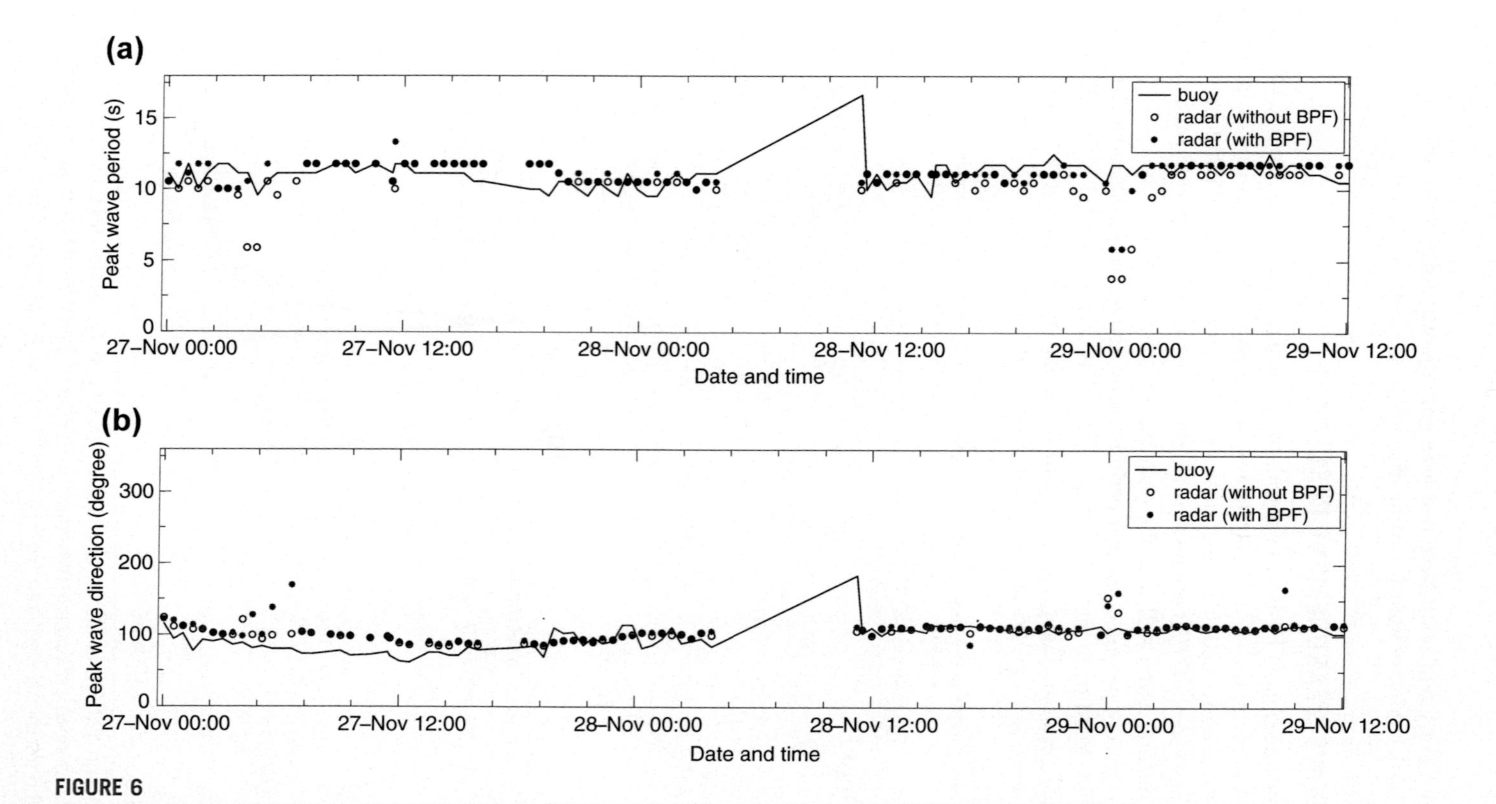

FIGURE 6

Comparison of radar-derived and buoy-recorded (a) peak wave period, and (b) peak wave direction.

hour for better comparison with the buoy data. From Figure 6, it can be seen that results from both algorithms agree well with the buoy result during most of the period. However, large deviation between the radar-derived and the buoy-recorded is observed for the time around 4:00 on November 27 and midnight of November 28. This is due to the low wind speed for these two periods (e.g., as shown in Figure 8(b), the wind speed is lower than 3 m/s from 23:40 on November 28 to 00:40 on November 29). It has been shown that a minimum sea surface wind speed of 3 m/s is required for HH-polarized operation to generate a usable level of backscatter for wave measurements.[37] As shown in Figure 8(c), the low-clutter direction percentages (LCDP, i.e., the number of low-clutter directions divided by the number of pulses as a percentage) of the images collected during that period are approximately 100%. Taking the buoy results as ground truth, the standard deviations (STDs) of the peak wave period obtained using the simplified and traditional wave algorithms are found to be 1.76 s and 2.19 s, respectively. Figure 6(b) illustrates that the peak wave direction results obtained using the simplified wave algorithm agree better with the buoy data than those from the traditional method. The corresponding STDs are 14.7° and 18.6°, respectively. The scatter plot in Figure 7 also shows the performances of these wave algorithms are comparable.

4.3 WIND RETRIEVAL RESULTS

Both the curve-fitting-based single-model and two-model wind algorithms described in Section 14.3 were implemented. The radar data considered here is classified as being contaminated by rain if the ZPP is below an empirical value of 10%. By utilizing this threshold, 9924 out of 49,182 images (i.e., 312 out of 1540 files) were identified as rain cases from the four-day dataset.

Figure 8 shows a typical comparison of the corresponding radar-derived and anemometer-measured wind results that were averaged over a 10-min interval. From Figure 8, it can be seen that both radar-deduced wind directions and speeds are affected by rain, and for the wind direction, the effect is relatively small. However, the radar-derived wind speed using the single-model algorithm was significantly overestimated as compared with the anemometer results. As mentioned earlier, the single-model method only involves the wind speed model (see Figure 3) obtained using the rain-free data. This means that the model is not applicable to rain-contaminated radar data. By using only the rain-contaminated data, the fitted wind speed model is depicted in Figure 9. From Figure 8, it is clear that the radar-derived wind speed results are improved significantly using the aforementioned rain mitigation scheme with two wind speed models. The error statistics are summarized in Table 1. By taking the anemometer wind data as ground truth, the rain mitigation scheme using two wind speed models (described in Step three) in Section 14.3.2 can reduce the wind speed STD and bias by 1.2 m/s and 1.3 m/s, respectively.

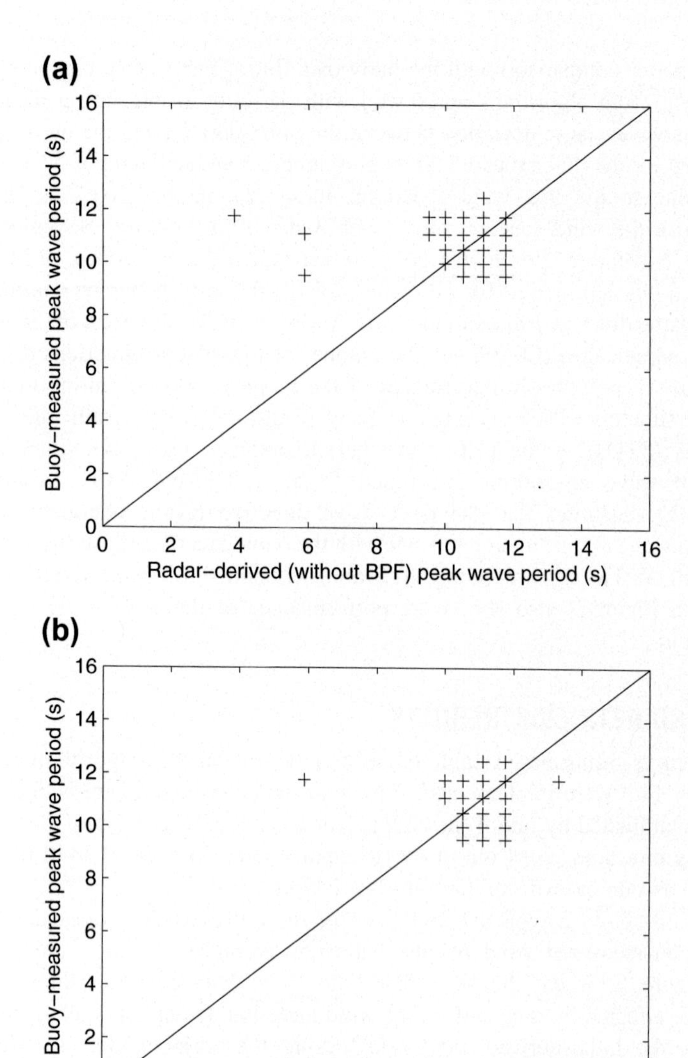

FIGURE 7

Scatter plot of (a) peak wave period: buoy-measured versus radar-derived (without BPF); (b) peak wave period: buoy-measured versus radar-derived (with BPF); (c) peak wave direction: buoy-measured versus radar-derived (without BPF); and (d) peak wave direction: buoy-measured versus radar-derived (with BPF).

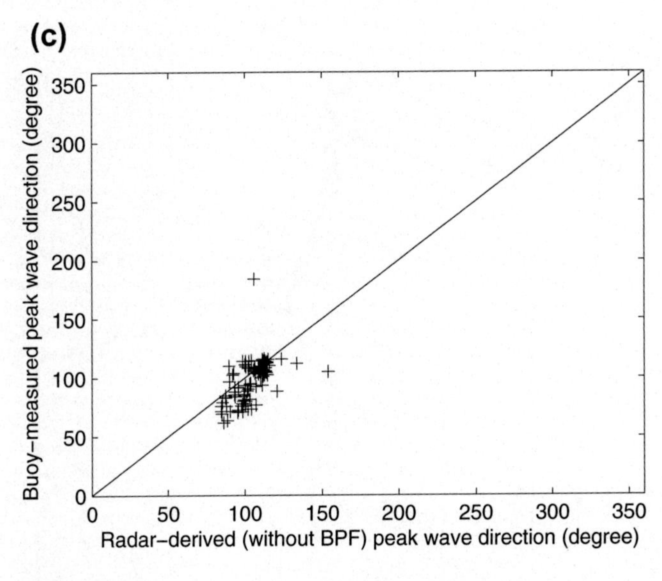

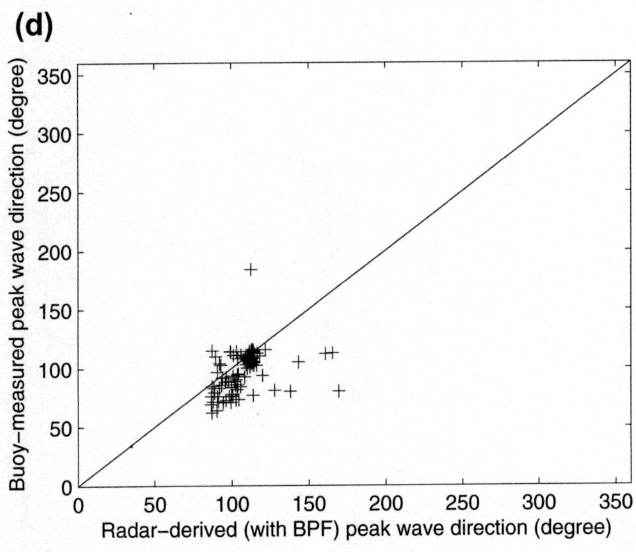

FIGURE 7 Continued

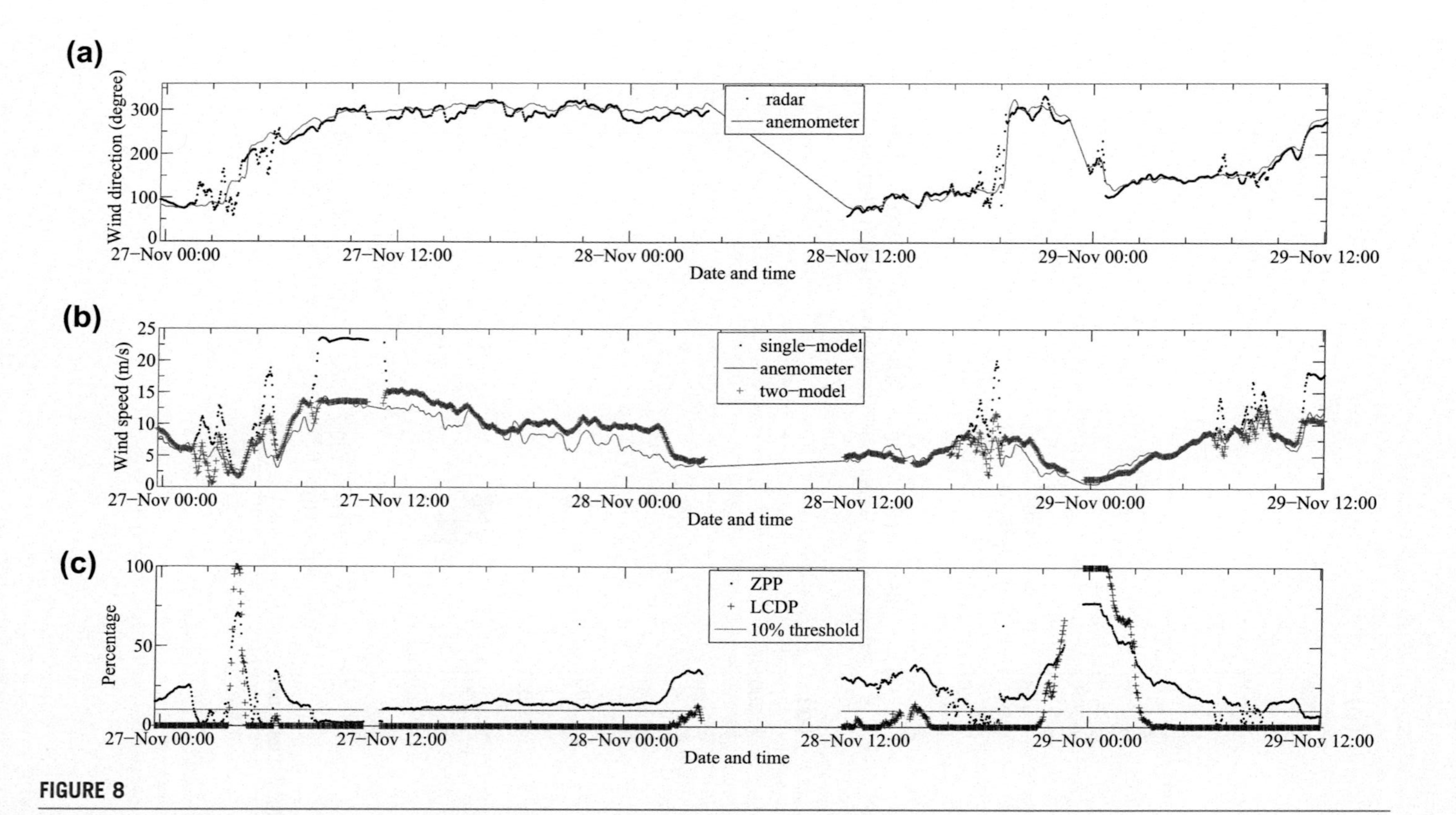

FIGURE 8

Results: (a) wind direction, (b) wind speed, and (c) zero pixel percentage (ZPP) and LCDP.

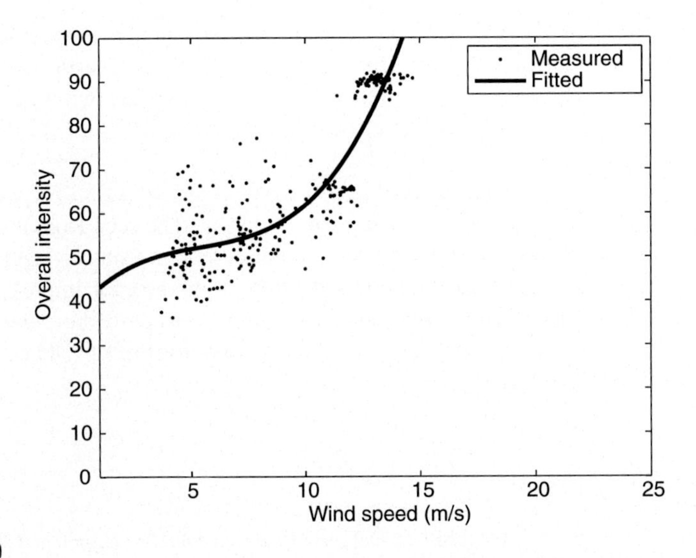

FIGURE 9

Scatter plot showing the anemometer wind speed, the corresponding radar backscatter intensity, and the best-fit curve based on a third-order polynomial function using rain-contaminated data.

Table 1 Wind Speed and Direction Recovery Error Statistics: Bias and Standard Deviation (STD)

Wind Algorithm	Wind Direction (degree)		Wind Speed (m/s)	
	Bias	STD	Bias	STD
Single-model	−3.8	18.3	2.1	2.9
Two-model	–	–	0.8	1.7

5. CONCLUSION

Recent research work involving sea surface remote sensing using X-band nautical radar in Eastern Canada has been reviewed. Traditional and simplified FFT-based wave algorithms as well as the curve-fitting-based wind parameter extraction method have been investigated. In addition, a scheme incorporating two wind speed models (one for rain-free and one for rain-contaminated data) has been proposed to improve the wind speed retrieval results under rain conditions. These methods have been tested using shipborne nautical radar data collected during a recent sea trial in the Northwest Atlantic Ocean. On comparing the radar-deduced wave measurements with the buoy record, it is seen that the performance of the simplified iterative FFT-based algorithm is comparable to that of the traditional method. Also, a comparison between the radar-measured wind results and the anemometer data confirms that the nautical radar can be a reliable tool for measuring sea surface wind parameters. The experimental results also show that the wind speed STD can be reduced by about 1.2 m/s using the rain-mitigation scheme with two speed models.

Although, here, the intention has been to develop an algorithm for improving wind speed from rain-contaminated radar data, it may also be seen from the foregoing analysis that the accuracy of wind direction measurement using X-band nautical radar decreases under rain conditions. In the future, a method for improving wind direction extraction from rain-contaminated radar data needs to be pursued. The work achieved here is expected to augment the existing utilities of X-band nautical radar for ocean remote sensing.

ACKNOWLEDGMENTS

The authors thank Dr Eric Thornhill at Defence Research and Development Canada for providing the field data. This work was supported in part by the Research and Development Corporation (RDC) IRIF Ignite grant (207765) and a Natural Sciences and Engineering Research Council of Canada grant (NSERC 402313-2012) to Dr W. Huang, an NSERC grant (NSERC 238263-2010) to Dr Eric Gill and an Atlantic Innovation Fund (AIF) award to Memorial University (E. Gill: principal investigator), a Newfoundland and Labrador Innovation Business and Rural Development grant (IBRD 30-10921-008) to Dr Gill and Dr Huang.

REFERENCES

1. Lipa BJ, Barrick DE. Least-squares methods for the extraction of surface currents from CODAR crossed-loop data – application at ARSLOE. *IEEE J Ocean Eng* 1983;**8**: 226–53.
2. Gurgel KW, Antonischki G, Essen HH, Schlick T. Wellen radar (WERA): a new ground-wave based HF radar for ocean remote sensing. *Costal Eng* 1999;**37**:219–34.
3. Liu Y, Weisberg RH, Merz CR. Assessment of CODAR SeaSonde and WERA HF radars in mapping surface currents on the west Florida Shelf. *J Atmos Ocean Technol* 2014;**31**: 1363–82.
4. Wyatt LR, Jaffres BD, Heron ML. Spatial averaging of HF radar data for wave measurement applications. *J Atmos Ocean Technol* 2013;**30**:2216–24.
5. Paduan JD, Washburn L. High-frequency radar observations of ocean surface currents. *Annu Rev Mar Sci* 2013;**5**:115–36.
6. Huang W, Gill EW, Wu X, Li L. Measurement of sea surface wind direction using bistatic high-frequency radar. *IEEE Trans Geosci Rem Sens* 2012;**50**:4117–22.
7. Young IR, Rosenthal W, Ziemer F. A three-dimensional analysis of marine radar images for the determination of ocean wave directionality and surface currents. *J Geophys Res* 1985;**90**(C1):1049–59.
8. Plant WJ, Keller WC. Evidence of Bragg scattering in microwave Doppler spectra of sea return. *J Geophys Res* 1990;**95**(C9):16,299–310.

9. Lee PHY, Barter JD, Beach KL, Hindman CL, Lake BM, Rungaldier H, et al. X-band microwave backscattering from ocean waves. *J Geophys Res* 1995;**100**(C2):2591−611.
10. Nieto-Borge JC, Rodriguez RG, Hessner K, Gonzales IP. Inversion of marine radar images for surface wave analysis. *J Atmos Ocean Technol* 2004;**21**:1291−300.
11. Izquierdo P, Guedes-Soares C, Nieto-Borge JC, Rodriguez GR. A comparison of sea-state parameters from nautical radar images and buoy data. *Ocean Eng* 2004;**31**: 2209−25.
12. Nieto-Borge JC, Hessner K, Jarabo-Amores P, de la Mata-Moya D. Signal-to-noise ratio analysis to estimate ocean wave heights from X-band marine radar image time series. *IET Radar Sonar Navig* 2008;**2**:35−41.
13. An J, Huang W, Gill EW. A self-adaptive wavelet-based algorithm for wave measurement using nautical radar. *IEEE Trans Geosci Rem Sens* 2015;**53**:567−77.
14. Dankert H, Horstmann J, Rosenthal W. Wind- and wave-field measurements using marine X-band radar-image sequences. *IEEE J Ocean Eng* 2005;**30**:534−42.
15. Dankert H, Horstmann J, Rosenthal W. Ocean wind fields retrieved from radar-image sequences. *J Geophys Res Oceans* 2003;**108**(C11):16-1−16-11.
16. Lund B, Graber HC, Romeiser R. Wind retrieval from shipborne nautical X-band radar data. *IEEE Trans Geosci Rem Sens* 2012;**50**:3800−11.
17. Vicen-Bueno R, Horstmann J, Terril E, de Paolo T, Dannenberg J. Real-time ocean wind vector retrieval from marine radar image sequences acquired at grazing angle. *J Atmos Ocean Technol* 2013;**30**:127−39.
18. Gangeskar R. Ocean current estimated from X-band radar sea surface images. *IEEE Trans Geosci Remote Sens* 2002;**40**:783−92.
19. Serafino F, Lugni C, Soldovieri F. A novel strategy for the surface current determination from marine X-band radar data. *IEEE Geosci Remote Sens Lett* 2010;**7**:231−5.
20. Senet CM, Seemann J, Ziemer F. The near-surface current velocity determined from image sequences of the sea surface. *IEEE Trans Geosci Remote Sens* 2001;**39**:492−505.
21. Huang W, Gill EW. Surface current measurement under low sea state using dual polarized X-band nautical radar. *IEEE J Sel Top Appl Earth Obs Remote Sens* 2012;**5**: 1868−73.
22. WaMoS II. *Wave and surface current monitoring system operating manual.* May 2012. version 4.0. [Online]. Available: oceanwaves.org.
23. Dobson FW, Smith SD, Anderson RJ, Vachon PW, Vandemark D, Buckley JR, et al. The Grand Banks ERS-1 SAR wave spectra validation experiment. In: *Proc. European Space Agency ERS-1 Symposium, Cannes, France*; 1992.
24. Buckley JR, Allingham M, Michaud R. On the determination of directional spectra from marine radar imagery of the sea surface. *Atmos Ocean* 1994;**32**:195−213.
25. Stredulinsky DC, Thornhill EM. Ship motion and wave radar data fusion for shipboard wave measurement. *J Ship Res* 2011;**55**:73−85.
26. Gangeskar R. An algorithm for estimation of wave height from shadowing in X-band radar sea surface images. *IEEE Trans Geosci Remote Sens* 2014;**52**:3373−81.
27. Huang W, Gill EW, An J. Iterative least-squares-based wave measurement using X-band nautical radar. *IET Radar Sonar Navig* 2014;**8**:853−63.
28. Hatten H, Ziemer F, Seemann J, Nieto-Borge J. Correlation between the spectral background noise of a nautical radar and the wind vector. In: *Proc. 17th Int. Conf. Offshore Mechanics and Arctic Eng. (OMAE), Lisbon, Portugal*; 1998.
29. Trizna D, Carlson D. Studies of dual polarized low grazing angle radar sea scatter in nearshore regions. *IEEE Trans Geosci Remote Sens* 1996;**34**:747−57.

30. Lund B, Graber HC, Horstmann J, Terrill E. Ocean surface wind retrieval from stationary and moving platform marine radar data. In: *Proc. IEEE IGARSS, Munich, Germany*; 2012. p. 2790–3.
31. Tournadre J, Quilfen Y. Impact of rain cell on scatterometer data: 1. Theory and modeling. *J Geophys Res* 2003;**108**(C7):18-1–18-14.
32. Contreras RF, Plant WJ. Surface effect of rain on microwave backscatter from the ocean: measurements and modeling. *J Geophys Res* 2006;**111**(C08):C08019.
33. Melsheimer C, Alpers W, Gade M. Investigation of multifrequency/multipolarization radar signatures of rain cells over the ocean using SIR-C/X-SAR data. *J Geophys Res* 1998;**103**(C9):18867–84.
34. Braun N, Gade M, Lange PA. Radar backscattering measurements of artificial rain impinging on a water surface at different wind speeds. In: *Proc. IEEE IGARSS, Hamburg, Germany*; 1999. p. 200–2.
35. Liu Y, Huang W, Gill EW. Analysis of the effects of rain on surface wind retrieval from X-band marine radar images. In: *Proc. MTS/IEEE oceans*. Canada: St John's; 2014.
36. Liu Y, Huang W, Gill EW, Peters DK, Vicen-Bueno R. Comparison of algorithms for wind parameters extraction from X-band shipborne marine radar images. *IEEE J Sel Top Appl Earth Obs Remote Sens* 2015;**8**:896–906.
37. Nieto-Borge JC, Guedes-Soares C. Analysis of directional wave fields using X-band navigation radar. *Coast Eng* 2000;**40**:375–91.

CHAPTER

Estimating Nearshore Bathymetry from X-Band Radar Data

15

Giovanni Ludeno[1,2], Ferdinando Reale[3], Fabio Dentale[3], Eugenio Pugliese Carratelli[3,4,*], Antonio Natale[1], Francesco Serafino[1]

Institute for Electromagnetic Sensing of the Environment (IREA), Italian National Research Council (CNR), Napoli, Italy[1]; Department of the Industrial and Information Engineering, Second University of Naples, Aversa, Italy[2]; Maritime Engineering Division University of Salerno (MEDUS), University of Salerno, Fisciano, Italy[3]; CUGRI—University Centre for Research on Major Hazard, Fisciano, Italy[4]

**Corresponding author: E-mail: epc@unisa.it*

CHAPTER OUTLINE

1. INTRODUCTION: THE RADAR IMAGING OF SEA WAVES

Sea state monitoring by land- and ship-based X-band marine radars provides obvious advantages: high temporal and spatial resolution can be achieved without physical contact with the sea surface and at a comparatively low cost, especially if the raw data are provided by a standard navigation radar.[1–3]

This chapter focuses on the determination of the sea depth by making use of images derived from X-band marine radar devices. This problem is of great interest because the possibility to get continuous measurements of the bathymetry and the sea state represents a key point for the monitoring of a number of coastal phenomena, including the reflection, diffraction, and refraction of the sea waves, as well as the changes of the sea bottom.

The basic principles involved in the analysis of the sea surface from radar data have been described and discussed by many authors,[4–7] and a concise review on this topic is also provided in Chapter 14 of this book.[8] The physical possibility to extract information on the sea surface from a radar signal is due to the interaction of the transmitted electromagnetic (EM) waves with the sea waves of similar wavelength (Bragg resonance). In particular, X-band EM waves are a few centimeters long and interact with the sea capillary-gravity waves (sometimes named "short" waves, "ripples," or "roughness"), which ride on the gravity waves (sometimes called "long" waves or, inappropriately, "swell"). Bragg scattering has been known for many years, and it represents an unwanted feature (clutter) when the radar is used for surveillance purposes. Research work over the years has shown that the signal backscattered from the sea surface (clutter) can be processed to derive important information about the wave motion. However, such a signal is not the direct representation of the sea surface because a number of effects are introduced during the radar imaging process. Indeed, due to the mechanisms that rule the EM scattering from rough surfaces,[3–7,9] the radar echo does not represent the sea elevation profile, but it is tied to the slope of the long waves (tilt modulation) and to the roughness of the riding ripples (hydrodynamic modulation). In addition, radar signals are also subject to the shadowing phenomenon, according to which no information is received from sea spots that are not in line of sight, and at grazing angle, this is the predominant effect.[3,9,10] Finally, for very strong winds and sea states, or in the immediate vicinity of beaches, breaking waves and floating foam should have a relevant impact on the response; however, this occurrence has only recently received the attention of researchers, and no useful results are yet available for grazing angles.[11,12] For these reasons, estimating sea state parameters from radar data is far from simple. At the present time, this task is treated as an inversion problem, and it is based on the knowledge of the fundamental laws that rule the dynamics of the sea gravity waves. According to the linear (Airy) wave theory, the elevation profile $\eta(\underline{r}, t)$ can be regarded as the superposition of monochromatic sea waves with different amplitudes A, periods T, directions $\widehat{k}$, and wavelengths λ, i.e., the following:

$$\eta(\underline{r}, t) = \iint_{K} \int_{\Omega} Z(\underline{k}, \omega) e^{j(\underline{k}\cdot\underline{r} - \omega t)} d\underline{k} d\omega, \tag{1}$$

where t is the observation time, and $\underline{r} = (x, y)$ represents the position vector, while $\omega = 2\pi/T$ is the angular frequency (pulsation), $\underline{k} = k\widehat{k} = (k_x, k_y)$ is the wave vector being $k = |\underline{k}| = 2\pi/\lambda$ the wave number, and $Z(\underline{k}, \omega) = A(\underline{k}, \omega) e^{j\varphi(\underline{k}, \omega)}$ is the complex amplitude of each component.[13,14]

These quantities cannot be arbitrarily chosen because their characteristics are strictly constrained by the physical properties of sea gravity waves. In particular, the dispersive behavior of sea gravity waves that propagate over a uniform sea bottom of depth, d, is ruled by the dispersion relation[15]:

$$\omega(\underline{k}) = \sqrt{kg \tanh(kd)}, \tag{2}$$

where g is the acceleration due to gravity. According to such an equation, each monochromatic component involved in the spectral representation of Eqn (1) propagates with a wave celerity (phase speed) $C = \omega(\underline{k})/k$ that depends on the wavelength, λ, as well as on the sea depth, d. Moreover, Eqn (2) states that not all the period/wavelength pairs are allowed for the gravity waves; in fact, the wavelength, λ, is uniquely determined by the period, T, for any given depth, d. This property plays a fundamental role on the location of the sea signal in the spectral (frequency-wave number) domain because the energy of the sea waves cannot lie anywhere in the ω-$\underline{k}$ space, but it concentrates over the ω-$\underline{k}$ locus of points defined by the dispersion relation. Accordingly, the dependence of the angular frequency on the sea depth offers the opportunity to retrieve the bathymetry from the spectrum of the sea signal. However, several factors contribute to complicate this task. First of all, the presence of a surface current field $\underline{U} = (U_x, U_y)$ on the sea surface modifies the dispersion equation in the following way:

$$\omega(\underline{k}) = \sqrt{kg \tanh(kd)} + \underline{k} \cdot \underline{U}, \tag{3}$$

That is, the surface current represents a Doppler shift on the wave spectrum. Therefore, any attempt to estimate the bathymetry should take into account the surface current. This means that if no a priori information is available about the sea surface current, it represents a further parameter to be recovered from the observation of the wave motion. The behavior of the angular frequency in function of the wave number for different values of the bathymetry and surface current is plotted in Figure 1. In addition, the estimation of the sea depth through radar data should also take into account all the distortions introduced on the radar signal. Due to the modulation effects, the energy of the radar spectrum is spread over the wave number domain according to the energy of the sea spectrum and, moreover, it is affected by undesired replicas introduced by the aforementioned nonlinear effects. An example of these distortions is depicted in Figure 2, where the spectrum of a simulated sea signal (left) is compared with its radar counterpart, obtained by numerically inserting the modulation effects (right). However, these distortions mainly affect the retrieval of sea state parameters, whose accomplishment requires a quite difficult equalization step involving the linear part of the Modulation Transfer Function (MTF),[13] while they can be easily handled if the main task is the reconstruction of the bathymetry and the surface current fields.

2. SEA SURFACE CURRENT AND BATHYMETRY RECONSTRUCTION FROM RADAR DATA

In the previous section, it was shown that the energy of the sea signal concentrates in a two-dimensional frequency-wave number domain $\omega(\underline{k})$, whose shape depends on the values of the sea surface current and bathymetry through the dispersion relation. Therefore, the estimation of these parameters from time

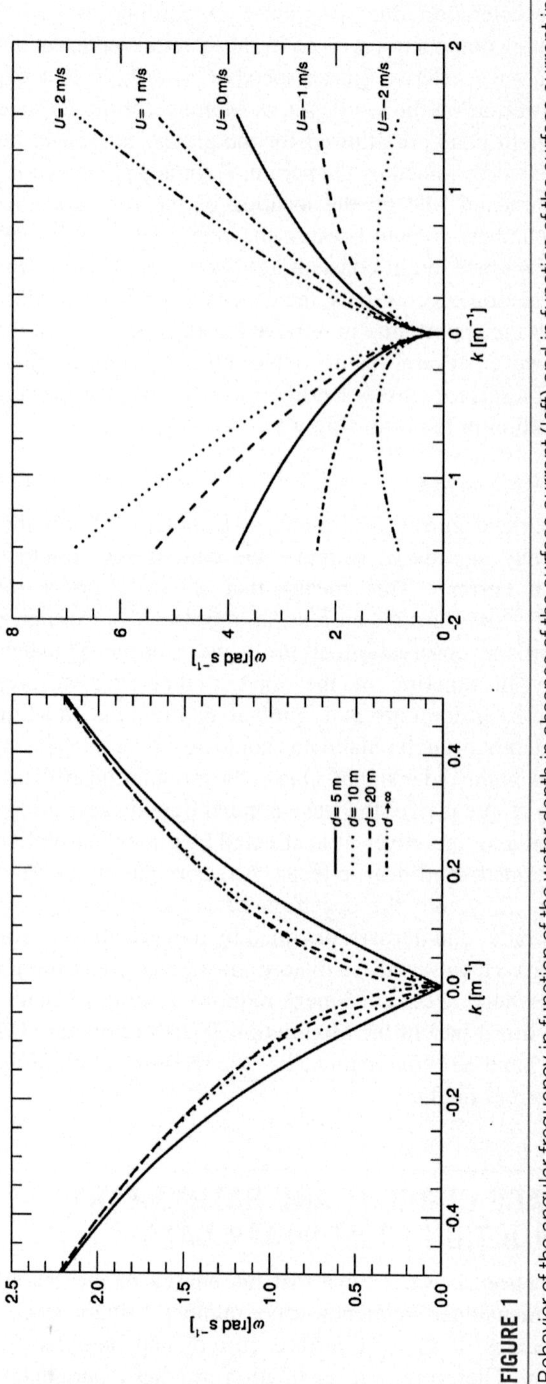

FIGURE 1

Behavior of the angular frequency in function of the water depth in absence of the surface current (left) and in function of the surface current in the deep water ($d \to \infty$) case (right).

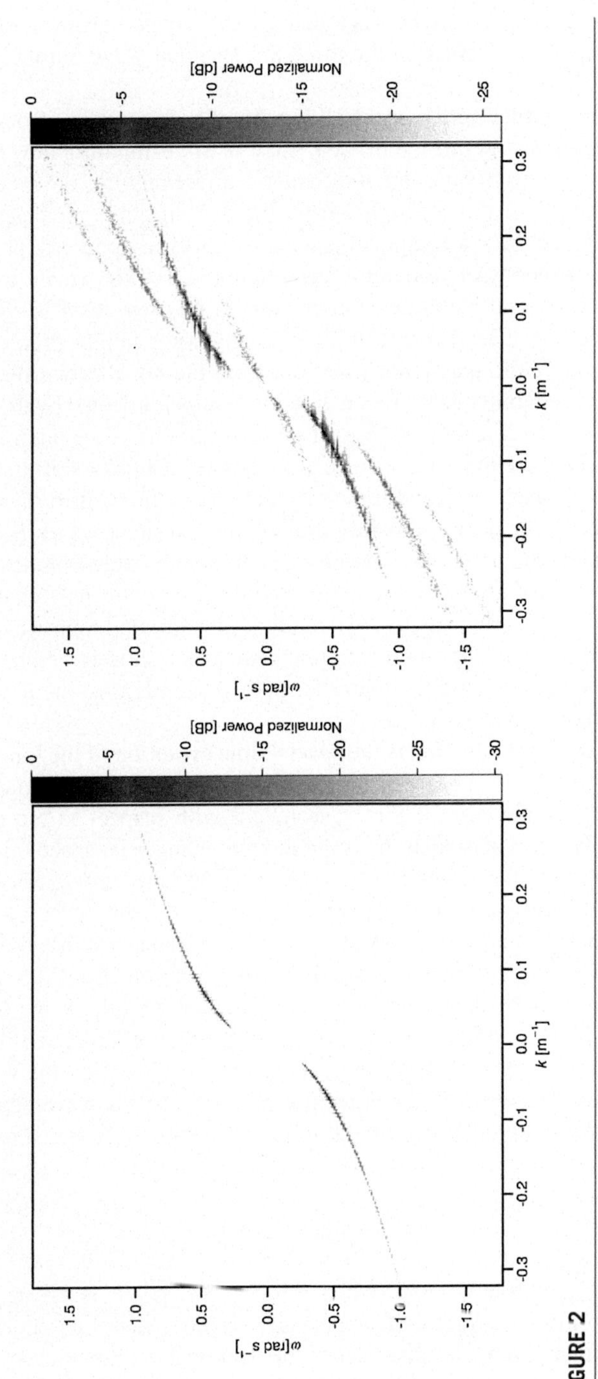

FIGURE 2

Spectrum of a simulated sea signal (left) and the corresponding radar spectrum (right).

sequences of radar data can be regarded as the shape retrieval of the two-dimensional dispersion surface in the three-dimensional wave number-frequency spectrum.

However, reconstructing the spectral domain of the sea waves from radar signals is anything but a straightforward task. First of all, estimating how the angular frequency $\omega(\underline{k})$ behaves in function of the wave number requires the retrieval of the spectral components of the sea signal from the overall noisy radar spectrum. In particular, confusing the sea components with the noise components, or vice versa, turns into a wrong reconstruction of the spectral domain of sea waves, thus causing poor surface current and bathymetry estimates. In this regard, it is worth noting that a wind speed of at least 3 m/s or a significant wave height greater than 1 m are required to make the sea signal stand out from the noise background.[16,17]

Moreover, the reliability of the reconstruction results is affected by the sensitivity of the angular frequency $\omega(\underline{k})$ with respect to the variations of the parameters d and $\underline{U}$. In particular, because the surface current, $\underline{U}$, causes a Doppler shift in the angular frequency, $\omega(\underline{k})$, a change in the values of $\underline{U}$ turns into a linear drift on $\omega(\underline{k})$. Information about the water depth is mainly provided by the long waves (small k) and about the current by the short waves (large k and, therefore, large Doppler shift). Significant variations of the sea surface current give rise to noticeable changes in the behavior of the angular frequencies, above all at medium and high wave numbers. However, the accuracy of the surface current estimates is closely tied to the resolution of the ω-$\underline{k}$ space, because the finer the spectral sampling is, the more accurate the surface current estimates can be.

The bathymetry strongly affects the space–time evolution of the long wave motion. In shoaling zones, for instance, the decreasing depth causes the reduction of the wavelength and the increase of the wave height with respect to the deep water ($d \rightarrow \infty$) case. The way in which the wave motion changes its space–time characteristics in function of the sea bottom[15] represents a key property to retrieve the bathymetry from the observation of the sea waves.

However, the sea bottom, d, affects the angular frequency in a quite complicated manner because it appears in the dispersion relation (Eqn (3)) through the hyperbolic tangent. The dependence of the wave wavelength on the bathymetry disappears for higher depths, thus resulting in the impossibility to retrieve sea depths greater than a critical depth, d^*, whose value depends on the wavelengths of the sea waves. The analytical connection between the sea wavelength and the sea depth can be computed by manipulating the dispersion relation, thus obtaining the following:

$$\lambda = \lambda_\infty \tanh\left(\frac{2\pi}{\lambda} d\right). \tag{4}$$

From Eqn (4), it follows that, when the sea bottom, d, is greater than half the wavelength of the sea waves, the hyperbolic tangent reaches its saturation level, and the wavelength practically approaches the deep water value λ_∞ ($\lambda > 0.99\lambda_\infty$).

Loosely speaking, this means that, in theory, sea waves begin to feel the influence of the bottom when the depth is less than half their wavelength. From a practical point of view, the bathymetry inversion procedures can provide a meaningful estimate on condition that the actual depth is smaller than the critical value $d^* = \lambda/4$, whereas $\lambda \geq 10 * d$ is normally assumed to be necessary to achieve a good accuracy.

The loci of points in the (λ, d) plane that satisfy Eqn (4) are plotted in Figure 3 for different values of the wave period, T (i.e., for different deep water wavelengths λ_∞). In that figure, the dashed and the dashed-dotted lines represent the pairs ($\lambda = 0.92\lambda_\infty$, $d = \lambda/4$) and ($\lambda = 0.99\lambda_\infty$, $d = \lambda/2$) corresponding to the practical and theoretical limits for the bathymetry inversion, respectively. Such a figure clearly shows that sea wavelengths are insensitive to the bathymetry values lying in the region to the right of the dashed-dotted line. Conversely, a noticeable impact of the sea bottom on the wavelengths can be observed in the left-hand region delimited by the dashed line, which represents the feasibility region for the bathymetry inversion purposes.[17] However, even if the sea depths belong to the feasibility region, the accuracy of the inversion results depends upon the sampling rates used to get the radar signals, as already stated for the surface current.

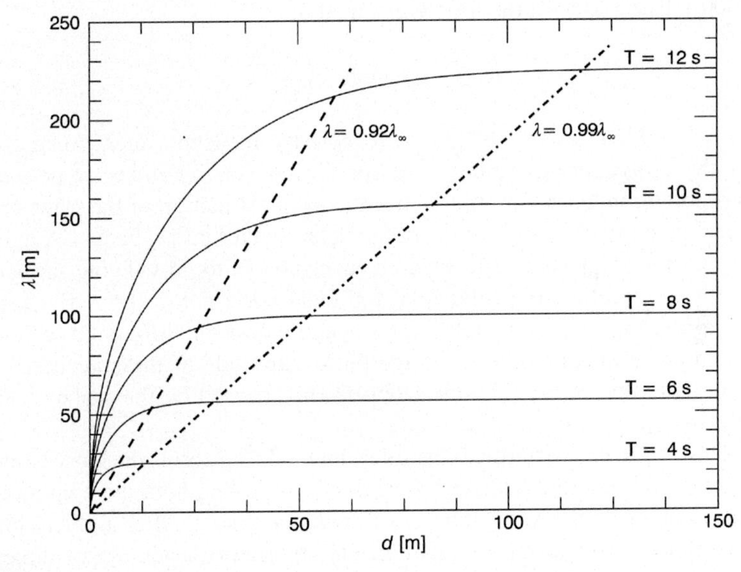

FIGURE 3

Sea wavelengths in function of bathymetry (solid curves) for different values of the sea period, T, together with the theoretical (dashed-dotted line) and practical (dashed line) limits for the bathymetry inversion.

Finally, it must be noted that previous discussion as well as the following inversion strategy are founded upon the linear wave theory, which fails to describe the sea evolution in shore areas where nonlinear effects rule the dynamics of the wave motion. Accordingly, the dispersion relation of Eqn (2) becomes inaccurate in very shallow water, and this gives rise to a slight overestimation of the sea depth. To improve the bathymetry estimation results, it is possible to resort to the nonlinear wave theory and introduce a correction term on the angular frequency. However, such a correction term depends upon the sea wave height, which, in absence of ancillary data, is a further parameter to be estimated from sea radar signals.

3. INVERSION PROCEDURES

A number of inversion techniques aimed at retrieving the bathymetry and the sea surface currents from radar data have been developed in last two decades. In this section, we briefly summarize the most used techniques available in literature to get such parameters.

One of the approaches to reconstruct the underwater topography from time sequences of radar data is described in Ref. 18. This method is founded on the connection between the wave period, T, the bathymetry, d, and wave celerity, C, which can be explicitly expressed via the following equation:

$$d = \frac{\lambda_\infty}{2\pi} \frac{C}{C_\infty} \tanh^{-1}\left(\frac{C}{C_\infty}\right), \tag{5}$$

where $C_\infty = gT/2\pi$ represents the wave celerity in deep water. According to Eqn (5), the bathymetry inversion turns into a wave celerity and wave period estimation problem. In particular, in Ref. 18, the local estimation of the wave celerity is founded on the spatial partitioning of the radar images into subareas. A temporal correlation analysis is then performed on each subarea to provide the surface current vector, $\underline{U}$, and then the celerity of the local wave field. The retrieval of the wave period, T, instead, is computed in the spectral domain, through the estimation of the angular frequency relevant to the peak amplitude of the considered radar spectrum. Note that in Eqn (5), the Doppler shift caused by the surface currents is neglected.

The DiSC (Dispersive Surface Classifier) method[19] performs the local estimation of the surface current and bathymetry fields by exploiting both the amplitude and phase information of the three-dimensional radar spectrum. After a global filtering on the three-dimensional radar spectrum, the algorithm makes use of directional dispersion frequency (DDF) filters to extract, for each angular frequency of the radar data, the wave numbers and directions that mainly contribute to the observed wave motion. The local estimates of the single-frequency and direction-dependent wave number require the knowledge of the phase of the radar spectrum because they are carried out through a classification method based on the AM/FM representation of the DDF filtered signals. This is necessary because of the inhomogeneity of the bathymetry. After estimating the wave number vectors with high spatial resolution from the spatial phase pattern, the water depth is retrieved by inversion of the dispersion relation using an iterative algorithm.

A least squares (LS) method is then performed on the effective spectral components reconstructed through the latter procedure to provide the estimation map of both the surface current and bathymetry fields. It is worth noting that the DiSC method can handle the spatial variations of the sea wave and bathymetry fields because it just requires the spatial homogeneity at the wavelength scale.

The most recent approach based on the NSP (normalized scalar product) method[20,21] performs the local surface current and bathymetry joint estimation by maximizing the normalized scalar product between the amplitude of the three-dimensional radar spectrum $F_I(\underline{k}, \omega)$ and a real characteristic function $G(\underline{k}, \omega, \underline{U}, d)$ based on the dispersion relation, such as shown by the following equation:

$$G(\underline{k}, \omega, \underline{U}, d) = \delta\left(\sqrt{kg\tanh(kd)} + \underline{k}\,\underline{U} - \omega\left(\underline{k}\right)\right), \tag{6}$$

where δ is the Dirac delta. Therefore, the retrieval of the surface current and bathymetry values is carried out through the following estimator:

$$\left(\widehat{\underline{U}}, \widehat{d}\right) = \underset{\underline{U},d}{\arg\max} \frac{\left\langle F_I\left(\underline{k}, \omega\right), G\left(\underline{k}, \omega, \underline{U}, d\right)\right\rangle}{\sqrt{P_F P_G}}, \tag{7}$$

where $\langle F_I, G\rangle$ represents the scalar product between the functions F_I and G, and P_F and P_G are the powers associated to the functions F_I and G, respectively. Because the estimation is founded on the correlation operator, it takes into account the integral strength of the sea signal over the spectral domain. This feature allows the NSP to outperform in terms of noise robustness (i.e., in terms of accuracy of the estimation results) the LS-based methods, where the reconstruction is carried out just exploiting the spectral location of the sea components.[22] Moreover, unlike the LS method, the NSP procedure does not require the critical step of defining an optimal threshold, whose value strongly affects the estimation results.

Recall that the NSP strategy relies on the spatial homogeneity of both the bathymetry and surface current fields, but, as stated, this condition is usually not satisfied in a nearshore area. Accordingly, the NSP local estimates require the spatial partitioning of the radar images into partially overlapping subareas. The subareas should be large enough to guarantee a reliable estimation but, at the same time, sufficiently small to avoid the spatial inhomogeneity of the parameters to be retrieved.[23–26] Note that the extent of each subarea represents the spatial resolution of the bathymetry and surface current retrieval maps.

4. ESTIMATION RESULTS ON REAL-WORLD DATA

In this section, we show the reconstruction of two inshore bathymetry fields from X-band incoherent radar data by exploiting the NSP strategy. To demonstrate the performance of X-band radars in reconstructing the underwater topography, in both cases, the estimates have been compared with some echo-sounder measurements available for the investigated areas.

The first reconstruction regards the bathymetry of the Salerno harbor, in Italy.[26] In this case, the estimation results have been obtained through the processing of 832 radar images acquired on March 27, 2013 (images courtesy of Remocean S.p.A.,[27]). The details of the acquisition system are summarized in Table 1.

In particular, the sequence of 832 consecutive radar images has been split into 12 partially overlapping (in time) subsets, each containing 128 radar images (with an overlap of 64 radar images between two consecutive subsets). Accordingly, the bathymetry reconstruction procedure (or, more accurately, the surface current and bathymetry joint estimation procedure) has been performed on these subsets, thus obtaining the 12 corresponding individual bathymetry maps, each having a pixel spacing equal to 50 m in both the coordinates. A subarea size of 500 m and spatial overlapping interval of 50 m have been used, in this test case. The temporal average of the latter maps represents the final outcome of the bathymetry reconstruction procedure and is shown in Figure 4, together with the geographic location of the Salerno harbor.

The reconstructed mean bathymetry field has been compared to the in situ data obtained through a survey performed by the Port Authority in the area bounded by the black contour in Figure 4, in February 2011. The survey was carried out with an Elac Hydrostar 4300 single-beam echo-sounder and a Leica System 1200 GPS.

Table 1 Parameters of the Acquisition System (Salerno)

Parameter	Value
Antenna rotation period	2.4 s
Range resolution	10.5 m
Azimuth resolution	1.2°
Radar range	2555 m
Radar's field of view	120°
Polarization	HH

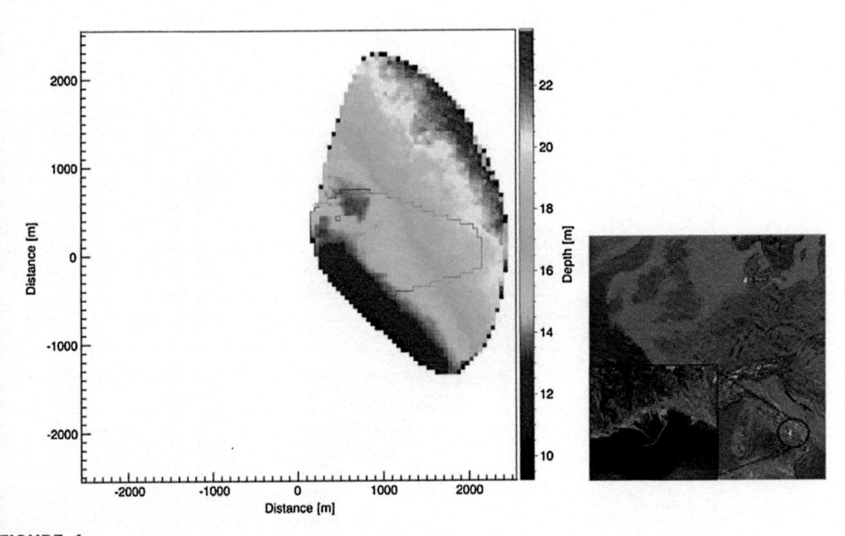

FIGURE 4

Reconstruction of the bathymetry field relevant to the Salerno harbor, in Italy. The black contour bounds the area for which the bathymetry estimates have been compared with the echo-sounder measurements.

The quantitative comparison between the ground truth and the bathymetry reconstructed from radar data is shown in Figure 5, which depicts the scatterplot of the radar depths versus the echo-sounder measurements as well as the histogram of the reconstruction error. In particular, the distribution of the estimate/measurement pairs represented in the scatterplot leads to a highly significant correlation coefficient ($R^2 = 0.95$) between the radar measured depth and that from the nautical chart. Moreover, by analyzing the histogram, the low effective error between the radar estimates and the ground truth can be easily appreciated (the standard deviation of the differences is about 0.52 m), even though a bias of about 1 m in the differences can be also observed. Part of this systematic error can be explained recalling that the NSP inversion procedure is founded on the linear wave theory, which leads to an overestimation of the water depth,[23–26] above all in shallow water.

The same approach has been applied on a dataset collected on August 27, 2003 at the experimental and monitoring radar station of HZG—Center for Materials and Coastal Research (formerly known as GKSS), which is located at the northern tip of the Sylt Island, in Germany.[24] The parameters relevant to this acquisition system are listed in Table 2.

The dataset consisted of 12 subsets, each composed by 256 consecutive radar images, to cover a 12-h tidal cycle. Therefore, 12 individual bathymetry maps, each

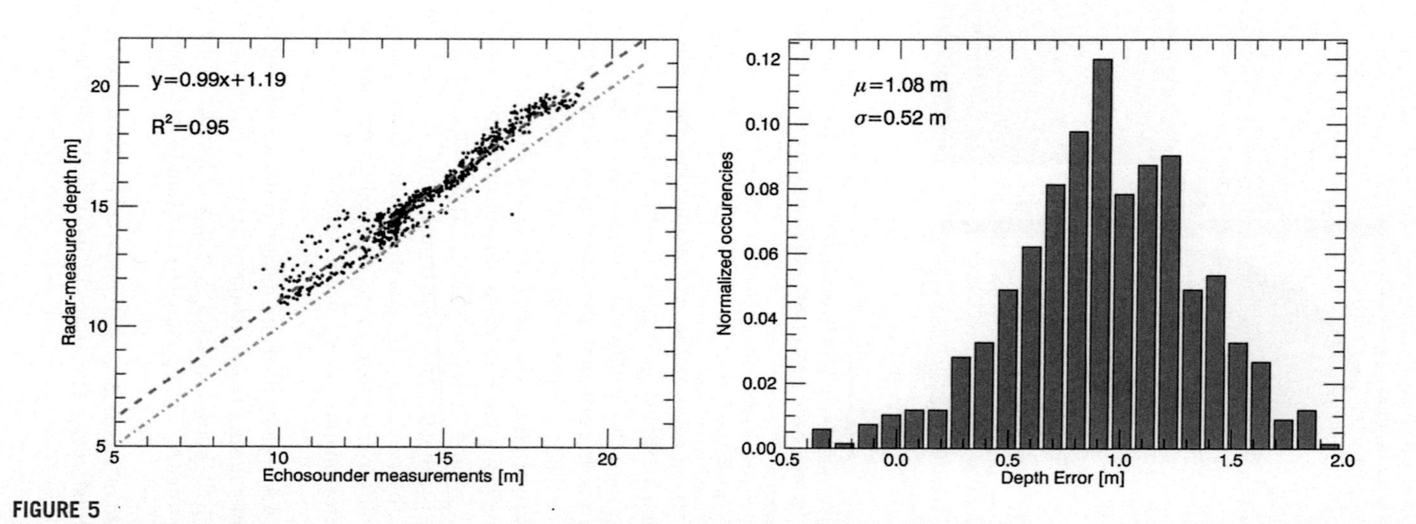

FIGURE 5

Left: radar measured depths versus echo-sounder measurements. The dashed line denotes the regression line, whose equation is given on the left-top side. Right: histogram of the reconstruction error. The mean value (μ) and the standard deviation (σ) of the differences are given on the left-top side.

Table 2 Parameters of the Acquisition System (Sylt Island)

Parameter	Value
Antenna rotation period	1.8 s
Range resolution	6.8 m
Azimuth resolution	0.95°
Radar range	1960 m
Radar's field of view	215°
Polarization	HH

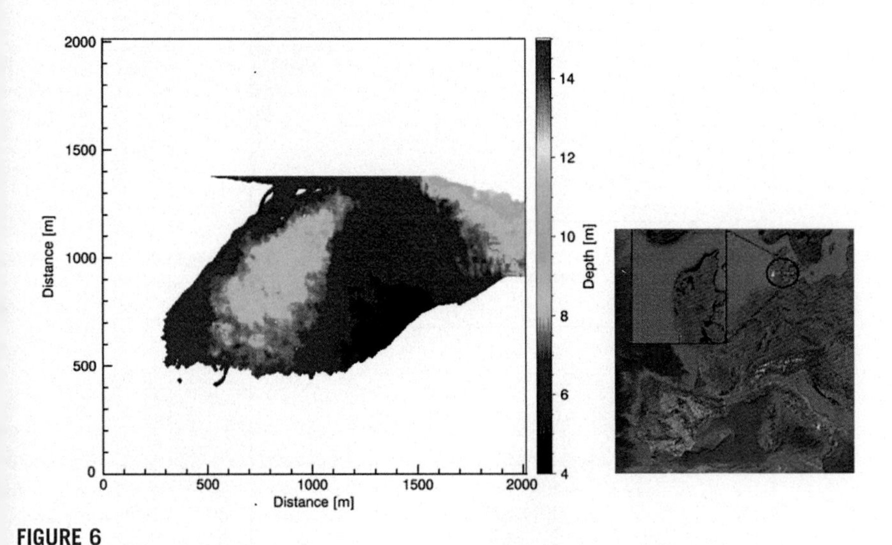

FIGURE 6

Reconstruction of the bathymetry field relevant to Sylt Island, in Germany.

having a pixel spacing equal to 3.4 m in both the coordinates, have been reconstructed by processing such subsets. Also, in this case, the final outcome of the reconstruction procedure is represented by the temporal average of the latter maps, to reduce the statistical uncertainty of the final estimates and compensate the errors induced by the tide levels.

The mean bathymetry map (see Figure 6) has been compared with the ground truth obtained two days before the radar data acquisition by a multibeam echo-sounder, and the scatterplot and the histogram of the depth error provided by this comparison are shown in Figure 7. An important issue is the inhomogeneity in this coastal area.

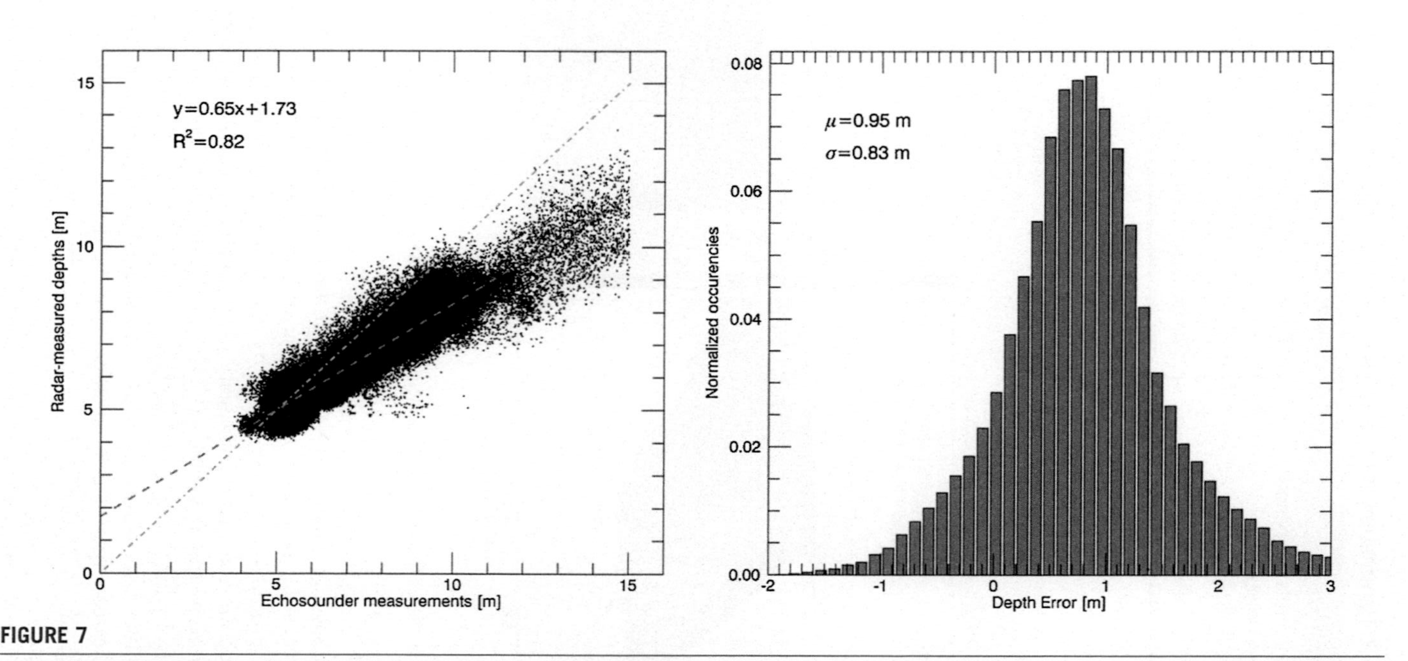

FIGURE 7

Left: radar depths versus echo-sounder measurements. The dashed line denotes the regression line, whose equation is given on the left-top side. Right: histogram of the reconstruction error. The mean value (μ) and the standard deviation (σ) of the differences are given on the left-top side.

5. CONCLUSIONS

The chapter has dealt with the problem of the sea depth estimation starting from X-band radar measurements. First, a simple analysis of the mathematical features of the problem has been performed, after which the most recent estimation strategy has been presented. The numerical analysis has provided results coherent with the theoretical expectations and pointed out how the intrinsic ill-conditioning of the problem makes it inapplicable for large values of the sea depth. In addition, the results showed that the methodology is accurate and independent from the type of input data captured by the radar; good performances of the approach were observed for a range of sea depths up to about 20 m. The main contribution of the work was the adoption of a "correlation" procedure to estimate the sea depth. Future work should be addressed to the analysis of radar data in the very shallow water scenario, where the wave linear theory fails, and more complex phenomena have to be taken into account.

ACKNOWLEDGMENTS

The authors are grateful to the Helmholtz-Zentrum Geesthacht Zentrum für Material-und Küstenforschung (Centre for Materials and Coastal Research) for providing the radar data set of the Sylt Island experiment.

REFERENCES

1. Young R, Rosenthal W, Ziemer F. Three-dimensional analysis of marine radar images for the determination of ocean wave directionality and surface currents. *J Geophys Res* 1985; **90**:1049–59.
2. Ziemer F, Rosenthal W. Directional spectra from ship board navigation radar during LEWEX. In: Beal RC, editor. *Directional ocean wave spectra: measuring, modeling, predicting, and applying*. Baltimore (MD, USA): The Johns Hopkins University Press; 1991. p. 80–4.
3. Plant WJ, Keller WC: Evidence of Bragg scattering in microwave Doppler spectra of sea return. *J Geophys Res* 1990;**95**:16299–310.
4. Pugliese Carratelli E, Dentale F, Reale F. Numerical pseudo-random simulation of SAR sea and wind response. In: Lacoste H, editor. *"Proceedings of SeaSAR 2006 (ESA SP-613): advances in SAR oceanography from Envisat and ERS missions", Frascati (RM), Italy, 23–26 January 2006*. Noordwijk (The Netherlands): ESA Publications Division (ESTEC); 2006.
5. Pugliese Carratelli E, Dentale F, Reale F. Reconstruction of SAR wave image effects through pseudo random simulation. In: Lacoste H, Ouwehand L, editors. *"Proceedings of Envisat symposium 2007 (ESA SP-636)", Montreux, Switzerland, 23–27 April 2007*. Noordwijk (The Netherlands): ESA Communication Production Office (ESTEC); 2007.
6. Clarizia MP, Gommenginger C, Di Bisceglie M, Galdi C, Srokosz M. Simulation of L-band bistatic returns from the ocean surface: a facet approach with application to ocean GNSS reflectometry. *IEEE Trans Geosci Remote Sens* 2012;**50**(3):960–71.
7. Reale F, Dentale F, Pugliese Carratelli E. Numerical simulation of whitecaps and foam effects on satellite altimeter response. *Remote Sens* 2014;**6**(5):3681–92.
8. Huang W, Gill EW. Ocean remote sensing using X-band shipborne nautical radar – applications in Eastern Canada. In: Liu Y, Kerkering H, Weisberg RH, editors. *Coastal ocean observing systems: advances and syntheses*. Elsevier; 2015.

9. Lee PHY, Barter JD, Beach KL, Hindman CL, Lake BM, Rungaldier H, et al. X band microwave backscattering from ocean waves. *J Geophys Res* 1995;**100**:2591–611. http://dx.doi.org/10.1029/94JC02741.
10. Wetzel LB. Electromagnetic scattering from the sea at low grazing angles. In: Plant WJ, Geernaert GL, editors. *Surface waves and fluxes*, vol. 2. Norwell (MA, USA): Kluwer Academic Publisher; 1990. p. 109–71.
11. Pugliese Carratelli E, Chapron B, Dentale F, Reale F. Simulating the influence of wave whitecaps on SAR images. In: Lacoste H, Ouwehand L, editors. *"Proceedings of SeaSAR 2008 (ESA SP-656)", Frascati (RM), Italy, 21–26 January 2008*. Noordwijk (The Netherlands): ESA Communication Production Office (ESTEC); 2008.
12. Debojyoti G, Mishra MK, Chauhan P. Deriving sea state parameters using RISAT-1 SAR data. *Adv Space Res* 1 January 2015;**55**(1):83–9. Available online 08.10.14, http://dx.doi.org/10.1016/j.asr.2014.09.033.
13. Nieto Borge JC. *Analisis de Campos de Oleaje Mediante Radar de Navegacion en Banda X* [Ph.D. thesis]. Spain: University of Madrid; 1997.
14. Nieto Borge JC, RodrÍguez GR, Hessner K, Gonzáles PI. Inversion of marine radar images for surface wave analysis. *J Atmos Oceanic Technol* 2004;**21**:1291–300.
15. Holthuijsen LH. *Waves in oceanic and coastal waters*. New York (NY, USA): Cambridge University Press; 2007.
16. Nieto Borge JC, Guedes Soares C. Analysis of directional wave fields using X-band navigation radar. *Coastal Eng* 2000;**40**:375–91.
17. Bell PS, Osler JC. Mapping bathymetry using X-band marine radar data recorded from a moving vessel. *Ocean Dyn* 2011;**61**:2141–56.
18. Bell PS. Shallow water bathymetry derived from an analysis of X-band marine radar images of waves. *Coastal Eng* 1999;**37**:513–27.
19. Senet CM, Seemann J, Flampouris S, Ziemer F. Determination of bathymetric and current maps by the method DiSC based on the analysis of nautical X-band radar image sequences of the sea surface (November 2007). *IEEE Trans Geosci Remote Sens* 2008;**46**:2267–79.
20. Serafino F, Lugni C, Soldovieri F. A novel strategy for the surface current determination from marine X-band radar data. *IEEE Geosci Remote Sens Lett* 2010;**7**:231–5.
21. Serafino F, Lugni C, Nieto Borge JC, Zamparelli V, Soldovieri F. Bathymetry determination via X-band radar data: a new strategy and numerical results. *Sensors* 2010;**10**(7):6522–34.
22. Huang W, Gill E. Surface current measurement under low sea state using dual polarized X-band nautical radar. *IEEE J Sel Top Appl Earth Obs Remote Sens* 2012;**5**(6):1860–73.
23. Serafino F, Lugni C, Ludeno G, Arturi D, Uttieri M, Buonocore B, et al. Remocean: a flexible X-band radar system for sea-state monitoring and surface current estimation. *IEEE Geosci Remote Sens Lett* 2012;**9**(5):822–6.
24. Ludeno G, Flampouris S, Lugni C, Soldovieri F, Serafino F. A novel approach based on marine radar data analysis for high-resolution bathymetry map generation. *IEEE Geosci Remote Sens Lett* 2014;**11**(1):234–8.
25. Ludeno G, Brandini C, Lugni C, Arturi D, Natale A, Soldovieri F, et al. Remocean system for the detection of the reflected waves from the Costa Concordia ship wreck. *IEEE J Sel Top Appl Earth Obs Remote Sens* 2014;**7**(7). 1939-1404.
26. Ludeno G, Reale F, Dentale F, Pugliese Carratelli E, Natale A, Soldovieri F, Serafino F. An X-band radar system for bathymetry and wave field analysis in harbor area. *Under review on Sensors* 2015;**15**(1):1691–707.
27. http://www.remocean.com/.

CHAPTER

16

Wind, Wave, and Current Retrieval Utilizing X-Band Marine Radars

Jochen Horstmann[1,*], Jose Carlos Nieto Borge[2], Jorg Seemann[1], Ruben Carrasco[1], Bjoern Lund[3]

Institute of Coastal Research, Helmhotlz-Zentrum Geesthacht, Germany[1]; Universidad de Alcalá, Spain[2]; Rosenstiel School of Marine and Atmospheric Science, University of Miami, Coral Gables, FL, USA[3]

**Corresponding author: E-mail: Jochen.Horstmann@hzg.de*

CHAPTER OUTLINE

1. INTRODUCTION

Marine radars, which have been developed for navigational purposes, have shown to be very useful systems for active oceanographic microwave remote sensing purposes. They operate in X-band (9.5 GHz) at grazing incidence where the backscatter of the ocean surface is primarily caused by small-scale surface roughness ($\sim$3 cm), which in turn is strongly dependent on the local wind speed[1] and wind direction.[2] At grazing incidence (>85°), the normalized radar cross-section (NRCS) is proportional to the spectral density of the surface roughness on scales comparable to the radar wavelength (Bragg scattering), which in the case of X-band is $\sim$1.5 cm. In addition, at grazing incidence, radar backscatter is induced by other scattering mechanisms, e.g., wedge scattering or small-scale wave breaking. Long surface waves modulate the small-scale surface roughness that, in turn, modulates the radar backscatter. At grazing incidence, the modulation originates from tilt and hydrodynamic modulation, as well as geometrical shadowing of the radar beam due to the ocean waves.[3] These modulation mechanisms lead to the imaging of surface waves whose wavelengths are greater than two times the radar resolution. The modulation of the NRCS is mathematically described in the linear approximation by the modulation transfer function

(MTF), which is a sum of the four contributing processes: shadowing, tilt modulation, hydrodynamic modulation, and wind modulation. For a detailed description of scattering and modulation mechanisms at low grazing incidence, refer to a special issue on "Low-grazing-angle Backscatter from Rough Surfaces."[4]

Today, marine radar image sequences are operationally used to determine two-dimensional wave spectra[5] and significant wave height.[6] Furthermore, marine radars have been used to measure individual waves,[7] wave groups,[8] near-surface currents,[9] bathymetry,[10] bathymetry and current in combination,[11,12] as well as monitoring of surface features.[13] Marine radars have also shown to be very capable of estimating wind speed and direction.[14]

2. WIND MEASUREMENTS

Ocean surface wind measurements are crucial for the study of energy and momentum transfer from the atmosphere to the ocean as well as for gas-exchange processes at the air–sea interface. On ships and offshore platforms, wind measurements are important for operation and safety and sometimes also reported to weather services for utilization in numerical weather prediction. In general, wind measurements are collected by wind vanes and wind-cup anemometers (in situ) or ultrasonic wind anemometers, which are normally mounted on a mast. In particular on ships, but also other platforms, these sensors are biased by the distortion of the airflow around the mast and superstructure, which is highly dependent upon the sensor's position, platform shape, and motion, as well as relative wind direction. Contrary to the traditional in situ sensors, marine X-band radars retrieve the wind vector in some distance to the platform where the wind is not influenced by flow distortion such as blockage and shadowing. Furthermore, these measurements do not have biases due to motion and height variations of the wind sensor. Last, marine radar has the capability of measuring the backscatter from the ocean surface in space and time under most weather conditions and independent of lighting conditions.

For studying the dependencies of the radar backscatter on wind speed and direction, marine radar data collected during the Office of Naval Research (ONR) field experiment on High Resolution Air–Sea Interaction (Hi-Res) off the southern coast of California in June 2010 were utilized. The data were acquired from the Research Platform (RP) Flip using a standard Furuno marine X-band radar operating at 9.4 GHz with HH-polarization with an output power of 12 kW. The radar was operated in short pulses (50 ns) with a pulse repetition frequency (PRF) of 1600 Hz. With this setup, the radar was installed at the top of the mast at a height of 30 m and was operated in continuous mode, collecting a radar image every 1.5 s (rotating speed of 40 rpm). The radar covers a range of 120 to 3950 m, representing incidence angles between 76° and 89.5°. For comparison, wind data were collected with an ultrasonic anemometer, which was located on the mast of the RP Flip. All wind measurements were converted to a 10-min mean wind covering wind speeds between 4 and 22 m/s.

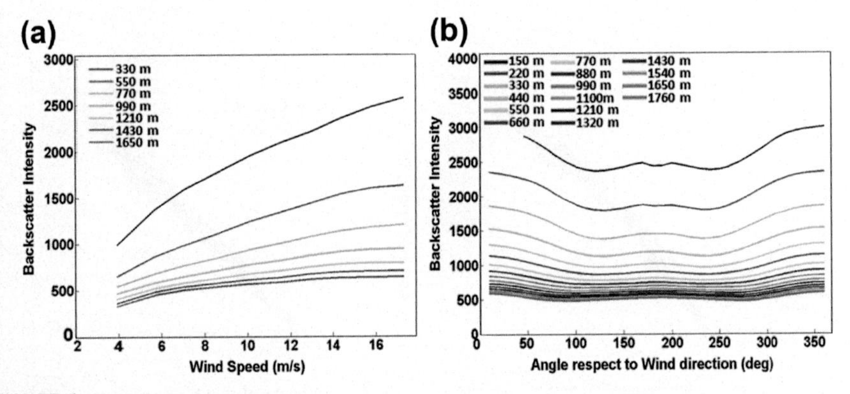

FIGURE 1

Dependency of the radar backscatter on wind speed (a) and wind direction (b). The color-coded lines represent the mean at different range distances between 330 and 1650 m and 110 and 1760 m, respectively.

In Figure 1(a), the wind speed dependency of the radar backscatter is plotted for wind speeds between 2 and 18 m/s. For this plot, the mean backscatter at range distances between 330 and 1650 m was retrieved considering the entire azimuth of the radar, which is equivalent to all wind directions. It can be seen that the radar backscatter increases with wind speed and that the dependence decreases significantly with range. At larger distances, there is only a weak dependency. In Figure 1(b), the dependency of the backscatter on wind direction (azimuth) with respect to the radar look direction is plotted. Here upwind corresponds to 0° (wind blows toward the radar) and 180° to downwind, respectively. Just as for Figure 1(a), the mean backscatter was retrieved for ranges between 110 and 1760 m. There is a clear wind direction dependence with the largest backscatter for upwind and lower backscatter for the other wind directions. These investigations show that the radar backscatter is strongly dependent on the wind speed and direction but also on the range distance to the radar. In addition to the dependencies shown in Figure 1, the radar backscatter has a dependence on the stability of the marine atmospheric boundary layer.[15]

The previous investigations show that marine radar is well suited to retrieve ocean surface winds. Therefore, the mean surface radar backscatter from the entire image (range of 4 km) is retrieved, which is related to the wind speed (Figure 2(a)) and converted via an empirical geophysical model function (GMF) to the surface wind speed (grey line in Figure 2(a)). Unfortunately, the GMF has to be fitted for each individual radar, as the systems are not radiometricaly calibrated. Figure 2(a) shows the wind speed measured at RP Flip versus the mean radar backscatter retrieved from a radar image sequence consisting of 32 individual radar images (equivalent to ~50 s of radar data). The correlation of the mean radar backscatter is excellent, and it is straightforward to fit a GMF, which is suited to convert the mean backscatter to wind speeds (typically the GMF can be fitted by a second- or third-degree polynomial function). Of particular interest is that the scatter of the

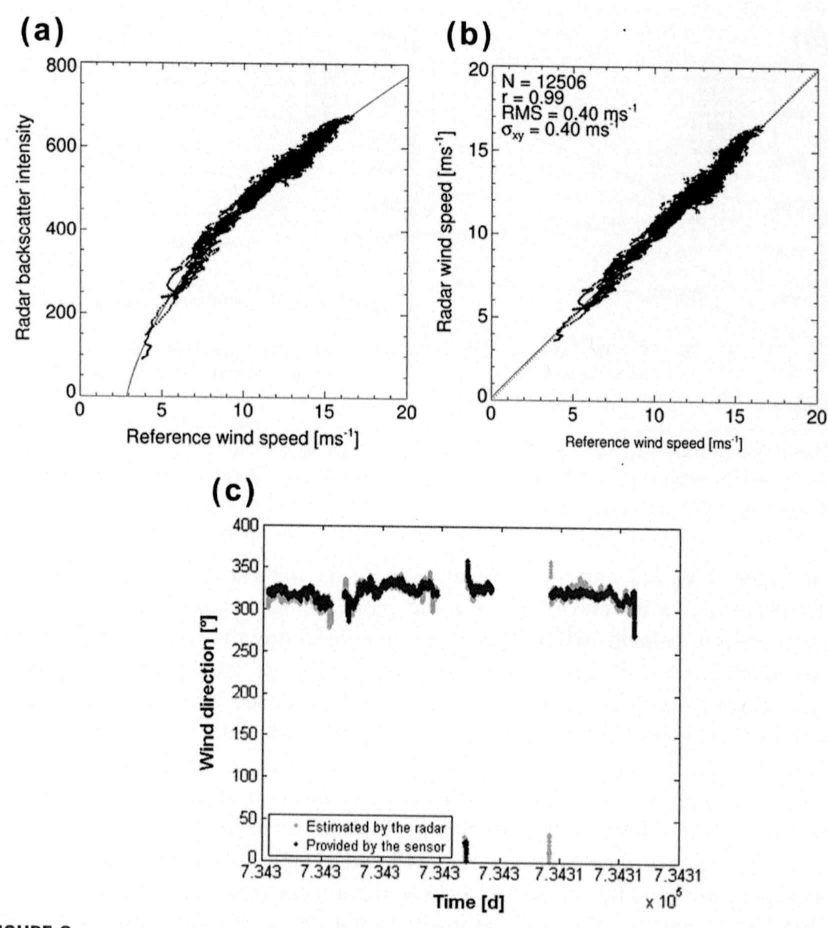

FIGURE 2

Depicted are scatter plots of wind speed versus mean radar backscatter of the entire image (a) and versus radar retrieved wind speeds (b). The gray line on the left-hand side shows the fitted empirical geophysical model function (GMF) converting backscatter to wind speeds. In (c), wind directions resulting from the radar (retrieved via the FFT-Method) are compared to in situ measurements.

data around the fitted GMF (gray line in Figure 2(a)) is very low, and therefore, only a small amount of data is needed for fitting the GMF. However, to get a representative GMF, it is important to have data covering a large range of wind speeds. In the case of RP Flip, the GMF was fitted by a second-order polynomial. Figure 2(b) shows a scatter plot of in situ wind speeds versus marine radar-retrieved wind speeds using the empirical GMF. The standard deviation of 0.42 m/s with a negligible bias

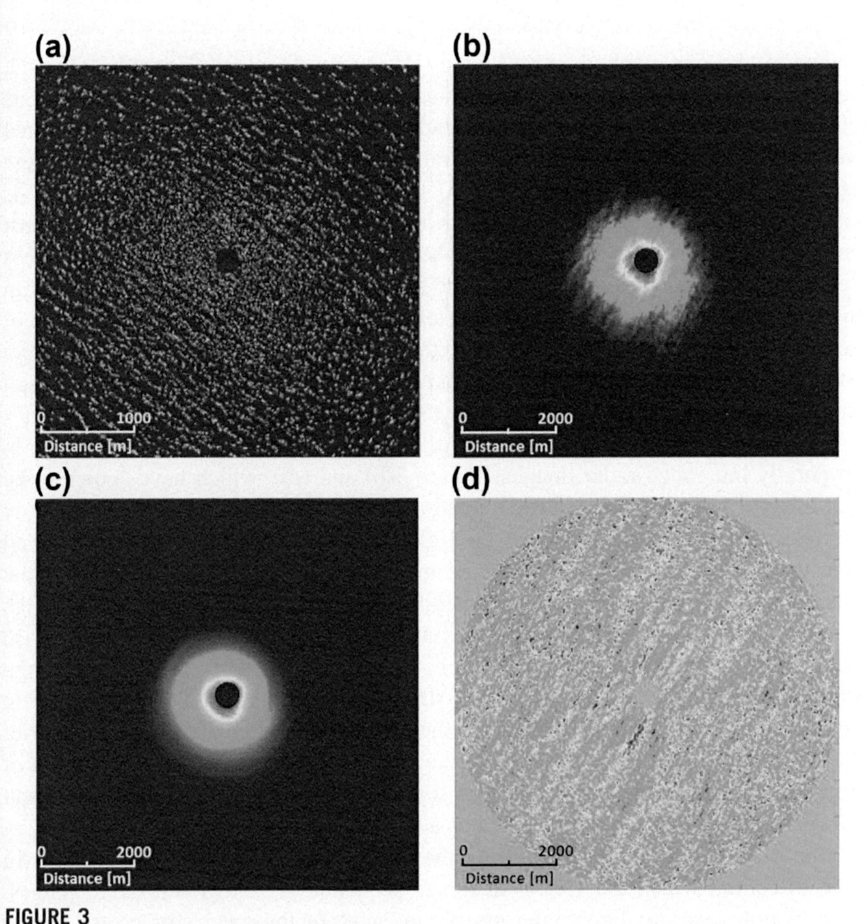

FIGURE 3

(a) Depicts an individual marine radar image that was acquired during the HiRes experiment in May 2010 from the RP Flip, (b) the radar backscatter after integrating an image sequence (64 radar images) over time, (c) the two-dimensional ramp resulting from the marine radar image sequence, and (d) subtracting the two-dimensional ramp and normalization of the image results.

demonstrates the high accuracy of the marine radar-retrieved wind speeds and compares well to results of other groups, which have an error of ~1 m/s.[14,16,17]

In marine radar data, there are three independent phenomena that are well suited to retrieve the mean surface wind directions. The first phenomenon is based on the wind directional dependence of the radar backscatter shown in Figure 1(b) and also mentioned in Refs. 2,15. In general, the radar cross-section is larger for wind blowing toward the radar than for the other directions. To enhance this artifact, the radar image sequence is integrated over time (~100 s), which significantly reduces the noise in the radar image due to speckle and also removes nonstatic patterns such as ocean surface waves (Figure 3(a)). Different methods have been developed to detect the upwind peak. The method suggested by Vicen-Bueno et al.[16] searches for the maximum range distance to a preselected intensity level. The intensity level is selected by searching for the maximum radar cross-section, which can be measured over the entire azimuth of the temporal integrated and smoothed radar image sequence. This method requires full 360° coverage of the radar and, therefore, is not suited for coastal stations as well as most shipborne systems. A fairly robust method was suggested by Lund et al.[17] where all radar pulses of an image sequences are integrated over range and then a cosine-squared function is fitted to the data. With this approach, the upwind peak can be determined even if it is located in a section of the radar without information. However, the shadowed or masked areas should not extend over too large a radius (<90°), and therefore, it is also not suited for coastal stations, which are typically limited to a 180° coverage.

The second phenomenon is based on streak-like features that are visible in temporally integrated radar images (Figure 3(b) and (d)), which have shown to be well aligned with the mean surface wind direction.[14,15] To enhance the streaks in the integrated marine radar images (~100 s), a two-dimensional ramp is removed from these images, which describes the mean range and azimuth behavior of the radar cross-section. This ramp is obtained by smoothing the integrated radar image in range and azimuth (Figure 3(c)). To further enhance the signal, in particular in large range distances, the backscatter is normalized with respect to range (Figure 3(d)). The orientation of these streaks can be estimated by transforming the integrated ramp-removed image into the wave number domain (via a Fast Fourier Transformation) where the direction of the wind is perpendicular to the direction of the main energy within wave numbers that correspond to scales between 30 and 400 m. The resulting wind direction has a 180° directional ambiguity.

The third phenomenon is based on the motion of radar backscatter patterns in wind direction, which are visible in image sequences of integrated radar images. These features are very likely imprints of wind gusts on the ocean surface and, therefore, can be used to estimate the mean surface wind direction with a 180° directional ambiguity. The direction of the movement of these patterns can be retrieved from two consecutive integrated and smoothed images by the Optical Flow (OF) method.[18] In a first step, the radar images are processed in the same manner as before for the streak detection (integration over time, removal of two-dimensional ramp, and normalization), which results in images such as the one depicted in Figure 3(d). This is repeated for another radar sequence acquired 60 s later. As the intensity changes that are to be investigated must be within the evaluated pixel neighborhood, the image resolution is resized considering the maximum feature speed to be resolved under consideration of radar resolution and image time step. By analyzing the optical flow in x and y directions, the complete flow field is obtained. The resulting mean direction of the flow field is considered to be the mean surface wind direction.

The resulting wind directions from each of the aforementioned methodologies are compared to the wind direction measured at RP Flip. The comparison statistics

Table 1 Statistics of the Different Wind Direction Retrieval Methods Applied to the Data Collected during the HiRes Experiment in May 2010 from the RP Flip

	RMSE (°)	Stdev (°)	Bias (°)
Upwind peak	6.4	6.2	−1.5
Fast Fourier transform	9.9	9.9	−0.4
Optical flow	7.5	7.1	−2.4

are listed in Table 1. Independent of the utilized methodology, the wind direction errors are below 10° with a negligible bias, showing that all three methods are well suited to retrieve the mean wind direction. A comparison of radar-retrieved wind directions resulting from the FFT Method and the in situ measurements is depicted in Figure 2(c).

3. WAVE AND CURRENT MEASUREMENTS

Wave information is usually derived from time series of the sea surface elevation measured at a specific position. These measurements are carried out by in situ sensors such as anchored buoys, wave lasers, and pressure sensors. However, deployments of such sensors are limited by the local water depth, as well as the mooring facilities. Furthermore, the use of point measurements assumes that the obtained wave information is representative for a particular area, which is often not the case, particularly in coastal waters where coastal effects like wave refraction, diffraction, and shoaling take place. Under these conditions, the sea state can vary significantly in the area of interest. The imaging of the sea surface with marine radars enables one to analyze the spatial and temporal behavior of ocean wave fields.[5] Therefore, sequences of consecutive sea clutter images are analyzed to estimate wave field spectral properties[6] as well as related sea state parameters.[19] Figure 4 shows an example of a radar images sequence of the ocean surface taken from the RP Fino-3 located in the North Sea.

Sea states are regarded as wind-generated wave fields with invariant statistical properties at a given sea surface position, $\boldsymbol{r} = (x, y)$, and at time, t. Hence, those wave fields are homogeneous in the spatial dependence and stationary in their temporal evolution. Under these assumptions, and taking into account a Eulerian description of the wave field,[20] the sea surface elevation, $\eta(\boldsymbol{r}, t)$, is described by using the following spectral representation:

$$\eta(R, T) = \int_{\Omega_{k,\omega}} e^{i(\boldsymbol{k}\cdot\boldsymbol{r}-\omega t)} dZ(\boldsymbol{k}, \omega), \tag{1}$$

where $\boldsymbol{k} = (k_x, k_y)$ is the two-dimensional wave number, ω is the angular frequency, and $dZ(\boldsymbol{k}, \omega)$ are the spectral amplitudes. The integration domain $\Omega_{\boldsymbol{k},\omega} = [-k_{x_c}, k_{x_c}]\times$

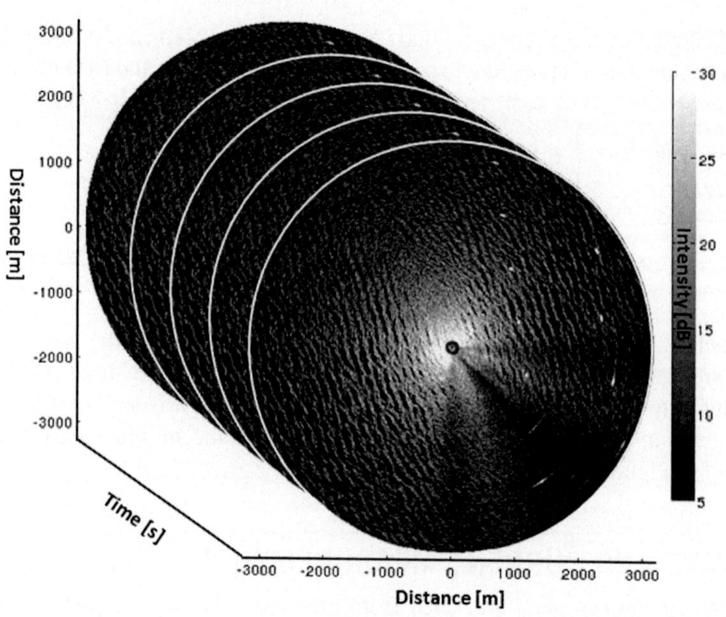

FIGURE 4

Example of a temporal sequence of radar images of the sea surface measure by a marine radar. The measurement was taken at the German research platform of Fino-3 in the North Sea.

$[-k_{y_c}, k_{y_c}] \times [-\omega_c, \omega_c]$ is defined from the range of wavelengths $\lambda = \frac{2\pi}{k}$ and periods $T = \frac{2\pi}{\omega}$ that form the wind-generated waves. In practice, the values of k_{x_c}, k_{y_c}, and ω_c are limited by the spatial and temporal resolution and sampling rate of the measuring sensor. The amplitudes $dZ(\boldsymbol{k},\omega)$ are the so-called random spectral measures. Hence, assuming the expression (1), the three-dimensional wave spectrum is defined as follows:

$$F^{(3)}(\boldsymbol{k},\omega)d^2kd\omega = \mathbb{E}[dZ(\boldsymbol{k},\omega)dZ^*(\boldsymbol{k},\omega)], \tag{2}$$

where $\mathbb{E}$ is the expectation operator. Taking into account that $\eta(\boldsymbol{r}, t)$ is a field with a real value, the three-dimensional spectrum (Eqn (2)) presents a point symmetry to the origin of its spectral variables:

$$F^{(3)}(\boldsymbol{k},\omega) = F^{(3)}(-\boldsymbol{k},-\omega), \tag{3}$$

Ocean waves are dispersive, and under the assumptions of the linear wave theory, the dispersion relation is given by the following:

$$\omega = \overline{\omega}(\boldsymbol{k}) = \sqrt{gk\tanh(kd)} + \boldsymbol{k}\cdot\boldsymbol{U} = \overline{\omega}_0(k) + \boldsymbol{k}\cdot\boldsymbol{U}, \tag{4}$$

where g is the acceleration of gravity, d is the water depth, and $\boldsymbol{U} = (U_x, U_y)$ is the two-dimensional near surface current, commonly known as the current of encounter.[9] The current vector, $\boldsymbol{U}$, induces a Doppler shift in the ocean wave frequency in Eqn (4) given by the dot product $\boldsymbol{k} \cdot \boldsymbol{U}$.

Integrating $F^{(3)}(\boldsymbol{k},\omega)$ over all the positive wave frequencies, the unambiguous two-dimensional wave number spectrum $F^{(2)}(\boldsymbol{k})$ is derived as follows:

$$F^{(2)}(\boldsymbol{k}) = 2 \int_0^{\omega_c} F^{(3)}(\boldsymbol{k}, \omega) d\omega \tag{5}$$

Under the frame of the linear wave theory, taking into account the dispersion relation (Eqn (4)), the three-dimensional wave spectrum (Eqn (2)) can be derived from $F^{(2)}(\boldsymbol{k})$ (Eqn (5)) as (see Figure 5)[20]:

$$F^{(3)}(\boldsymbol{k}, \omega) = \frac{1}{2} \left[F^2(\boldsymbol{k}) \delta(\omega - \overline{\omega}(\boldsymbol{k})) + F^2(-\boldsymbol{k}) \delta(\omega + \overline{\omega}(-\boldsymbol{k})) \right]. \tag{6}$$

It can be seen that Eqn (3) holds. Taking into account the dispersion relation (Eqn (4)), and the two-dimensional wave number spectrum, $F^{(2)}(\boldsymbol{k})$, given by Eqn (5), other spectral densities can be derived. One of the most common spectral representation used is the directional spectrum, $E(\omega,\theta)$, which is given by the following:

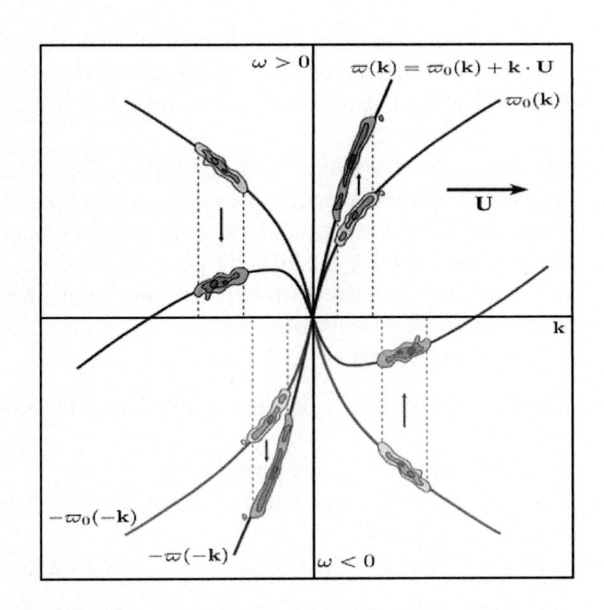

FIGURE 5

Distribution of the spectral wave components of the three-dimensional wave spectrum $F^{(3)}(k, \omega)$ given by Eqn (6) and the modifications due to the existence of a current of encounter U.

$$E(\omega,\theta)=F^{(2)}(\mathbf{k}(\omega,\theta))\mathbf{k}(\omega,\theta)\frac{\partial k(\omega,\theta)}{\partial\omega}, \tag{7}$$

where $\theta=\tan^{-1}\left(\frac{k_y}{k_x}\right)$ is the wave propagation direction. The term $k(\omega,\theta)\cdot\frac{\partial k(\omega,\theta)}{\partial\omega}$ in Eqn (7) is the Jacobian needed to change the coordinates from the (k_x, k_y) space into the (ω, θ) space.[6] $E(\omega, \theta)$ is usually factorized as follows:

$$E(\omega,\theta)=D(\omega,\theta)S(\omega), \tag{8}$$

where $D(\omega, \theta)$ is the so-called directional spreading function, and $S(\omega)$ is the frequency spectrum that is usually estimated from wave elevation records measured by anchored buoys. From Eqn (8), $S(\omega)$ is obtained as follows:

$$S(\omega)=\int_{-\pi}^{\pi}E(\omega,\theta)d\theta. \tag{9}$$

The directional spreading function $D(\omega, \theta)$ describes the statistical properties of the directional behavior of the sea state. Hence, the function $D(\omega, \theta)$ is the circular probability function in the interval $[-\pi,\pi]$, and therefore, it can be expressed as the following Fourier expansion:

$$D(\omega,\theta)=\frac{1}{2\pi}+\frac{1}{\pi}\sum_{n=1}^{\infty}a_n\cos n\theta+b_n\sin n\theta \tag{10}$$

where a_n and b_n are directional Fourier coefficients that are functions of the frequency ω. From directional buoy records, only a limited number of Fourier coefficients can be estimated. Thus, standard directional buoys can only measure the first four of them (e.g., a_1, a_2, b_1, and b_2).

As already mentioned, the measurement of sea states using marine radars is based on backscatter of the electromagnetic waves by the ripples and the roughness of the free sea surface due to the local wind. Hence, the pattern of electromagnetic energy backscattered by the ripples is modulated by the larger ocean surface structures, such as swell and wind sea waves, which, therefore, can be detected on the radar screen. However, there are various other phenomena besides the ocean waves that appear in the marine radar images:

- Range dependence, due to tilt and shadowing modulation and loss of power with range (Figure 1(a))
- Azimuthal dependence due to the wave propagation
- Shadowing modulation, which occurs when higher waves hide lower waves to the radar antenna[7,21]
- Hydrodynamic modulation due to the orbital motion of the waves[22]

Because of their nonlinearity, all these phenomena contribute with additional spectral components to the image spectrum of the radar image sequence. These components do not belong to the ocean wave spectra and, therefore, are not described by the linear modulation transfer function (MTF).[21] Hence, the image spectrum has to be processed to derive reliable estimation of the wave spectrum and the related sea state parameters.

The first step in this processing chain is to apply a three-dimensional Fast Fourier Transformation (3D FFT) of the image sequence to estimate the image spectrum $I^{(3)}(\boldsymbol{k},\omega)$. Within the image spectrum, the following contributions can be summarized[19,21]:

- Static and quasi-static spectral components due to the range dependencies
- Wave field components
- Higher harmonics of the wave components due to nonlinear imaging of marine radar and nonlinearities of the waves
- Background noise due the interaction of the microwave with the sub-resolution-scale roughness of the sea surface, called speckle. This roughness is generated by the local wind stress.
- Additional contributions in lower frequency planes, such as subharmonics of the nonlinear imaging of wave groups and subharmonics of the nonlinear wave field. This structure is termed group line.

Figure 6 shows a wave number frequency slice through the wave number frequency cube resulting from the 3D FFT. This example illustrates the energy contribution of the spectral components mentioned. This measurement was taken by a land-based radar station located on top of a cliff at the northern coast of Spain

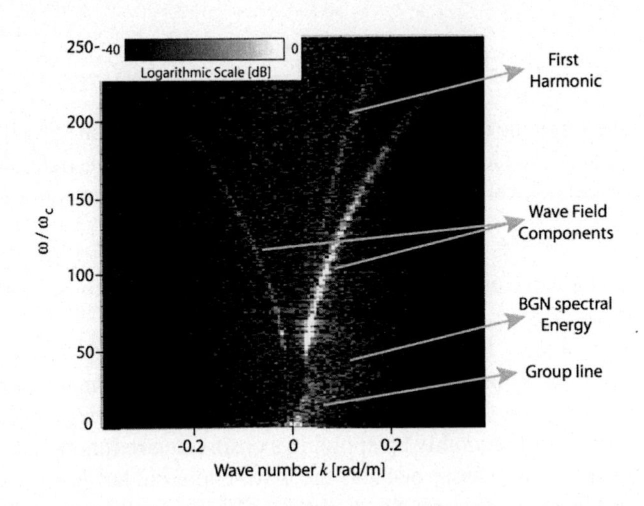

FIGURE 6

Example of a two-dimensional transect of an image spectrum $I^{(3)}(k,\omega)$ (see text for explanation). The transect is carried out in the (k_p,ω) domain, where k_p is the wave number vector along the spectral-peak wave propagation direction. The value of the cut-off angular frequency is $\omega_c = 1.78$ rad/s. Only the positive domain of angular frequencies ω is represented.

(Bay of Biscay). This is a swell-dominated area where the long waves reach the cliff walls and some part of the incoming wave energy is reflected. Therefore, in this case, the radar is measuring a bimodal sea state. This bimodality can be seen by the energy in the two opposite branches of the dispersion relation, where the weaker branch corresponds to the reflected wave components, which enables one to quantify the ratio of the reflected energy to the incoming energy.

To derive the estimation of the wave spectrum, it is necessary to apply an inversion modeling technique. The basic assumption of this inversion modeling algorithms is the existence of the dispersion relation $\overline{\omega}(\boldsymbol{k})$ given by Eqn (4). In the following, the different steps that conform the inversion modeling needed to derive sea state information are described.

3.1 CURRENT ESTIMATION

Once the image spectrum $I^{(3)}(\boldsymbol{k}, \omega)$ is obtained, the next step is to apply the inversion model to derive the wave spectrum and the related sea state parameters. This modeling technique is composed of the following steps.

In the first step, a low-pass filter is applied to $I^{(3)}(\boldsymbol{k}, \omega)$ to remove the static and quasi-static patterns that are induced by non-stationary and nonhomogeneous trends in the radar image sequence. A good empirical threshold for the frequency of the high-pass filter is $\omega_{th} = 2\pi \cdot 0.03$ rad/s (as lower frequencies are not due to the sea state). Hence, the transfer function of filter $\mathcal{F}_{th}(\boldsymbol{k}, \omega)$ is defined as follows:

$$\mathcal{F}_{th}(\boldsymbol{k}, \omega) = \begin{cases} 0 & \text{if}(\boldsymbol{k}, \omega) \in \Omega_{th} \\ 1 & \text{otherwise} \end{cases} \tag{11}$$

where Ω_{th} is the set in the three-dimensional spectral domain defined by the very low wave numbers and frequencies that do not belong to the wave field. Taking into account the dispersion relation (Eqn (4)), Ω_{th} is defined as follows:

$$\Omega_{th} = \{\boldsymbol{k} : |\boldsymbol{k}| \leq k(\omega_{th})\} \times [-\omega_{th}, \omega_{th}]. \tag{12}$$

Here, $k(\omega_{th})$ is the wave number solution of the dispersion relationship without current. In this case, $|\boldsymbol{k}|$'s are very small. Hence, the dot product $\boldsymbol{k} \cdot \boldsymbol{U}$ is small as well, and it can be eliminated from the dispersion relationship expression (4). The filter (Eqn (11)) could be considered too sharp in the spectral domain ($\boldsymbol{k}$,ω). In this case, it is possible to use a transition volume inside the domain Ω_{th}, where the filter takes values from 0 to 1 smoothly by applying a cosine square function, for example. In general, this is not necessary because the wave energy is far enough, in the $\Omega_{\boldsymbol{k},\omega}$ domain, from the static pattern. So, a simple filter, as indicated in Eqn (11), is enough to analyze the wave energy distribution in the image spectrum.[6]

In the second step, the velocity of encounter is estimated by matching the information of $\boldsymbol{k}$ and ω space by analyzing the distribution of the energy in the domain of $\Omega_{\boldsymbol{k},\omega}$. The physical model applied is the linear wave theory and the dispersion relation (Eqn (4)). Thus, assuming that the wave energy is a first-order contribution to the total image spectral energy, which can be proved using simulated radar images as well as real measurements,[5,6] the current of encounter $\boldsymbol{U} = (U_x, U_y)$ can be estimated by minimizing the following functional (according to the parameters U_x and U_y)[5]:

$$\mathcal{V} = \sum_{j=1}^{N_r} \left[\omega_j - \overline{\omega}_0\left(k_j\right) - k_{x_j} U_x - k_{y_j} U_y\right]^2 \tag{13}$$

where $k_j = \sqrt{k_{x_j}^2 + k_{y_j}^2}$, and $\overline{\omega}_0(k)$ is the dispersion relation (Eqn (4)) without presence of velocity of encounter $\boldsymbol{U}$. N_r is the number $(\boldsymbol{k}, \omega)$ points whose spectral energy is due to the linear wave field and not due to other nonlinear effects in the radar imaging. Using numerical simulation methods of sea states and their associated radar images, as well as real measurements, N_r is estimated from all points with energy higher than 20% of the maximum value of $I^{(3)}(\boldsymbol{k}, \omega)$.

The computation of $\boldsymbol{U}$ from the functional (Eqn (13)) can be improved using an additional iterative scheme, where the estimation of $\boldsymbol{U}$ from Eqn (13) represents a first guess of the current fit.[9] The iterative scheme decreases the threshold in energy below the value of 20% mentioned previously, permitting one to increase the number of fitting points and the accuracy of the fit. To apply this iterative method, the resolutions of the spectral variables $(\Delta\omega, \Delta k_x, \Delta k_y)$ given by the discrete Fourier transform theory in three dimensions has to be taken into account.

Within the iterative scheme, the first guess current estimation is utilized to localize the energy of the first harmonics and the aliased modes within the image spectrum. The coordinates of the signal of the first harmonic are obtained from the coordinates of the signal from the fundamental mode by scaling with a factor of two:

$$(k_1, \omega_1) = 2(k_0, \omega_0). \tag{14}$$

Because of the location of the spectral energy on a dispersion surface, the aliased spectral energy can be de-aliased over the Nyquist frequency limit according to the following signal coordinate mapping:

$$\left(-k_A, \omega_{Ny} - \omega_A\right) \Rightarrow \left(k, \omega_{Ny} + \omega\right). \tag{15}$$

Taking into account these additional structures of the radar image spectrum, the number of spectral signal coordinates is increased by an order of magnitude, which increases the accuracy of the current estimation by approximately a factor of three.

The accuracy of the current fit can be described by a confidence ellipse in the (u_x, u_y) plane. The size of the two half axes and the orientation of the ellipse depends on the shape of the wave spectral coordinate distribution in the wave number plane. In general, the accuracy of the current fit depends, in addition to the number of spectral signal coordinates, on the following factors:

- The current estimation accuracy increases with wave number, because the Doppler frequency shift of the waves depends linearly on the wave number.

- Only the current component parallel to the wave direction results in a Doppler frequency shift. Therefore, only the directional spread of the sea state allows the estimation of the current component perpendicular to the main wave direction.

In summary, a wind sea with short waves and a large directional spread allows a more precise current estimation than a swell with long waves and a small directional spread.

The current estimation is also influenced by the vertical gradient of the current, because with increasing distance z to the sea surface, the amplitude of the wave-induced orbital paths are damped by the factor e^{2kz}. Effectively, only the current up to a depth of one-tenth of the dominant wave length influences the frequency of the surface waves.[23]

The current estimation method was validated using radar data acquired at the research platform FINO-3 in the German Bight, where the German Federal Maritime and Hydrographic Agency operates a bottom-mounted Acoustic Doppler Current Profiler (ADCP) as well as a directional wave rider buoy. In this area, the surface currents can be assumed to be homogeneous, and therefore, the point sensors can be compared to the radar measurements, which represent a spatial average. Close to the sea surface, the ADCP measurements, especially during higher sea states, are influenced by the wave-induced turbulence. Therefore, the 6-m water depth data were utilized for the comparison. Figure 7 shows the scatter plot of radar-retrieved currents (using the iterative scheme[9]) to the ADCP data. Within this data set, three data points were removed as their *SNR* was too low, indicating the

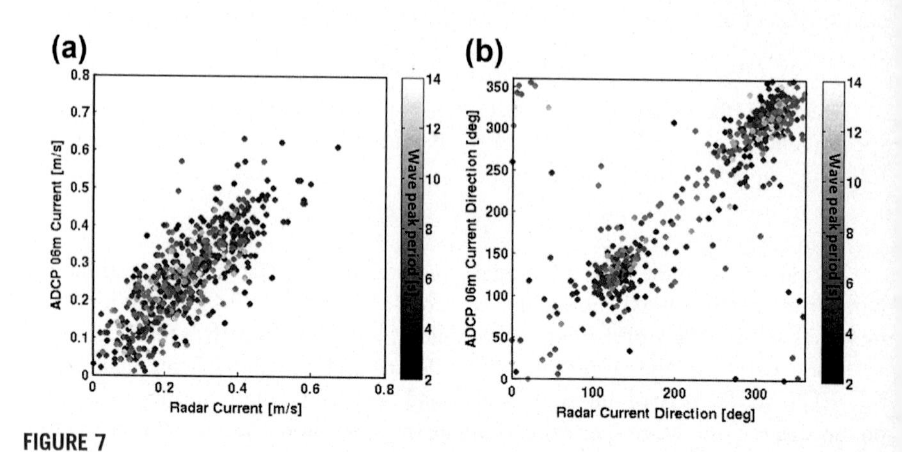

FIGURE 7

Scatter plot of radar retrieved current speeds (a) and current directions (b) versus currents measured by an Acoustic Doppler Current Profiler (ADCP) at 6-m water depth. The gray scale represents the wave peak period as measured by a directional wave rider buoy located in vicinity of Fino-3. The comparison was performed with data collected at the German offshore platform Fino-3 located 80 km west of the island of Sylt.

lack of waves for retrieving the currents. The correlation coefficient between the two data sets was 0.76, and the rms error was 0.07 m/s with a negligible bias.

In shallow water, the intrinsic frequency of the waves depends on the water depth. To fit the current, or to use the dispersion relation as a spectral filter (Section 4), the water depth has to be known. If the information about the water depth is not available, this magnitude can be estimated together with the two components of the near surface current. In this case, due to the tanh, the cost function (Eqn (13)) depends nonlinearly on the water depth. Therefore, least squares optimization cannot be carried out by solving a system of linear equations. Instead, the cost function is minimized numerically as a function of the two current components and the water depth using the multidimensional downhill simplex method. The water depth has an influence on the dispersion relation if it is less than 25% of the dominant wave length of the sea state. This influence is strong enough to obtain significant accuracy for the depth estimation if the ratio $\frac{d}{\lambda_{max}} \leq 0.1$.[24]

In coastal areas, the water depth and near surface current are often inhomogeneous. Analyzing radar image sequences of these areas, the inhomogeneity causes a broadening of the dispersion surface in the spectral wave number frequency domain. In the spatial domain, each directional frequency component has a varying wavelength. After a directional frequency decomposition in the complex-valued wave number-frequency image spectrum of the radar, the wave number vectors are estimated with a high resolution in the spatial domain, and afterward, minimizing the cost function (Eqn (13)) maps of water depth and current can be retrieved with a spatial resolution in the order of the dominant wave length.[11]

Note, that the retrieval of currents and bathymetry only depends on the dynamics of the sea surface. Therefore, every sensor that can capture the wave field in space and time at appropriate scales can be used to retrieve these parameters. Piotrowski and Dugan[25] have retrieved current and water depth maps of a shoaling zone using optical image sequences acquired from an aircraft. Abileah[26] demonstrates the feasibility of bathymetric retrieval from optical satellite images of the coastal sea.

3.2 WAVE SPECTRUM AND PARAMETER ESTIMATION

Once the surface current is derived, the estimation of the wave spectrum is needed to obtain sea state information. As the typical X-band marine radar data sets are sampled in space, (x, y), and time, t, the corresponding spectral information is described in the wave number, $\boldsymbol{k} = (k_x, k_y)$, and frequency, ω, domain. Therefore, after a three-dimensional Fourier decomposition of the sea surface radar image time series, the so-called image spectrum, $I^{(3)}(\boldsymbol{k}, \omega)$, is estimated. The estimation of the wave spectrum requires an inverse modeling scheme that is composed of the following steps.

Removal of the static and quasi-static pattern in the $(\boldsymbol{k}, \omega)$ domain: A high-pass filter is applied to remove all those low-frequency spectral components that are induced by the long-range modulation of the radar image due to the radar equation. All the $(\boldsymbol{k}, \omega)$ spectral components of $I^{(3)}(\boldsymbol{k}, \omega)$ with $\omega < \omega_{th}$ are suppressed.

A typical threshold cut-off frequency is $\omega_{th} = 0.04\pi$ rad/s. Hence, the high-pass filtered image spectrum is obtained $I_{th}^{(3)}(\boldsymbol{k}, \omega)$.

In the next step, the $(\boldsymbol{k}, \omega)$ spectral components that hold the dispersion relation of linear gravity waves are selected. All spectral energy outside of the dispersion shell is filtered by applying a three-dimensional band-pass filter in the $(\boldsymbol{k}, \omega)$ domain to keep only those wave field components that belong to the wave field.[5] Hence, imposing the dispersion relation (Eqn (4)) $\omega = \overline{\omega}(\boldsymbol{k})$ as a spectral filter, a new three-dimensional spectral density $F_f^{(3)}(\boldsymbol{k}, \omega)$ is obtained as follows:

$$F_f^{(3)}(\boldsymbol{k}, \omega) = \int\limits_{\Omega_{\boldsymbol{k},\omega}} I_{th}^{(3)}(\boldsymbol{k}', \omega')\delta(\boldsymbol{k}' - \boldsymbol{k})\cdot\delta(\omega' - \omega(\boldsymbol{k}'))d^2\boldsymbol{k}'d\omega' \tag{16}$$

where $I_{th}^{(3)}(\boldsymbol{k}', \omega) \equiv I^{(3)}(\boldsymbol{k}, \omega)\cdot\mathcal{F}_{th}(\boldsymbol{k}, \omega)$, and $\delta(\cdot)$ is the Dirac's delta. The expression (16) represents a continuous formulation of the three-dimensional band-pass filter. In practice, a discrete numerical technique has to be applied, taking into account the spectral resolution of the wave number vector components and the angular frequency components (e.g., Δk_x, Δk_y, and $\Delta\omega$).

In the next step, a modulation transfer function (MTF) has to be applied that depends on the incidence angle and polarization of the radar. The MTF describes the differences between the radar-retrieved image spectra and the corresponding spectra from the in situ sensors. This deviation is due to the radar wave imaging mechanisms, in particular shadowing and/or tilt modulation[21] that are not considered in the band-pass filter (Eqn (14)). By using a so-called modulation transfer function,[27] this deviation can be minimized. As the radar imaging mechanism for marine radar are still not well understood, the modulation transfer function $\mathcal{T}(\boldsymbol{k})$ at grazing incidence and horizontal polarization has been derived empirically. It can be parameterized with a power law:

$$\mathcal{T}(\boldsymbol{k}) = k^{\beta} \tag{17}$$

where the exponent β has been obtained empirically, as well as from numerical simulations. For operational purposes, the value $\beta \approx -1.2$ has been obtained.[28] From Eqn (17), and taking into account Eqn (16), the estimation of the three-dimensional wave spectrum $\tilde{F}^{(3)}(\boldsymbol{k}, \omega)$ is given by the following:

$$\tilde{F}^{(3)}(\boldsymbol{k}, \omega) = \mathcal{T}(\boldsymbol{k})\cdot F_f^{(3)}(\boldsymbol{k}, \omega) \tag{18}$$

In the final step, the significant wave height is estimated. Marine radars provide images coded in relative values of gray scales, $\zeta(\boldsymbol{r}, t)$, rather than values of physical parameters related to the backscattering phenomenon, such as the radar cross-section. Hence, the spectral estimation of the wave field (Eqn (18)) is not properly scaled in the sense that its integral over all the domain $\Omega_{\boldsymbol{k},\omega}$ does not provide an estimation of the variance of the sea surface, and therefore, $\tilde{F}^{(3)}(\boldsymbol{k}, \omega)$ does not inform directly about the significant wave height, H_s. Analyzing the structure of the image spectrum, $I^{(3)}(\boldsymbol{k}, \omega)$, it can be seen that H_s can be estimated in a similar way to synthetic aperture radar (SAR) systems.[22] This method is based on the occurrence of multiplicative speckle noise with radar imaging. The speckle noise results in a white noise floor in the image spectrum, which enables the calibration of the modulation variance spectrum. For the particular case of radar image sequence analysis, the method has to be extended to the three dimensions of the image spectrum. Under these conditions, the significant wave height has a linear dependence with the root-squared of the signal-to-noise ratio (SNR)[28]:

$$H_s = c_0 + c_1\sqrt{SNR} \tag{19}$$

where c_0 and c_1 are calibration constants. The factor SNR is defined as follows:

$$SNR = \frac{\int\limits_{\Omega_{k,\omega}} \tilde{F}^{(3)}(\boldsymbol{k}, \omega)\delta(\omega - \overline{\omega}(k))d^2\, kd\omega}{\int\limits_{\Omega_{k,\omega}} F^{(3)}_{BNG}(\boldsymbol{k}, \omega)d^2kd\omega}, \tag{20}$$

where $F^{(3)}_{BNG}(\boldsymbol{k}, \omega)$ is the spectral energy of the background noise (see Figure 6). From H_s derived from expression (19), the three-dimensional wave spectrum can be derived as follows:

$$F^{(3)}(\boldsymbol{k}, \omega) = C\cdot\Omega_{\boldsymbol{k},\omega}, \tag{21}$$

where C is the constant needed to rescale the spectrum $\tilde{F}^{(3)}(\boldsymbol{k}, \omega)$,

$$C \equiv \frac{H_s^2}{16\cdot \int\limits_{\Omega_{k,\omega}} \Omega_{\boldsymbol{k},\omega}d^2kd\omega} \tag{22}$$

From the wave spectrum, the different sea state parameters, e.g., significant wave height and peak period, can be deduced.

Some of the data shown in this section were taken from the EXBAYA 95 oceanographic campaign carried out in February of 1995.[19] During this campaign, a shipborne marine radar was used in the vicinity of the mooring position of a pitch–roll directional buoy in deep water conditions (600 m depth). During the experiment, several cases of swell coming from northwest were measured, which is the typical sea state situation in the Bay of Biscay. These incoming wave fields are generated by storms in northern Atlantic Ocean arriving to the north of the Iberian Peninsula as very long, grouped waves. The sampling time in the time series is given by the antenna rotation period, which was 2.6 s for this setup. The used buoy is a conventional moored directional pitch–roll buoy, which provides three time series: heave, pitch, and roll. The sampling time of the time series is 1 s, and 2048 s is the total duration of each buoy measurement. The directional spectrum $E(f, \theta)$, where $f = \omega/(2\pi)$, was estimated from these data using the extended maximum likelihood method (EMLM) and maximum entropy method (AR(2)). These two methods

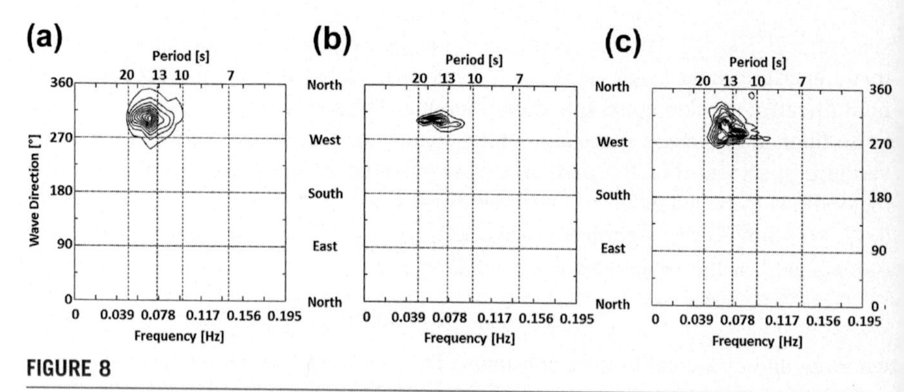

FIGURE 8

Estimation of $E(f, \theta)$ from extended maximum likelihood method (EMLM) (a), maximum entropy method (AR(2)) (b), and the marine radar (c).

provide different estimations of $E(f, \theta)$, keeping the first four directional Fourier coefficients (e.g., $a_1(f)$, $a_2(f)$, $b_1(f)$, and $b_2(f)$) as seen in Eqn (10).

Some examples of interpolation for the $E(f, \theta)$ from the buoy records and the corresponding estimation from the sea clutter time series can be seen in Figure 8. This measurement corresponds to a swell wave field coming from north-northwest, approximately. It is well known that the correct estimation of the $E(f, \theta)$ is impossible when measuring only a limited number of sea state features at a fixed sea surface location (e.g., the mooring position). Most of the commercial buoys measure three time series: heave, and the two horizontal wave displacements along the east–west and north–south directions, or, alternatively, the wave slopes at the mooring point along the east–west and north–south directions. Therefore, only the first four directional Fourier coefficients $a_1(f)$, $a_2(f)$, $b_1(f)$, and $b_2(f)$ can be estimated. This fact can be a serious limitation in cases of multimodal sea states (i.e., superposition of one or two swells and a wind sea).

Each estimation of $E(f, \theta)$ obtained from buoy records can provide quite different results depending on the behavior of the analyzed sea state but keeping the spectral properties of measured time series. EMLM in Figure 8(a) provides more background directional noise than AR(2) in Figure 8(b). AR(2) presents some problems in the estimation of the directional spectrum for very focused sea states, such as a very long swell. This behavior is due to the intrinsic bimodality of this method.

The estimation from the radar sea clutter is illustrated in Figure 8(c). The marine radar has more directional resolution than a buoy because this system provides spatial information about the wave fields. An example of bimodal sea state can be seen in Figure 9. This measurement was taken in the North Sea during the ERS-1/ERS-2 Tandem campaign. Hence, Figure 9 illustrates the wave number spectrum $F^{(2)}(\boldsymbol{k})$ of the bimodal sea state. Therefore, a swell component, located at a low wave number region (long wave lengths) with low angular spreading, is clearly visible. In addition, the wind sea contribution is visible as well, presenting a large angular spreading for higher wave numbers (shorter wave lengths).

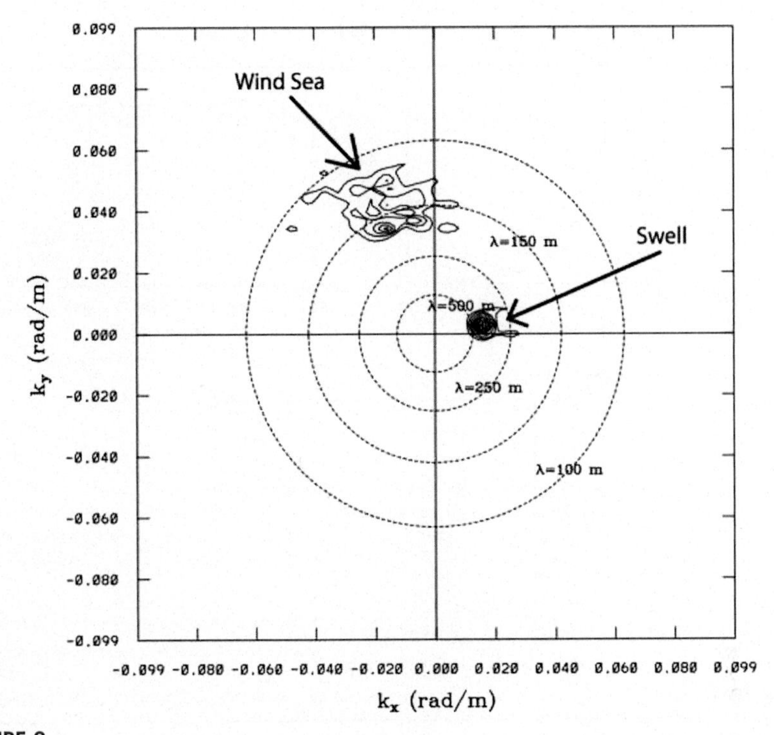

FIGURE 9

Bimodal wave number spectrum, $F^{(2)}(k)$, measured by a marine radar close to the Norwegian island of Utsira in the North Sea.

Once the directional frequency spectrum $E(f, \theta)$ is obtained, the one-dimensional frequency spectrum $S(f)$ as well as other sea state parameters depending on the frequency can be computed. As mentioned previously, standard directional buoys only measure three geometric properties of the wave field at a fixed point. So, to compare the results from the radar sea clutter analysis with directional buoy data, it is necessary to take into account the limitations of directional buoys in the measurement of the sea state directionality. Figure 10(a) shows one example of frequency spectrum $S(f)$ obtained in the Bay of Biscay at the same location where a pitch–roll buoy was moored. It can be observed that there is a good agreement in the shape of the two spectral estimations.

The estimation of coming-from mean wave propagation direction, depending on the frequency

$$\overline{\theta}(f) \equiv \overline{\theta}_1(f) = \tan^{-1}\left[\frac{b_1(f)}{a_1(f)}\right] \tag{23}$$

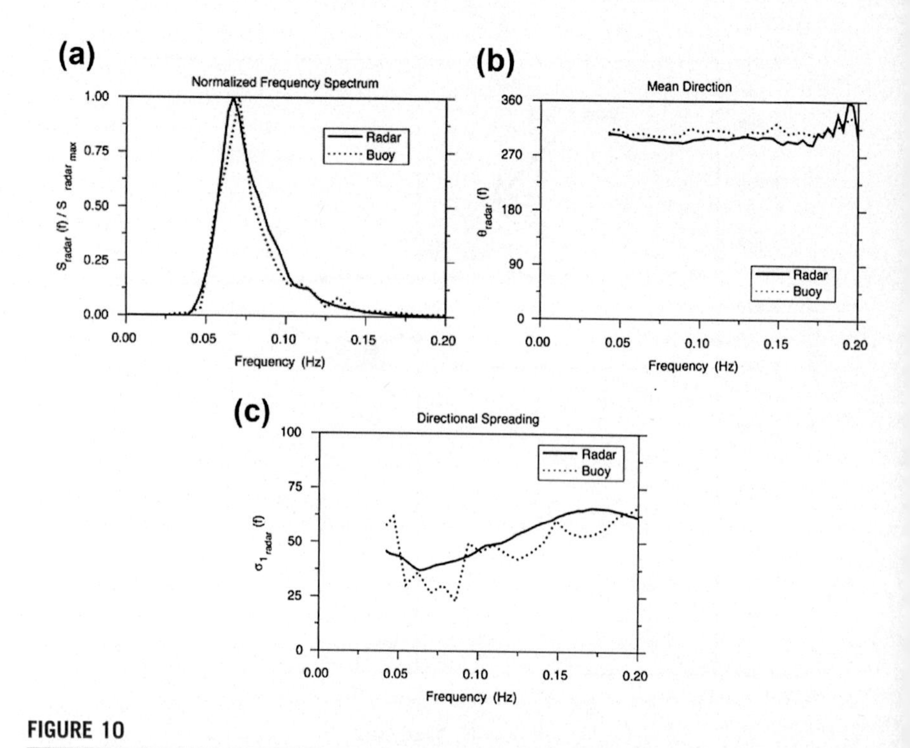

FIGURE 10

(a) Frequency spectrum S(f), (b) mean wave direction $\overline{\theta}(f)$(coming-from convention), and (c) angular spreading depending on the frequency $\sigma_1(f)$, all retrieved from the radar (solid line) and the buoy (dotted line).

from both sensors appears in Figure 10(b). As in the case of the one-dimensional spectrum, the comprised $\overline{\theta}(f)$ from the radar image analysis is close to the estimation from the buoy records.

Figure 10(c) illustrates the estimation of the angular spreading $\sigma_1(f)$, which is defined from the first two directional Fourier coefficients as follows:

$$\sigma_1(f) = \sqrt{2\left[1 - \sqrt{a_1^2(f) + b_1^2(f)}\right]}. \tag{24}$$

Once the different wave spectral representation are derived, most of the related sea state parameters can be obtained, such as peak and mean periods, mean wave lengths, etc.[29] All those parameters depend on the spectral shape rather the amount of energy of the spectrum (e.g., the variance of the sea surface). Nevertheless, as was mentioned before, the parameters that depend on the sea surface variance, as the significant wave height, H_s, need a proper scaling of the wave spectral estimation. Hence, for a proper spectral rescaling, after correcting the spectral shape with a power law (Eqn (17)), it is necessary to use Eqn (19):

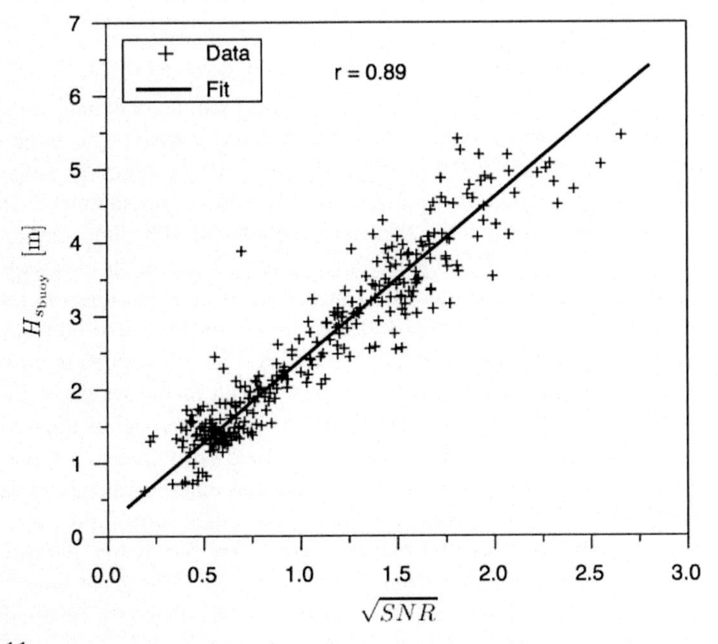

FIGURE 11

Scatter plot of the $\sqrt{SNR}$ and the buoy H_s estimation using wave energy and the *BGN* energy outside the dispersion shell. A high-pass filter is applied to the radar data to avoid quasi-static patterns due to radar imaging mechanisms ($f > 0.04$ Hz).

$$H_s = c_0 + c_1\sqrt{SNR}. \tag{25}$$

Figure 11 illustrates an example of correlation between H_s measured by a buoy and $\sqrt{SNR}$ derived by the radar using Eqn (20). The data were acquired onboard the *FPSO Norne* in the northern North Sea. The good agreement between the two parameters can be seen. The linear fit, as it is shown in Figure 11, permits one to obtain the calibration constants c_0 and c_1. These constants depend on each radar installation (e.g., type of radar transmitter, antenna, and location). Hence, once the field campaign of the calibration phase is finished and the constants c_0 and c_1 are known, the marine radar is properly calibrated and can be used to measure wave heights as well. In the same way as other sea state parameters, the estimation of H_s from the analysis of marine radar data sets depends on the quality of the signal. Therefore, at least a minimum amount of local wind is necessary to produce the ripples in the sea surface to obtain a reliable radar intensity. The minimum amount of wind speed depends on each radar installation and radar polarization; a typical value is 3 m/s. In addition to modulate the radar backscatter a sea state with a significant wave height of at least 0.5 m is required to retrieve oceanographic information from the sea clutter pattern.

4. SUMMARY

The marine X-band radar is a powerful tool for monitoring the ocean surface winds, waves, and currents in vicinity of the radar. The radar can be operated from coastal stations, offshore platforms, as well as from moving vessels. The wind can be measured with an accuracy of 0.5 m/s in speed and 10° in direction, respectively. The major advantage to traditional wind sensors is that the measurements are taken in some distance to the radar and are, therefore, not influenced by flow distortion such as blockage and shadowing of the measurement platform as well as biases due to the sensors motion and height variation. Ongoing work is concentrating on retrieving wind fields as well as estimating gust and their properties in space and time.

Furthermore, the radar can be utilized to retrieve the ocean surface currents by estimating the difference of the measured phase speed of the waves to the phase speed according to linear wave theory. The error of this approach is approximately 0.07 m/s with a negligible bias. However, the method is limited to a homogenous current within the acquired area. In the case of spatial inhomogeneous surface currents, other methods have to be applied, such as described in Ref. 30.

In contrast to other sensors, the marine radar offers the unique opportunity to measure surface wave properties from moving vessels with an accuracy similar to a directional wave rider buoy. The main parameters that can be obtained by the radar are the wave spectra, significant wave height, peak period, and angular spreading. In contrast to buoys, the full directionality of crossing seas can also be resolved. In addition, the radar has shown to be suited to retrieve individual waves and their properties, such as individual wave height, wave grouping group and phase speed, and other related parameters. Recent research is focusing on the investigation of coherent marine radars, which measure, in addition to the intensity of the radar backscatter, the speed of the scatterers. The main objectives of these activities are to overcome the individual calibrations of the radars with respect to the significant wave height, as well as to get direct measurements of the waves orbital velocities, which would enable a further estimation possibility of individual wave heights.[31]

REFERENCES

1. Chaudhry A, Moore R. Tower-based backscatter measurements of the sea. *IEEE J Oceanic Eng* 1984;**9**:309–16.
2. Trizna DB, Carlson D. Studies of dual polarized low grazing angle radar sea scatter in nearshore regions. *IEEE Trans Geosci Remote Sens* 1996;**34**:747–57. http://dx.doi.org/10.1109/36.499754.
3. Wetzel L. Electromagnetic scattering from the sea at low grazing angles, in surface waves and fluxes. In: Geernaert GL, Plant WL, editors. *Remote sensing*, vol. 2. Kluwer Academic; 1990. p. 109–71.
4. Brown GS. Low-grazing-angle backscatter from rough surfaces. *IEEE Trans Ant Prop* 1998;**46**(1). http://dx.doi.org/10.1109/TAP.1998.655445. special issue.

5. Young I, Rosenthal W, Ziemer F. A three-dimensional analysis of marine radar images for the determination of ocean wave directionality and surface currents. *J Geophys Res* 1985;**90**(C1):1049−59.
6. Nieto-Borge JC, Reichert K, Dittmer J. Use of nautical radar as a wave monitoring instrument. *Coastal Eng* 1999;**3−4**(37):331−42.
7. Nieto-Borge JC, Rodríguez G, Hessner K, González P. Inversion of marine radar images for surface wave analysis. *J Atmos Oceanic Technol* 2004;**21**:1291−300.
8. Dankert H, Horstmann J, Lehner S, Rosenthal W. Detection of wave groups in SAR images and radar image sequences. *IEEE Trans Geosci Remote Sens* 2003;**41**: 1437−46. http://dx.doi.org/10.1109/ TGRS.2003.811815.
9. Senet C, Seemann J, Ziemer F. The near-surface current velocity determined from image sequences of the sea surface. *IEEE Trans Geosci Remote Sens* 2001;**39**:492−505.
10. Bell P. Shallow water bathymetry derived from an analysis of X-band marine radar images of waves. *Coastal Eng* 1999;**37**:513−27.
11. Seemann J, Senet CM, Ziemer F. Local analysis of inhomogeneous sea surfaces in coastal waters using nautical radar image sequences. In: *Mustererkennung*. Springer; 2000. p. 179−86.
12. Flampouris S, Ziemer F, Seemann J. Accuracy of bathymetric assessment by locally analyzing radar ocean wave imagery. *IEEE Trans Geosci Remote Sens* 2008;**46**: 2906−13. http://dx.doi.org/10.1109/TGRS.2008.919687.
13. Horstmann J, Vicen-Bueno R, Coffin M. *Reading the ocean with marine radars*. Defence Global; 2011. pp. 58−60.
14. Dankert H, Horstmann J, Rosenthal W. Ocean wind fields retrieved from radar-image sequences. *J Geophys Res* 2003;**108**:3352. http://dx.doi.org/10.1029/2003JC002056.
15. Dankert H, Horstmann J. A marine radar wind sensor. *J Atmos Oceanic Technol* 2007;**24**: 1629−42.
16. Vicen-Bueno R, Horstmann J, Terril E, de Paolo T, Dannenberg J. Real-time ocean wind vector retrieval from marine radar image sequences acquired at grazing angle. *J Atmos Oceanic Technol* 2013;**30**:127−39. http://dx.doi.org/10.1175/JTECH-D-12-00027.1.
17. Lund B, Graber HC, Romeiser R. Wind retrieval from shipborne nautical X-band radar data. *IEEE Trans Geosci Remote Sens* 2012;**99**:1−12.
18. Lucas BD, Kanade T. An iterative image registration technique, with an application to stereo vision. In: *Int'l joint conference artificial intelligence*; 1981. p. 121−30.
19. Nieto-Borge JC, Guedes-Soares C. Analysis of directional wave fields using X-band navigation radar. *Coastal Eng* 2000;**4**(40):375−91.
20. Krogstad HE, Trulsen K. Interpretations and observations of ocean wave spectra. *Ocean Dyn* 2010;**60**(4):973−91.
21. Seemann J, Ziemer F. Computer simulation of imaging ocean wave fields with a marine radar. In: *Proc. oceans, challenges of our changing global environment*; 1995. p. 1128−33.
22. Alpers W, Hasselmann K. Spectral signal to clutter and thermal noise properties of ocean wave imaging synthetic aperture radars. *Int J Remote Sens* 1982;**3**:423−46.
23. Stewart RH, Joy JW. HF radio measurements of surface currents. *Deep Sea Res* 1974;**21**: 1039−49.
24. Flampouris S, Seemann J, Senet CM, Ziemer F. The influence of the inverted sea wave theories on the derivation of coastal bathymetry. *IEEE Trans Geosci Remote Sens* 2011; **8**(3):436−40.

25. Piotrowski CC, Dugan WL. Accuracy of bathymetry and current retrievals from airborne optical time-series imaging of shoaling waves. *IEEE Trans Geosci Remote Sens* 2002; **40**(12):2606–18.
26. Abileah R. Mapping shallow water depth from satellite. In: *ASPRS 2006 annual conference*; 2006.
27. Plant WH. In: Komen GJ, Oost WA, editors. *The modulation transfer function: concept and applications, radar scattering from modulated wind waves*. Kluwer Academic; 1988. p. 155–72.
28. Nieto Borge JC, Hessner K, Jarabo Amores P, de la Mata Moya D. Signal-to-noise ratio analysis to estimate ocean wave heights from X-band marine radar image time series. *IET Radar Sonar Navig* 2008;**2**(1):35–41.
29. Izquierdo P, Guedes Soares C, Nieto Borge JC, Rodrguez GR. A comparison of sea-state parameters from nautical radar images and buoy data. *Ocean Eng* 2004;**31**:2209–25.
30. Senet CM, Seemann J, Flampouris S, Ziemer F. Determination of bathymetric and current maps by the method DiSC based on the analysis of nautical X-band radar image sequences of the sea surface. *IEEE Trans Geosci Remote Sens* 2007;**46**(8):2267–79.
31. Seemann J, Stresser M, Ziemer F, Horstmann J, Wu LC. Coherent microwave radar backscatter from shoaling and breaking sea surface waves. In: *Proc. oceans 2014*; 2014.

CHAPTER

Glider Salinity Correction for Unpumped CTD Sensors across a Sharp Thermocline

17

Yonggang Liu*, Robert H. Weisberg, Chad Lembke

College of Marine Science, University of South Florida, St. Petersburg, FL, USA

**Corresponding author: E-mail: yliu18@gmail.com, yliu@mail.usf.edu*

CHAPTER OUTLINE

1. INTRODUCTION

By offering economical platforms for interdisciplinary ocean observations, autonomous underwater gliders (hereafter "gliders") are becoming important assets of ocean observing systems.[1–3] Gliders are robots that propel themselves through the water using changes in buoyancy and adjustments in attitude to allow them to soar on wings. By cyclically falling and floating, they progress forward in a sawtooth fashion while transiting from near the surface to the seafloor or desired depth. Those powered on batteries are capable of being deployed weeks to months, allowing 100–1000s of km transects of water column data to be collected. Several commercial products are available, for example, the Slocum glider created by and offered commercially by Teledyne Webb Research Corporation,[4,5] the Seaglider developed at the Applied Physics Laboratory at the University of Washington and now available through Kongsberg Maritime,[6] and the Spray glider of Scripps Institution of Oceanography and Bluefin Robotics.[7] Their applications are found in many regions of the world's oceans, e.g., the US West Coast,[8,9] the US East Coast,[10,11] the Gulf of Mexico,[3,12–16] the Mediterranean Sea,[17] and the Australia coast.[18–20]

For salinity determinations, some of the gliders use passive (unpumped) conductivity cells to conserve power and extend the range of the glider mission. As with conventional CTDs, there is a mismatch between the temperature and the conductivity measurements primarily due to the thermal inertia of the conductivity sensor. It takes time for the conductivity sensor to adjust to surrounding water, e.g., to diffuse its heat stored when it moves from hot to cold water, whereas the temperature sensor responds more rapidly to the ambient temperature change. This temperature–conductivity response time mismatch leads to erroneous salinity calculations, referred to as thermal lag effects. Such effects were examined by Gregg and Hess[21] and Lueck,[22] and a numerical algorithm for thermal lag correction was proposed by Lueck and Picklo.[23] A practical method for determining the thermal lag correction parameters was proposed by Morison et al.[24] based on minimizing the salinity separation of temperature–salinity (T–S) curves from the upcasts and downcasts of a yo-yo sequence of conductivity–temperature–depth (CTD) profiles. The thermal lag corrections for two Sea-Bird CTD instruments (SBE-41CP and SBE-41) were reported by Johnson et al.[25] based on screening thousands of profiles from Argo profiling floats. Mensah et al.[26] revisited the thermal mass inertia correction of SBE4 conductivity sensors for the calculation of salinity, and they proposed an empirical method, also based on Morison et al.[24] to determine optimal values for the correction parameters.

Unlike conventional CTD casts from a surface vessel where the ascent/descent speeds may be controlled to be constant and slow enough, a glider's speed may vary through the water depending on the glider's buoyancy manipulation, altitude, and depth. The glider moves both horizontally and vertically. A complete descent and ascent course of the glider is called a yo-yo. The conductivity measurement relies on the glider's motion through the water to passively flush the conductivity cell. The rate of flushing is related with glider altitude and speed, and varies during the course of a yo-yo. Also, the glider's CTD sampling frequency is often set to be lower ($\sim$0.5 Hz) than that of the high-resolution sampling of the CTD operated on a surface vessel, and the sampling interval may be irregular. In particular, some glider data acquisition systems do not sample the CTD sensors at a constant rate, nor do they collect data continuously, depending on how the glider firmware was programmed on a mission and how many sensors it has reporting to it in a given sequence. The glider data output may not be a continuous and evenly spaced time series, even though the glider is set up to sample at a constant rate. There can be gaps that are substantial. The users have no way of knowing the exact time stamp placed on the data (personal communication, Carol Janzen, Seabird Scientific). These features render glider CTD data different from traditional CTD data, and the thermal lag correction is more difficult. This issue is discussed in recent publications.[27–29]

For salinity data from unpumped CTD sensors installed on gliders, proper corrections are critical prior to their applications in physical oceanography, e.g., hydrographic analysis,[30,31] mixed layer depth estimation,[32] assimilation into ocean circulation models,[33–35] multiplatform analysis,[36] etc. Salinity data also affect the calculation of other variables and derivatives, e.g., dissolved oxygen, chlorophyll, and spiciness.[8,37–39] Careful calibration of salinity data is, therefore, of primary importance in glider data processing.

Thermal lag corrections for glider data have gained increased attention in recent years. For example, Bishop[27] corrected the thermal lag effects in Slocum glider observations based on the methods of Lueck and Picklo[23] and Morison et al.[24] He used the mean vertical speed of the Slocum glider to calculate the correction parameters. Ericksen[40] reported salinity estimation using an unpumped conductivity cell on a Seaglider. Frajka-Williams et al.[28] briefly described the thermal inertia corrections of Seaglider data based on the flight model of Eriksen et al.[6] and the thermal lag theory of Lueck.[22] Their relaxation constants were determined by minimizing the along-isopycnal difference between salinities of successive climb–dive near the surface or dive–climb at depth. Recently, Garau et al.[29] proposed a thermal lag correction method for Slocum CTD glider data, also based on the work of Morison et al.[24] but using the variable speed of the glider. Another novel part of that method is that the four correction parameters are determined by minimizing an objective function that measures the area between two T–S curves formed by two CTD profiles, one upcast and one downcast. This method has been used in routine glider data processing.[41,42]

Thermal lag effects for glider data may vary with water column stratification. For a well-mixed water column, the thermal lag effects may not be a problem at all, because the water properties are about the same throughout the water column, and the time lags in temperature and conductivity sensors may not be noticeable. However, in a stratified water column, the thermal lag effects may become an issue.[43] The stronger the stratification, the larger the expected thermal lag effects.[26] The importance of correcting the thermal lag effects was demonstrated for the SBE-25 CTD data collected in the Mediterranean summer thermocline.[44]

In this paper, we examine the performance of the thermal lag correction methods in correcting the salinity errors in the glider data collected on the eastern Gulf of Mexico, West Florida Shelf (WFS), with emphasis on a strong thermocline. The purpose is two-fold: (1) to test the existing thermal lag correction methods in the case of a strong thermocline, and (2) to improve the glider salinity correction results empirically.

2. A SHARP THERMOCLINE

Since 2009, Webb Slocum gliders (model G1 with a 200-m buoyancy engine) have been added to integrated ocean observation systems on the WFS[45] that include arrays of moored Acoustic Doppler Current Profilers,[46] high-frequency radars,[47,48] and satellite-tracked drifters.[49] These glider data have been used in oceanographic applications and data assimilation experiments on the WFS.[13–15] Unpumped glider

CTD assemblies of Seabird Electronics Inc. are used on Slocum gliders up through 2010. The sensors are electronically derived from the SBE41CP Argo float CTD. It is important to recognize that they are distinctly very different from the instrument/CTD as used on the Argo profiling floats. This means processing applied to SBE41CP Argo profiling float data (or any pumped CTD data set) cannot be directly transferred to the glider unpumped CTD data processing (personal communication, Carol Janzen).

As a low-energy shelf,[47,50] the eastern Gulf of Mexico water tends to be stratified in spring and summer,[51] which is related to the seasonal variations of the wind forcing and heating/cooling.[52–55] Interactions between the Loop Current eddies and the shelf slope could ventilate the WFS with cold waters of deep ocean origin[13,14,56] and generate a strong thermocline on the shelf.

A particular example can be seen from the glider data of mission 16 between the 25- and 50-m isobaths (Figure 1). Even though this glider mission was of short duration (aborted due to an air bladder leak after three days of survey in April of 2009), a sharp thermocline was observed in the area offshore of the 40-m isobath. Both the temperature and conductivity data show a distinct difference between the upper layer and the near bottom layer of the water column with a sharp thermocline around the 30-db level (Figure 2(a) and (b)). The calculated salinity data show a thin lens of saline water in the transect plot (Figure 2(c)), which corresponds to the vertical level of the thermocline. This feature cannot be interpreted as a slope water lens transported on-shelf within the thermocline as found by Hopkins et al. (2012) in the Celtic Sea[57];

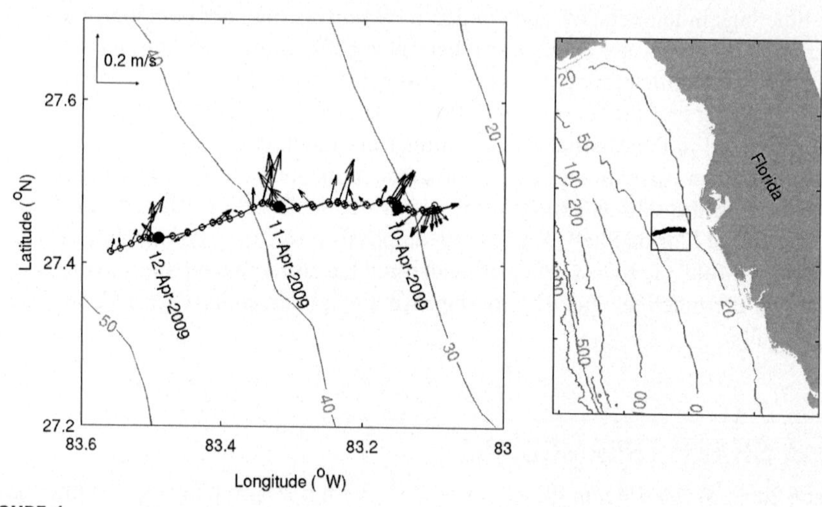

FIGURE 1

Glider track and the surface velocity estimated from the glider observations (left panel), and the relative position of the track on the West Florida Shelf (right panel). Bottom bathymetry contours are in meters.

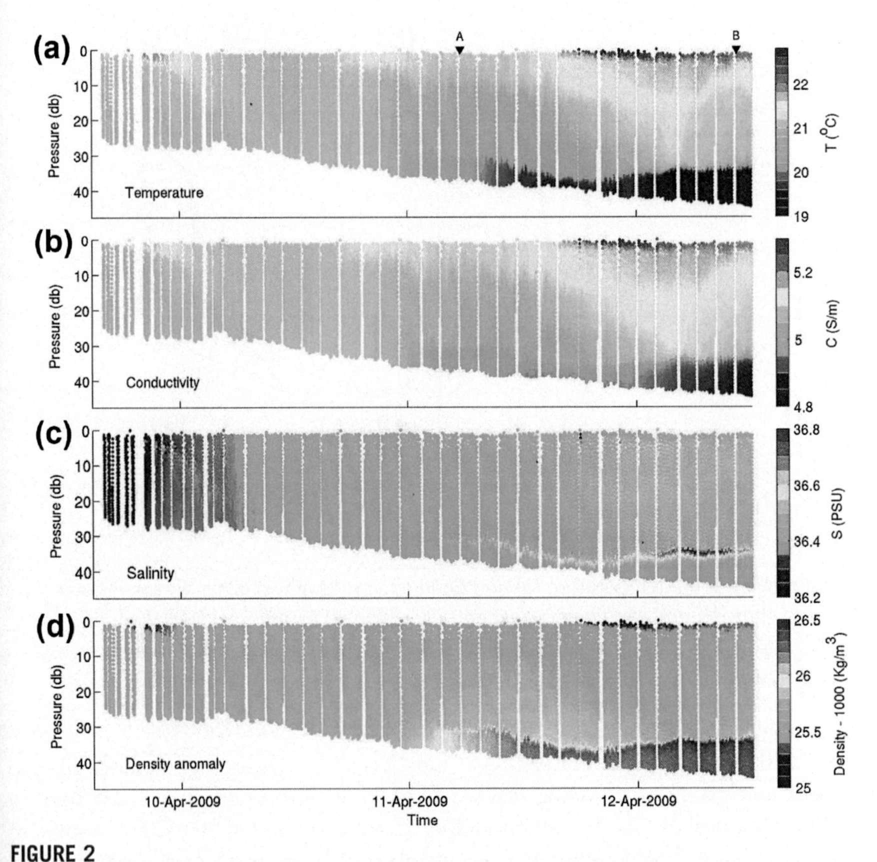

FIGURE 2

Water property distributions along the Figure 1 glider transect: (a) temperature, (b) conductivity, (c) salinity, and (d) density anomaly. Note the false high salinity lens around the 30-m depth level (c) where the thermocline is located, and the associated vertical density inversion (d). The upside-down triangles in panel (a) denote the locations of two yo-yos in weak and sharp thermocline cases, respectively. The gaps between the profiles correspond to the time window for data transfer when the glider floated on the surface.

rather, it is due to persistent thermal lag errors on the glider salinity calculations in the presence of a strong thermocline. This can be evidenced as follows: (1) The density transect shows vertical inversions (heavier water on top of lighter water, Figure 2(d)) as a result of this salinity artifact; (2) The vertical profiles of temperature show mismatches around the thermocline between the downcast and upcast of a glider yo-yo (Figure 3(a)); (3) The vertical profiles of conductivity show even more obvious mismatches at those levels between the downcast and upcast of a glider yo-yo (Figure 3(b)); and (4) The ascending and descending vertical profiles of salinity exhibit errors of opposite sign near the thermocline (Figure 4). These are typical features related to the thermal lag effects of CTDs across a sharp thermocline.[26]

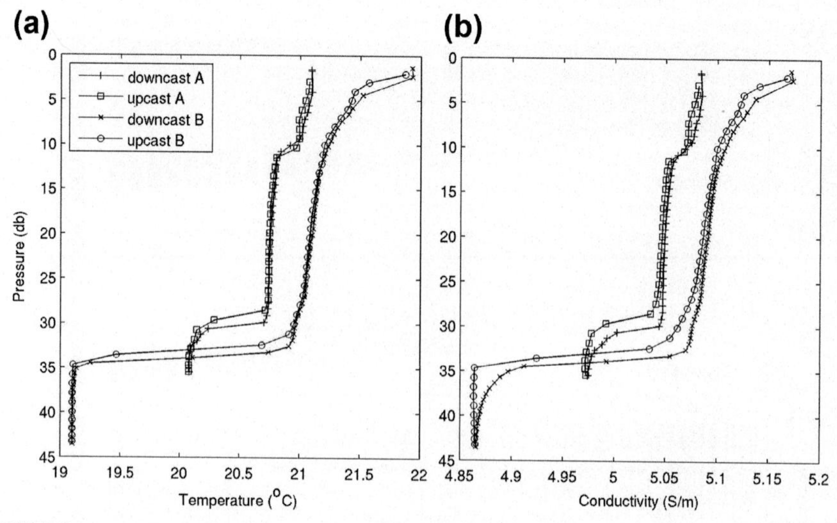

FIGURE 3

Vertical profiles of the temperature (a) and conductivity (b) of two glider yo-yos representing weak and sharp thermocline cases, respectively (denoted as the upside-down triangles in Figure 2). Note the sharp thermocline near ~34 m depth with a temperature change of about 2 °C within 3 m of the water column.

We will examine the effectiveness of the thermal lag correction methods in both the weak and strong thermocline cases. The glider data from mission 16 are used in this study because of this sharp thermocline located around the 30 to 35 m depths of the water column. Temperature changes are about 2 °C within 3 m of the water column at the 45-m site (Figure 3(a)). In terms of the vertical temperature gradient, this case may be referred to as a "strong thermocline" according to Mensah et al.[26] Along the transect, the strength of the thermocline decreases toward the land. Two yo-yos (points A and B in Figure 2(a)) are chosen to represent a weak and a strong thermocline, respectively. The horizontal distances the glider maneuvered during the two yo-yos are 60 and 84 km, respectively.

3. METHODS

As mentioned, a mismatch in the temporal response of a pair of temperature and conductivity sensors can lead to significant errors in the calculated salinity. So, the thermal lag correction methods, based on theoretical model of Lueck,[22] are generally categorized into two approaches: to estimate the conductivity data measured inside the conductivity cell, or to estimate the temperature of water inside the conductivity cell, with the sole purpose of calculating salinity. This has been well documented in literature.[24,26,29]

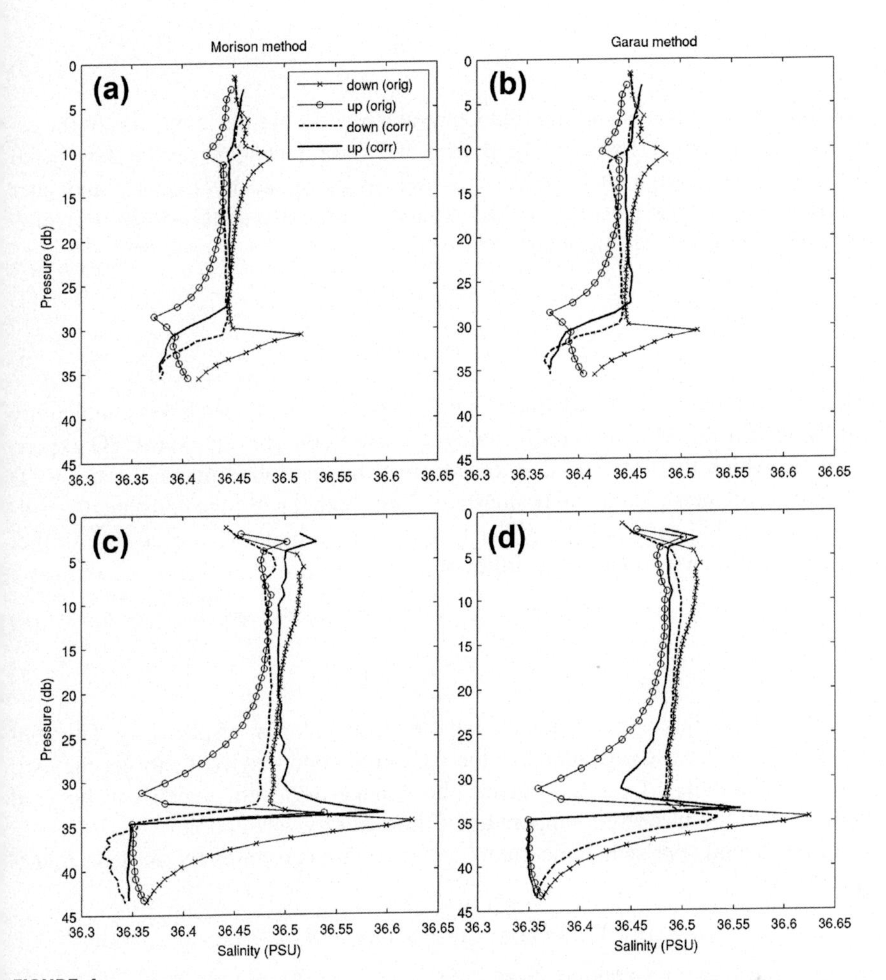

FIGURE 4

Vertical profiles of the original and corrected salinity of one glider yo-yo (denoted as the upside-down triangles in Figure 2) using different methods: (left panels) Morison et al. (1994) and (right panels) Garau et al. (2011). Thermal lag correction is successful for the weak thermocline profile (panels (a) and (b)), but not for the sharp thermocline case (panels (c) and (d)). Salinity spikes near the sharp thermocline are still seen in the corrected salinity profiles.

In the first approach, a conductivity correction (C_T) may be expressed as follows:

$$C_T(n) = -bC_T(n-1) + \gamma a[T(n) - T(n-1)], \tag{1}$$

where n is the sample index, T is the temperature, and γ is the sensitivity of conductivity to temperature. The coefficients a and b are given as follows:

$$a = \frac{4f_n\alpha\tau}{1 + 4f_n\tau}, \tag{2}$$

$$b = 1 - \frac{2a}{\alpha}, \quad (3)$$

where f_n is the Nyquist sampling frequency, and α and τ are the amplitude of the error and time constant, respectively. Both α and τ are dependent on the flow speed (flushing rate) through the conductivity cell, as predicted by Lueck[22] and later confirmed by Morison et al.[24] Based on the empirical results, Morison et al.[24] suggested the following formulas:

$$\alpha = 0.0135 + 0.0264/V, \quad (4)$$

$$\tau = 7.1499 + 2.7858\Big/\sqrt{V}, \quad (5)$$

where V is the average velocity (units in meters per second) through the conductivity cell. Note that the Morison et al.[24] study was based on conventional CTD experiments, in which the flow speed is known (may be controlled manually in a CTD operation) and then assumed to be constant. Thus, both α and τ are constant. Considering the variable speed of the glider in the water column, Garau et al.[29] further expressed these two equations as follows:

$$\alpha(n) = \alpha_o + \alpha_s/V(n), \quad (6)$$

$$\tau(n) = \tau_o + \tau_s\Big/\sqrt{V(n)}. \quad (7)$$

where the subscripts o and s indicate the offsets and slopes, respectively. The conductivity correction, C_T, is added to the measured conductivity, C, to get the estimated conductivity, $(C_T + C)$, outside the conductivity cell, which can be used together with the measured temperature, T, for salinity calculation.

In the second approach, temperature correction was proposed by Morison et al.[24] as follows:

$$T_T(n) = -bT_T(n-1) + a[T(n) - T(n-1)], \quad (8)$$

where a and b are the same as those defined in Eqns (2) and (3). The temperature correction T_T is added to the measured temperature T to get the estimated temperature $(T_T + T)$ inside the conductivity cell, which can be used together with the measured conductivity C for salinity calculation.

The conductivity correction was used by some researchers,[23,26] whereas the temperature correction was preferred by others.[24,29] Note that Eqn (8) is simpler than Eqn (1) because it does not need the estimated sensitivity γ. Also, it avoids the error induced by implicitly linearizing the equation of state in the first approach (by effectively assuming a uniform ratio between temperature and conductivity).

The process for choosing the parameters α_o, α_s, τ_o, and τ_s is critical in the thermal lag correction. Morison et al.[24] proposed a practical (empirical searching) method to determine these parameters by minimizing the salinity separation of T–S curves from upcasts and downcasts of a yo-yo sequence of CTD profiles. The hypothesis is that both the upcast and downcast CTD profiles should measure the same water mass. The empirical results are given in Eqns (4) and (5). This empirical searching method was also used in Mensah et al.[26] Garau et al.[29] proposed another method to estimate these parameters based on the same hypothesis. They minimize an objective function that measures the area between two T–S curves from upcasts and downcasts of a yo-yo sequence of CTD profiles. The minimization is an iterative process using the optimization toolbox from MATLAB. The minimum of the constrained nonlinear multivariable function is found using a sequential quadratic programming (SQP) method.

The method of Garau et al.[29] has received attention in the glider community. A MATLAB toolbox containing the thermal lag correction code is also available freely online (http://www.socib.es/~glider/doco/gliderToolbox/ctdTools/thermalLagTools). Its applications are found in many studies.[41,58] However, to our knowledge, the correction for salinity data in a strong thermocline has not been reported.

The thermal lag correction code provided by Garau et al.[29] allows for a choice between two methods: (1) constant correction parameters for Eqns (6) and (7) as recommended by Morison et al.[24] in Eqns (4) and (5), and (2) variable correction parameters that are determined from a downcast–upcast pair of CTD profiles of each glider yo-yo.[29] In this paper, they are referred to Morison and Garau methods, respectively. Note that the velocity in Eqns (6) and (7) is not constant, which is different from that of the original method developed for conventional CTD casts.[24] Regardless, it is still referenced as the Morison method in this chapter. So, the only difference between these two methods is the choice of the four correction parameters (α_o, α_s, τ_o, and τ_s).

4. THERMAL LAG CORRECTION RESULTS

4.1 SUCCESSES AND LIMITATIONS

The two correction methods mentioned previously were applied to the two CTD profiles shown in Figure 3, and the results of thermal lag correction are shown in Figure 4. Both the original and corrected salinity profiles are shown in the same panels so that the successes and limitations of the corrections may be seen.

For the weak thermocline, both methods successfully adjusted the salinity profiles (Figure 4(a) and (b)). The differences between the downcast and upcast salinity profiles are significantly reduced, especially for the two (weak) thermocline layers around the 12- and 30-m levels, respectively. The corrected downcast and upcast salinity profiles tend to align with each other throughout the water column. Two haloclines, one weaker around the 12-m level and the other stronger around the 30-m level, are evident in the corrected salinity profiles. These features correspond well with the weak thermoclines (Figure 3(a)) and the vertical gradients of conductivity at those two levels (Figure 3(b)), respectively. Despite the minor differences

(<0.02 PSU) between the upcast and downcast profiles, the thermal lag corrections for the weak thermoclines are satisfactory.

For the strong thermocline, the differences between the downcast and upcast salinity profiles are also reduced by both methods (Figure 4(c) and (d)). For example, the salinity spikes are reduced by 0.08 to 0.1 PSU using the Garu method (Figure 4(d)). However, salinity spikes are still seen in the corrected profiles near the strong thermocline around the 30- to 35-m levels. Note that both the temperature and conductivity profiles are reasonably smooth at these levels (Figure 3(a) and (b)). These salinity spikes are due to the short-term mismatch between the temperature and conductivity sensors.[26] The examples of salinity spikes, as shown in Figure 4(c) and (d), are not isolated cases. Persistent spikes are seen in the calculated salinity in this strong thermocline area. For some glider yo-yos, the salinity spikes of the downcast and upcast profiles point to different directions (i.e., one in $S+$ direction, the other in $S-$ direction) instead of one direction as shown in Figure 4(c) and (d). After the preliminary thermal lag correction using the Morison method, the salinity spikes in the downcast profiles are removed, whereas those in the upcast profiles are only slightly reduced (Figure 4(c)). However, using the Garau method, both salinity spikes are still present, though largely reduced compared with the original salinity data (Figure 4(d)). These preliminary experiments show that thermal lag effects are still a problem for the strong thermocline, and further improvements are needed for the thermal lag corrections.

It is worthy to point out that proper visualization is important in the examination of the salinity errors due to the thermal inertia effects. If shown in a transect plot as Figure 2(c), the preliminarily corrected salinity data are almost "acceptable" because the reduced salinity spikes are hardly seen from such a color plot (figure not shown). However, if shown in line plots (Figure 4(c) and (d)), these salinity errors are more evident. Another example is the very weak thermocline around the 12-m level at point A. It is not eye-catching at all in either the temperature or the conductivity profiles (Figure 3(a) and (b)). However, the original salinity data, directly calculated from these temperature and conductivity data, show a noticeable salinity difference (~0.07 PSU) between the upcast and downcast profiles (Figure 4(a) and (b)). Such salinity differences are not easily seen from the scattered color dots (Figure 2(c)) generated by the MATLAB internal function "scatter," which is widely adopted by the glider community. These differences are usually too small to be resolved by the colors over a large range. These small features are not easily seen in color contour plots because the color contour plots are a smoothed version of the scattered dots. As a result, one should be cautious in interpreting the color scatter or contour plots of salinity for a halocline or for thermal lag effects. It is better to examine them in line plots as in Figure 4.

4.2 IMPROVEMENTS WITH A MEDIAN FILTER

Salinity spikes were seen from CTD profiles through sharp thermoclines/haloclines in early studies.[59] These salinity spikes were in error and cannot be eliminated by the regular smoothing (e.g., low-pass filtering) technique that usually spreads the error and leaves its integral effect unchanged.[22] Emery and Thomson[60] suggested some methods for detecting and removing large errors or spikes from data. A standard method for isolating large errors is to compute a histogram of the sample values, and see if the divergent values fit into the assumed probability distribution function for the assumed variable. Another method, which is more automatic and objective, is to identify and eliminate all values that exceed a specified standard deviation (e.g., $\pm$ three standard deviations). However, these approaches have the weakness that they must first consider all the data points, including the extreme values, as valid in order to determine which data points are outliers.[60] These conventional approaches are not effective or convenient for ridding glider data of salinity spikes. Mensah et al.[26] proposed a median filter to deal with these spikes in CTD profiles. Salinity profiles are slightly corrected by replacing, for a centered window of N points, the value at the center point by the median value of this window:

$$S(n) = median\left(S_{n-(N-1)/2}, \cdots, S_{n-1}, S_n, S_{n+1}, \cdots, S_{n+(N-1)/2}\right). \tag{9}$$

The one-dimensional median filter is a nonlinear digital filtering technique, often used to remove noise from a sequence of data. Using this technique for temperature, conductivity, and salinity, Mensah et al.[26] found that the spikes were effectively corrected so that errors spanning across a wide range of depth through the profile were identified.

To see the effect of de-spiking by this median filter, the salinity profile at point A is used as an example. Different filter window lengths (N) are used and the results are shown in Figure 5. The median filter is effective in removing the peaks from the profile data. A 5-point median filter ($N = 5$) can largely reduce the peaks of the spikes near the haloclines. When using 11 points or more, almost all the peaks are removed (actually replaced with the median values). To be conservative, a 7-point median filter ($N = 7$) is chosen for the following calculations. This corresponds to about 7 m in the vertical water column for this glider yo-yo (Figure 5). The depth range may vary as the speed of the glider changes.

The median filter is applied to the salinity data at point B (in Figure 2). A simple application of the median filter to the original salinity data can effectively remove the sharp spike in the upcast profile, and it can significantly reduce the large peak in the downcast profile. The median filter is further applied to the preliminarily corrected salinity profiles using the Morison method, and the spike in the corrected upcast profile is completely removed (Figure 6(a)). The improvement of the results is significant, which can be seen by comparing Figure 6(a) (filtered profiles) with Figure 4(c) (not filtered profiles).

Application of the median filter in conjunction with the Garau method is more complicated than with the Morison method. To obtain the optimal correction parameters, the temperature and salinity profiles are used to compute the area encompassed by the downcast and upcast profiles in the T–S diagram (e.g., Figure 7), which is to be minimized in an iterative process. Thus, for each iteration, the median filter is

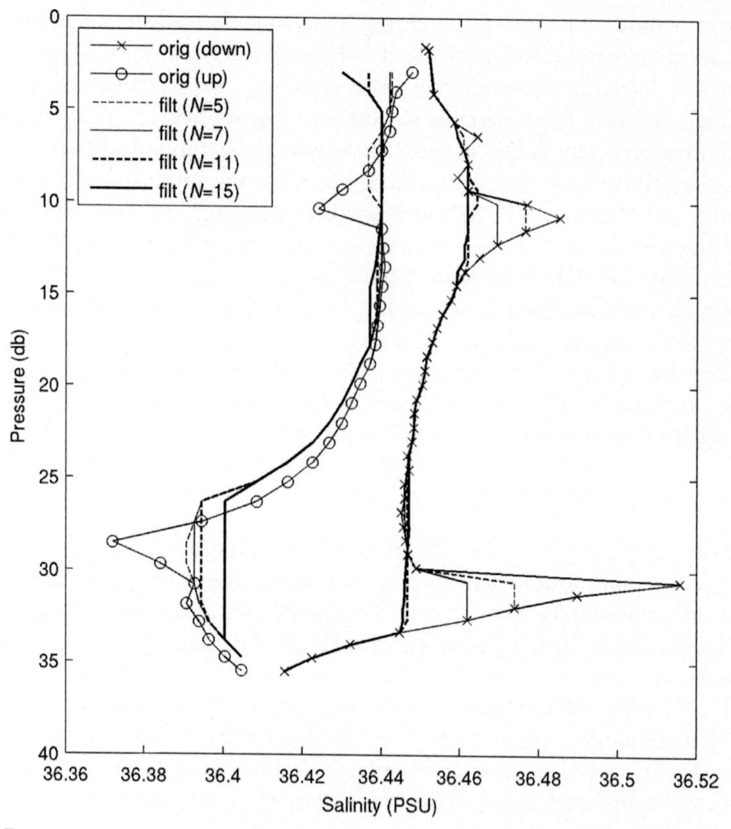

FIGURE 5

Vertical profiles of the salinity of one glider yo-yo (indicated as point A in Figure 2) before and after removing the salinity spikes using the median filter (N is the number of data points in a median filter window).

applied to the salinity data before they are used to calculate the T–S area. After the correction, the median filter is applied to the salinity profile one more time to remove the spikes, if still present. The improved results are shown in Figure 6(b). The differences between the downcast and upcast salinity profiles are minimized, especially those spikes in the strong thermocline (with salinity errors of 0.13 PSU), which are effectively removed. Significant improvements are seen in the salinity correction.

The difference of the T–S diagrams with and without the application of the median filter is evident. The area between the downcast and upcast profiles (using the corrected salinity) becomes smaller when the median filter is used (Figure 7). That is to say, the differences between the downcast and upcast water property (T and S) are reduced when the median filter is used in conjunction with the Garau method.

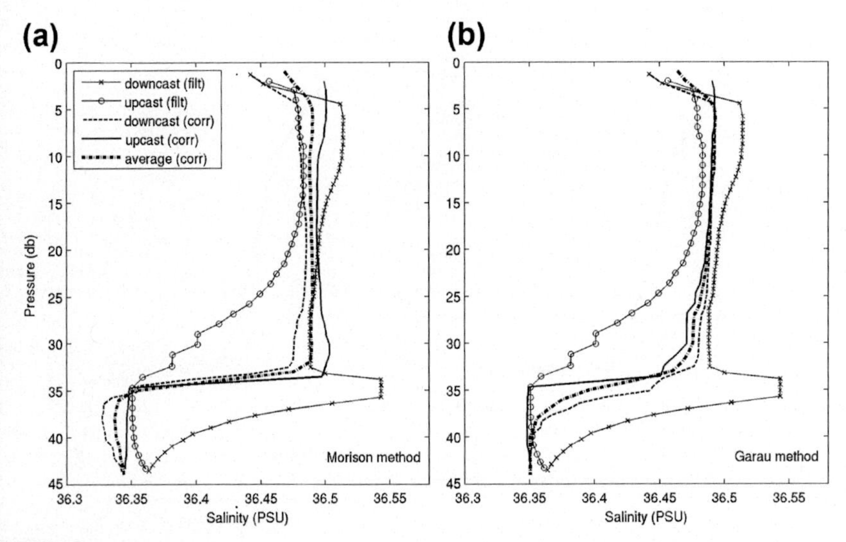

FIGURE 6

Vertical profiles of the de-spiked and the corrected salinity of one glider yo-yo (denoted as point B in Figure 2) using different methods: (a) Morison et al. (1994) and (b) Garau et al. (2011). An average of upcast and downcast salinity profiles is also shown.

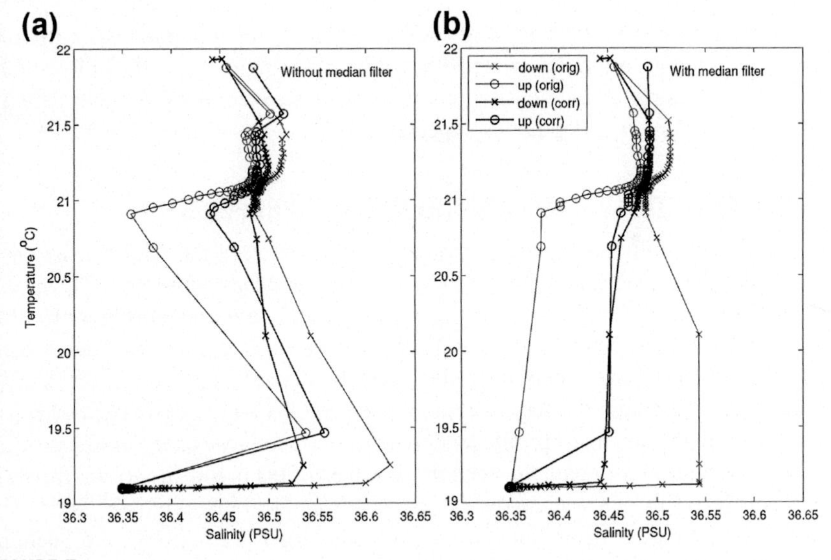

FIGURE 7

Temperature–salinity diagrams of the original and the thermal lag adjusted glider CTD data using Garau et al. (2011) method before (a) and after (b) removing the salinity spikes near the sharp thermocline using the median filter.

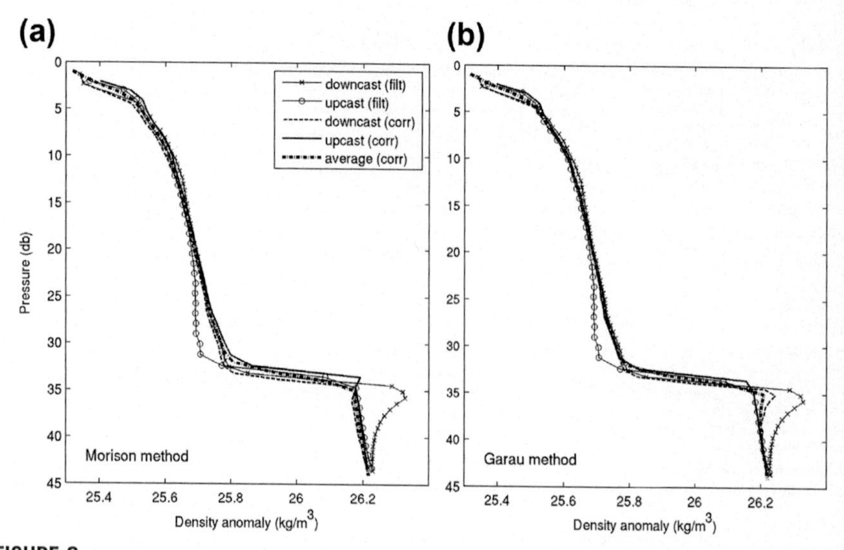

FIGURE 8

Vertical density anomaly (defined as density − 1000) profiles of one glider yo-yo (denoted as point B in Figure 2) after the salinity data are de-spiked and corrected using different methods: (a) Morison et al. (1994) and (b) Garau et al. (2011). An average of upcast and downcast density profiles is also shown for each case.

Density anomaly profiles are also shown (Figure 8) for both methods with and without the median filter. The water density becomes more hydrostatically stable after the corrections, as the density generally increases with depth (higher density values are seen in lower depths) as expected.

4.3 AVERAGE OF UPCAST AND DOWNCAST PROFILES

The corrected salinity data are now more consistent between the downcast and upcast profiles than the original data. However, the mismatches between the downcast and upcast profiles of a glider yo-yo are not completely eliminated (Figure 6). For the salinity data corrected using the Morison method, larger mismatches (up to 0.02 to 0.03 PSU) are found near the surface, immediately above the strong thermocline, and below the thermocline, respectively (Figure 6(a)). The corrected salinity data in the downcast profile tend to be lower than those of the upcast profile. Lower corrected salinity data are seen in the water levels below the strong thermocline. This seems to be a systematic bias, as seen from the transect plot (Figure 9(a)). For the salinity data using the Garau method, larger mismatches are seen in the strong thermocline (Figure 6(b)). These minor issues remain in the corrected salinity profiles.

A simple average between the downcast and upcast profiles seems to have a more acceptable salinity profile for the glider yo-yo (Figure 6). To facilitate the averaging,

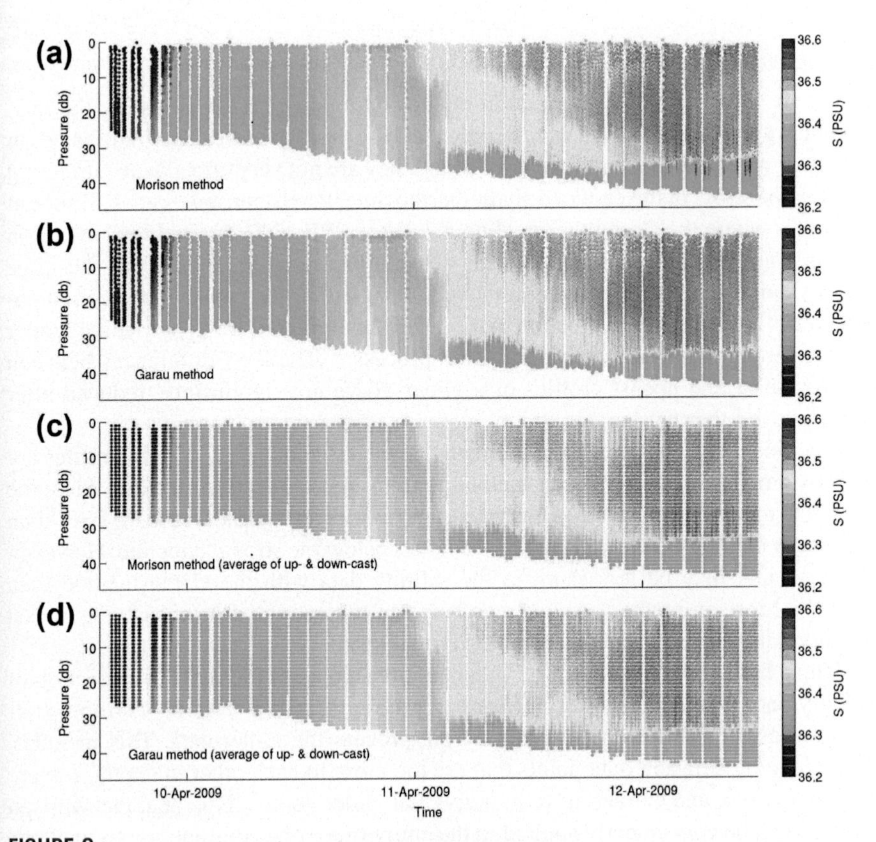

FIGURE 9

Distribution of salinity along the glider transect after correction of the thermal lag effect using different methods: (a) Morison et al. (1994) and (b) Garau et al. (2011). Salinity spikes near the thermocline depths are removed after applying the median filter. The averages of upcast and downcast salinity profiles are also shown (c & d).

the corrected downcast and upcast salinity profiles are first linearly interpolated to the same vertical levels at integer decibars, respectively. The two profiles are then averaged into one single profile. The averaged data, one profile for one yo-yo, would still be dense enough for most oceanographic studies.[61,62] These data are more redundant than the conventional CTD casts from vessels.

The averaged salinity data are shown in Figure 6(a) and (b) as line profiles, and in Figure 9(c) and (d) as transect plots. The main differences between the two types of salinity correction are that the Morison method–corrected salinity data tend to be systematically lower by 0.01 PSU in the water levels below the strong thermocline, while the Garau method–corrected data are closer to the true data in the layer below the strong thermocline. However, using the Garau method, the halocline may be a bit smoothed in the salinity data corrected (Figure 6(b)).

5. SUMMARY AND DISCUSSIONS

Although new thermal lag correction methods are powerful for adjusting the mismatches of the downcast and upcast glider salinity profiles in weakly stratified ocean waters or weak thermoclines, we have found they are not very effective in correcting the salinity errors in the case of a sharp thermocline. Persistent spikes are still seen at the levels where the strong thermocline is located. When the thermal lag correction methods are used in conjunction with a one-dimensional median filter, the large salinity spikes can be effectively removed, using either the constant correction parameters[24] or variable correction parameters determined from individual glider yo-yos through an iterative minimization process.[29] The T–S differences between the downcast and upcast profiles of a glider yo-yo are significantly reduced after the improved thermal lag correction.

The mismatches between the corrected downcast and upcast salinity profiles are slightly smaller using the Garau method than using the Morison method. Also, the Morison method–corrected salinity tends to be lower by 0.01 to 0.02 PSU than the Garau method–corrected data in the levels below the strong halocline. However, the halocline may be less sharp in the salinity data with the Garau method than with the Morison method. An average between the corrected downcast and upcast salinity data yields a more acceptable vertical profile.

Thus, based on the studies,[24,29] practical procedures of thermal lag correction of salinity data from unpumped CTD sensors are suggested for waters of strong stratification as the following: (1) Properly pre-process the glider data. This includes, e.g., reducing adjacent data points that are too close to each other in depth/pressure within a yo-yo, and extracting valid individual glider yo-yos. Note that the Morison method can be conveniently applied to the entire time series of glider mission at one time, whereas the Garau method can only be applied to one glider yo-yo at a time. In the Garau method, the correction parameters, estimated from a glider yo-yo, may vary for different yo-yos; (2) For each yo-yo containing both downcast and upcast profiles of CTD data, apply the Garau method in conjunction with the median filter; (3) Concatenate the time series of individual yo-yos to form a long time series for a glider mission or transect; and (4) Further average between the downcast and upcast salinity profiles to get a single vertical profile for each glider yo-yo, depending on the applications of the glider data.

Some glider manufacturers may have already switched for pumped CTDs, which could improve the quality of salinity calculations.[43,63] However, unpumped CTDs are still widely used in oceanographic observations. Our proposed method may help to better quality control the archived and real-time glider salinity data, as sharp thermoclines may widely exist in the world's oceans. The salinity correction was based on a Slocum glider data set. The proposed technique is also likely useful for other applications in which unpumped CTD measurements are made, for example, underway CTD.[64]

Underwater gliders have been increasingly employed as an important component of coastal ocean observing systems. A vision of developing a sustained glider network for the US coastal oceans can be seen in the IOOS National Underwater Glider Network Plan.[65,66] It is critical to perform a proper quality assurance and quality control to the collected real-time data.[67] Our proposed method can be implemented as an additional improvement to this process before the real-time data are provided to users for decision-making.[68]

ACKNOWLEDGMENTS

Support was by ONR Grant #N00014-10-1-0785, NOAA grant #s NA06NOS4780246 and NA07NOS4730409 (the first being for the ECOHAB program and the second being through the NOAA IOOS Office for the SECOORA program), NSF grant # OCE-0741705, NASA Ocean Surface Topography Science Team (OSTST) grant # NNX13AE18G, and the Gulf of Mexico Research Institute through the Florida State University Deep-C Program. A University of South Florida (USF) College of Marine Science (CMS) internal award also helped facilitate this work. The CMS-USF glider group staff were responsible for the maintenance and deployment of the gliders as well as preliminary data processing. Dr. Carol Janzen at Sea-Bird Electronics provided enlightening discussions on the glider data acquisition system and the CTD sensors. This is CPR Contribution 37.

REFERENCES

1. Rudnick D, Davis R, Eriksen C, Fratantoni D, Perry M. Underwater gliders for ocean research. *Mar Technol Soc J* 2004;**38**:73–84.
2. Fratantoni DM, Haddock SHD. Introduction to the autonomous ocean sampling network (AOSN) program. *Deep Sea Res Part II* 2009;**56**:61.
3. Robbins C, Kirkpatrick GJ, Blackwell SM, Hillier J, Knight CA, Moline MA. Improved monitoring of HABs using autonomous underwater vehicles (AUV). *Harmful Algae* 2006;**5**:749–61.
4. Stommel H. The SLOCUM mission. *Oceanography* 1989;**19**:22–5.
5. Webb DC, Simonetti PJ, Jones CP. SLOCUM: an underwater glider propelled by environmental energy. *IEEE J Ocean Eng* 2001;**26**:447–52.
6. Eriksen CC, Osse TJ, Light RD, Wen T, Lehman TW, Sabin PL, et al. Seaglider: a long-range autonomous underwater vehicle for oceanographic research. *IEEE J Ocean Eng* 2001;**26**:424–36.
7. Sherman J, Davis RE, Owens WB, Valdes J. The autonomous underwater glider "Spray". *IEEE J Ocean Eng* 2001;**26**(4):437–46.
8. Davis RE, Ohman MD, Rudnick DL, Sherman JT, Hodges B. Glider surveillance of physics and biology in the southern California current system. *Limnol Oceanogr* 2008; **53**:2151–68.
9. Pierce SD, Barth JA, Shearman RK, Erofeev AY. Declining oxygen in the northeast Pacific. *J Phys Oceanogr* 2012;**42**:495–501.
10. Glenn S, Jones C, Twardowski M, Bowers L, Kerfoot J, Kohut J, et al. Glider observations of sediment resuspension in a Middle Atlantic bight fall transition storm. *Limnol Oceanogr* 2009;**53**:2180–96.

11. Schofield O, Kohut J, Glenn S, Morell J, Capella J, Corredor J, et al. A regional slocum glider network in the Mid-Atlantic bight leverages broad community engagement. *Mar Technol Soc J* 2010;**44**:185–95.
12. Zhao J, Hu C, Lenes JM, Weisberg RH, Lembke C, English D, et al. Three-dimensional structure of a Karenia brevis bloom: observations from gliders, satellites, and field measurements. *Harmful Algae* 2013;**29**:22–30. http://dx.doi.org/10.1016/j.hal.2013.07.004.
13. Weisberg RH, Zheng L, Liu Y, Murawski S, Hu C, Paul J. Did Deepwater horizon hydrocarbons transit to the west Florida continental shelf? *Deep Sea Res Part II* 2014. http://dx.doi.org/10.1016/j.dsr2.2014.02.002.
14. Weisberg RH, Zheng L, Liu Y, Lembke C, Lenes JM, Walsh JJ. Why a red tide was not observed on the west Florida continental shelf in 2010. *Harmful Algae* 2014;**38**:119–26. http://dx.doi.org/10.1016/j.hal.2014.04.010.
15. Pan C, Zheng L, Weisberg RH, Liu Y, Lembke C. Comparisons of different ensemble schemes for glider data assimilation on West Florida shelf. *Ocean Model* 2014;**81**: 13–24. http://dx.doi.org/10.1016/j.ocemod.2014.06.005.
16. Perry RL, DiMarco SF, Walpert J, Guinasso Jr NL, Knap A. *Glider operations in the northwestern Gulf of Mexico*. San Diego: MTS/IEEE Oceans 2013; 2013. www.mtsjournal.org/Papers/PDFs/130503-198.pdf.
17. Ruiz S, Pascual A, Garau B, Faugere Y, Alvarez A, Tintoré J. Mesoscale dynamics of the Balearic Front, integrating glider, ship and satellite data. *J Mar Syst* 2009;**78**:S3–16. http://dx.doi.org/10.1016/j.jmarsys.2009.01.007.
18. Baird ME, Suthers IM, Griffin DA, Hollings B, Pattiaratchi C, Everett JD, et al. The effect of surface flooding on the physical–biogeochemical dynamics of a warm-core eddy off southeast Australia. *Deep Sea Res Part II* 2011;**58**:592–605. http://dx.doi.org/10.1016/j.dsr2.2010.10.002.
19. Pattiaratchi C, Hollings B, Woo M, Welhena T. Dense shelf water formation along the south-west Australian inner shelf. *Geophys Res Lett* 2011;**38**:L10609. http://dx.doi.org/10.1029/2011GL046816.
20. Roughan M, Schaeffer A, Suthers I. *Sustained ocean observing along the coast of southeastern Australia: NSW-IMOS 2007–2014*. 2015 [Chapter 6 of this book].
21. Gregg MC, Hess WC. Dynamic response calibration of sea-bird temperature and conductivity probes. *J Atmos Ocean Technol* 1985;**2**:304–13.
22. Lueck RG. Thermal inertia of conductivity cells: theory. *J Atmos Ocean Technol* 1990;**7**: 741–55.
23. Lueck RG, Picklo JJ. Thermal inertia of conductivity cells: observations with a sea-bird cell. *J Atmos Ocean Technol* 1990;**7**:756–68.
24. Morison J, Andersen R, Larson N, D'Asaro E, Boyd T. The correction for thermal-lag effects in sea-bird CTD data. *J Atmos Ocean Technol* 1994;**11**:1151–64.
25. Johnson GC, Toole JM, Larson NG. Sensor corrections for sea-bird SBE-41CP and SBE-41 CTDs. *J Atmos Ocean Technol* 2007;**24**:1117–11301.
26. Mensah V, Le Menn M, Morel Y. Thermal mass correction for the evaluation of salinity. *J Atmos Ocean Technol* 2009;**26**:665–72.
27. Bishop CM. *Sensor Dynamics of autonomous underwater gliders* [Masters thesis], Memorial University of Newfoundland; 2008.
28. Frajka-Williams E, Eriksen CC, Rhines PB, Harcourt RR. Determining vertical water velocities from seaglider. *J Atmos Ocean Technol* 2011;**28**:1641–56.
29. Garau B, Ruiz S, Zhang WG, Pascual A, Heslop E, Kerfoot J, et al. Thermal lag correction on slocum CTD glider data. *J Atmos Ocean Technol* 2011;**28**:1065–71.

30. Castelao R, Glenn S, Schofield O, Chant R, Wilkin J, Kohut J. Seasonal evolution of hydrographic fields in the central Middle Atlantic bight from glider observations. *Geophys Res Lett* 2008;**35**:L03617. http://dx.doi.org/10.1029/2007GL032335.
31. Hodges BA, Fratantoni DM. A thin layer of phytoplankton observed in the Philippine sea with a synthetic moored array of autonomous gliders. *J Geophys Res* 2009;**114**:C10020. http://dx.doi.org/10.1029/2009JC005317.
32. Chu PC, Fan CW. Optimal linear fitting for objective determination of ocean mixed layer depth from glider profiles. *J Atmos Ocean Technol* 2010;**27**:1893–8.
33. Shulman I, Rowley C, Anderson S, DeRada S, Kindle J, Martin P, et al. Impact of glider data assimilation on coastal model predictions. *Deep Sea Res Part II* 2009;**56**: 188–98.
34. Zhang WG, Wilkin JL, Arango HG. Towards an integrated observation and modeling system in the New York Bight using variational methods. Part I: 4DVAR data assimilation. *Ocean Model* 2010;**53**(3):119–33.
35. Pan C, Yaremchuk M, Nechaev D, Ngodock H. Variational assimilation of glider data in Monterey Bay. *J Mar Res* 2011;**69**:331–46.
36. Troupin C, Pascual A, Valladeau G, Pujol I, Lana A, Heslop E, et al. Illustration of the emerging capabilities of SARAL/AltiKa in the coastal zone using a multi-platform approach. *Adv Space Res* 2014. http://dx.doi.org/10.1016/j.asr.2014.09.011.
37. Perry MJ, Sackmann BS, Eriksen CC, Lee CM. Seaglider observations of blooms and subsurface chlorophyll maxima off the Washington coast. *Limnol Oceanogr* 2008;**53**: 2169–79.
38. Alvarez A, Mourre B. Oceanographic field estimates from remote sensing and glider fleets. *J Atmos Ocean Technol* 2012. http://dx.doi.org/10.1175/JTECH-D-12-00015.1.
39. Send U, Regier L, Jones B. Use of underwater gliders for acoustic data retrieval from subsurface oceanographic instrumentation and bidirectional communication in the deep ocean. *J Atmos Ocean Technol* 2013;**30**:984–98.
40. Eriksen CC. Salinity estimation using an un-pumped conductivity cell on an autonomous underwater glider. In: *Extended abstracts, fourth EGO conf., Larnaca, Cyprus*. EGO; 2009. p. 9.
41. Ruiz S, Renault L, Garau B, Tintoré J. Underwater glider observations and modeling of an abrupt mixing event in the upper ocean. *Geophys Res Lett* 2012;**39**:L01603. http://dx.doi.org/10.1029/2011GL050078.
42. Bouffard J, Renault L, Ruiz S, Pascual A, Dufau C, Tintoré J. Sub-surface small scale eddy dynamics from multi-sensor observations and modeling. *Prog Oceanogr* 2012. http://dx.doi.org/10.1016/j.pocean.2012.06.007.
43. Janzen CD, Creed EL. Physical oceanographic data from seaglider trials in stratified coastal waters using a new pumped payload CTD. In: *Proceedings of oceans 2011 MTS/IEEE, Kona, Hawaii, USA, September 19–23, 2011*, ISBN 978-1-4577-1427-6. p. 1–7.
44. Pinot J-M, Velez P, Tintore J, Lopez-Jurado JL. The thermal-lag effect in SBE-25 CTDs: importance of correcting data collected in the Mediterranean summer thermocline. *Sci Mar* 1997;**61**:221–5.
45. Weisberg RH, He R, Liu Y, Virmani JI. West Florida shelf circulation on synoptic, seasonal, and inter-annual time scales. In: Sturges W, Lugo-Fernandez A, editors. *Circulation in the Gulf of Mexico: observations and models*, vol. 161. Washington, D.C: AGU; 2005. p. 325–47.

46. Weisberg RH, Liu Y, Mayer DA. West Florida shelf mean circulation observed with long-term moorings. *Geophys Res Letts* 2009;**36**:L19610. http://dx.doi.org/10.1029/2009GL040028.
47. Liu Y, Weisberg RH, Merz CR, Lichtenwalner S, Kirkpatrick GJ. HF radar performance in a low-energy environment: CODAR seasonde experience on the West Florida Shelf. *J Atmos Ocean Technol* 2010;**27**:1689–710. http://dx.doi.org/10.1175/2010JTECHO720.1.
48. Merz CR, Weisberg RH, Liu Y. Evolution of the USF/CMS CODAR and WERA HF radar network. *Proc. MTS/IEEE Oceans* 2012;**12**:1–5. http://dx.doi.org/10.1109/OCEANS.2012.6404947.
49. Liu Y, Weisberg RH, Hu C, Kovach C, Riethmüller R. Evolution of the loop current system during the deepwater horizon oil spill event as observed with drifters and satellites. In: Liu Y, et al., editors. *Monitoring and modeling the deepwater horizon oil spill: a record-breaking enterprise. Geophys. Monogr. Ser.*, vol. 195. Washington, D.C: AGU; 2011. p. 91–101. http://dx.doi.org/10.1029/2011GM001127.
50. Weisberg RH, Liu Y, Merz CR, Virmani JI, Zheng L. A critique of alternative power generation for Florida by mechanical and solar means. *Mar Technol Soc J* 2012;**46**(5):12–23. http://dx.doi.org/10.4031/MTSJ.46.5.1.
51. Liu Y, Weisberg RH. Ocean currents and sea surface heights estimated across the West Florida shelf. *J Phys Oceanogr* 2007;**37**(6):1697–713. http://dx.doi.org/10.1175/JPO3083.1.
52. Virmani JI, Weisberg RH. Features of the observed annual ocean-atmosphere flux variability on the West Florida Shelf. *J Clim* 2003;**16**:734–45.
53. He R, Weisberg RH. West Florida shelf circulation and temperature budget for the 1999 spring transition. *Cont Shelf Res* 2002;**22**:719–48.
54. He R, Weisberg RH. West Florida shelf circulation and temperature budget for the 1998 fall transition. *Cont Shelf Res* 2003;**23**:777–800.
55. Liu Y, Weisberg RH. Seasonal variability on the West Florida shelf. *Prog Oceanogr* 2012;**104**:80–98. http://dx.doi.org/10.1016/j.pocean.2012.06.001.
56. Weisberg RH, He R. Local and deep-ocean forcing contributions to anomalous water properties on the West Florida shelf. *J Geophys Res* 2003;**108**(C6):3184. http://dx.doi.org/10.1029/2002JC001407.
57. Hopkins J, Sharples J, Huthnance JM. On-shelf transport of slope water lenses within the seasonal pycnocline. *Geophys Res Lett* 2012;**39**:L08604. http://dx.doi.org/10.1029/2012GL051388.
58. Bouffard J, Pascual A, Ruiz S, Faugère Y, Tintoré J. Coastal and mesoscale dynamics characterization using altimetry and gliders: a case study in the Balearic sea. *J Geophys Res* 2010;**115**:C10029. http://dx.doi.org/10.1029/2009JC006087.
59. Bray NA. Salinity calculation techniques for separately digitized fast response and platinum resistance CTD temperature sensors. *Deep Sea Res* 1987;**34**:627–32.
60. Emery WJ, Thomson RE. *Data analysis methods in physical oceanography.* 2nd Rev ed. Elsevier; 2001. 658p.
61. Pascual A, Ruiz S, Tintoré J. Combining new and conventional sensors to study the Balearic current. *Sea Technol* 2010;**51**:32–6.
62. Helber RW, Kara AB, Richman JG, Carnes MR, Barron CN, Hurlburt HE, et al. Temperature versus salinity gradients below the ocean mixed layer. *J Geophys Res* 2012;**117**:C05006. http://dx.doi.org/10.1029/2011JC007382.

63. Alvarez A, Stoner R, Maguer A. *Performance of pumped and un-pumped CTDs in an underwater glider.* San Diego, CA: IEEE Oceans 2013; 2013.
64. Ullman DS, Herbert D. Processing of underway CTD data. *J Atmos Ocean Technol* 2014; **31**:984–98. http://dx.doi.org/10.1175/JTECH-D-13-00200.1.
65. U.S. IOOS. *National underwater glider network Plan.* 2014. http://www.ioos.noaa.gov/glider/strategy/natl_glider_ntwrk_plan_draft_v9.pdf.
66. Willis S. *National observing systems in a global context.* 2015 [Chapter 2 of this book].
67. U.S. IOOS. *Manual for real-time quality control of in-situ temperature and salinity data.* 2014. QARTOD, http://www.ioos.noaa.gov/qartod/temperature_salinity/qartod_temperature_salinity_manual.pdf.
68. Porter DE, Dorton J, Leonard L, Kelsey H, Ramage D, Cothran J, et al. *Integrating environmental monitoring and observing systems in support of science to inform decision making: case studies for the Southeast.* 2015 [Chapter 22 of this book].

CHAPTER 18

New Sensors for Ocean Observing: The Optical Phytoplankton Discriminator

Justin Shapiro[1], L. Kellie Dixon[1], Oscar M. Schofield[2], Barbara Kirkpatrick[1,3], Gary J. Kirkpatrick[1,*]

Mote Marine Laboratory, Sarasota, FL, USA[1]; Institute of Marine and Coastal Sciences, Rutgers University, New Brunswick, NJ, USA[2]; Gulf of Mexico Coastal Ocean Observing System, College Station, TX, USA[3]

**Corresponding author: E-mail: gkirkpat@mote.org*

CHAPTER OUTLINE

1. INTRODUCTION

Phytoplankton are integral to complex natural processes such as the carbon cycling, food web dynamics, coastal hypoxia events, and harmful algal blooms (HABs). Identification and quantification of phytoplankton are listed as high-priority measurements needed to address six of the seven societal goals identified in the Integrated Ocean Observing System (IOOS) Summit[1] and were listed as core variables for observatory systems.[2,3] Similarly, chromophoric dissolved organic matter (CDOM) is the primary constituent that is absorbing light in the ocean and often exceeds even the light absorbed by phytoplankton.[4] As a result, CDOM dominates ocean color, plays a critical role in photobiology and photochemistry, photoproduction of CO_2,[5] as well as controlling the absorption of light energy and subsequent impacts on heat flux[6] and other ocean–climate interactions. The IOOS Summit[1] included CDOM among its 26 high-priority variables required to address three of its seven societal goals. The Optical Phytoplankton Discriminator (OPD) is an automated, in situ instrument that directly measures both phytoplankton and CDOM.

Traditional phytoplankton sampling in the ocean has been anchored by net tows or grab water sampling and subsequent identification of cells using microscopy in a laboratory. Those methods are time intensive, require a skilled microscopist and taxonomist, and they are, therefore, costly. Samples are usually examined for a small number of target taxa, limiting the view of the community structure. Net tows and grab sampling schemes usually result in undersampling due to chronic limitations of the method including vessel time, adequate sample storage, unforeseen conditions in boat operations such as weather/sea state, and the inability to sample continuously throughout the water column. There also is the likelihood of biased sampling when sampling sites are not assigned a priori, particularly when dealing with HABs that have visible surface expressions. These are limitations that make traditional phytoplankton sampling and analysis techniques unsuitable for widespread use in ocean observing systems, and they are limitations overcome by use of automated techniques.

Several automated phytoplankton taxonomic sensors have been developed to supplement traditional field/lab-based (net tow/grab samples/microscopy) methods. In addition to the light absorption-based OPD, other in situ phytoplankton taxonomic sensors include imaging-based and molecular probe-based devices. Both the Imaging FlowCytobot (IFCB, McLane Research Laboratories, Inc.) and the FlowCam (Fluid Imaging Technologies) use microscope imaging approaches. The IFCB is an automated underwater imaging microscope that uses flow cytometry technology to isolate particles for imaging. A computer is "trained" to identify the acquired phytoplankton images.[7,8] The FlowCam is a benchtop instrument that microscopically measures plankton and particles in a continuous fluid flow. Size, shape, fluorescence, and concentration statistics are provided.[9,10] The Environmental Sample Processor (ESP, McLane Research Laboratories, Inc.) implements a DNA support membrane for a particular target species and provides in situ species-specific identification using molecular probes.[11,12] These sensors are rapidly maturing; however, they often are large and power intensive, which limits their utility for integrations into autonomous systems.

The other key approach used extensively by the community is through satellite and airborne remote sensing, which has advanced due to the availability of multiple ocean color satellites with a variety of sensors and development of new ocean color algorithms that target specific phytoplankton taxa.[13] These platforms, sensors, and algorithms provide relevant spatial resolution to detect and map phytoplankton abundance over large areas.[14] Remote sensing will continue to fill important roles in ocean observing systems due to the synoptic views it provides. For some applications, it is necessary to partner satellite remote sensing with in situ observations to fully characterize subsurface features.[15] Observation of phytoplankton populations below the observable depth range of remote sensing require in-water sensors such as the OPD.

Development of the optics-based forerunner to the OPD was initiated specifically to detect the toxic dinoflagellate, *Karenia brevis*, which is responsible for nearly annual blooms in the Gulf of Mexico.[16] Blooms of this organism have been observed along the coasts of all the US Gulf states, and many of coastal states of the southeastern United States and the Mexican Gulf region. These blooms have caused massive fish kills, contaminated bivalves such as clams and oysters, caused deaths of other marine life,[17] and produced toxic aerosols.[18] These events led to closure of commercial seafood production, disruption of recreational fishing enterprises, human respiratory illness, despondent tourists due to the dead fish and toxic aerosols at the beach, and associated economic losses.[19] It is difficult, time-consuming, and expensive to manually monitor coastal waters at temporal and spatial resolution appropriate for detection of developing blooms. Historically, HAB monitoring programs typically detected the presence of their target species by visual confirmation (water discoloration and fish kills), illness to fish consumers, microscopic identification of the HAB organism, chemical analysis for toxin levels in shellfish samples and mouse bioassays. The inadequacy of those methods for HAB monitoring were well recognized, and in a 10-year national plan dealing with harmful algae[20] that identified research and information needs for addressing the issues surrounding harmful algae, it was recommended that the community "develop methodologies for rapid field-based detection of HABs and toxins." Fortunately, biological oceanographers had developed optical instrumentation (including light absorption, light scattering, and fluorescence) that could collect data over biologically relevant spatial/temporal scales in a nonintrusive manner (cf. Refs. 21,22). The need for automation of HAB monitoring provided the incentive for OPD development, and the well-developed field of bio-optics provided the means to accomplish that development. The OPD was designed to be an early warning sensor to indicate when and where waters needed to be evaluated for *K. brevis* by accepted traditional methods.

The OPD development was guided by several methods that used light absorption characteristics of photopigment complements of aquatic microalgae to discriminate taxonomic composition of natural communities. Faust and Norris[23] demonstrated the utility of derivative analysis of in vivo particle absorbance spectra to determine the concentration of major photopigment groups (chlorophylls and carotenoids). Hoepffner and Sathyendranath[24] formulated an absorption spectrum decomposition approach, through Gaussian curve fitting, to elucidate the pigment types present in a sample and from that deduce the dominant algal classes present. Johnsen et al.,[25] used a stepwise discriminant analyses to classify absorption spectra among 31 bloom-forming phytoplankton (representing the four main groups of phytoplankton with respect to accessory chlorophylls). Due to the inherent design of the OPD, it has the ability to measure light absorption by dissolved optically active carbon or chromophoric dissolved organic matter (CDOM).[26] Levels of oceanic CDOM, and the accuracy with which absorption is quantified, can substantially affect estimates of global-scale chlorophyll and primary production rates[27] and photochemical production rates,[28] with subsequent effects on the accuracy of global climate models. Characteristics of the exponential shape of CDOM spectral absorption (spectral slope) can also provide insight into CDOM age and photodegradation history.[29,30] At the present time, in situ measurement typically utilizes the fluorescence response of organic matter (FDOM) as a proxy for CDOM. Fluorescence-based CDOM instruments, however, measure only a small subset of DOM molecules that have the aromaticity and conjugation to fluoresce. The FDOM:CDOM relationship varies with the source of CDOM, its lability, and its light exposure history. Significantly, FDOM cannot provide spectral slope information.

Unlike FDOM determinations, spectrophotometric CDOM absorption measurements directly quantify the desired light absorption properties in both coastal and oceanic waters, provided adequate sensitivity and accuracy are obtained. Spectral slope data can be readily derived from full spectrum absorption. Conventional path lengths of both laboratory spectrophotometers and field absorption instrumentation are typically limited to 10 or 25 cm, respectively, which limits some open ocean applications. Corrections for salinity, temperature, and scattering[31] are also applied for the most exacting work. The most sensitive commercially available CDOM absorption instrument and the OPD utilize liquid-core waveguide (LCW) technology, in which the difference in refractive index between sample and waveguide wall results in a highly efficient internal reflection, permitting an illumination of the core to be transmitted through a coiled waveguide and resulting in path lengths of 200 cm or more. Operational issues common to all LCW instruments include fragility of silica capillary, bubble artifacts, condensation, and clogging of small lumen apertures. Additionally, because light transmission through the LCW is a function of the sample refractive index in addition to the sample absorption,[32] data collection requires careful accounting for temperature and salinity differences between references and samples. The commercial LCW instrument, though limited to manual benchtop operation, has advanced the sensitivity of CDOM analysis and resulting knowledge.[4,32,33] The OPD provides similar absorption measurement sensitivity with the additional feature of unattended, in situ, automated operation.

2. HISTORY OF THE OPD

In a set of laboratory experiments, Millie et al.[34] utilized in vivo absorbance spectra to discriminate different light acclimation states of *K. brevis* cultures grown under differing light levels. Results from those experiments on a single species provided evidence that there might be utility in the use of absorbance spectra to discriminate multiple taxonomic groups of phytoplankton. Subsequently, taxonomic groups were discriminated in theoretical mixes of absorbance spectra collected from multiple monospecific cultures.[35] A stepwise discriminant analyses was used to differentiate mean-normalized absorbance spectra for laboratory cultures of *K. brevis* from absorbance spectra of a diatom, a prasinophyte, and peridinin-containing dinoflagellates. Wavelengths delineated by the stepwise techniques were associated with the accessory carotenoids. Unfortunately, the comparative absorption by the carotenoids in the green, yellow, and orange wavelengths was much less than the absorption by chlorophyll in the blue and red wavelengths, limiting the sensitivity of that approach. Furthermore, the absorbance attributable to class-specific groupings of accessory pigments was difficult to discern due to the dampening of the shoulders on absorbance spectra due to pigment packaging effects.[36] To maximize the minor inflections in spectral absorbance, fourth derivative analysis[37] was used to resolve the positions of absorbance maxima attributable to class-specific photosynthetic pigments groupings.[38,39] The fourth derivative was chosen to minimize the influence of noise, the absorbance attributable to nonpigmented particles and the concurrent light loss due to light scatter. The application of derivative analysis and comparison to a standard derivative spectrum for *K. brevis* proved successful at not only discriminating the presence of *K. brevis*, but also at providing an estimate of the fraction of the chlorophyll *a* biomass attributable to *K. brevis* in laboratory cultures.[35]

With these laboratory results, the OPD theory of operation was in place. In 1996, one of the authors was gaining experience with miniature fiberoptic spectrometers (Ocean Optics, Inc.) while working on a phytoplankton photo-physiology project. At the same time, a local laboratory device manufacturer came to him with a product they believed would improve the sensitivity of high-performance liquid chromatography (HPLC) pigment analyses. That product was the Liquid-Waveguide Capillary Cell (LWCC, World Precision Instruments, Co.). Those components were recognized as the basis for an instrument to carry out the processing necessary to apply the OPD principles.

The application of the OPD at sea began in 1997 and continues to this day (Table 1). Major events along the way are reported below. A collection of components needed to manually perform the phytoplankton discrimination process were taken to sea in August 1997. Results from that cruise demonstrated the linear relationship between the *Karenia* sp. similarity indexes (SI) produced by the OPD and *Karenia* sp. chlorophyll *a* biomasses.[40] The multiple separate components were cumbersome for field work because each trip required assembly and disassembly of the apparatus. Additionally, the need for automation was evident because continuous unattended operation was the goal. Construction of the first prototype BreveBuster (predecessor to the OPD) began in the spring of 1998. Initially, it was manually operated, but in the summer of 1999, it was upgraded to automated, computer-controlled operation. A second, parallel, light absorption sampling channel, consisting of LWCC, light source, and spectrometer, was added in the winter of 1999–2000 to enable dual measurements of particle and CDOM absorption and minimize downtime while channels alternately went through cleaning cycles. The automated prototype OPD was deployed on board ships during process cruises of ECOHAB: Florida in 1999, 2000, 2001, and 2002. During a diel study in October 2001 tracking a surface drifter, the OPD verified the increase in abundance of *Karenia* sp. in surface waters in the morning and the subsequent decrease in abundance in the evening. Accompanying the increase in abundance in the morning was an increase in CDOM absorption. However, after the *Karenia* sp. left the surface layer in the evening, the CDOM absorption did not drop.[41] The prototype OPD was deployed for continuous underway CDOM absorption measurement during the HyCODE project at the Rutgers University Cabled Underwater Observatory, LEO-15 in the summers of 2000 and 2001. Those deployments demonstrated the OPDs ability to map surface CDOM

Table 1 Listing of OPD Applications, Locations and Supporting Agencies

Project/Location	Platform				
	Bench	Buoy	Pier/ Piling	Ship	AUV
Gulf of Mexico					
OSV Anderson HAB cruise (EPA)				X	
ECOHAB: Florida (EPA)				X (3)	
HAB Monitoring for Aquaculture (FL SG)			X (3)		
AUV Payload Integration (NSF)					X
ECOHAB: AUV HAB Detection (NOAA/CSCOR)					X (2)
MERHAB: GoMx HAB Sentinel (NOAA/CSCOR)		X (2)		X	X
FWRI/Mote HAB Monitoring (FL FWC)		X (4)	X (2)	X	X
GLIMO: Red Tide Project (NOAA/CCMA)			X		X (2)
Binational HAB Observing System (NOAA, EPA)[a]			X (3)		
HAB Detection Methods (EPA)	X				
HAB Bulletin Support (NOAA/CO-OPS)					X
Coastal Ocean Observing (NOAA/GCOOS)		X (2)	X (2)		X
Other Locations					
New Zealand (Mote)	X				
Pacific NW (OSU)				X	
Galapagos (NCSU)				X	
Mid-Atlantic Bight (Rutgers)				X	X
Great Lakes (SUNY)	X (2)			X	
Freshwater Blue-Greens (USDA)	X				
Mediterranean (Rutgers)				X	
Puget Sound (NRL)					X
Arctic (UMaine)[b]				X	

[a] *Installed in Veracruz, Mexico.*
[b] *CDOM only.*
Updated from Kirkpatrick GJ, Millie DF, Lohrenz SE, Moline MA, Schofield OM, 2011. Automated, in-water determination of colored dissolved organic material and phytoplankton community structure using the optical phytoplankton discriminator. Proc. SPIE 8030, Ocean Sensing and Monitoring III, 80,300E (May 04, 2011). *2011; http://dx.doi.org/10.1117/12.884250.*

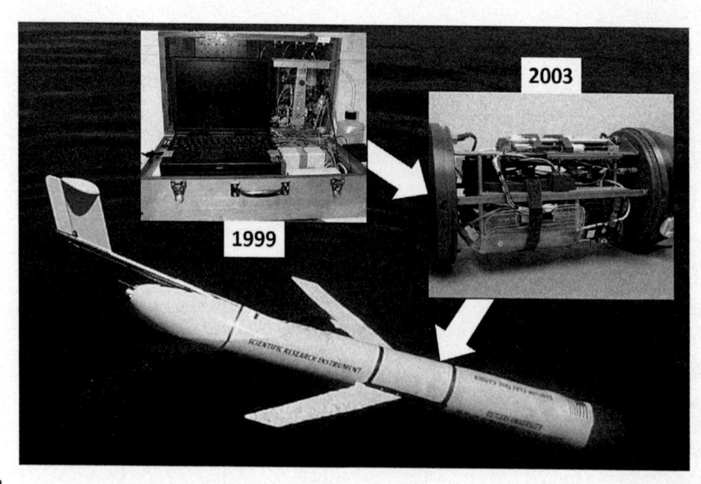

FIGURE 1

Chronological progression of the OPD configuration from benchtop prototype to AUV.

absorption.[26] In 2002, an NSF-supported project facilitated the conversion of the prototype OPD to an AUV payload for both the Slocum Glider and the REMUS. The first OPD-equipped Slocum glider open water deployment took place May 20, 2003 in Tuckerton, New Jersey, at the Rutgers University LEO-15 site (Figure 1). The first Gulf of Mexico deployment of OPDs, supported by the NOAA ECOHAB program, took place November 11–13, 2003, with OPDs in a Slocum glider and in a REMUS. The first OPD for shallow, fixed installations was deployed May 5, 2004, on a channel marker in Pine Island Sound, Florida, under a Sea Grant project examining the utility of the OPD for early warning of HAB presence for shellfish mariculture. The first extended OPD-equipped Slocum glider mission to successfully elucidate phytoplankton community structure throughout the water column ran September 28 thru October 12, 2004.[42] From 2003 to 2008 the OPD was adapted to and deployed on a Bottom-Stationed Ocean Profiler (BSOP) and an offshore oceanographic buoy under the NOAA MERHAB program.[43] During that project, a series of research cruises were undertaken to compare *Karenia* sp. detection methods including microscopic enumeration, molecular probe assay, HPLC pigment analyses and the OPD (Figure 2). On January 21 and 26, 2005, an OPD-equipped REMUS was used to examine biophysical interactions within a significant *K. brevis* bloom.[44] A *K. brevis* bloom in the fall of 2011 provided an opportunity to conduct a multidisciplinary, multi-institutional study of its three-dimensional structure. Two gliders (one equipped with an OPD) were deployed simultaneously on October 12, 2011. A University of South Florida glider examined the far-field physical structure with two transects through the bloom. A Mote Marine Laboratory glider equipped with an OPD operated mainly within the bloom. Satellite remote sensing, field measurements, and numerical models were used to put the bloom in environmental context and characterize the dynamics involved.[15]

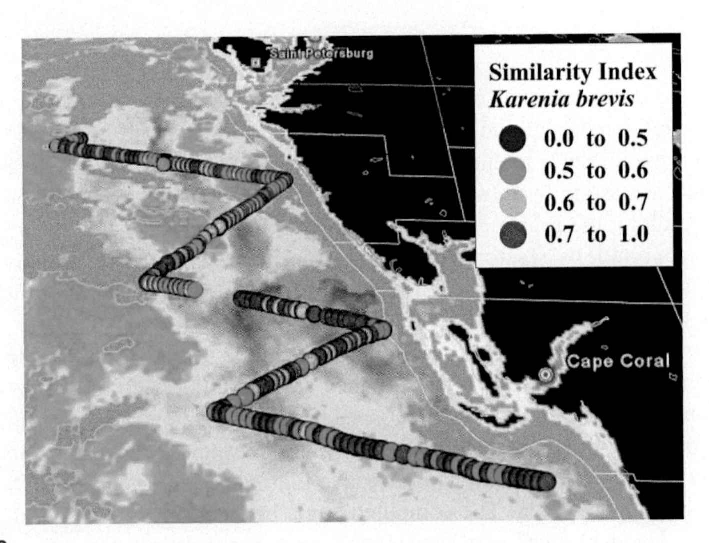

FIGURE 2

The near surface *Karenia* sp. similarity indexes (SI) determined by a shipboard OPD on November 8, 2005. Background image is MODIS remote sensing fluorescence line height.

Remote sensing image courtesy of USF-IMARS.

3. METHODOLOGY

Photopigments of plants and algae are light-harvesting molecules that function to channel light energy into the photochemical pathway for photosynthesis or to shunt excess light energy away from the photochemical pathway when there is a risk of damage from too much light energy.[45] There are approximately 45 known plant pigments found in marine microalgae, each with a slightly different molecular structure.[46] These differences in molecular structure yield differences in the shapes of light absorption spectra for each pigment. The absorption spectrum of an individual plant pigment can be modeled as the sum of a set of Gaussian curves centered at wavelengths of maximum absorption by the light-absorbing chemical structures. The absorption spectrum of any phytoplankton cell is the sum of the absorption spectra of all the pigments making up the cells pigment complement modified by factors such as cell size and the concentration of pigments within the cell (pigment-packaging effects).

The OPD method is a computational means of highlighting the absorption characteristics of plant photopigments, removing or minimizing the absorption and scattering characteristics of nonpigmented components of the bulk sample, and then fitting a set of known taxonomic class photopigment signatures to the highlighted photopigment absorption characteristics. To accomplish this, the bulk water particle absorbance spectrum is subjected to derivative analysis, and then that derivative spectrum is compared to the derivative spectrum of the known target taxa yielding a similarity index (SI).

The even derivatives of a Gaussian curve have Gaussian characteristics. Hence, no information about the pigments contained in the particles is lost during derivative analysis. However, the interfering absorbance components resulting from light loss due to particle scattering and detrital absorption are monotonic functions of wavelength and are strongly attenuated by derivative analysis. The fourth derivative of the particle scattering term (apparent absorbance) is attenuated by a factor $\approx 10^{-8}$. Similarly, taking the fourth derivative of the detrital component of absorbance reduces its magnitude by a factor $\approx 2 \times 10^{-8}$. This processing method strongly reduces the undesirable signals of particle scattering and detrital absorption while the pigment signal is conserved and effectively amplified.

The spectral SI is related to the angle between the vectors (300 element) representing the fourth derivative of absorbance spectra (400–700 nm) of a "known" phytoplankton taxon (culture or monospecific bloom sample) and the "unknown" natural, mixed-population sample. As the angle between vectors gets smaller, the similarity gets larger between the absorbance fourth derivative spectra of the known taxon and unknown taxon. The computed angle between vectors is scaled to range between -1 and 1. An $SI = -1$ means that absorbance fourth derivative spectrum of the unknown is the mirror image (about the wavelength axis) of the known (an impossibility for two phytoplankton taxa). An $SI = 1$ indicates a perfect match between the absorbance fourth derivative spectra of the known and unknown taxa (only occurs in practice when one spectrum is compared to itself).

A reference taxa spectrum or standard species spectrum is generated from individual species of phytoplankton cultured in a laboratory environment. A sample of culture at a detectable concentration (typically >20,000 cells/L) is passed through the OPD, and several characteristic absorbance spectra are captured. The spectra are converted to a reference taxa spectrum in a custom software package called "OPD Analysis." That conversion process averages the multiple absorbance spectra collected from the single species, assigns a user-specified name and the culture sample chlorophyll *a* concentration, then it stores the "species" spectrum. Multiple reference species spectra can be stored in a "library" onboard the OPD for automated processing or kept on a desktop machine for offline use. The OPD is capable of evaluating SIs for up to 16 reference spectra simultaneously while the instrument is deployed. Once the instrument is recovered, the spectral files stored during each sampling instance can be post-processed using OPD Analysis and SI for any number of phytoplankton species can be generated.

Multiple regression analysis simultaneously finds the least squares best fit of a set of known taxa absorbance fourth derivative spectra to the unknown sample, bulk water absorbance fourth derivative spectrum obtained by the OPD.[42,47] The known taxa absorbance fourth derivative spectra are quantified by the concentration of chlorophyll *a* in the culture samples used to create the known spectra. The resultant best fit taxa are estimates of the chlorophyll *a* biomasses of the dominant phytoplankton taxa in the sampled phytoplankton community. Typically, three to five taxa (of 20 included in the known set) are selected by this fitting process to represent the community structure.

Absorption by CDOM is determined using the standard approach where the natural logarithm of the ratio of the light transmission through the CDOM containing water sample to the light transmission through "pure" water is scaled by the optical path length of the water-containing cell. CDOM absorption is an exponential function of wavelength and can be expressed as follows:

$$a_{CDOM}(\lambda) = a_{CDOM}(\lambda_S) * e^{-S*(\lambda-\lambda_s)} \tag{1}$$

where $a_{CDOM}(\lambda_S)$ is the absorption value at a "standard" wavelength (λ_S), typically 400 or 440 nm), and S is the exponential slope of the CDOM absorption spectrum at λ_S. By accepting the standard form of the CDOM absorption spectrum (Eqn (1)), it is possible to completely describe a CDOM absorption spectrum by reporting just $a_{CDOM}(\lambda_S)$ and S. To determine those two parameters for any CDOM absorption spectrum, first, the spectral absorption values are transformed by the natural log (ln), and then a least squares linear regression is fit to the transformed absorption values over the wavelength range from 380 to 500 nm. The best fit linear coefficients, intercept and slope, then represent $\log_e(a_{CDOM}(\lambda_S))$ and S, respectively. During every sample cycle, the CDOM absorption is calculated.

4. SYSTEMS LEVEL INTEGRATION

The OPD is a system of fluidic, optical, and computational systems that obtains a water sample, illuminates the sample with a calibrated light source, and measures the transmission spectrum through the water sample (Figure 3). There are two

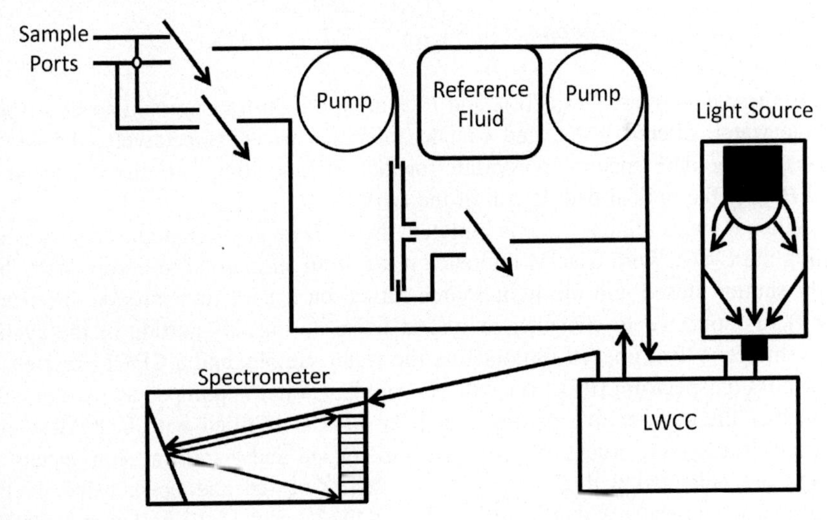

FIGURE 3

Schematic of major OPD components including the fluidic pathways.

standard OPD models, a low-pressure (LP) version rated for operation in less than 20 m of water, and a high-pressure (HP) version rated for use up to 100 m depth. Samples are drawn into the device through two intake/exit ports to enable continuous sampling and port flushing. Samples are initially filtered through coarse stainless-steel filters to inhibit clogging of the optical system. Fluids are drawn into the device by either peristaltic pumps (LP) or a custom syringe pump (HP) and routed through a tubing path controlled by pinch valves. Samples are initially pumped through a 0.2 μm cross-flow membrane filter, which is composed of a bundle of cylindrical filter elements sealed in a cylindrical housing with fluid ports on both the ends of sides of the cartridge. This enables access to filtered water pumped into the end of the filter and out of the side ports, and to whole water flushed directly through the cylindrical membranes. Those samples are routed into an LWCC. In the LWCC, an optical fiber inserts into a hollow, coated, quartz flow cell that allows a wavelength-calibrated light source (Heraeus GmbH) to illuminate the sample fluid in the LWCC. The spectrum (350–800 nm) of light transmitted through the fluid is measured by a miniature fiberoptic spectrometer (Ocean Optics, Inc.). During the sampling cycle, three water types are measured, including whole water (ambient particulates and CDOM), filtered water (CDOM only), and CDOM reference water (chosen to represent pure water). Dark spectra are measured with the shutter of the light source closed to quantify system noise. The light transmission spectra from these samples are used to calculate particle absorbance (Eqn (2)) and CDOM absorption (Eqn (3)) spectra.

$$A(\lambda)_{particle} = -\log\left(\frac{I(\lambda)_w - D(\lambda)_w}{I(\lambda)_f - D(\lambda)_f}\right) \tag{2}$$

$$a(\lambda)_{CDOM} = -\ln\left(\frac{I(\lambda)_f - D(\lambda)_f}{I(\lambda)_r - D(\lambda)_r}\right) \Big/ PL_{lw} \tag{3}$$

In these equations, $I(\lambda)_w$, $I(\lambda)_f$, and $I(\lambda)_r$ are the light transmission spectra from whole water, filtered water, and CDOM reference water, respectively; the $D(\lambda)_x$ terms are the dark spectra appropriate for the corresponding transmission spectra, and PL_{lw} is the optical path length of the LWCC.

When a measurement cycle is initiated, the OPD begins its sample sequence by filling the LWCC with CDOM reference water from an internal reservoir. With the light shutter closed and the light source turned on a dark transmission spectrum ($D(\lambda)_r$) is collected to characterize system noise during this portion of the cycle. The shutter is then opened illuminating the water sample, and a CDOM reference transmission spectrum ($I(\lambda)_r$) is captured. Ambient water is pumped across the hollow fiber filter to remove particles, and the LWCC is filled with CDOM-laden, particle-free sample water. A dark spectrum ($D(\lambda)_f$) and a transmission spectrum ($I(\lambda)_f$) are collected with the light-source shutter closed and open, respectively. Ambient whole water is then pumped through the interior of the hollow membrane filter tubes without being filtered, filling the LWCC with the particle-laden sample, and a whole water transmission spectrum ($I(\lambda)_w$) is collected. For this sample, a dark spectrum is not collected because it follows immediately after the previous filtered (CDOM) sample. Calculation of SI and CDOM absorption spectra are completed, and results are stored and transmitted as specified by the user. If the OPD is configured to continuously cycle, only the portion of cycle that comes after the CDOM reference is repeated. The CDOM reference cycle is repeated on an adjustable schedule, but usually every 8 to 10 cycles to account for the development of fouling in the LWCC, changes in the light source spectrum, and drift in the spectrometer. Additionally, if there will be a delay before the next cycle, a small volume of CDOM reference water is pumped into the LWCC to displace fouling organisms and compounds and to inhibit growth.

Computational, electronic, fluidic, and optical systems are controlled by a low-power Persistor Instruments, Inc. CF2 microcontroller. This processor handles all data management, processing, and communications capabilities of the OPD. Communications with host systems, which can include external communications systems (LP units) or independent control and communications systems (HP units on AUVs), are handled by RS232 serial standard protocol.

The Optical Phytoplankton Detector provides a relatively low-cost way to monitor phytoplankton community structure and CDOM. On a systems level, an LP OPD capable of operating unattended for approximately one month has a purchase price of under $30,000. Operating costs are on the order of $1000 per deployment, including costs of operating marine vessels to deploy and recover units. A single trained operator is capable of maintaining approximately four instruments. An HP OPD for deployment on an AUV has a purchase cost on the order of $40,000, in addition to the cost of the AUV, and it can run with minimal user intervention on the order of two weeks. In comparison, shipboard survey work costs on the order of $10,000–20,000 per day for vessel operations and personnel, plus the cost of scientific staff on board, and sample processing costs once grab samples are returned to shore. For the purpose of regional HAB monitoring, the OPD is capable of identifying domains of interest for shipboard surveys without the cost associated with large-scale surveys.

5. APPLICATIONS

The OPD is a modular instrument, in that components can be adapted for different water types with varying optical transmission properties and phytoplankton concentrations. In offshore environments, namely oligotrophic waters, planktonic and CDOM concentrations can be extremely low, requiring an increase in instrument sensitivity. This can be achieved by increasing the sample volume and path length, for planktonic and CDOM detection, respectively, to result in a measurable change in absorbance. Similarly, in extremely turbid waters, absorbance by CDOM in the water can extinguish the characteristic transmission spectra of planktonic cells contained within the sample volume, necessitating a shortened waveguide.

FIGURE 4

OPD deployment modes.

The OPD is capable of being implemented in different configurations for different measurement platforms (Figure 4). Typical applications include mounting an LP instrument to a dock piling in an inlet or to the submerged subframe of a buoy deployed offshore. LP instruments are also connected to shipboard flow-through systems during oceanographic cruises for continuous measurement while underway. Depending on the available communications and electrical infrastructure of the deployment, LP units transmit their measurements to a central database in real time directly via TCP/IP network connection, network enabled high-frequency radio, cellular, or satellite phone (Figure 5). High-pressure variants of the OPD are designed for full system integration in autonomous underwater vehicles (AUVs) in operation up to 100 m. By utilizing AUVs, phytoplankton community structure and CDOM can be measured along track, throughout the operating depth range of the AUV. HP OPDs have been developed for the Slocum glider AUV (Teledyne-Webb Research Corporation) and the REMUS 100 AUV (Hydroid). In the Slocum glider, the OPD is integrated into the existent science bay, and it is controlled by the computer that regulates the collection of onboard scientific data. A subset of OPD data is returned every time the glider surfaces, allowing for near real-time monitoring of phytoplankton community structure over the range of depths that the glider undulates. Full spectral data is accessible when the Glider is recovered. In the REMUS AUV, the OPD is integrated as a custom modular hull section. Data is available once the AUV is recovered. The OPD has also been integrated into the Bottom-Stationed Ocean Profiler (BSOP) vehicle. The BSOP is a buoyancy-controlled underwater platform with a 200 m operating depth limit that spends the

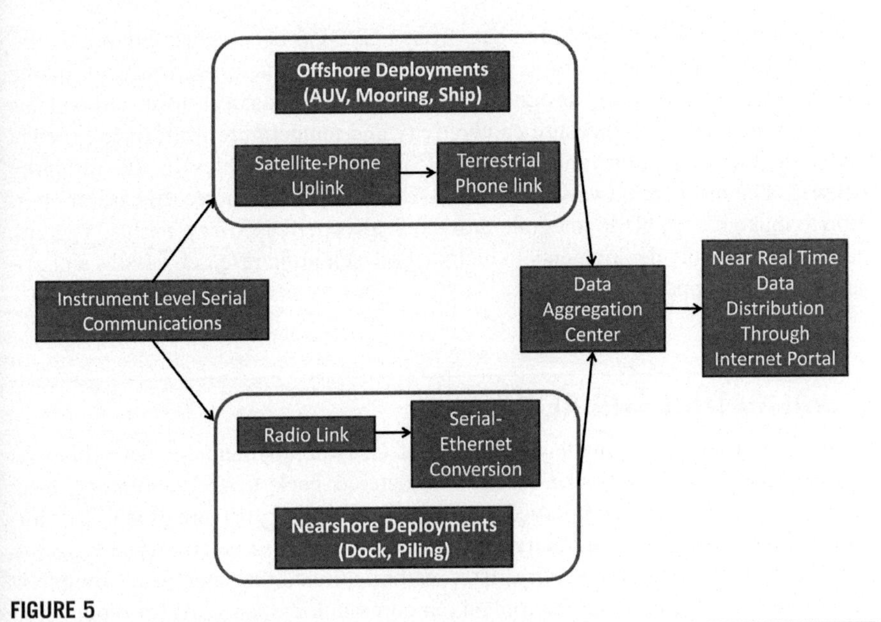

FIGURE 5

OPD data telemetry schemes.

majority of its time on the sea floor. It travels vertically through the water column generating no horizontal speed of its own in an attempt to limit horizontal displacement without using ground tackle. During a sampling cycle, it rapidly ascends and descends to minimize horizontal displacement by water currents. Ascent and descent rates are adjustable to enable sampling by sensors with low cycling rate. A benchtop variant of the OPD is available for manual or automated operation in the laboratory or on board ships. The water-resistant benchtop version has the same components and functionality as the in-water versions, but with a few additional features including a laptop computer for human interface, a GPS interface for recording position, and additional inlet ports to accommodate several cleaning solutions.

Utilizing the core design of the OPD, a spectrophotometric device named the CDOM Mapper has emerged during the development of the OPD. CDOM mapper is optimized to provide CDOM measurements with high accuracy and sensitivity in a variety of water masses during long-term unattended operations. Similar to the benchtop OPD, the CDOM Mapper includes fluid stores and plumbing pathways implemented so that it is able to manage fouling by periodically pumping cleaning and rinsing solutions through optical pathways. CDOM sensitivity can be increased by implementing an LWCC with a longer path length. The LWCCs in these instruments are readily replaceable, making it relatively easy for a user to select a path length to suit their changing needs. Working in estuarine waters might require an LWCC path length of 25 cm, whereas work in open ocean waters might call for a

1-m or longer LWCC. Geolocation data from a GPS has been integrated into the instrument for continuous position information in time and space. Variants of the CDOM mapper have been designed that incorporate updated sensor components to prevent obsolesce, to measure conductivity and temperature for refractive index modeling and correction, and to prevent and remove bubbles in the sampling pathway. The most recent variant of the CDOM mapper has increased ease of operation to make it possible to integrate into the SeaKeeper Discovery Yachts Program, increasing available deployments to include both scientific research vessels and private vessels of opportunity.

6. VALIDATION AND RESULTS

Because of the remote, unattended nature of OPD deployments, it was often not feasible to directly verify the results telemetered back to the laboratory. Data from several projects that employed the OPD (Table History1) were used in the validation of the method. To accurately conduct a validation exercise, it was necessary to assure that the OPD and the comparison method utilized the same water sample at very close to the same time. Phytoplankton communities, especially at bloom concentrations, can be spatially very patchy. Photo-acclimation of photopigments can change pigment compliments and the resulting absorbance signature on minute time scales. Fortunately, there have been studies, both laboratory and field, that conducted simultaneous sampling and processing. Additionally, comparison methods are subject to their own inaccuracies, making it necessary to place caveats on validation results. For instance, the use of optical microscope enumeration of phytoplankton taxonomic classes has been the de facto standard method for many years. Although the optical microscope is a powerful tool when used by skilled taxonomists, it is very time consuming and problematic when identifying very small cells. Because *Karenia* sp. cells are large and very distinctly shaped, optical microscope enumeration provides very accurate data for comparison to the OPD estimations of *Karenia* sp. However, the diatom class, for instance, includes a wide range of species with varying sizes and shapes. Some are large and uniquely shaped, making practical the use of microscopic enumeration for numerous samples. Conversely, very small-sized diatom species with nondescript shapes (at optical microscope resolution) are difficult to enumerate accurately by optical microscopes in large numbers of samples. The upshot of these issues in optical microscope enumeration is that there were no complete taxonomic enumerations of community structure to use in validation of the OPD community structure estimates. There are molecular techniques for identification of taxonomic groups, but few simultaneously provide comprehensive coverage of all the possible groups. Chemotaxonomic classification of class-level taxonomy, utilizing high-performance liquid chromatography (HPLC), is a widely accepted approach to dealing with relatively large numbers of samples and for including the full complement of taxonomic classes in natural water samples.

The MERHAB Gulf of Mexico HAB Sentinel project conducted cruises off the southwest coast of Florida from 2003 thru 2008. During those cruises, several methods of taxonomic classification of phytoplankton (with emphasis on *Karenia* sp.) were compared. All of the methods mentioned previously were applied to the natural phytoplankton communities present in each of the water samples collected during the cruises. Special attention was paid to utilizing aliquots from the same, well-mixed containers for all four classification methods. In addition, the time between sample collection from the water column and onboard analysis or preservation were kept as short as possible and as similar as possible between methods. Data from those MERHAB cruises are used here to provide validation of the OPD taxonomic classification method. Haywood et al.[48] did collect molecular probe-based estimates of *Karenia* sp. abundance simultaneously with OPD operation. However, because their results were *Karenia* sp. specific and, thus, did not include quantitation of other taxonomic groups, they were not included in this chapter. Chemotaxonomic classification was conducted on all of the samples, and those results are used here for comparison to OPD results. Several caveats are necessary before the validation results are presented. First, *Karenia* sp. was the taxonomic group that received the lion's share of the calibration effort over the development of the OPD. Second, *Karenia* sp. is a group of species in the same genus, and they are very similar to each other relative to similarity between genera and classes. Third, the taxonomic class "standards" library of absorbance spectra was made up from individual species of the class and does not represent all species in that class. Development of a more inclusive standards library will improve on the results presented here. Figure 6 presents a comparison of the taxonomic classification results between the OPD and the chemotaxonomic method (ChemTax) using HPLC-derived photopigments. The three most common taxonomic groups present were included because OPD estimates of minor taxonomic groups in any sample are highly variable. The F test was used to test if the slope of the linear regression was different from zero for each of taxonomic groups and depths. All of the tests showed that the slope was significantly different from zero ($p < 0.0001$). The slopes of the relationships between the estimates of chlorophyll *a* concentration from the OPD and those from ChemTax for each taxon and depth are not equal to the value one. This indicates the two methods do not agree. Furthermore, the disagreement is different for each taxon. There are a number of reasons for this disagreement. The first is that both methods are based on numerical estimation models, not direct quantification. Another factor that separates the two techniques is that the OPD concentrates particles while it filters water for the "filtered" (CDOM) measurement immediately before the particle absorbance measurement. Setup for the particle absorbance measurement flows water through the hollow filter elements and, in doing so, sweeps a concentrated collection of particles out of the filter and into the LWCC. This concentrating mechanism works differently for different taxa, so the rate at which the different taxa reach the LWCC for measurement varies based on the particle size and "stickiness." Different levels of taxon concentration were detected during the brief measurement period done at a fixed time following the

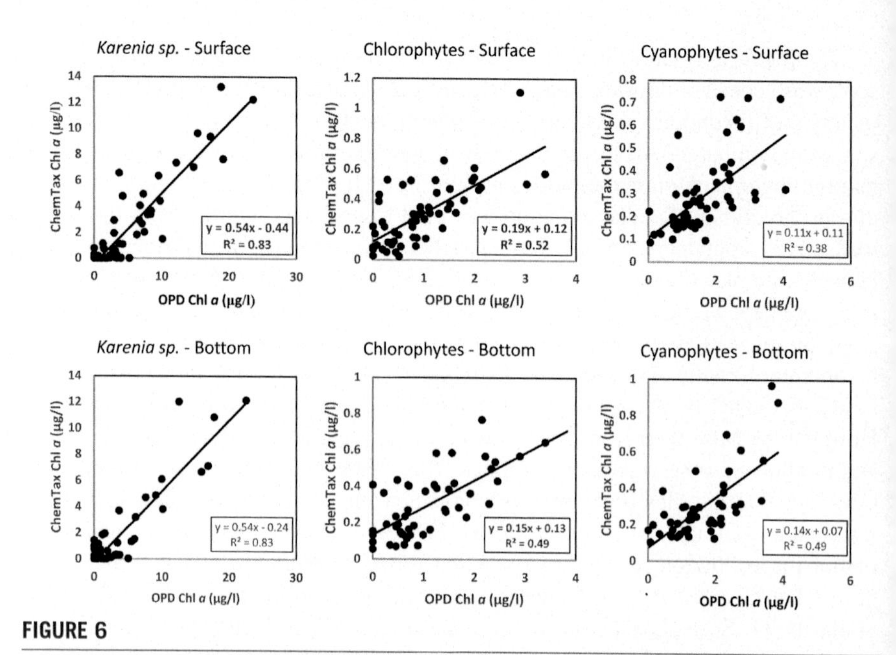

FIGURE 6

ChemTax versus OPD taxon-specific chlorophyll *a* biomass. Samples for each method were collected simultaneously from the same aliquots of water sampled from 18 to 22 September, 2006, during a MERHAB cruise off the southeast coast of Florida. Note that the slopes of the regression lines are statistically different from zero, and they are not equal to the value one (see text).

initiation of water flow through the filter. This concentrating action can be compensated for if a second particle absorbance measurement (a repeat) is made after sufficient time to flush the concentrated particles through the LWCC. The particle concentration at that time should be very close to the ambient particle concentration. If a taxon is present in a high enough abundance in the ambient water to be detected without the concentrating mechanism, then repeat values from several cycles can be used to determine the relationship between concentrated and nonconcentrated absorbance measurements. That relationship can be used to scale down the concentrated particle absorbance measurement. This was not done for the data presented here. An additional factor influencing the comparison of the methods involves the difference between species selected for calibration of the OPD and ChemTax to represent the taxonomic classes. All of the differences and their impacts on the different measurement techniques are too extensive to discuss in detail here. In brief, the representative species are different, and the growth conditions for the calibration efforts were probably different from those of the natural populations tested here, leading to different physiological photo-acclimation states between calibration and field sample species. This could lead to different ratios of accessory pigments between the representative species in the ChemTax calibration relative to those found in the test samples, leading to some over- or underestimation of that taxonomic group in the estimated community structure. The same discrepancy between photo-acclimation state of calibration and field sample species would impact the OPD's match of absorbance spectra, but in addition, the size difference between species used to represent taxonomic classes during calibration relative to those in the field samples would impact the pigment packaging effect[36] influence on absorbance spectra. Improvements in the performance of both taxonomic estimation techniques would result from additional calibration work that would include more species representatives for each taxonomic class and include several different photo-acclimation states.

Because AUVs by nature operate in a manner that obscures their exact sampling locations from real-time observation by humans, collecting verification samples during OPD-equipped AUV operations was impractical. Therefore, these results rely on prior validation to support the taxonomic classifications estimates. An OPD-equipped Slocum glider mission conducted in the fall of 2004 is presented here to illustrate the capability of the OPD to provide phytoplankton community structure in four dimensions. Figure 7 presents taxonomic distribution on two dimensions from one leg of four legs across the inner West Florida continental shelf (WFS).

The measurement of in situ CDOM in marine environments by the OPD and CDOM Mapper is complicated by systematic issues. In utilizing an LWCC, mismatch between the refractive index (RI) of the capillary tube and the fluid that it contains induces apparent absorbance without any change in CDOM. Because salinity has a nonlinear impact on the RI of water, calibration of CDOM measurements requires measurement of the salinity of the sampled water and a mathematical model of the nonlinear change in RI due to salinity. Additionally, CDOM absorbance measured utilizing an LWCC varies with the difference in RI between filtered sample water and CDOM reference solution. Unless there is an exact match between the salinity of the reference solution and that of the sample water, the difference in RI will induce an apparent gain or drop in absorption that may require correction.

In the best case scenario, that being an exact match in salinity between the CDOM reference solution and the sample fluid, there is a linear difference between CDOM absorbance measurements (A440) obtained by the OPD and a PerkinElmer Lambda 650 UV spectrophotometer, across a range of CDOM concentrations. In experiments with increasing sample salinity, a nonlinear difference between the CDOM intercept measured by the benchtop spectrometer and the OPD can be seen. Because the offset between the OPD and the benchtop spectrophotometer has a linear difference, a linear correction factor can be implemented to maximize the agreement with the calibrated laboratory instrument (Figure 8).

Low-pressure OPDs on fixed platforms measure CDOM every time a phytoplankton discrimination sample is processed. This is typically on a 2-h sampling period. The OPDs that are located on pilings in estuaries have shown tidally driven variation in CDOM absorption, both the absorption value at 440 nm (Figure 9) and the exponential slope of the absorption spectrum.

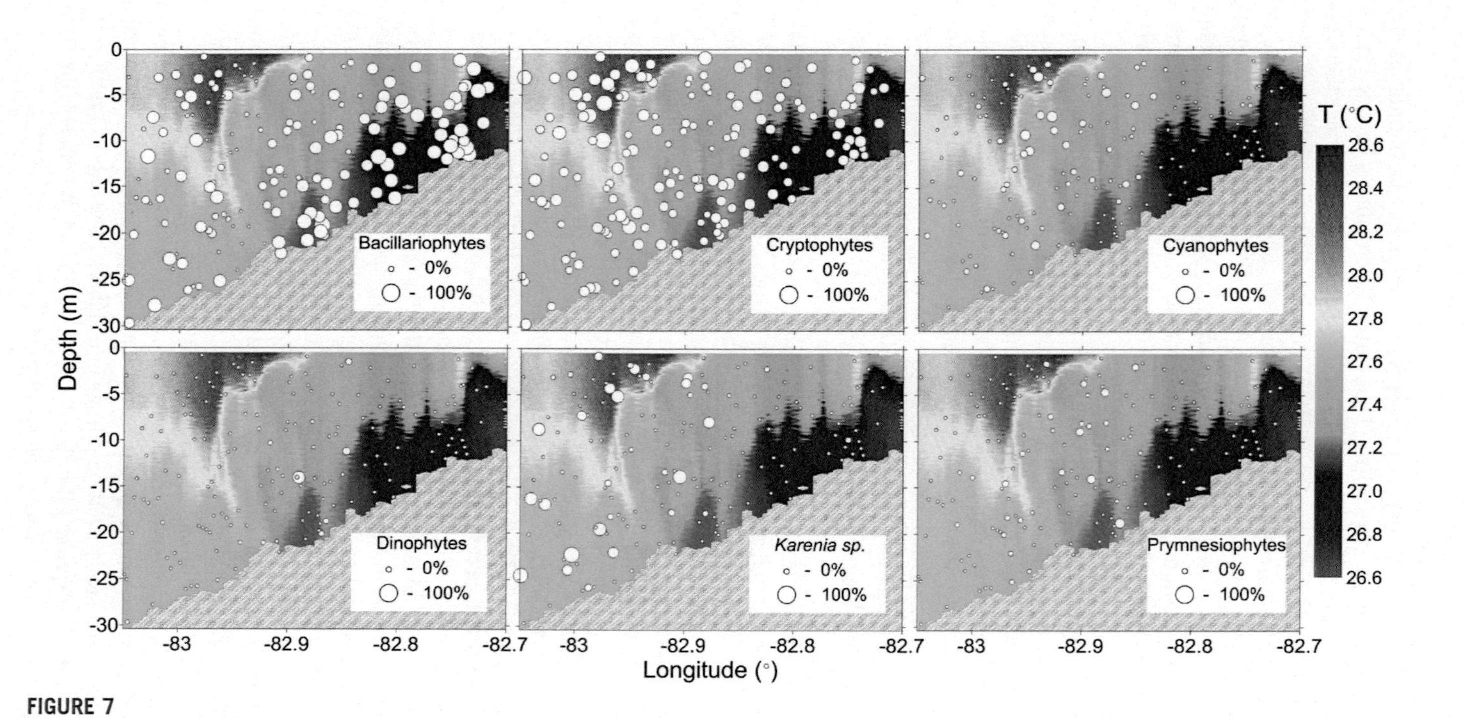

FIGURE 7

Two-dimensional distribution of each taxon percentage of the phytoplankton community structure determined by an OPD-equipped Slocum glider, 28–30 September, 2004. Background colors represent water temperature measured by the glider CTD.

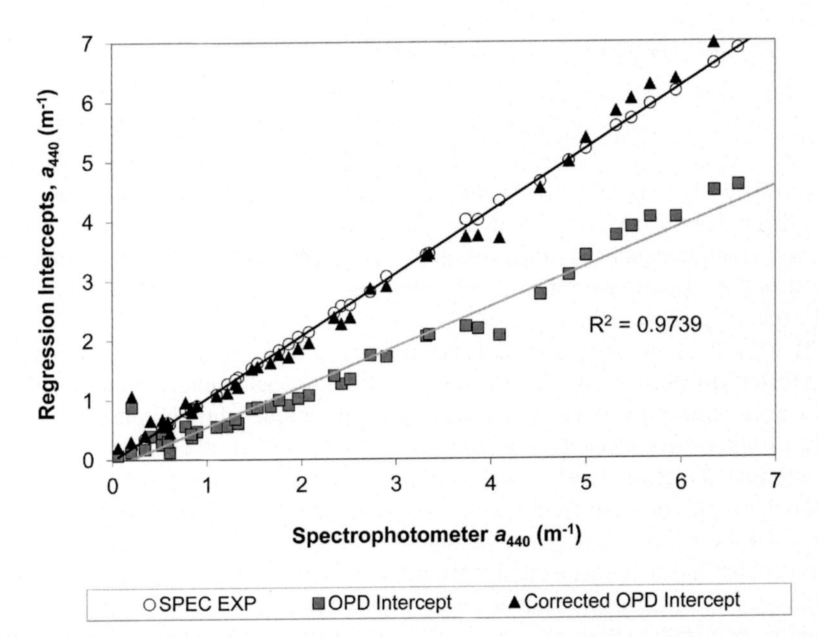

FIGURE 8

Comparison of CDOM measurements from a benchtop spectrophotometer and the OPD. The OPD consistently underestimated (accepting that the benchtop spec was the standard method for this measurement) the CDOM. A linear correction term was applied to the OPD values to yield a good matchup with the spectrophotometer values.

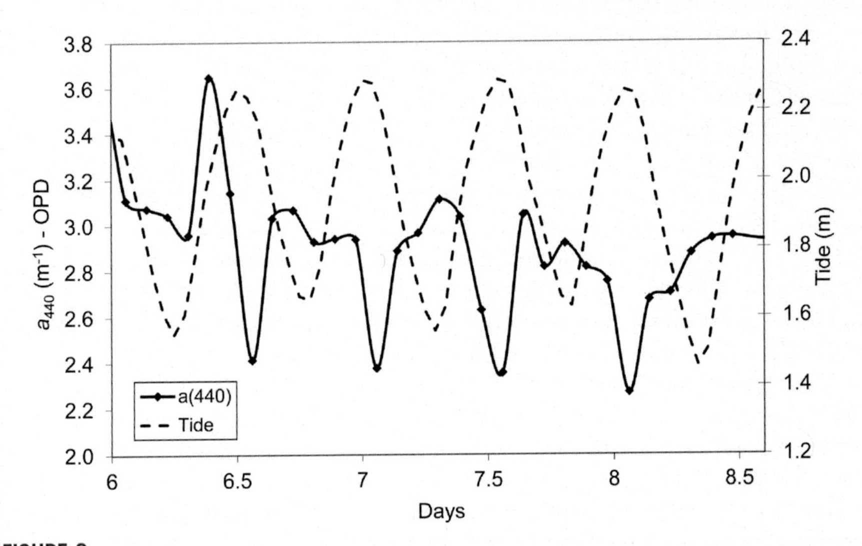

FIGURE 9

OPD CDOM absorption measurements relative to tidal water level variation in Tarpon Bay, Florida (Sanibel-Captive Conservation Foundation Marine Laboratory dock). Note that the two parameters are highly correlated, but confounded by the strong influence of the nearby mouth of the Caloosahatchee River.

7. FUTURE DEVELOPMENT/PLANS

The OPD has been developed for over a decade and has been implemented across a broad geographic range including North America, from the Eastern Pacific to the Western Atlantic, around the Gulf of Mexico, and in the Great Lakes. An OPD has made its way across the Mediterranean, around the Arctic, and there have been deployments in Mexico. With this experience, there are several developmental pathways that will enhance future development. Improvements fall into the domains of expanding species identification capabilities, enhancing CDOM measurement, and expanding the user base.

To identify plankton species and allocate species to a community structure using the OPD, characteristic fourth derivative spectra must be maintained on file. These spectra are obtained by running isolated plankton samples through the OPD. Typically, plankton are collected during bloom conditions, or they are isolated and grown in culture to detectable levels. A rigorous effort to generate absorption spectra from cultured samples must be conducted and verified by testing plankton from both culture and wild blooms. Additional validation of OPD results against known samples of mixed cultures and against ambient samples determined through more complete molecular and microscopic methods is desirable. The present library of species files would benefit from including phytoplankton from different geographic regions, and a database of libraries would enable researchers and operators of monitoring stations to access and share species files.

The OPD has demonstrated success in responding to naturally varying levels of CDOM in the natural sampling environment. For OPD to provide CDOM measurements as absorption coefficients (m^{-1}) for research purposes, remote sensing validation, and generating hybrid in situ remote sensing products, a thorough validation and calibration of CDOM measurement must be demonstrated. Simultaneous analyses of CDOM via OPD and benchtop spectroscopy would be performed on estuarine samples that cover a range of both CDOM and salinity values, as well as a matrix of constructed, fixed CDOM-varying salinity samples. The resulting dataset would allow an algorithm to correct measured absorbance with actual CDOM absorption coefficients by accounting for variations in refractive index between sample and reference salinity. The successful algorithm to compensate CDOM absorbance for changing salinity would motivate that integration of a conductivity cell within the OPD, similar to that designed as part of CDOM mapper, as well as the integration of this algorithm into the automated sampling sequence.

To integrate additional sensors, such as a conductivity cell and thermistor into the OPD, as well as to maintain compatibility with the upcoming generations of ocean sampling platforms, it will eventually be desirable to convert the OPD hardware from the Persistor CF2 processor to an ARM Linux architecture. This would enable a more flexible computing infrastructure and additional computing power to calculate plankton community structure and chlorophyll *a* biomass contributions in real time. It has already been determined that the Slocum Glider will make a similar transition from the Persistor series of processors to the more flexible Linux architecture.

In the Gulf of Mexico, the Gulf of Mexico Coastal Ocean Observation System (GCOOS) currently supports a number of OPD installations in the eastern Gulf. In addition, GCOOS is currently seeking to expand observing systems including, among other systems, an extended harmful algal bloom monitoring system, of which the OPD could be an integral component. The Sarasota Operations of the Coastal Ocean Observation Laboratories (SO-COOL) model of combining fixed and mobile HAB monitoring instruments with real-time data telemetry and distribution to end users can easily be extend to accommodate additional sampling sites. An effective model would be an instrument exchange program that would support sites around the Gulf, maintaining continuous OPD operations with minimal instrument/site downtime and minimal replication of technical expertise.

REFERENCES

1. *U.S. IOOS summit report: a new decade for the integrated ocean observing system.* Interagency Ocean Observation Committee; Copyright © 2013.
2. U.S. Commission on Ocean Policy. *An ocean blueprint for the 21st century.* 2004. Final Report. Washington, DC. ISBN: 09759462-X.
3. Ocean.US. *The first U.S. integrated ocean observing system (IOOS) development plan.* No. 9. The National Office for Integrated and Sustained Ocean Observations; 2006.
4. Nelson NB, Siegel DA. The global distribution and dynamics of chromophoric dissolved organic matter. *Ann Rev Mar Sci* 2013;**5**:447–76. http://dx.doi.org/10.1146/annurev-marine-120710-100751.
5. Clark CD, Hiscock WT, Millero FJ, Hitchcock G, Brand L, Miller WL, et al. CDOM distribution and CO_2 production on the Southwest Florida Shelf. *Mar Chem* 2004;**89**. http://dx.doi.org/10.1016/j.marchem.2004.02.011.
6. Hill V. Impacts of chromophoric dissolved organic material on surface ocean heating in the Chukchi Sea. *J Geophys Res* 2008;**113**(C7). http://dx.doi.org/10.1029/2007JC004119.
7. Sosik HM, Olson RJ. Automated taxonomic classification of phytoplankton sampled with imaging-in-flow cytometry. *Limnol Oceanogr Methods* 2007;**5**. http://dx.doi.org/10.4319/lom.2007.5.204.
8. Campbell L, Olson RJ, Sosik HM, Abraham A, Henrichs DW, Hyatt CJ, et al. First harmful *Dinophysis* (Dinophyceae, Dinophysiales) bloom in the U.S. is revealed by automated imaging flow cytometry. *J Phycol* 2010;**46**. http://dx.doi.org/10.1111/j.1529-8817.2009.00791.x.
9. See JH, Campbell JL, Richardson T, Pickney JL, Shen R, Guinasso NL. Combining new technologies for determination of phytoplankton community structure in the Northern Gulf of Mexico. *J Phycol* 2005;**41**. http://dx.doi.org/10.1111/j.1529-8817.2005.04132.x.
10. Buskey EJ, Hyatt CJ. Use if the FlowCAM® for semi-automated recognition of red tide cells (*Karenia brevis*) in natural plankton samples. *Harmful Algae* 2006;**5**(6). http://dx.doi.org/10.1016/j.hal.2006.02.003.
11. Greenfield D, Marin III R, Doucette GJ, Mikulski C, Jones K, Jenson S, et al. Field applications of the second generation environmental sample processor (ESP) for remote detection of harmful algae: 2006–2007. *Limnol Oceanogr Methods* 2008;**6**. http://dx.doi.org/10.4319/lom.2008.6.667.

12. Scholin C, Doucette G, Jensen S, Roman B, Pargett D, Marin III R, et al. Remote detection of marine microbes, small invertebrates, harmful algae, and biotoxins using the environmental sample processor (ESP). *Oceanography* 2009;**22**. http://dx.doi.org/10.5670/oceanog.2009.46.
13. Moore TS, Dowell MD, Franz BA. Detection of coccolithophore blooms in ocean color satellite imagery: a generalized approach for use with multiple sensors. *Remote Sens Environ* 2012;**117**. http://dx.doi.org/10.1016/j.rse.2011.10.001.
14. Tomlinson MC, Stumpf RP, Ransibrahmanakul V, Truby EW, Kirkpatrick GJ, Pederson BA, et al. Evaluation of the use of SeaWiFS imagery for detecting *Karenia brevis* harmful algal blooms in the eastern Gulf of Mexico. *Remote Sens Environ* 2004;**91**. http://dx.doi.org/10.1016/j.hal.2010.02.002.
15. Zhao J, Hu C, Lenes JM, Weisberg RH, Lembke C, English D, et al. Three-dimensional structure of a *Karenia brevis* bloom: observations from gliders, satellites, and field measurements. *Harmful Algae* 2013;**29**. http://dx.doi.org/10.1016/j.hal.2013.07.004.
16. Steidinger KA. Historical perspective on *Karenia brevis* red tide research in the Gulf of Mexico. *Harmful Algae* 2009;**8**. http://dx.doi.org/10.1016/j.hal.2008.11.009.
17. Landsberg JH, Flewelling LJ, Naar J. *Karenia brevis* red tides, brevetoxins in the food web, and impacts on natural resources: decadal advancements. *Harmful Algae* 2009;**8**. http://dx.doi.org/10.1016/j.hal.2008.11.010.
18. Kirkpatrick B, Fleming LE, Squicciarini D, Backer LC, Clark R, Abraham W, et al. Literature review of Florida red tide: implications for human health. *Harmful Algae* 2004;**3**. http://dx.doi.org/10.1016/j.hal.2003.08.005.
19. Hoagland P, Jin D, Polansky L, Kirkpatrick B, Kirkpatrick G, Fleming L, et al. The costs of respiratory illnesses arising from Florida gulf coast *Karenia brevis* blooms. *Environ Health Perspect* 2009;**117**(8). http://dx.doi.org/10.1289/ehp.0900645.
20. HARRNESS. In: Ramsdell JS, Anderson DM, Glibert PM, editors. *Harmful algal research and response: a national environmental science strategy 2005–2015*. Washington DC: Ecological Society of America; 2005. p. 96.
21. Special issue. *Opt Limnol Oceanogr* 1989;**34**(8)
22. *J Geophys Res* 1995;**100**(C7).
23. Faust MA, Norris KH. In vivo spectrophotometric analysis of photosynthetic pigments in natural populations of phytoplankton. *Limnol Oceanogr* 1985;**30**(6). http://dx.doi.org/10.4319/lo.1985.30.6.1316.
24. Hoepffner N, Sathyendranath S. Determination of the major groups of phytoplankton pigments from the absorption spectra of total particulate matter. *J Geophys Res* 1993;**98**(C12). http://dx.doi.org/10.1029/93JC01273.
25. Johnsen G, Samset O, Granskog L, Sakshaug E. In vivo absorption characteristics in 10 classes of bloom-forming phytoplankton: taxonomic characteristics and responses to photoadaptation by means of discriminant and HPLC analysis. *Mar Ecol Prog Ser* 1994;**105**. http://dx.doi.org/10.3354/meps105149.
26. Kirkpatrick GJ, Orrico C, Moline MA, Oliver M, Schofield OM. Continuous hyperspectral absorption measurements of colored dissolved organic material in aquatic systems. *App Opt* 2003;**42**(33). http://dx.doi.org/10.1364/AO.42.006564.
27. Siegel DA, Maritorena S, Nelson NB, Behrenfeld MJ, McClain CR. Colored dissolved organic matter and its influence on the satellite-based characterization of the ocean biosphere. *Geophys Res Lett* 2005;**32**. http://dx.doi.org/10.1029/2005GL024310.
28. Reader HE, Miller WL. Effect of estimations of ultraviolet absorption spectra of chromophoric dissolved organic matter on the uncertainty of photochemical production calculations. *J Geophys Res* 2011;**116**. http://dx.doi.org/10.1029/2010JC006823.

29. Nelson NB, Siegel DA, Carlson CA, Swan CM, Smethie WM, Khatiwala S. Hydrography of chromophoric dissolved organic matter in the North Atlantic. *Deep Sea Res I* 2007;**54**. http://dx.doi.org/10.1016/j.dsr.2007.02.006.
30. Helms JR, Stubbins A, Ritchie JC, Minor EC, Kieber DJ, Mopper K. Absorption spectral slopes and slope ratios as indicators of molecular weight, source, and photobleaching of chromophoric dissolved organic matter. *Limnol Oceanogr* 2008;**53**. http://dx.doi.org/10.4319/lo.2008.53.3.0955.
31. Ohi N, Makinen CP, Mitchell R, Moisan TA. *Absorption and attenuation coefficients using the WET labs ac-s in the mid-Atlantic bight: field measurements and data analysis, wallops coastal ocean observation laboratory project. Document series. NASA/TM–2008–214157*, vol. 3. Greenbelt (MD): NASA; 2008. http://ntrs.nasa.gov/archive/nasa/casi.ntrs.nasa.gov/20080023441_2008022754.pdf.
32. Miller RL, Belz M, Del Castillo C, Trzaska R. Determining CDOM absorption spectra in diverse coastal environments using a multiple pathlength, liquid core waveguide system. *Cont Shelf Res* 2002;**22**. http://dx.doi.org/10.1016/S0278-4343(02)00009-2.
33. D'Sa EJ, Steward RG, Vodacek A, Blough NV, Phinney D. Determining optical absorption of colored dissolved organic matter in seawater with a liquid capillary waveguide. *Limnol Oceanogr* 1999;**44**. http://dx.doi.org/10.4319/lo.1999.44.4.1142.
34. Millie DF, Kirkpatrick GJ, Vinyard BT. Relating photosynthetic pigments and in vivo optical density spectra to irradiance for the Florida red-tide dinoflagellate *Gymnodinium breve*. *Mar Ecol Prog Ser* 1995;**120**. http://dx.doi.org/10.3354/meps120065.
35. Millie DF, Schofield OM, Kirkpatrick GJ, Johnsen G, Tester PA, Vinyard BT. Detection of harmful algal blooms using photopigments and absorption signatures: a case study of the Florida red-tide, *Gymnodinium breve*. *Limnol Oceanogr* 1997;**42**. http://dx.doi.org/10.4319/lo.1997.42.5_part_2.1240.
36. Morel A, Bricaud A. Inherent optical properties of algal cells including picoplankton theoretical and experimental results. *Can Bull Fish Aquat Sci* 1986;**214**:521–60.
37. Butler WL, Hopkins DW. Higher derivative analysis of complex absorption spectra. *Photochem Photobiol* 1970;**12**. http://dx.doi.org/10.1111/j.1751-1097.1970.tb06076.x.
38. Bidigare RR, Morrow JH, Keifer DA. Derivative analysis of spectral absorption by photosynthetic pigments in the western Sargasso Sea. *J Mar Res* 1989;**47**. http://dx.doi.org/10.1357/002224089785076325.
39. Smith CM, Alberte RS. Characterization of in vivo absorption features of chlorophyte, phaeophyte, and rhodophyte algal species. *Mar Biol* 1994;**118**. http://dx.doi.org/10.1007/BF00350308.
40. Kirkpatrick GJ, Millie DF, Moline MA, Schofield O. Optical discrimination of a phytoplankton species in natural mixed populations. *Limnol Oceanogr* 2000;**45**(2). http://dx.doi.org/10.4319/lo.2000.45.2.0467.
41. Schofield O, Kerfoot J, Mahoney K, Moline M, Oliver M, Lohrenz S, et al. Vertical migration of the toxic dinoflagellate *Karenia brevis* and the impact on ocean optical properties. *J Geophys Res* 2006;**111**(C06009). http://dx.doi.org/10.1029/2005JC003115.
42. Kirkpatrick GJ, Millie DF, Moline MA, Lohrenz SE, Schofield OM. Phytoplankton community composition observed by autonomous underwater vehicle. In: Moestrup Ø, Tester PA, Enevoldsen H, editors. *Proceedings of the XIIth international conference on harmful algae*. Copenhagen: ISSHA and IOC UNESCO; 2008.
43. Bendis B, Fries D, Haywood A, Kirkpatrick G, Millie D, Orsi T, et al. Integrating autonomous data acquisition and forecasting into regional monitoring efforts; a synopsis of the MERHAB 2002: eastern Gulf of Mexico sentinel program. *J Phycol* 2004;**39**(s1). http://dx.doi.org/10.1111/j.0022-3646.2003.03906001_7.x.

44. Robbins IC, Kirkpatrick GJ, Blackwell SM, Hillier J, Knight CA, Moline MA. Improved monitoring of HABs using autonomous underwater vehicles (AUV). *Harmful Algae* 2006;**5**(6). http://dx.doi.org/10.1016/j.hal.2006.03.005.
45. Falkowski PG, Raven JA. *Aquatic photosynthesis*. 2nd ed. Princeton University Press; 2007. 484 pp.
46. Jeffrey SW, Mantoura RFC, Wright SW, editors. *Phytoplankton pigments in oceanography.* Paris: United Nations Educational, Scientific and Cultural Organization (UNESCO) Publishing; 1997.
47. Stæhr PA, Cullen JJ. Detection of *Karenia mikimotoi* by spectral absorption signatures. *J Plankton Res* 2003;**25**. http://dx.doi.org/10.1093/plankt/fbg083.
48. Haywood AJ, Scholin CA, Marin III R, Steidinger KA, Heil C, Ray J. Molecular detection of the brevetoxin-producing dinoflagellate *Karenia brevis* and closely related species using rRNA-targeted probes and a semiautomated sandwich hybridization assay. *J Phycol* 2007;**43**. http://dx.doi.org/10.1111/j.1529-8817.2007.00407.x.

CHAPTER

Observing System Impacts on Estimates of California Current Transport

19

Andrew M. Moore*, Christopher A. Edwards, Jerome Fiechter, Michael G. Jacox

Department of Ocean Sciences, University of California Santa Cruz, Santa Cruz, CA, USA

**Corresponding author: E-mail: ammoore@ucsc.edu*

CHAPTER OUTLINE

1. INTRODUCTION

Assessment of observing system impacts on state estimates and forecasts derived from data assimilation is routine practice in numerical weather prediction, and it has proved to be a valuable tool for monitoring the performance of global data assimilation systems and for identifying problems with specific platforms in the observing network. Similar practices are now starting to be used in ocean analysis systems, where they also provide valuable information about the efficacy of ocean state estimates and ocean observing systems. It is anticipated that in the near future the routine monitoring of observation impacts on ocean analyses and forecasts will form an integral component of activities in support of the US national coastal ocean observing systems. To date, most efforts to quantify observation impacts in oceanography have been based on traditional Observing System Experiments (OSEs) (e.g., Refs. 1–3). However, more recent efforts have included other novel techniques such as spectral analysis of the representer matrix,[4] quantification of the number of degrees of freedom of the observing system,[5] and observation footprints.[6] An extensive review of some very recent efforts to quantify the impact of observing systems on ocean state estimates and forecasts using a variety of methods can be found in Refs. 7,8.

In the work presented here, we have followed the method developed by Langland and Baker[9] that is commonly used in numerical weather prediction at several operational centers in which the impact of the observations on an ocean analysis or forecast can be readily computed from the output of the data assimilation system.[10–12] This technique has the tremendous advantage over more traditional OSEs in that only minimal additional computational effort is required, whereas in OSEs, the entire analysis must be recomputed when observations are withheld. For routine sequential data assimilation systems, the analysis minus background differences (i.e., the analysis increments) are often small enough that the linear assumption can be invoked to expand the increments directly in terms of the contributions associated with each observation. The linear map from state space into observation space by which this decomposition is achieved is given by the transpose (or adjoint) of the Kalman gain matrix derived during the analysis stage. Therefore, all that is required to compute the observation impacts is the a posteriori reconstruction of the gain matrix using archived information from the data assimilation system.

In this chapter, the method of Langland and Baker[9] has been applied to two sequences of historical ocean circulation estimates for the California Current System (CCS) spanning the last three decades. The model and data assimilation system used to compute the CCS analyses are described in Section 2. The focus of this work is on the transport of the CCS in the vicinity of the central California coast, and the method used to quantify the impact of the observing system on CCS transport is described in Section 3. Using this approach, two different aspects of the analysis system are illustrated here. In Section 4, we monitor the performance of the data assimilation system in terms of the contribution of the individual components of the control vector to the ocean circulation estimate, and in Section 5, the impact of each individual observing platform on the central California CCS transport is quantified. A summary of important findings is presented in Section 6.

2. HISTORICAL ANALYSES OF THE CALIFORNIA CURRENT SYSTEM

2.1 ROMS 4D-VAR

The model implemented in this study is the Regional Ocean Modeling System (ROMS). ROMS is a hydrostatic primitive equations model with orthogonal curvilinear coordinates in the horizontal and terrain following coordinates in the vertical making it ideally suited for modeling coastal and near shore circulation environments.[13] In addition, ROMS supports an impressive suite of different numeric algorithms, mixing schemes and open boundary conditions that provide the user with considerable flexibility.

The configuration of ROMS used here spans the west coast of North America from near the Canadian border to mid-way down the Baja Peninsula, as illustrated in Figure 1, which shows the model domain and the ocean bathymetry. The model resolution is 1/10° in the horizontal, and there are 42 terrain following coordinates in the vertical, which yields a vertical resolution ranging from ~0.3 to 8 m on the continental shelf and ~7 to 100 m in the deep ocean. Observational estimates based on satellite altimetry indicate that eddies range in size between ~40 and 350 km, with a mean size of ~165 km,[14] so the horizontal resolution used here is sufficient to resolve most of the CCS mesoscale eddy field.[15,16] Though higher resolution may be desirable, data assimilation imposes a heavy computational burden as described later in the chapter, so 1/10° resolution is a trade-off between computational effort and the degree of realism of the simulations. Nevertheless, despite the relatively low horizontal resolution, the model configuration used here (hereafter referred to as ROMS-CCS) yields a simulation of the CCS that compares favorably with observations as shown by Veneziani et al.[17] where further details of the model configuration can also be found.

ROMS also supports a suite of state-of-the-art four-dimensional variational (4D-Var) data assimilation algorithms.[18] Though a detailed description of the

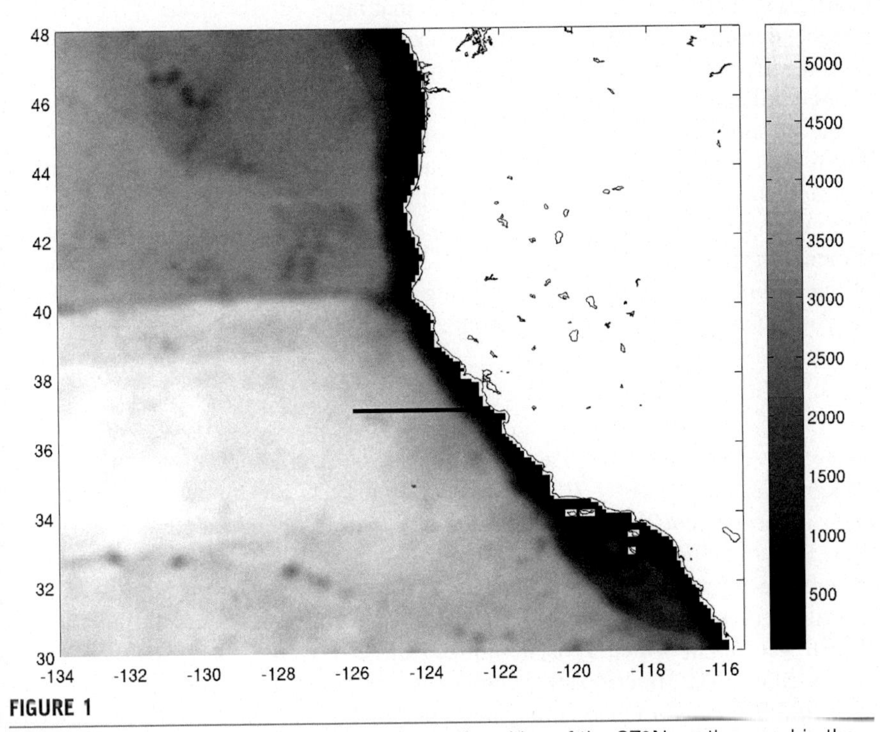

FIGURE 1

The ROMS domain showing the bathymetry and position of the 37°N section used in the observation impact calculations.

ROMS 4D-Var system is beyond the scope of this chapter, a brief description of the important features that are relevant to quantification of the observation impacts described in Section 3 will be given here. With this in mind, we will follow the standard notation established by Ide et al.[19] and expanded for more oceanographically relevant cases by Daget et al.[20]

Let $\mathbf{x}$ denote the state-vector of the ocean, namely all of the model grid point values of temperature, salinity, two components of velocity, and free surface displacement. The time evolution of the state vector over the time interval (t_i, t_{i+1}) can then be represented as $\mathbf{x}(t_{i+1}) = \mathcal{M}(\mathbf{x}(t_i), \mathbf{f}(t_i), \mathbf{b}(t_i))$ where $\mathbf{f}(t_i)$ and $\mathbf{b}(t_i)$ represent the surface forcing and open boundary conditions, respectively, over the interval (t_i, t_{i+1}), and $\mathcal{M}$ denotes the nonlinear model. The model solution $\mathbf{x}(t_{i+1})$ is uniquely determined by the initial condition $\mathbf{x}(t_i)$, ocean surface forcing, and open boundary conditions, which collectively are referred to as the control vector denoted by $\mathbf{z} = (\mathbf{x}^T(t_i), \mathbf{f}^T(t_i), \mathbf{b}^T(t_i))^T$. The goal of 4D-Var is to find the control vector $\mathbf{z}$ that yields the "best" ocean circulation estimate, $\mathbf{x_a}$. The best circulation estimate is that associated with the $\mathbf{z}$ that minimizes the cost function given by $J_{NL} = (\mathbf{z} - \mathbf{z_b})^T\mathbf{B}^{-1}(\mathbf{z} - \mathbf{z_b}) + (\mathbf{y}^o - H(\mathbf{z}))^T\mathbf{R}^{-1}(\mathbf{y}^o - H(\mathbf{z}))$, where $\mathbf{z_b}$ is a *prior* or background estimate of the control vector, $\mathbf{y}^o$ is the vector of observations, $\mathbf{B}$ and $\mathbf{R}$ are, respectively, the background error and observation error covariance matrices, and H is the observation operator that maps $\mathbf{z}$ to the space−time observation locations. In the case of 4D-Var, the operator H includes $\mathcal{M}$, and information is dynamically interpolated in time via the model dynamics.

According to Bayes' theorem, the cost function J_{NL} corresponds to the logarithm of the posterior probability distribution of $\mathbf{z}$, so identification of the minimum of J_{NL} is tantamount to identifying the maximum likelihood circulation estimate, corresponding to $\mathbf{z_a}$, given the prior control vector $\mathbf{z_b}$ and the observations $\mathbf{y}^o$.[21] The topology of J_{NL} may be very complicated, so in general, the minimum of J_{NL} is found using an iterative truncated Gauss-Newton method, which solves a sequence of linear minimization problems.[22] During each sequence of linear minimizations, the cost function that is actually minimized is given by the following:

$$J_k = \delta\mathbf{z}_k^T\mathbf{B}^{-1}\delta\mathbf{z}_k + (\mathbf{H}_{k-1}\delta\mathbf{z}_k - \mathbf{d}_{k-1})^T\mathbf{R}^{-1}(\mathbf{H}_{k-1}\delta\mathbf{z}_k - \mathbf{d}_{k-1}) \tag{1}$$

where $\delta\mathbf{z}_k = \sum_{i=1}^{k-1}\delta\mathbf{z}_i = \mathbf{z}_k - \mathbf{z_b}$ represents the departure of the kth iterate from the background, $\mathbf{d}_{k-1} = (\mathbf{y}^o - H(\mathbf{z}_{k-1}))$ is referred to as the innovation vector, and $\mathbf{H}_{k-1}$ is the tangent linearization of the observation operator H linearized about $\mathbf{z}_{k-1}$. Linearization of the minimization problem in this way is referred to as the incremental approach[23] and is used in ROMS 4D-Var. The primary workhorse algorithm that will be used here is the strong constraint, dual formulation of ROMS 4D-Var in which J_k is minimized directly in the space spanned by the observations. The minimization proceeds via a Lanczos formulation of the restricted $\mathbf{B}$-preconditioned conjugate gradient (CG) method.[24] Each sequence of linear minimizations of Eqn (1) proceeds via so-called inner-loops. When the $\delta\mathbf{z}_k$ that minimizes J_k has been identified, the estimate $\mathbf{z}_k$ about which $\mathbf{H}$ is linearized is updated (a so-called outer-loop), and minimization of Eqn (1) proceeds again. The inner- and outer-loops are continued until J_{NL} is reduced to an acceptable level or until the iterative procedure has converged to the point where further reductions in J_{NL} are negligible. In the 4D-Var calculations presented here, a single outer-loop and 15 inner-loops were used, so the subscript k will be dropped for convenience. This configuration represents a trade-off between the further error reduction afforded by multiple outer-loops and the considerable increase in computational cost when $k > 1$, and it is based on previous experience with the same system configuration.[18,25]

In most operational meteorological and oceanographic data assimilation systems, errors in the observations are assumed to be independent, in which case, **R** takes the form of a diagonal matrix. Though this is a reasonable assumption for independent in situ observations, it is certainly not the case for gridded and post-processed satellite products. However, using a nondiagonal **R** is computationally very challenging and is currently an active area of research in data assimilation. Consequently, we have used here a diagonal **R**, where the leading diagonal elements are the sum of the instrument error variance, σ_o^2, and the variance associated with the error of representativeness, σ_r^2.

The other important component of J_{NL} that requires comment is the background error covariance matrix **B**. Because of the very large dimension of the control vector $\mathbf{z}$ ($\sim 10^6$–10^7), **B** is never handled explicitly as a matrix, but instead a model is used for **B**. Specifically, the state-vector increments $\delta\mathbf{x}$ are decomposed into a balanced (e.g., geostrophic) component and an unbalanced (e.g., ageostrophic) component, and the component of **B** for the initial conditions, $\mathbf{B_x}$, is factorized according to the following:

$$\mathbf{B_x} = \mathbf{K_b}\boldsymbol{\Sigma}\mathbf{C}\boldsymbol{\Sigma}^T\mathbf{K_b}^T \tag{2}$$

where **C** is the univariate correlation matrix of the errors in the unbalanced increments, $\boldsymbol{\Sigma}$ is the diagonal matrix of unbalanced error standard deviations, and $\mathbf{K_b}$ is a multivariate balance operator that accounts for the cross-covariances between errors in different state variables.[26] In ROMS 4D-Var, the action of **C** is modeled as the solution of a pseudo-heat diffusion equation following the seminal work of Weaver and Courtier.[27] The matrix of standard deviations, $\boldsymbol{\Sigma}$, was computed from the differences between random draws of the model solution from a long run without data assimilation. The estimates for the appropriate error correlation lengths for **C** are based on the experiments of Broquet et al.[25,28] using 4D-Var applied to the same model. The standard ROMS multivariate balance operator is based on Weaver et al.[26] and uses the hydrostatic balance, geostrophic balance, and appropriate T–S relationships to inform the cross-covariances imposed by $\mathbf{K_b}$, but it was not used in the calculations presented here (i.e., $\mathbf{K_b} = \mathbf{I}$) because of some issues that were discovered in estimating the ocean mixed layer depth. Unlike 3D-Var, 4D-Var very naturally propagates uncertainty information between variables via the model dynamics, so the lack of a formal balance operator in the calculations presented here is not considered to be a limiting factor. A sequence of similar pilot observation impact calculations by Moore et al.[29] that employ $\mathbf{K_b}$ but span a much shorter period (2002–2004) yield quantitatively similar results to those presented here, indicating that the omission of a formal balance operator is unlikely to significantly influence the overall conclusions of this work.

As described in Section 2.2, two sequences of historical analyses were computed using ROMS 4D-Var, one spanning the period 1980–2010 and the other spanning the period 1999–2012. In both cases, the analysis period was divided into 8-day assimilation windows that overlap by 4 days, where the starting time of one assimilation cycle corresponds to the midpoint of the previous cycle. A full description of the data assimilation system and configuration as applied during both analysis periods can be found in Neveu et al.[30]

2.2 OBSERVATIONS AND PRIORS

As noted in Section 2.1, two sequences of CCS circulation analyses were computed. The first spans the period 1980–2010, hereafter referred to as WCRA31, and the second spans the shorter period 1999–2012, hereafter WCRA14. The configuration of ROMS 4D-Var was identical for both WCRA31 and WCRA14, and the two sequences of analyses differ only in the prior estimates that were used to compute the ocean surface forcing.

In the case of WCRA31, the surface forcing was computed using a combination of different atmospheric products. The surface winds were taken from the cross-calibrated multiplatform (CCMP) product of Atlas et al.[31] which is a 2D-Var analysis of all available surface wind observations, using the ECMWF ERA40 reanalysis as the prior estimate for the period 1987–1999, and the operational ECMWF analysis after 1999. Consequently, we used sea level pressure, radiation fluxes, precipitation, and standard height temperature and humidity sampled every six hours from the ERA40 reanalysis so that the ROMS-derived heat and fresh water fluxes are consistent with the prior used for the CCMP winds. Prior to 1987, the ERA-40 reanalysis fields were used, and after 1987, the ERA-40 reanalysis fields were employed in conjunction with CCMP winds. After 2001, ERA-Interim reanalysis fields were used in conjunction with CCMP winds. The horizontal resolution of the various atmospheric products ranges from 25 km for the CCMP winds to 2.5° for ERA-40 reanalysis fields, whereas ERA-Interim reanalysis fields have a resolution of 0.7°. Despite the coarse resolution of ERA-40, it was felt that maintaining consistency between the CCMP wind field (for which ERA40 was the prior) and the other components of surface forcing during the 4D-Var analyses was more important than using the higher resolution ERA-Interim product throughout. Nonetheless, the circulation analyses of WCRA31 appear to be very good despite the relatively coarse resolution of the prior heat and freshwater flux forcing estimates.

Many details of the near shore CCS circulation depend on the fine-scale structure of the surface wind, particularly in the vicinity of prominent coastal capes.[32,33] Given the potential limitations due to low resolution of some components of the surface forcing used during WCRA31, the second sequence of analyses WCRA14 were computed using prior surface forcing derived from the Naval Research Laboratory's Coupled Ocean–Atmosphere Mesoscale Prediction System (COAMPS[34]), which has only been available since January 1999 for the CCS region. The standard height atmospheric variables were derived from four different nests of COAMPS with horizontal resolution ranging from 3 to 81 km from the inner to the outer nest. Only data from the three innermost nested grids were used in ROMS and yield surface fields with a resolution of 3 to 9 km near the coast. This model provides the highest resolution atmospheric forcing data set currently available for the CCS region, and COAMPS verifies well against independent observations, indicating that it is a high-quality product. Doyle et al.[34] demonstrate that COAMPS also captures many of the important regions of orographically enhanced wind stress curl along the coast in the lee of major coastal capes.

The prior fields for the open boundary conditions for both WCRA31 and WCRA14 were taken from the global Simple Ocean Data Assimilation (SODA) product of Carton and Giese.[35]

The observations assimilated into the model were collected from a variety of platforms and include the following: SST from AVHRR (1980–2012), AMSR-E (2002–2011), and MODIS (2000–2012); daily SSH maps from Aviso (1992–2012); and all available in situ hydrographic observations from CTDs, XBTs, MBTs, and Argo profiling floats (1980–2012) from the quality-controlled EN3 archive maintained by the UK Meteorological Office.[36] Because of land contamination effects,[37] SSH observations were not assimilated within 50 km of the coast. A secondary quality control check based on departures of the observations from the prior circulation $\mathbf{x_b}$ was also used to screen observations prior to assimilating them into the model.[38] However, this procedure typically results in less than 0.1% of the observations being rejected.[30]

To avoid data redundancy, multiple observations of a single variable type within a single grid cell and within a 6-h window were combined to form super observations.[39] This is particularly important for SST observations during the last decade when observations from three separate satellite observing systems were used. Super observations were computed based on the standard formula for a minimum variance estimate by combining SST observations using the generally accepted measurement errors of each instrument.

2.3 VALIDATION OF THE 4D-VAR ANALYSES

Various aspects of the 4D-Var circulation estimates WCRA31 and WCRA14 circulation estimates have been explored and validated against independent observations that were not assimilated into the model. For example, Schroeder et al.[40] have performed a detailed analysis of the root mean square (rms) error, bias, and correlation between temperature (T), salinity (S), the $\sigma = 26$ kg/m^3 isopycnal depth, and stratification for WCRA31 along the central California coast using hydrographic observations collected by the U.S. National Marine Fisheries Service between 1987 and 2010 as part an annual Rockfish Recruitment and Ecosystem Assessment Survey

(RREAS), and using observations from the Monterey Bay Aquarium Research Institute (MBARI) M2 mooring (36°N 41′N, 122° 24′W) spanning the period 1998–2010. WCRA31 compares favorably with these independent coastal observations with typical rms (bias) errors in T and S averaged throughout the entire water column ~0.6 °C (~0.5 °C) and 0.1 (~0.05), respectively, with correlations between modeled and observed time series in the range 0.6–0.91. Neveu et al.[30] have examined the differences between the two sequences of analyses and hydrographic observations collected mainly in the Southern California Bight (equatorward of 35°N and offshore to 124°W in Figure 1) during quarterly survey cruises as part of the California Cooperative Fisheries Investigation (CalCOFI). Neveu et al.[30] find that the rms error and bias in T and S over the upper 150–200 m of the water are significantly reduced in the 4D-Var analyses compared to a forward run of the model where there is no data assimilation. In this case though, some of the CalCOFI observations were assimilated into the model. Jacox et al.[41] have explored variability in coastal upwelling along the entire US West Coast, and they show that the analyses exhibit considerable skill in the upwelling response to climate forcing, such as the El Niño Southern Oscillation (ENSO). Further evidence of the ability of WCRA31 to resolve coastal variability associated with the Pacific Decadal Oscillation (PDO), the North Pacific Gyre Oscillation (NPGO), and ENSO is provided by Crawford et al.[42]

3. QUANTIFYING THE IMPACT OF THE OBSERVATIONS ON OCEAN CIRCULATION ANALYSES

The control vector $\mathbf{z_a}$ (and corresponding ocean circulation estimate $\mathbf{x_a}$) that minimizes the cost function in Eqn (1) can be expressed as follows:

$$\mathbf{z_a} = \mathbf{z_b} + \mathbf{K}(\mathbf{y} - H(\mathbf{z_b})) \tag{3}$$

where $\mathbf{K} = \mathbf{BH}^T(\mathbf{HBH}^T + \mathbf{R})^{-1}$ is the Kalman gain matrix. The contribution of each observation y_i^o to any scalar function $I(\mathbf{x})$ of the state vector $\mathbf{x}$ is of particular interest, where $I(\mathbf{x})$ represents any metric of interest such as ocean transport, eddy kinetic energy, eddy fluxes, etc. Following Langland and Baker,[9] it is easy to show that the change in $I(\mathbf{x})$ due to assimilating the observations is given by the first-order Taylor expansion:

$$\Delta I = I(\mathbf{x_a}) - I(\mathbf{x_b}) \simeq \mathbf{d}^T\mathbf{K}^T(\partial I/\partial \mathbf{x})|_{\mathbf{x_b}}. \tag{4}$$

If we write instead $\Delta I = \mathbf{d}^T\mathbf{g}$, where $\mathbf{g} = \mathbf{K}^T(\partial I/\partial \mathbf{x})|_{\mathbf{x_b}}$, then it is clear that $\Delta I = \sum_{i=1}^{N} d_i g_i$ is simply the dot-product of two vectors, where each element of the dot-product is *uniquely* associated with each of the N observations, and it is the product $d_i g_i$ that quantifies the impact of the ith observation on ΔI. In any data assimilation system, the true gain matrix $\mathbf{K}$ will never be known, and in practice, an approximation $\tilde{\mathbf{K}}$ (hereafter referred to as the practical gain matrix) is computed by the assimilation algorithm. In ROMS 4D-Var, the practical gain matrix can be reconstructed from the CG matrix of Lanczos vectors, $\mathbf{V}_m$, according to $\tilde{\mathbf{K}} = \mathbf{B}\mathbf{H}^T\mathbf{V}_m\mathbf{T}_m^{-1}\mathbf{V}_m^T$, where m is the number of inner-loops employed, and $\mathbf{T}_m$ is a tridiagonal matrix. In this case, we have the following:

$$\Delta I = I(\mathbf{x_a}) - I(\mathbf{x_b}) \simeq \mathbf{d}^T\mathbf{V}_m\mathbf{T}_m^{-1}\mathbf{V}_m^T\mathbf{H}\mathbf{B}(\partial I/\partial \mathbf{x})|_{\mathbf{x_b}}. \quad (5)$$

The CG Lanczos vectors that comprise $\mathbf{V}_m$ are routinely archived as part of ROMS 4D-Var, so evaluation of Eqn (5) requires only a single integration of the tangent linear model $\mathbf{H}$ sampled at the observation points. If the definition of $I(\mathbf{x})$ includes a time integral (as is the case considered here), then an additional integration of the adjoint model is also required as described in Ref. 29.

The increment ΔI can also be decomposed as $\Delta I = \Delta I_\mathbf{x} + \Delta I_\mathbf{f} + \Delta I_\mathbf{b}$, where $\Delta I_\mathbf{x}$, $\Delta I_\mathbf{f}$, and $\Delta I_\mathbf{b}$ are the contributions to ΔI from 4D-Var corrections to the initial conditions, surface forcing, and open boundary conditions, respectively. For example, $\Delta I_\mathbf{x}$ can be computed by running the tangent linear model separately subject only to the 4D-Var initial condition increment $\delta\mathbf{x}(t_i)$, with $\delta\mathbf{f} = 0$ and $\delta\mathbf{b} = 0$. Similarly, $\Delta I_\mathbf{f}$ is computed with $\delta\mathbf{x}(t_i) = 0$ and $\delta\mathbf{b} = 0$, and so on. The contribution of each observation or observing platform to each of the control vector contributions to ΔI can be then be computed using Eqn (5) and the appropriate $\mathbf{H}$, which highlights the level of detail that is available using this type of analysis.

The impact of the observations on a metric $I(\mathbf{x})$ computed in this way is typically used in numerical weather prediction to quantify the impact of different observing systems on forecast errors.[11,43] When evaluating the observation impact on forecast errors, the verifying analysis at the forecast time is used as a measure of the truth. The application of Eqn (5) here, however, is somewhat different in that it is applied to the analysis increments, with the aim of quantifying the impact of each observing platform on the analysis itself. The skill of forecasts initialized at the end of each analysis cycle is not considered here, although Moore et al.[29] have applied the same method to forecast errors in ROMS-CCS to illustrate the impact of the observations on forecast skill.

4. CONTROL VECTOR IMPACTS ON ALONGSHORE TRANSPORT

The functional $I(\mathbf{x})$ used in this study is based on the transport crossing a vertical section at 37°N extending from the coast to 127°W, as illustrated in Figure 1. Specifically, $I(\mathbf{x})$ is defined as the total transport from the surface to 500 m and averaged in time over each 8-day 4D-Var analysis cycle. This metric is, therefore, a measure of the total transport carried by the equatorward-flowing California Current, the poleward undercurrent, and a near shore coastal jet that forms seasonally.

Figure 2 shows time series of the monthly-averaged 37°N transport increments $\Delta I = I(\mathbf{x_a}) - I(\mathbf{x_b})$ computed from WCRA31, which vary between $\sim\pm 0.5$ Sv. This compares with a total transport that varies seasonally between $\sim\pm 5$ Sv. Also shown in Figure 2 is a time series of ΔI based on the first-order Taylor

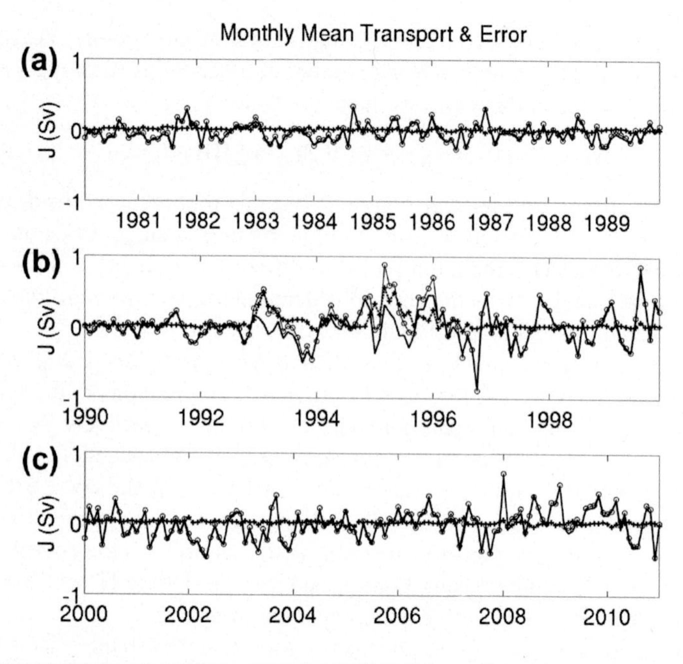

FIGURE 2

Time series of the monthly mean transport increments, ΔI, computed directly from the difference between the posterior x_a and prior x_b circulation estimates of WCRA31 from the nonlinear model (black), and from the first-order Taylor approximation in Eqn (4) (circles). The difference between the nonlinear and tangent linear transport increments is also shown (plus signs). (a) 1980–1989, (b) 1990–1999, and (c) 2000–2010.

expansion in Eqn (4), which will be limited by the tangent linear assumption inherent in $\mathbf{K}^T$ by virtue of $\mathbf{H}$. With the exception of the period 1993–1995 (Figure 2(b)), Eqn (4) provides a very good approximation for the transport increment ΔI over the 31-year period. The poor performance of Eqn (4) during 1993–1995 period is associated with the introduction of Aviso SSH observations into the 4D-Var analysis system toward the end of 1992, a period discussed more extensively later in the chapter. Therefore Figure 2 indicates that, apart from the 1993–1995 period, Eqn (4) will yield reliable measures of the impact of the control vector increments and observations on ΔI derived from WCRA31. The same is true of WCRA14 (not shown).

As shown in Section 3, the transport increment ΔI can be decomposed into the control vector contributions $\Delta I_{\mathbf{x}}$, $\Delta I_{\mathbf{f}}$, and $\Delta I_{\mathbf{b}}$ associated with the 4D-Var corrections to the initial conditions, surface forcing, and open boundary conditions, respectively. Figure 3 shows the annual rms increments associated with the three components of the control vector for WCRA31 and WCRA14. Figure 3 reveals that during most years the 4D-Var corrections to the initial conditions have the largest influence on ΔI in both WCRA31 and WCRA14, although the contribution by corrections to

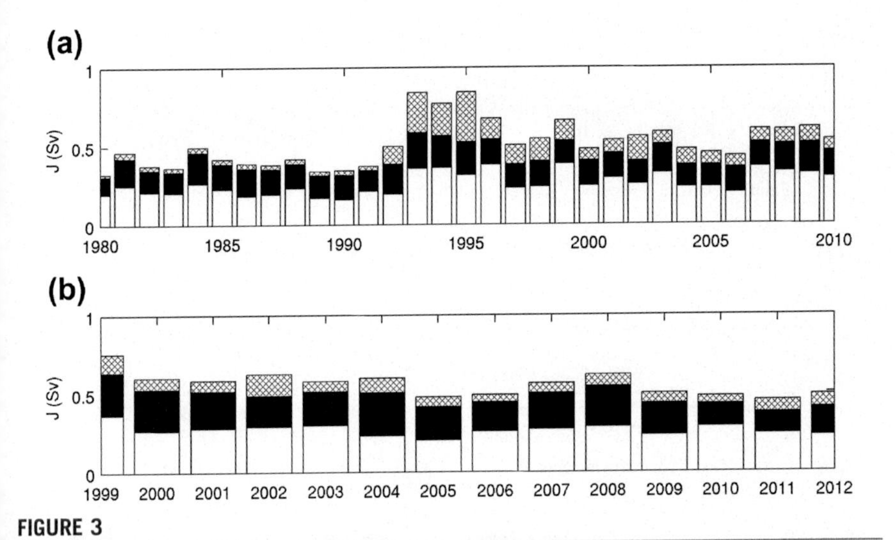

FIGURE 3

Time series of the annual rms contributions ΔI_x (white), ΔI_f (black), and ΔI_b (hatched) to the transport increment for (a) WCRA31 and (b) WCRA14.

the surface forcing is comparable, particularly during WCRA14. Prior to the introduction of SSH observations into the 4D-Var system in 1992, the contribution to ΔI by corrections to the open boundary conditions was relatively small. However, after 1992, $\Delta I_\mathbf{b}$ is significantly larger, particularly during the period of 1993–1995 identified previously; although, as shown in Figure 2, the transport increments ΔI based on the tangent linear approximation in Eqn (4) are generally not reliable at this time. Nevertheless, after 1995, the $\Delta I_\mathbf{b}$ are still substantially larger than during the pre-SSH era.

To illustrate the level of detail that is possible in the decomposition of $\Delta I_\mathbf{x}$, $\Delta I_\mathbf{f}$, and $\Delta I_\mathbf{b}$, it is instructive to drill down on a single year. As an example, we will focus on the control vector impacts on ΔI during the period of January 1998 to January 1999. Figure 4(a) shows time series of the control vector contributions to ΔI for each 4D-Var data assimilation cycle during this period, and it illustrates the changing interplay between the initial conditions, surface forcing, and open boundary conditions in controlling this aspect of the CCS circulation. Figure 4(a) also reveals that during many cycles different control variables can exert opposing influences on ΔI.

Figure 4(b) focuses on the impact of the surface forcing components of the control vector on ΔI, and it shows that almost all of the change in ΔI is due to corrections in $\mathbf{f}$ associated with the surface wind stress. Figure 4(c), on the other hand, shows the impact of the open boundary conditions that are imposed on the five state variables, and it reveals that sea surface elevation, ς, along the boundaries exerts the largest influence on ΔI. The contribution of ς to ΔI during each cycle is further decomposed in Figure 4(d) into the contribution from each of the three open boundaries, and it

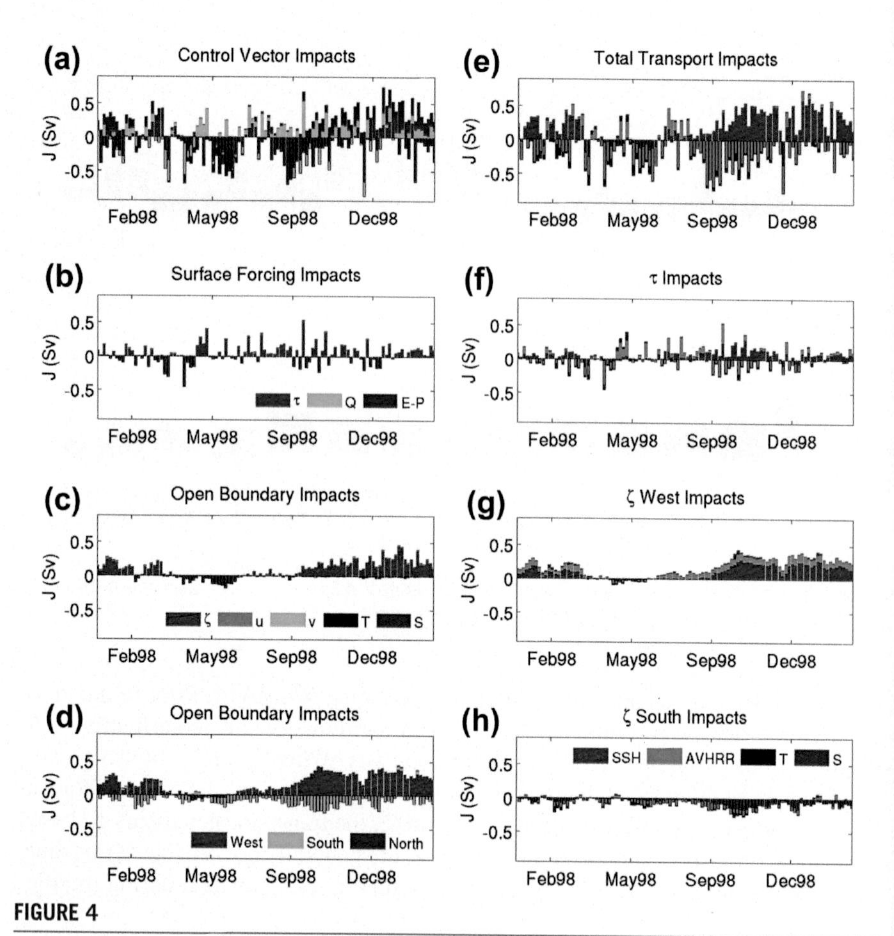

FIGURE 4

Time series of the contribution of each component of the control vector and observing platform to the transport increment, ΔI, during each 4D-Var data assimilation cycle for 1998. (a) Initial conditions, ΔI_x (red), surface forcing, ΔI_f (green), and open boundary conditions, ΔI_b (blue). (b) ΔI_f from (a) is here decomposed into the contribution due to wind stress τ (red), heat flux Q (green), and surface freshwater flux E-P (blue). (c) ΔI_b from (a) is here decomposed into the contribution of the open boundary conditions applied to each of the state variables ς, u, v, T, and S to ΔI. (d) Because ς dominates ΔI_b in (c), the contribution of the open boundary condition for ς at each of the three open boundaries is shown here. (e) The impact of individual observing platforms on the transport increments, ΔI, shown in (a): SSH (red), AVHRR SST (green), in situ profiles of temperature (brown) and salinity (blue). (f) The same as (e) but for the contribution of τ to ΔI_f, which dominates the surface forcing impacts in (b). (g) The same as (e) except for the contribution of the ς western open boundary condition to ΔI_b shown in (d). (h) The same as (e) except for the contribution of the ς southern open boundary condition to ΔI_b shown in (d).

shows that it is 4D-Var adjustments to ς along the western and southern open boundaries that have the greatest impact on adjustments to the central CCS transport, although the changes along the two boundaries always oppose each other.

Figure 4 is also an illustration of how Eqn (4) can be used to monitor the performance of the 4D-Var system. In this case, there is a significant change in the contribution of the open boundary conditions to $I(\mathbf{x})$, which is most likely the result of an inconsistency between the Aviso SSH product and the SODA open boundary condition (see Section 5). Figure 4 indicates that the 4D-Var algorithm corrects for this inconsistency by adjusting ς along the western and southern open boundaries. Nevertheless, as discussed previously, the resulting circulation increments $\delta\mathbf{x} = \mathbf{x_a} - \mathbf{x_b}$ are large enough to render Eqn (4) unreliable during 1993–1995 while the system readjusted to constraints imposed by the SSH observations.

5. OBSERVATION IMPACTS ON ALONGSHORE TRANSPORT

Following Section 3, the impact of each observation on ΔI, $\Delta I_\mathbf{x}$, $\Delta I_\mathbf{f}$, and $\Delta I_\mathbf{b}$ was computed according to Eqn (4). To distill what is a large amount of information, the impacts were grouped according to observing platform type, namely SSH from altimeters, SST from each of the three sensors used (AVHRR, AMSR, and MODIS) as well as SST super observations, and temperature and salinity profiles from all in situ instruments. To summarize the results from all 4D-Var cycles, the annual rms was computed for each observing platform type, as shown in Figure 5, which shows time series of the rms contribution of each observing platform to ΔI, $\Delta I_\mathbf{x}$, $\Delta I_\mathbf{f}$, and $\Delta I_\mathbf{b}$ for both WCRA31 and WCRA14.

Figure 5 shows that during 1980, the first year of WCRA31, only in situ observations were present and accounted for 100% of the impact on ΔI. During the period of 1981 to 1991, AVHRR SST observations were also available during WCRA31, and they clearly exert the largest influence on the ΔI, although the contribution from in situ is significant also. Between 1992 and 2000, the observing system was augmented by SSH, which has an impact on each ΔI that is comparable to that of AVHRR. During the last decade of both analyses, SST observations have dominated both in number and in the impact that they have on each ΔI. The SST super observations and MODIS SST generally have the largest impact on the total transport increment and the contribution associated with each control variable.

As noted in Section 4, Figure 5(g) shows that prior to 1992 the impact of the observations on ΔI via adjustments to the open boundary conditions was small. However, following the introduction of SSH data into the assimilation system, both SSH and AVHRR conspire to adjust the boundary condition influence on transport, even though the impact of AVHRR on $\Delta I_\mathbf{b}$ before the altimeter era was negligible. After the year 2000, the open boundary contributions to ΔI are smaller and spread across all satellite platforms in both WCRA31 (Figure 5(g)) and WCRA14 (Figure 5(h)). This illustrates the complex interplay between different observation types within the data assimilation system that can be partially

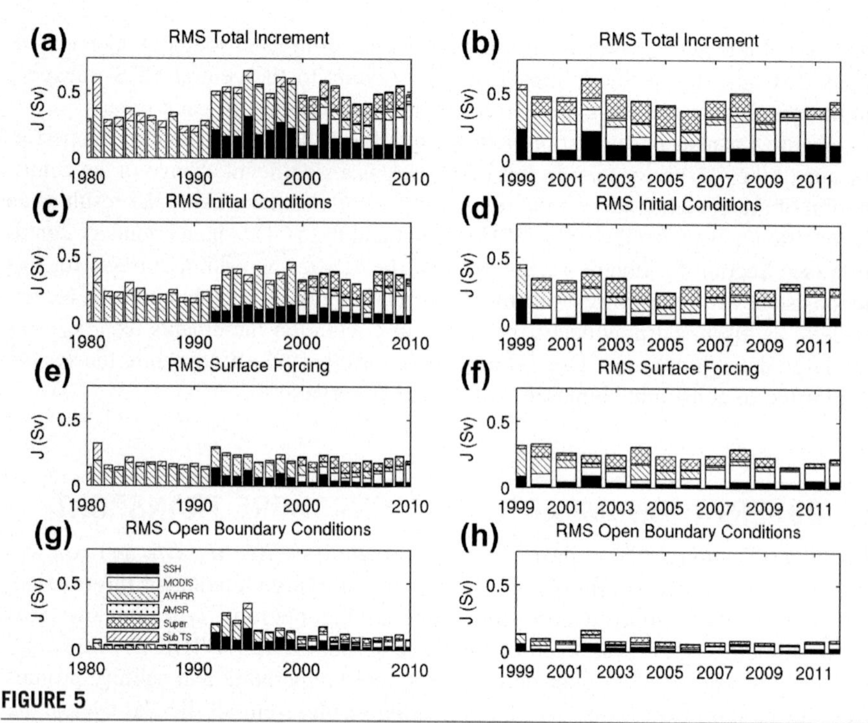

FIGURE 5

Time series of the annual rms impact of each observing platform on ΔI for (a) WCRA31 and (b) WCRA14. The rms impact of each platform on ΔI_x is shown for (c) WCRA31 and (d) WCRA14, and similarly for ΔI_f in (e) for WCRA31 and (f) for WCRA14, and for ΔI_b in (g) for WCRA31 and (h) for WCRA14. The impact associated with each platform is coded according to the following: SSH (black), MODIS SST (white), AVHRR SST (\\\), AMSR SST (dots), SST super observations (cross-hatched), and in situ hydrographic observations (///). Observations labeled as MODIS, AVHRR, and AMSR are from the respective platforms and have not been combined with data from other platforms in the super observation process.

unraveled using the observation impact tools used here. Figure 5 also shows that the impact of the in situ observations has been reduced dramatically in the recent decade due to the plethora of satellite observations. Moore et al.[5] demonstrate that there is a great deal of redundancy in the various satellite SST observing systems within the CCS, and they suggest that the satellite data should be decimated before assimilation to allow in situ observations to exert more influence on the analyses. Such a practice would be in line with that currently applied to satellite observations in numerical weather prediction.

There is some evidence of seasonal variations in the observation impacts. For example, Figure 6(a) shows time series of the rms observation impacts of the surface forcing contribution ΔI_f to ΔI averaged over each calendar month for WCRA14. The transport increments ΔI_f are clearly largest during the summer, mainly

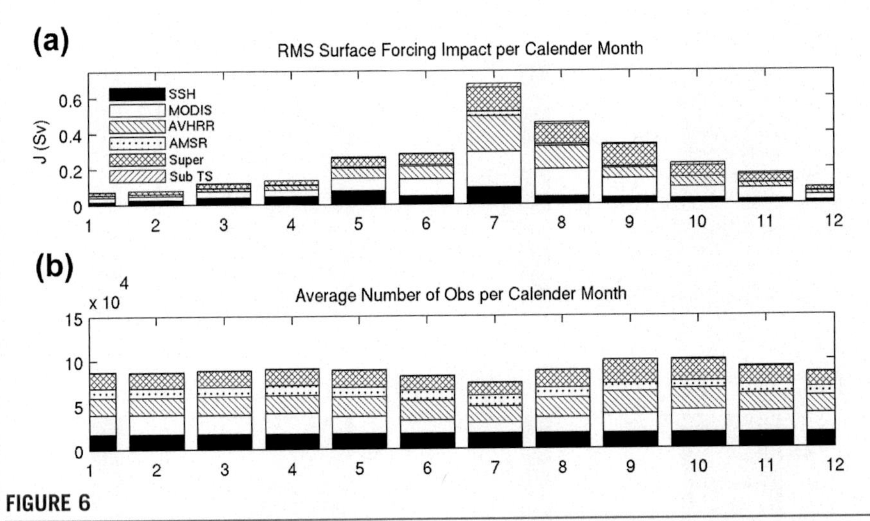

FIGURE 6

(a) Time series of the monthly rms observation impacts on the surface forcing contribution to ΔI for WCRA14. (b) Time series of the average number of observations from each platform during each calendar month.

associated with the impact of MODIS and AVHRR SST observations. The number of observations actually reaches a low point during July (when ΔI_f peaks) as shown in Figure 6(b) due to the extensive layer of marine stratus that is so prevalent along the California coast during summer. The summertime peak in ΔI_f is most likely associated with a systematic error in upwelling ocean temperatures in ROMS-CCS as discussed by Broquet at al.[28]

As in Section 4, we can drill down and look in more detail at the period of January 1998 to January 1999 and quantify the impact of the observations on the total transport increment and the contributions from each of the control variables. Figure 4(e) shows the observation impacts for the total transport increment, ΔI, and reveals that the contributions by AVHRR SST and Aviso SSH almost always oppose each other. The same is generally true for the wind stress contributions (Figure 4(f)). At the open boundaries, however, SSH and AVHRR SST reinforce each other, as shown in Figures 4(g) and (h), although SSH dominates and is largely responsible for the 4D-Var boundary adjustments as discussed in Section 4.

Geographic variations in the observation impacts on the alongshore transport increments, ΔI, are illustrated in Figure 7, which shows the rms impact associated with each observation location for each of the observing platforms. The impact of AMSR SST observations that are not associated with super observations was small and is not shown. Perhaps unexpectedly, the observations that have the largest impact on ΔI are not always those in the vicinity of the 37°N section. For example, SST observations that are primarily north of the transport section (i.e., upstream for the time-mean near surface circulation) have the largest impact. For AVHRR and

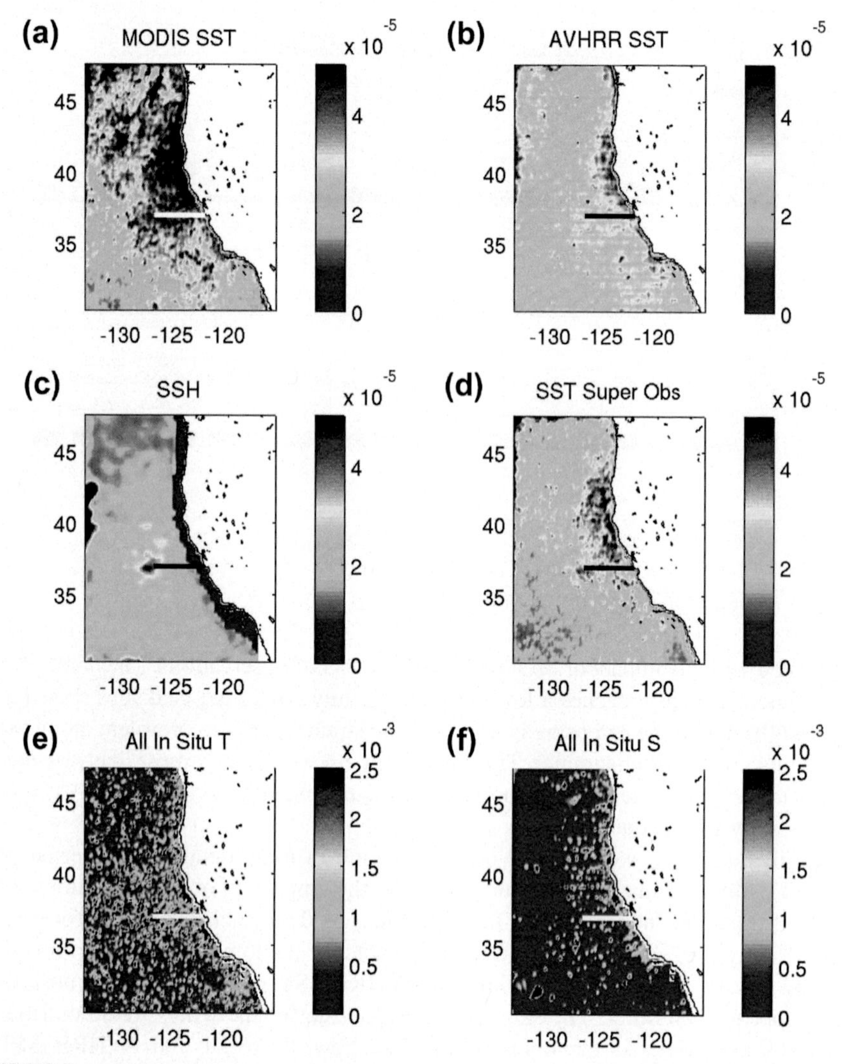

FIGURE 7

Maps of the rms impact of each observation location on the transport increment ΔI in Sv for WCRA31. (a) MODIS SST, (b) AVHRR SST, (c) SSH, (d) SST super observations, (e) all in situ temperature profiles, and (f) all in situ salinity profiles. All impact data were interpolated onto a regular grid and smoothed with a second-order Shapiro filter. The 37°N section along which ΔI was computed is also indicated. The impact of AMSR observations was small compared to MODIS, AVHRR and SST super observations and is not shown.

SST super observations (Figure 7(b) and (d)), data close to the coast have the largest impact, whereas in the case of MODIS (Figure 7(a)), ΔI is impacted by observations over much of the domain. For the in situ observations, Figure 7(e) and (f) reveal that observations close to the coast generally have the largest impact on ΔI both upstream and downstream of the 37°N section. Figure 7 also reveals that the impact of each individual in situ datum is typically one to two orders of magnitude larger than that associated with satellite data. Figure 7(c) shows that the transport increment ΔI is typically influenced by SSH observations offshore and in the deep ocean, and ΔI is particularly sensitive to SSH near a section of the western open boundary.

The geographic distributions of the observation impacts in Figure 7 can be more fully appreciated by appealing to the underlying dynamics of the circulation. Recall from Section 2.1 that information contained in the observations is dynamically interpolated in space and time by the tangent linear dynamics of ROMS. Specifically, information is carried backward in time by $\mathbf{H}^T$ to yield the Green's functions associated with each observation, and Green's function influences are carried forward in time by $\mathbf{H}$. Thus, the propagation of information through the system is controlled by familiar dynamic processes such as advection, diffusion, and wave propagation. The maximum distance that observational information can be transported by different processes can be estimated, which are referred to here as information horizons. Assuming a mean speed for the California Current ~0.1 m/s,[44] then the advection horizon for each 8-day 4D-Var cycle is ~70 km (i.e., ~1° of longitude at 37°N and ~0.6° of latitude). Therefore, only observations in the vicinity of the 37°N section can influence ΔI via horizontal advection. The fastest baroclinic waves that are available to carry information are nondispersive inertia-gravity waves. If we assume an average first baroclinic mode phase speed ~2.5 m/s for the NE Pacific,[45] the inertia-gravity wave horizon is ~1700 km (~19° of longitude and 15° of latitude), so information from observations anywhere in the domain can potentially influence ΔI when carried by inertia-gravity waves. Other important vehicles for information transfer are coastally trapped waves, which include coastal Kelvin waves, topographic Rossby waves (or continental shelf waves), and edge waves.[46] The horizon associated with each wave type will be different, but for coastal Kelvin, waves will be ~1700 km in the alongshore direction. Barotropic wave adjustment represents the fastest mode of information propagation, and the barotropic horizon will extend to the open boundaries, which is reflected in the boundary impacts evident in Figure 7.

For SSH, another dynamic aspect of the circulation is important, namely the establishment of a horizontal pressure gradient. Figure 7(c) shows that the zonal pressure gradient is altered by the assimilation of SSH observations, and the geostrophic adjustment time required to establish a new geostrophically balanced alongshore transport across 37°N will be less than a day.

Information can also be spread by the background error covariance matrix **B** in Eqn (5) according to the de-correlation lengths assumed for the errors in $\mathbf{z_b}$. The correlation lengths used here vary between 50 km for the initial conditions errors up to 300 km for errors in surface forcing.[25,28] The information horizons associated with the resulting regularization by **B** are, therefore, relatively small compared to the wave horizons.

6. SUMMARY AND CONCLUSIONS

Langland and Baker[9] have demonstrated that in the linear limit, the transpose of the practical gain matrix provides a linear map from state space to observation space that can be used to uniquely attribute increments in a scalar function of the state vector to each individual observation and to each element of the control vector, or both at the same time. The practical gain matrix can be conveniently reconstructed using the archived output from 4D-Var, and the observation and control vector impacts of scalar functions can be evaluated at modest additional computational cost. This benefit is in contrast to more traditional OSEs, where the data assimilation analyses must be recomputed to evaluate the impact of withholding observations. For example, to quantify the impact of five different platforms, a traditional OSE would require five additional analyses to be performed. The 4D-Var observation impact method has been applied here to two long sequences of 4D-Var circulation estimates for the CCS, in one case spanning the last three decades, to explore the impact of the observations and control vector increments on the net transport of the CCS off the central California coast.

The impacts of the control vector can be used to monitor the performance of the data assimilation system through time. In the case considered here, the resulting impacts suggest that a mismatch exists between the assimilated SSH and that prescribed at the open boundaries, an aspect of the 4D-Var calculations that requires further investigation. One novel aspect of this study is that we are able to monitor the impact of different ocean observing platforms during the past three decades. In particular, the observation impact calculations reveal the complex interplay that can occur between data as different observing systems come online in the 4D-Var system. Furthermore, the impact of the in situ observations can become swamped by the sheer volume of satellite data, although the impact per datum is typically one to two orders of magnitude larger for in situ observations than for satellite data. This result suggests that decimation of the satellite observations may be appropriate before data assimilation (standard practice in numerical weather prediction) to reduce the high degree of redundancy associated with the latter[5] and to allow the in situ observations to exert more influence on the analyses.

The geographic distribution of the observation impacts can be largely understood in terms of the underlying dynamics of the circulation, and approximate information horizons can be computed to delineate the control that different processes (e.g., horizontal advection and wave propagation) have on the transfer of observational information via the 4D-Var algorithm.

This study was limited to the impact of the data assimilation control vector and observations on the alongshore transport. However, any other differentiable function $I(\mathbf{x})$ can be considered in Eqns (4) and (5). For example, Fiechter et al.[47] explored the impact of different observing systems on eddy kinetic energy and surface chlorophyll concentrations in the Gulf of Alaska. The main limitation on the choice of $I(\mathbf{x})$ is the degree to which it is nonlinear, because Eqn (4) represents a first-order Taylor expansion. Therefore, if the linear assumption implicit in Eqn (4) is a poor approximation, the resulting estimate of ΔI may not be representative of $I(\mathbf{x_a}) - I(\mathbf{x_b})$, but this, of course, can always be checked a posteriori. In addition to the alongshore transport, the impact of the control vector and observations on other metrics $I(\mathbf{x})$ of the CCS circulation is also currently being explored by the authors, including upwelling transport, the transport of the California Undercurrent, and the depth of the pycnocline at different locations along the US West Coast. The results of that study will be the subject of a future publication.

It is anticipated that in the near future the routine monitoring of observation impacts on ocean analyses and forecasts, such as that demonstrated here, will form an integral component of activities in support of the U.S. Integrated Ocean Observing System (IOOS). Plans are underway to perform such monitoring of observation impacts in the near real-time analysis forecast system for the CCS in support of the Central and Northern California Ocean Observing System (CeNCOOS) described in Moore et al.[38] and Kourafalou et al.[48,49]

ACKNOWLEDGMENTS

The authors gratefully acknowledge the support of the National Science Foundation (OCE 1061434), the National Oceanic and Atmospheric Administration (NA13NOS0120139) via a subaward from the Southwest Universities Research Association (SURA), and the Office of Naval Research (N00014-10-1-0476).

REFERENCES

1. Balmaseda MA, Anderson DLT, Vidard A. Impact of Argo on analyses of the global ocean. *Geophys Res Lett* 2007;**34**:L16605. http://dx.doi.org/10.1029/2007GL0304452.
2. Oke PR, Schiller A. Impact of Argo, SST, and altimeter data on an eddy resolving ocean reanalysis. *Geophys Res Lett* 2007;**34**:L19601. http://dx.doi.org/10.1029/2007GL031549.
3. Smith GC, Haines K. Evaluation of the S(T) assimilation method with the Argo dataset. *Q J R Meteorol Soc* 2009;**135**:739–56.
4. Le Hénaff M, De Mey P, Marsaleix P. Assessment of observational networks with the representer matrix spectra method-application to a 3D coastal model of the Bay of Biscay. *Ocean Dyn* 2009;**59**:3–20.
5. Moore AM, Arango HG, Broquet G, Edwards C, Veneziani M, Powell BS, et al. The regional ocean modeling system (ROMS) 4-dimensional variational data assimilation

systems. Part II: performance and application to the California current system. *Prog Oceanogr* 2011;**91**:50–73.

6. Oke PR, Sakov P. Assessing the footprint of a regional ocean observing system. *J Mar Syst* 2012;**105**:30–51. http://dx.doi.org/10.1016/j.jmarsys.2012.05.009.
7. Oke PR, Larnicol G, Fujii Y, Smith GC, Lea DJ, Guinehut S, et al. Assessing the impact of observations on ocean forecasts and reanalyses: part 1, global studies. *J Oper Oceanogr* 2015. http://dx.doi.org/10.1080/1755876X.2015.1022067.
8. Oke PR, Larnicol G, Jones EM, Kourafalou V, Sperrevik AK, Carse F, et al. Assessing the impact of observations on ocean forecasts and reanalyses: part 2, regional applications. *J Oper Oceanogr* 2015. http://dx.doi.org/10.1080/1755876X.2015.1022080.
9. Langland RH, Baker NL. Estimation of observation impact using the NRL atmospheric variational data assimilation adjoint system. *Tellus* 2004;**56A**:189–201.
10. Daescu DN. On the sensitivity equations for four-dimensional variational (4D-Var) data assimilation. *Mon Weather Rev* 2008;**136**:3050–65.
11. Zhu Y, Gelaro R. Observation sensitivity calculations using the adjoint of the gridpoint statistical interpolation (GSI) analysis system. *Mon Weather Rev* 2008;**136**:335–51.
12. Gelaro R, Zhu Y. Examination of observation impacts derives from observing system experiments (OSEs) and adjoint models. *Tellus* 2009;**61A**:179–93.
13. Shchepetkin AF, McWilliams JC. The regional oceanic modeling system (ROMS): a split explicit, free-surface, topography-following-coordinate oceanic model. *Ocean Model* 2005;**9**:347–404.
14. Stegmann PM, Schwing F. Demographics of mesoscale eddies in the California current. *Geophys Res Lett* 2007;**34**:L14602. http://dx.doi.org/10.1029/2007GL029504.
15. Marchesiello P, McWilliams JC, Shchepetkin AF. Equilibrium structure and dynamics of the California current system. *J Phys Oceanogr* 2003;**33**:753–83.
16. Gruber N, Lachkar Z, Frenzel H, Marchesiello P, Munnich M, McWilliams JC, et al. Eddy-induced reduction of biological production in eastern boundary upwelling systems. *Nat Geosci* 2011;**4**:787–92.
17. Veneziani M, Edwards CA, Doyle JD, Foley D. A central California coastal ocean modeling study: 1. Forward model and the influence of realistic versus climatological forcing. *J Geophys Res* 2009;**114**:C04015. http://dx.doi.org/10.1029/2008JC004774.
18. Moore AM, Arango HG, Broquet G, Powell BS, Zavala-Garay J, Weaver AT. The regional ocean modeling system (ROMS) 4-dimensional variational data assimilation systems. Part I: system overview and formulation. *Prog Oceanogr* 2011;**91**:34–49.
19. Ide K, Courtier P, Ghil M, Lorenc AC. Unified notation for data assimilation: operational, sequential and variational. *J Meteorol Soc Jpn* 1997;**75**:181–9.
20. Daget N, Weaver AT, Balmaseda MA. Ensemble estimation of background-error variances in a three-dimensional variational data assimilation system for the global ocean. *Q J R Meteorol Soc* 2009;**135**:1071–94.
21. Wikle CK, Berliner LM. A Bayesian tutorial for data assimilation. *Phys D* 2007;**230**:1–16.
22. Lawless AS, Gratton S, Nichols NK. Approximate iterative methods for variational data assimilation. *Int J Numer Methods Fluids* 2005;**1**:1–6.
23. Courtier P, Thépaut J-N, Hollingsworth A. A strategy for operational implemenation of 4D-Var using an incremental approach. *Q J R Meteorol Soc* 1994;**120**:1367–88.
24. Gürol S, Weaver AT, Moore AM, Piacentini A, Arango HG, Gratton S. B preconditioned minimization algorithms for variational data assimilation with the dual formulation. *Q J Roy Meteorol Soc* 2014;**140**:539–56.

25. Broquet G, Edwards CA, Moore AM, Powell BS, Veneziani M, Doyle JD. Application of 4D-variational data assimilation to the California current system. *Dyn Atmos Oceans* 2009;**48**:69–92. http://dx.doi.org/10.1016/j.dynatmoce.2009.03.001.
26. Weaver AT, Deltel C, Machu E, Ricci S, Daget N. A multivariate balance operator for variational ocean data assimilation. *Q J R Meteorol Soc* 2005;**131**:3605–25.
27. Weaver AT, Courtier P. Correlation modelling on the sphere using a generalized diffusion equation. *Q J R Meteorol Soc* 2001;**127**:1815–46.
28. Broquet G, Moore AM, Arango HG, Edwards CA. Corrections to ocean surface forcing in the California current system using 4D-variational data assimilation. *Ocean Modell* 2011;**36**:116–32.
29. Moore AM, Arango HG, Broquet G, Edwards C, Veneziani M, Powell BS, et al. The regional ocean modeling system (ROMS) 4-dimensional variational data assimilation systems. Part III: observation impact and observation sensitivity in the California current system. *Prog Oceanogr* 2011;**91**:74–94.
30. Neveu E, Moore AM, Edwards CA, Fiechter J, Drake P, Jacox MG, et al. An historical analysis of the California current using ROMS 4D-Var. Part I: System configuration and diagnostics, submitted for publication.
31. Atlas R, Hoffman RN, Ardizzone J, Leidner SM, Jusem JC, Smith DK, et al. A cross-calibrated, multiplatform ocean surface wind velocity product for meteorological and oceanographic applications. *Bull Am Meteorol Soc* 2011;**92**:157–74.
32. Enriquez AG, Friehe CA. Effect of wind stress and wind stress curl variability on coastal upwelling. *J Phys Oceanogr* 1995;**25**:1651–71.
33. Pickett MH, Paduan JD. Ekman transport and pumping in the California current based on the US Navy's high-resolution atmospheric model (COAMPS). *J Geophys Res* 2003;**108**:3327. http://dx.doi.org/10.1029/2003JC001902.
34. Doyle JD, Jiang Q, Chao Y, Farrara J. High-resolution atmospheric modeling over the Monterey Bay during AOSN II. *Deep Sea Res II* 2009;**56**:87–99.
35. Carton JA, Giese BS. A reanalysis of ocean climate using simple ocean data assimilation (SODA). *Mon Weather Rev* 2008;**136**:2999–3017.
36. Ingleby B, Huddleston M. Quality control of ocean temperature and salinity profiles — historical and real-time data. *J Mar Syst* 2007;**65**:158–75.
37. Saraceno M, Strub PT, Kosro PM. Estimates of sea surface height and near surface along-shore coastal currents from combinations of altimeters and tide gauges. *J Geophys Res* 2008;**113**:C11013. http://dx.doi.org/10.1029/2008JC004756.
38. Moore AM, Edwards CA, Fiechter J, Drake P, Arango HG, Neveu E, et al. A prototype for an operational regional ocean data assimilation system. Chapter 14. In: Xu L, Park S, editors. *Data assimilation for atmospheric, oceanic and hydrological applications*, vol. 2. Springer; 2013. p. 345–66.
39. Daley R. *Atmospheric data analysis*. Cambridge University Press; 1991. 457 pp.
40. Schroeder ID, Santora JA, Moore AM, Edwards CA, Fiechter J, Hazen E, et al. Application of a data-assimilative regional ocean modeling system for assessing California current system ocean conditions, krill, and juvenile rockfish interannual variability. *Geophys Res Lett* **41**:5942–50.
41. Jacox MG, Moore AM, Edwards CA, Fiechter J. Spatially resolved upwelling in the California current system and its connections to climate variability. *Geophys Res Lett* 2014;**41**:3189–96.
42. Crawford WJ, Moore AM, Jacox MG, Neveu E, Edwards CA, Fiechter J. An historical analysis of the California current using ROMS 4D-Var. Part II: climate variability, submitted for publication.

43. Errico RM. Interpretations of an adjoint-derived observational impact measure. *Tellus* 2007;**59A**:273–6.
44. Hickey BM. Coastal oceanography of western North America from the tip of Baja, California to Vancouver Island. In: Robinson AR, Brink KH, editors. *The sea*, vol. 11. John Wiley and Sons; 1998. p. 345–93.
45. Chelton DB, deSzoeke RA, Schlax MG, El Naggar K, Siwertz N. Geographical variability of the first baroclinic Rossby radius of deformation. *J Phys Oceanogr* 1998;**28**: 433–60.
46. Gill AE. *Atmosphere-ocean dynamics*. Academic Press; 1982. 662 pp.
47. Fiechter J, Broquet G, Moore AM, Arango HG. A data assimilative, coupled physical-biological model for the coastal Gulf of Alaska. *Dyn Atmos Ocean* 2011;**52**:95–118.
48. Kourafalou VH, De Mey P, Le Hénaff M, Charria G, Edwards CA, He R, et al. Coastal ocean forecasting: system integration and validation. *J Oper Oceanogr* 2015. http://dx.doi.org/10.1080/1755876X.2015.1022336.
49. Kourafalou VH, De Mey P, Staneva J, Ayoub N, Barth A, Chao Y, et al. Coastal ocean forecasting: science drivers and user benefits. *J Oper Oceanogr* 2015. http://dx.doi.org/10.1080/1755876X.2015.1022348.

CHAPTER

20

Assimilation of HF Radar Observations in the Chesapeake–Delaware Bay Region Using the Navy Coastal Ocean Model (NCOM) and the Four-Dimensional Variational (4DVAR) Method

Hans Ngodock[1,*], Philip Muscarella[1], Matthew Carrier[1], Innocent Souopgui[2], Scott Smith[1]

Naval Research Laboratory, Stennis Space Center, Mississippi, USA[1]; Department of Marine Science, University of Southern Mississippi, Stennis Space Center, Mississippi, USA[2]

**Corresponding author: E-mail: hans.ngodock@nrlssc.navy.mil*

CHAPTER OUTLINE

1. INTRODUCTION

Consistent and accurate coastal ocean monitoring necessitates the availability of three key components: (1) an observing network that adequately samples the monitored domain, (2) a coastal ocean circulation model with a sufficiently high resolution that takes into account the often complex geometry and dynamics that occur near the coastline, and (3) an analysis system that is able to accurately assimilate the sampled observations to initialize the coastal model for forecasting. Modern analysis systems can also provide an observation impact assessment for the design, evaluation, and possibly reassignment of observing resources.

Coastal current measurement types are limited to moored buoys and ADCPs, which do not provide adequate spatial distribution/resolution, surface drifters, which tend to leave the deployment area relatively quickly, and shipboard ADCPs, which are relatively expensive to operate. The continuous monitoring of coastal waters for circulation properties requires long-term station observations. High-frequency (HF) radar units are unique observation platforms that provide surface current measurements at horizontal resolutions of 1 to 6 km ranging from 25 to 200 km off the coast. This amount of spatial coverage would be unattainable with current meter stations. In addition, HF radar units are installed by various universities and institutions along much of the coastline of the continental United States. HF radar observations have been assimilated into ocean models mostly using sequential methods (e.g., Refs 1–4). However, there are a few assimilation examples using a 4DVAR method.[5–7]

Regional ocean models for coastal circulation monitoring require initial and boundary conditions from larger or global domain models that are usually run with much coarser horizontal resolution, as well as surface forcing fields from atmospheric models. To a large extent, the accuracy of the coastal models depends on (1) the accuracy of the larger domain model providing initial and boundary conditions, (2) the accuracy of the atmospheric model providing surface forcing fields, and equally important, (3) the accuracy of the parameterization of the physics due to increased resolution. The increased resolution can also become a liability for the assimilation as the model resolves small-scale circulation features that cannot be constrained by the available observations. Usually, only coarse observation coverage is available for assimilation into the larger domain (with the exception of sea surface temperature (SST)), making it difficult to provide accurate initial and boundary conditions for the coastal model. Also, atmospheric models can contain errors in the coastal oceans due to the coarse resolution often used and the complex land–sea boundary, not to mention the lack of frequent feedback from the ocean to the atmosphere in these areas. Failure to do this also translates to errors in the atmospheric fields in these areas. In addition, the ocean model's horizontal resolution may not be high enough to capture all the details of the coastline and the bathymetry. All these elements contribute to the discrepancies that are seen when coastal ocean model solutions are compared to observations. This is where the data assimilation plays the critical role of combining the ocean model and available observations in a dynamically consistent way to not only provide a better initial condition for the prediction of the ocean environment, but also to correct at least some components of the model error, e.g., errors in the atmospheric forcing fields.

Due to the high temporal variability of surface currents in the coastal areas, a necessary requirement of the assimilation system is the ability to take into account the temporal dimension in the observations. Such capability is inherent to 4DVAR.

Contrary to the often used sequential methods that assimilate observations at a given time (thus correcting the model state at fixed time stamps), e.g., the three-dimensional variational data assimilation (3DVAR), or the methods based on the Kalman filter, 4DVAR seeks to correct the entire model trajectory for a given time window by assimilating all the observations (distributed in time and space) that were sampled during that time window. In this process, 4DVAR (1) uses observations at almost the exact times that they are sampled, which suits most asynoptic data, (2) implicitly uses flow-dependent background errors, which ensures the analysis quality for rapidly changing environments, and (3) uses a forecast model as a constraint, which ensures the dynamic balance of the final analysis.

It was recently shown that the assimilation of surface velocity observations derived from drifters, using a 4DVAR with the Naval Coastal Ocean Model (NCOM-4DVAR[8]), improved ocean model forecasts of sea surface height, surface and subsurface velocity, temperature, and salinity in the Gulf of Mexico.[9] Unfortunately, this study was limited in time due to the deployment and lifespan of the drifters. This paper aims to expand on the previous study by assimilating a sustained and dense source of surface velocity observations from HF radars in the Chesapeake–Delaware Bay region to show that they are a viable dataset for constraining and forecasting the coastal circulation.

2. HF RADAR OBSERVATIONS

The surface current observations used for this study come from a network of three SeaSonde HF radar units that are deployed in the mid-Atlantic region of the East Coast of the United States (black stars on Figure 1). The northernmost site is on Assateague Island, Maryland, the central site is at Cedar Island, Virginia, and the southernmost site is at Little Island Park, Virginia. Throughout the study period, data is available from these sites during 93%, 99.9%, and 100% of the time, respectively. These HF radar units produced by CODAR ocean sensors scatter radio waves off the ocean surface and infer movement of near surface currents. During July 2013, the three stations were operating at 4.5 MHz that resonantly scatter off surface gravity waves of approximately 30-m wavelength. For these so-called long-range site observations are provided hourly on a polar-coordinate grid with a range step of 6 km and a bearing step of 5°. Additionally, the horizontal range of a single site is approximately 200 km offshore.

Roarty et al.[10] discusses the operation and maintenance of the mid-Atlantic HF radar observing network; this includes the three sites used in this study. Because the HF radar data is used by the U.S. Coast Guard Search and Rescue Optimal Planning System (SAROPS), there is an implemented procedure for quality control and assurance of the observations collected by these sites. Additionally, they go on to report RMS differences between in situ measurements of both acoustic doppler current profilers (ADCPs) and drifters of 7.4 to 9.8 cm/s using an unweighted least squares method.[11] This unweighted least squares approach merges the single-site radial

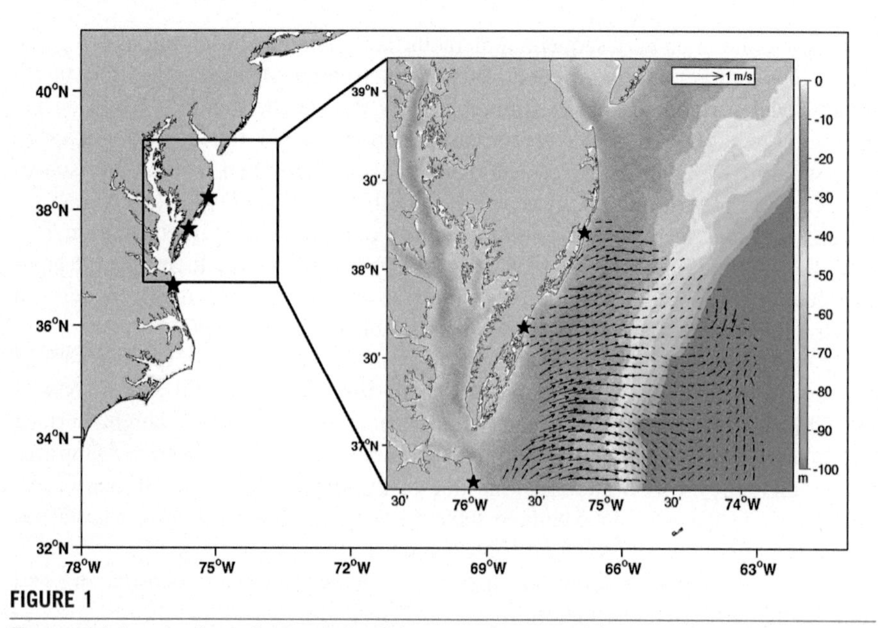

FIGURE 1

The model domain: the outer box is the 3-km parent nest, and the inner and expanded box is the actual 1-km domain with a sample coverage of HR radar observations on July 01, 2013, overlaid on the colored bathymetry. The radar stations are represented by the stars.

velocities located within a search radius around each grid point. This processing step uses a Matlab toolbox called HFR_Progs to create total current vectors. For the three-site setup used here, the search radius is 10 km with a minimum of two radials from at least two sites required to create a total. In theory, accurate surface velocities are recovered when two HR radar beams form an angle of 90°. In practice, however, accurate velocities can still be constructed from beams forming an angle as low as 15°, which has become a standard for operational processing of HF radar observations, at least in the mid-Atlantic Bight HF radar observing network.[10] The HF radar observations assimilated here were processed with the 15° minimum angle threshold. The HF radar measurements are sensitive to environmental factors that can affect the spatial extent of the velocity footprint. This usually results in occasional gaps within a coverage area. For more information see Ref. 12.

3. THE MODEL

The ocean model used in this study is the Navy coastal ocean model (NCOM). NCOM is a free-surface model that has been described in the literature.[13,14] The model domain (and bathymetry) shown in Figure 1 spans longitudes 76.6°W to 73.6°W and latitudes 36.75°N to 39.2°N at 1-km horizontal resolution with 50 vertical levels. The initial conditions were obtained from downscaling the operational 1/8° resolution global NCOM (GNCOM) to an intermediate model with horizontal resolution of 3 km, and then to a high-resolution 1-km model. Horizontal viscosities and diffusivities are computed using either the grid-cell Reynolds number (Re) or the Smagorinsky schemes, both of which tend to decrease as resolution is increased. The grid-cell Re scheme sets the mixing coefficient K to maintain a grid cell Re number below a specified value, e.g., if $\mathrm{Re} = u * dx/K = 30$, then $K = u * dx/30$. Hence, as dx decreases, K decreases proportionally. A similar computation is performed for the Smagorinsky scheme.

The surface atmospheric forcing, including wind stress, atmospheric pressure, and surface heat flux, is provided by the Navy Global Atmospheric Prediction System (NOGAPS[15–17]) with a horizontal resolution of 0.5°. River forcing is provided at all river in-flow locations in this mid-Atlantic domain. Additionally, eight tidal constituents (K1, O1, P1, Q1, K2, M2, N2, and S2) are forcing the domain through the open boundaries. Open boundary conditions use a combination of radiative models and prescribed values provided by the parent 3-km nest. Different radiative options are used at the open boundaries depending on the model state variables: a modified Orlanski radiative model is used for the tracer fields (temperature and salinity), an advective model for the zonal velocity (u), a zero gradient condition for the meridional velocity (v) as well as the barotropic velocities, and the Flather boundary condition for elevation.

4. THE ASSIMILATION SYSTEM

The assimilation system used here is described in more detail in Ref. 8. The brief presentation that follows only serves to elucidate the focus of this study. For a given model, the following is presented:

$$\begin{cases} \dfrac{\partial X}{\partial t} = F(X) + f, \; 0 \le t \le T \\ X(t=0) = I(x) + i(x) \end{cases} \tag{1}$$

where X stands for all the dependent model state variables, i.e., the two-dimensional SSH and barotropic velocities, and the three-dimensional temperature, salinity, and baroclinic velocities; F includes the model tendency and forcing terms, f is the model error with covariance C_f, $I(x)$ is the prior initial condition, and $i(x)$ is the initial condition error with covariance C_i; x and t represent the position in the three-dimensional space and time, respectively. Given a vector Y of M observations of the model state in the space–time domain, with the associated vector of observation errors ε (with covariance C_ε), the following is shown:

$$y_m = H_m X + \varepsilon_m, \quad 1 \le m \le M \tag{2}$$

where H_m is the observation operator associated with the mth observation. One can define a weighted cost function as follows:

$$J = \int_0^T \int_\Omega \int_0^T \int_\Omega f(x,t) W_f(x,t,x',t') f(x',t') dx' dt' dx dt + \int_\Omega \int_\Omega i(x) W_i(x,x') i(x') dx' dx + \varepsilon^T W_\varepsilon \varepsilon \tag{3}$$

where Ω denotes the model domain, the weights W_f and W_i are defined as inverses of C_f and C_i in a convolution sense, and W_ε is the matrix inverse of C_ε. Boundary condition errors are omitted from Eqns (1) and (3) only for the sake of clarity. The solution of the assimilation problem, i.e., the minimization of the cost function (Eqn (3)), is achieved by solving the following Euler-Lagrange (EL) system:

$$\begin{cases} \dfrac{\partial X}{\partial t} = F(X) + C_f \cdot \lambda, \ 0 \le t \le T, \\ X(t=0) = I(x) + C_i \circ \lambda(x,0) \\ -\dfrac{\partial \lambda}{\partial t} = \left[\dfrac{\partial F}{\partial X}(X)\right]^T \lambda + \displaystyle\sum_{m=1}^{M} \sum_{n=1}^{M} W_{\varepsilon,mn} (y_m - H_m X) \delta(x - x_m) \delta(t - t_m), \quad 0 \le t \le T \\ \lambda(T) = 0 \end{cases} \tag{4}$$

where λ is the adjoint variable defined as the weighted residual:

$$\lambda(x,t) = \int_0^T \int_\Omega W_f(x,t,x',t') f(x',t') dx' dt', \tag{5}$$

and δ denotes the Dirac delta function, $W_{\varepsilon,mn}$ are the matrix elements of W_ε, the superscript T denotes the transposition, and the covariance multiplication with the adjoint variable is the convolution:

$$C_f \cdot \lambda(x,t) = \int_0^T \int_\Omega C_f(x,t,x',t') \lambda(x',t') dx' dt', \tag{6}$$

and

$$C_i \circ \lambda(x,0) = \int_\Omega C_i(x,x') \lambda(x',0) dx' \tag{7}$$

for the model and initial condition errors, respectively.

It can be seen in Eqn (4) that the adjoint model is forced by the innovations (model-data misfits at the observation locations), and its solution initializes and/or forces the forward model, depending on whether a strong or weak constraints assumption is adopted.

A standard approach to solving the Euler-Lagrange system (Eqn (4)) is the strong constraints *4dvar* that assumes that only the initial condition is erroneous, i.e., the model has no errors ($C_f = 0$). The solution of Eqn (4) is found iteratively as follows: (1) a first guess initial condition is used to solve the nonlinear model, (2) the nonlinear solution is used to compute the model-data misfits that appear in the right-hand side of the adjoint model, (3) the adjoint model is solved and used to compute the correction to the initial condition, and (4) the process is repeated until the minimum of the cost function or a preselected convergence criterion is reached.

The weak constraints 4dvar approach takes into account the model errors and, thus, increases the dimension of the control space, which now becomes the entire model trajectory for the selected assimilation window. This rather huge control space also increases the computational cost of the assimilation, and it usually renders the minimization (of the cost function) process poorly conditioned. This difficulty can be avoided if the minimization is done in the data space, which does not depend on and is usually much smaller than the control space. That is possible through the representer algorithm,[18,19] which expresses the solution of Eqn (4) as the sum of a first guess and a finite linear combination of representer functions, one per datum. Being a linear expansion, the representer algorithm cannot be applied to Eqn (4) directly, mainly because of its nonlinear property. However, following Refs 8,20 the representer algorithm can be applied to a linearized form of Eqn (4).

5. EXPERIMENTS AND RESULTS

For this study, the initial condition error for the experiment is set as 1.0 °C for temperature, 0.1 practical salinity unit (psu) for salinity, and 0.5 m/s for velocity. These errors are set by examining the innovation values between a free-running NCOM and available observations. The error values are uniformly prescribed across the domain and reduced at depth. This is deemed acceptable because we are mostly interested in the accuracy of surface currents. It is also important to note the model errors in this study are attributed to errors in the specified atmospheric surface forcing. This a reasonable assumption as ocean surface currents are strongly influence by surface wind stress. The model errors are 0.05 °C for temperature, 0.005 practical salinity unit (psu) for salinity, and 0.05 m/s for velocity. The model errors represent 5% of the magnitudes of the atmospheric forcing in respective equations, with the exception of the free surface, and are converted from fluxes units to units of the ocean state variables using the relationships imposed by the discretization of the model (see Refs 21,8). The horizontal correlation scales of the initial and model errors are taken to be 20 km (approximately the Rossby radius of deformation in the domain) and fixed in time. When these isotropic covariances are convolved with the adjoint solution and included in the forward model as dictated by Eqn (4), the 4dvar system produces analysis increments that are flow dependent, thanks to the dynamics of the tangent linear and adjoint models, and also the nonlinear model trajectory around which the system is linearized.

Each of the observation data types is also assigned errors. The observation errors are usually a combination of the estimated instrument error and the representativeness error. Here, the temperature error is 0.35 °C, 0.035 practical salinity unit (psu), and 0.05 m/s for velocity. The experiment carried out here takes place from July 1 to 31, 2013, in sequential assimilation windows of three days, with observations binned hourly and subsampled to keep only one observation per 20-km correlation scale in both meridional and zonal directions. With the exception of the first cycle, the background (i.e., the solution that the assimilation is trying to correct) for each cycle is the forecast obtained by running the nonlinear with the final condition from the analysis in the previous cycle.

In order to assess the fit to the observations over time in the whole assimilation window, we define the following "fit to the observations" metric:

$$J_{FIT} = \frac{1}{M} \sum_{m=1}^{M} \frac{|y_m - H_m X^a|}{\sigma_m}. \tag{8}$$

In Eqn (8), y_m is the mth observation, M is the total number of observations, H_m is the observation operator, X^a is the assimilated solution or analysis, and σ_m is the observation error or standard deviation. The right-hand side of Eqn (8) can be computed as a time series and also evaluated for the free-run solution and the first guess. Because the assimilation is expected to fit the observations to within the observation standard deviation at the observation locations, the metric J_{FIT} in Eqn (8) is expected to be less or equal to one for the analysis. One only hopes that the same is true for the subsequent forecasts as a result of fitting the observations in previous cycles.

The results of this assimilation experiment show that the NCOM-4DVAR is capable of assimilating HF radar velocities by significantly reducing the discrepancies between the modeled and the observed surface velocities. It can be seen in Figure 2 that the free-running model for the 1-km resolution (black line) is in significant disagreement with the observations, having J_{FIT} values between 2 and 4 observation standard deviations. The assimilation (gray dashed line) is able to reduce those discrepancies, sometimes by as much as 2 standard deviations, with J_{FIT} values generally between 1 and 2.4 observation standard deviations. These values are still higher than the target value of 1, indicating that, although the assimilation has done a good job of reducing the discrepancies compared to the free-running model, the assimilated solution is still not fitting the observations accurately. On the other hand, the first guess solution (dashed black line), which consists of the forecast from analysis in the previous assimilation window, shows discrepancies of the same magnitudes as the free-running solution. This indicates that the gains of the

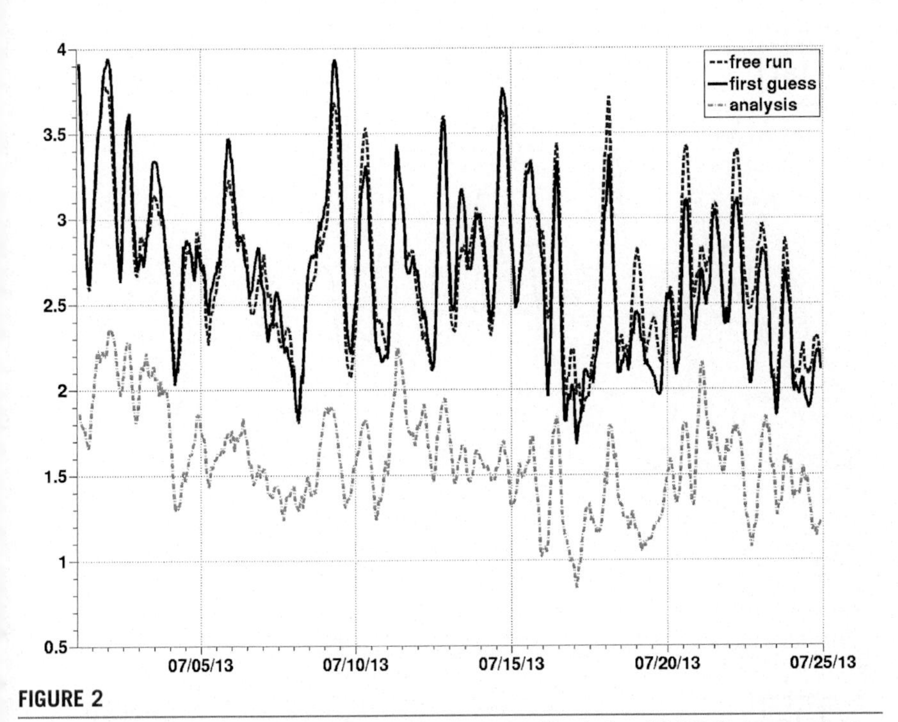

FIGURE 2

The J_{FIT} metric for a free-running model (black dashed line), the 4DVAR first guess fields (solid black line), and the 4DVAR analysis (gray dashed line) at the observation locations. This experiment assimilates observations hourly.

assimilation are quickly lost in the forecast, primarily due to erroneous surface forcing and the high resolution of the model that resolves circulation features that are not observed.

6. VALIDATION

The validation of any assimilation experiment requires independent observations against which the assimilation results can be compared. Those usually consist of buoys along the coast. However, those are not available during the time of this experiment. We carried out a second assimilation experiment where observations are assimilated every 3 h instead of every hour as in the previous experiment. The unassimilated observations, i.e., those that are left out every 2 h, are considered independent for the purpose of validation, even though they may be correlated with those that are assimilated, by reason of proximity in space and time. The same fit to the observations metric J_{FIT} is also used to evaluate this assimilation experiment, not only for the assimilated observations, but also for the withheld observations. Results in Figure 3(a) show the J_{FIT} values for the assimilated solution and the first guess

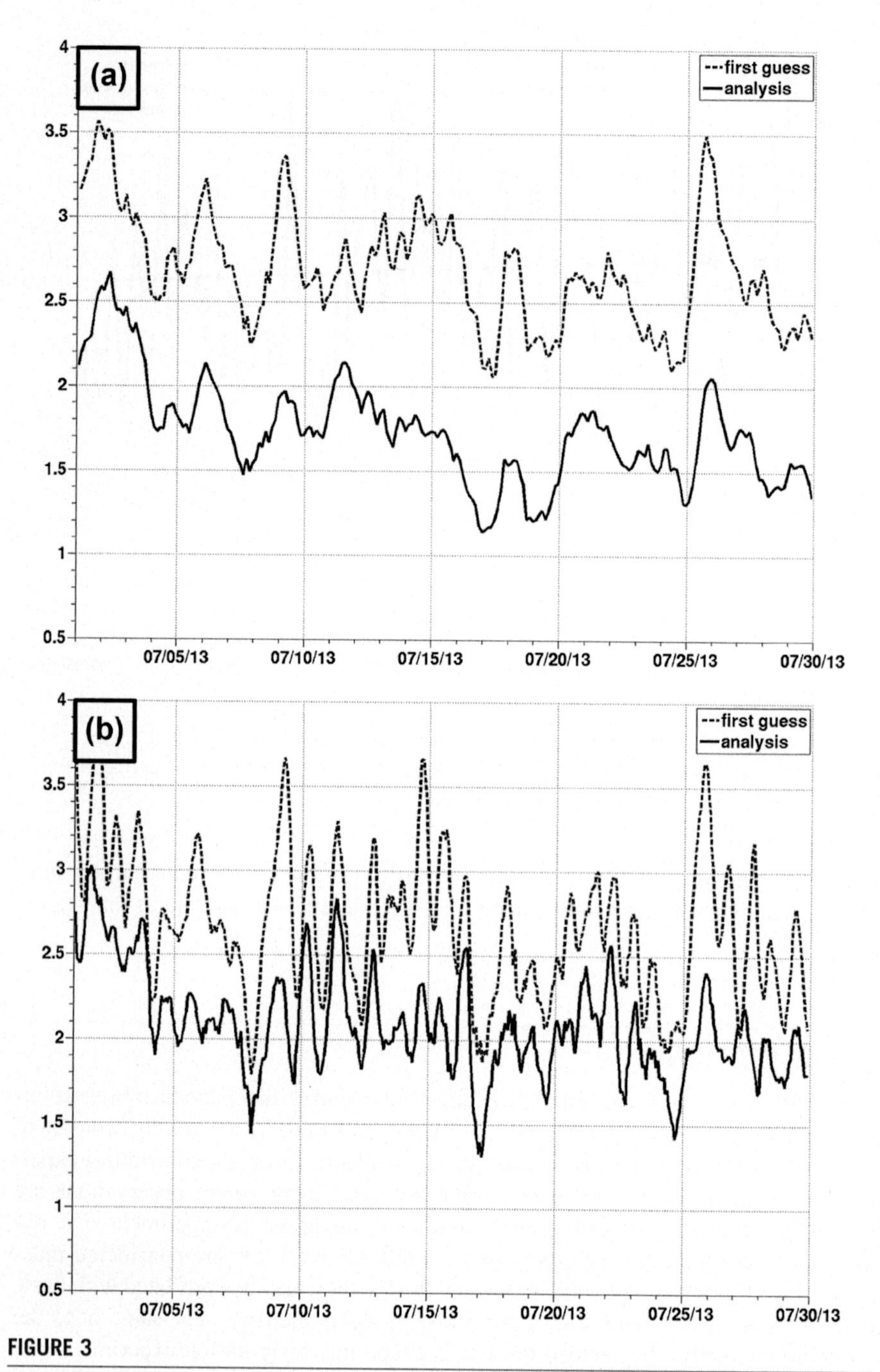

FIGURE 3

J_{FIT} metric for the first guess solution (dashed) and the new assimilated solution (solid) at the assimilated observations every third hour (a) and at the withheld observations (b).

compared to assimilated observations, whereas Figure 3(b) shows similar J_{FIT} values for the assimilated solution and the first guess compared to withheld observations. It can be seen that similar to the previous experiment, there is a significant reduction in the J_{FIT} values from the first guess (2–3.5 standard deviations) to the assimilated solution (1.2–2.7 standard deviations). More importantly, there is a good improvement of the assimilated solution versus the first guess when these two solutions are compared to the withheld observations, an improvement that sometimes exceeds a standard deviation.

Figure 4 shows a comparison of surface velocities maps from the observations, the first guess, and the analysis 5 and 16 days into the assimilation, respectively. We first note that at these two time levels, there is almost no agreement at all between the circulation patterns shown in the observations and those in the first guess, i.e., the model is significantly in error compared to the observations, even after being initialized by the assimilation. For example, on day 5, the observations describe an offshore surface circulation, whereas the model shows an alongshore circulation. This results from the atmospheric forcing fields being erroneous themselves, see Figure 5. The assimilation procedure alters the first guess enough to produce an analysis that fits (looks like) the observations, albeit not perfectly: The offshore current is reconstructed by the assimilation on day 5; and on day 16, the northeastward coastal current that turns offshore is also recovered, though these features were missing in the first guess. However, the analysis sometimes still has the patterns of the first guess. This indicates that the background and model errors prescribed for the assimilation are too small.

A major difficulty in coastal ocean modeling resides in the lack of high spatial resolution atmospheric forcing. Atmospheric models are usually run with coarser resolutions compared to the ocean models (especially for coastal applications), because resolving the rather fast motion of the atmosphere with high horizontal resolution would require very small time steps that would be computationally prohibitive. For the case at hand, the ocean model has a horizontal resolution of 1 km, while the atmospheric forcing fields are obtained from interpolating results from an atmospheric model that used a 0.5° resolution that does not capture the variability of the model domain. According to Ekman theory, a modest wind stress of 5 m/s would cause the surface velocity to deflect to the right of the wind stress direction by an angle of 45° if the water depth is at least 45 m. The topography shown in Figure 1 indicates that this would not apply to a significant portion of the domain where the depth is less than 45 m, and it is expected that the surface velocity be strongly correlated to the wind stress in that part of the domain. Figure 5 shows the direction of the wind stress compared to the direction of the surface currents from the HF radar stations. The wind stress is generally uniform as a result of interpolating from a few grid points of the atmospheric model. It can be seen that the direction of the wind is never aligned with the direction of the observations; they are quite different. This presents a significant challenge to the assimilation system and explains why the assimilation could not accurately fit the observations with small initial conditions and model errors.

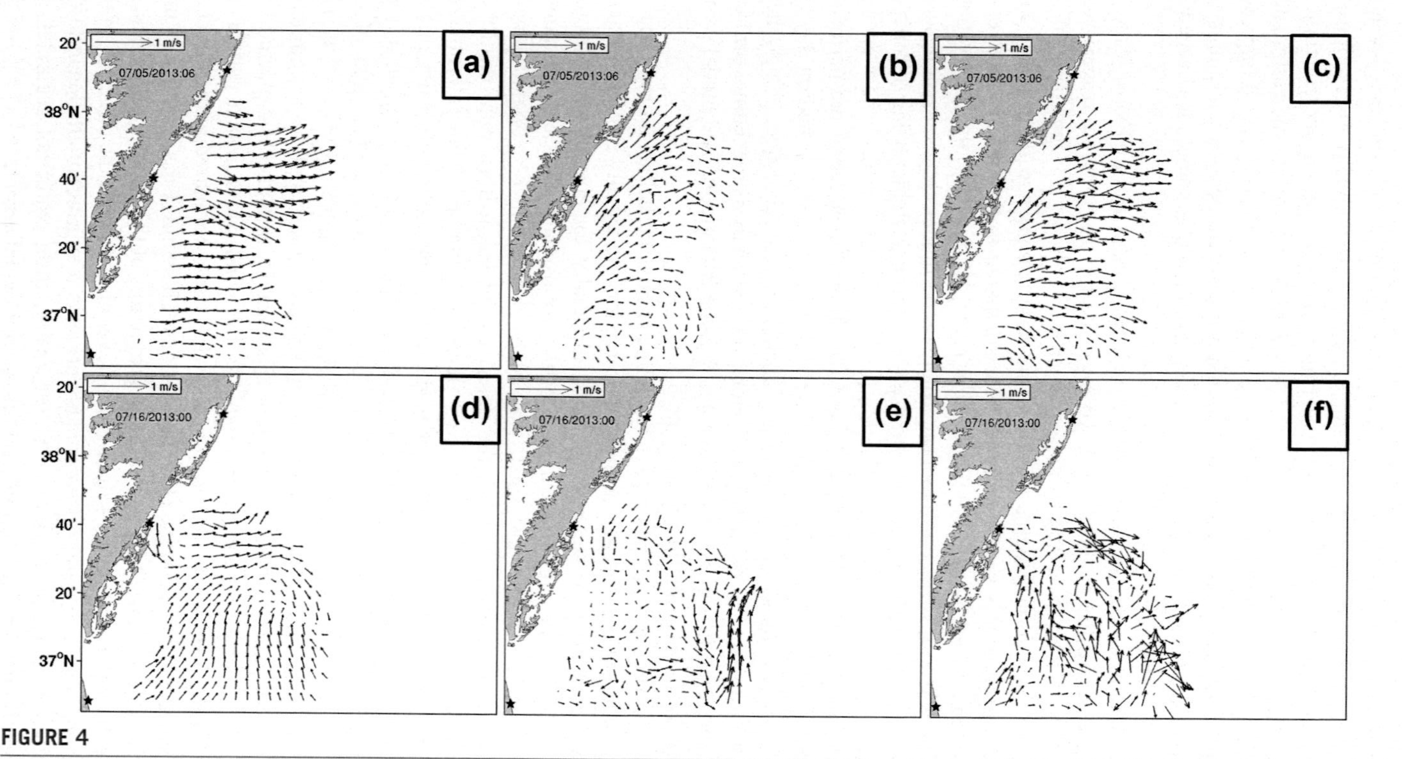

FIGURE 4

Surface velocities from the observations (a, d), the first guess (b, e), and the analysis (c, f) 5 and 16 days into the assimilation, respectively.

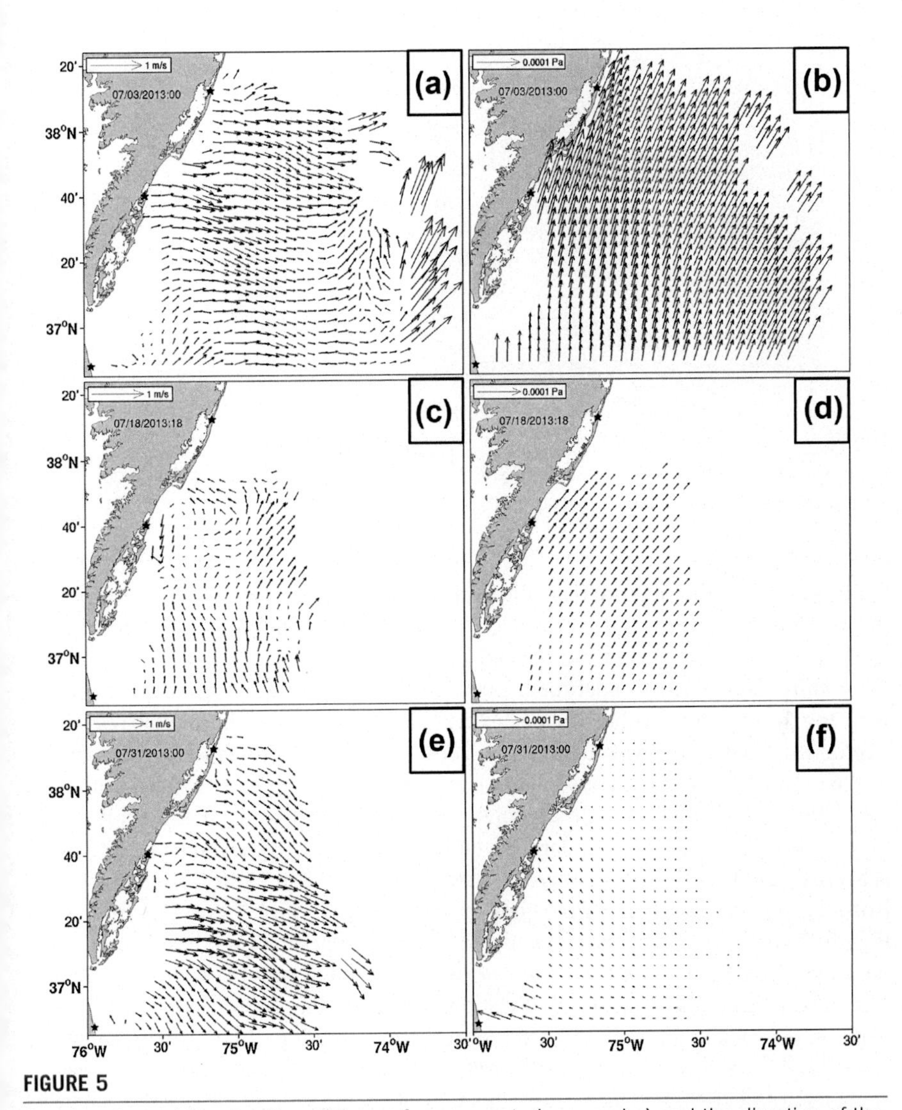

FIGURE 5

A comparison of the direction of the surface currents (a, c, and e) and the direction of the wind stress (b, d, and f) at the end of the first, the sixth, and the tenth assimilation window, respectively.

Another challenge to the assimilation resides in the resolution of the ocean model and that of the observations. The processed observations have a spatial resolution of about 6 km, whereas the model is run at a resolution of 1 km. As seen in Figure 6 in a subset of the domain, the model resolves multiple small-scale features that are absent from the observations. Also, the strong flow at the southeast corner of the domain reveals that the model boundary conditions contain an intrusion of the

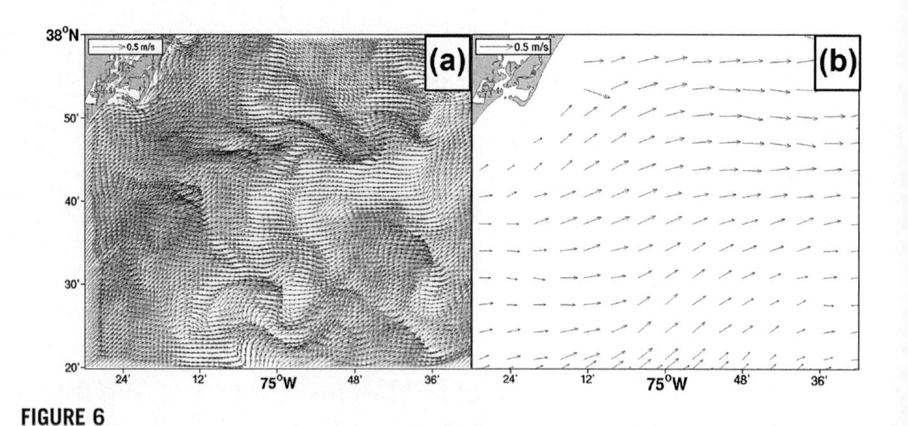

FIGURE 6

A comparison of the 1-km first guess velocity field (a) and the observations (b) on day 10 in a subset of the model domain.

Gulf Stream, another feature that is not present in the observations. Thus, the assimilation could benefit from more accurate boundary conditions and, perhaps, a slightly coarser resolution from the model, say 3 km.

A third assimilation experiment is carried out for three cycles of 3 days each, with the same setup as the original experiment, except that the model resolution is reduced from 1 to 3 km. Figure 7(a) shows a comparison between the analyses of the 1 and 3 km experiments, where significant improvements in the accuracy of the 3-km analysis can be seen, especially at the times when the 1-km analysis has J_{FIT} values exceeding 1.5 standard deviations. In general, J_{FIT} values for the 3-km analysis are lower than 1.5 standard deviations. On the other hand, a similar comparison of J_{FIT} values for the first guesses from both 1- and 3-km experiments in Figure 7(b) shows that the forecast from the 3-km analysis has significantly improved compared to the 1-km forecast, with noticeable gaps between the two lines sometimes exceeding 1 standard deviation. However, the 3-km first guess still displays large discrepancies with the observation, having J_{FIT} values generally exceeding 2 observation standard deviations, and only occasionally falling below 1.5 standard deviations. Once again, this loss of accuracy in the first guess is attributed to the erroneous atmospheric forcing (in this case the wind stress) because the model resolution is now closer to that of the observations coverage, and the analysis is also significantly closer to the observations.

Similar to Figure 4, Figure 8 shows a comparison of surface velocities maps from the observations, the first guess, and the analysis 3 and 5 days into the assimilation, respectively, for the 3-km model resolution. There is a significant difference in the direction of the velocity between the observations and the first guess on July 3, especially at the lower-right side of the domain showing an intrusion of the Gulf Stream in the first guess, whereas the observations locate this intrusion further to the east and slightly to the north, compared to its location in the first guess. The assimilation

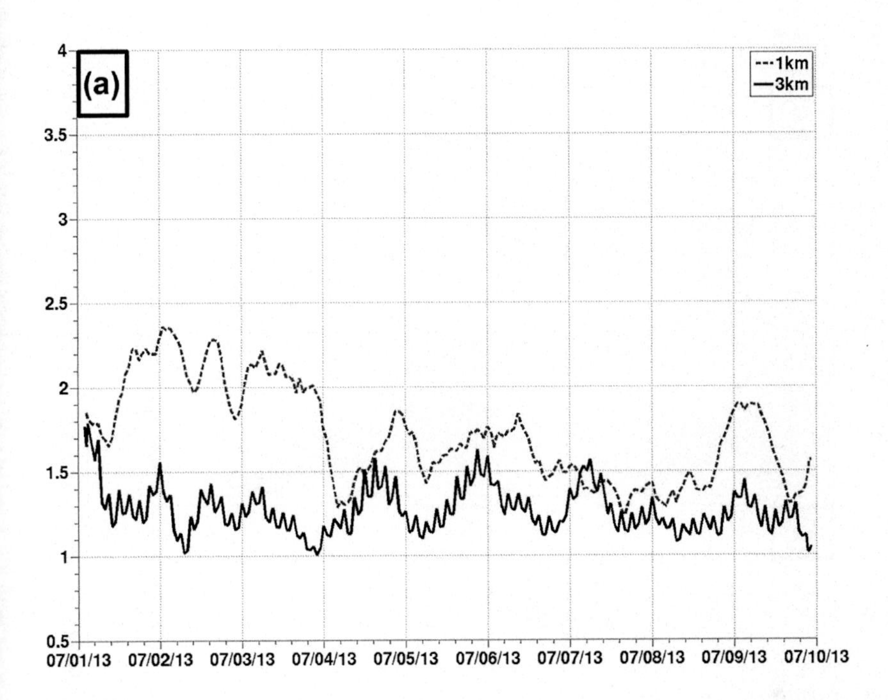

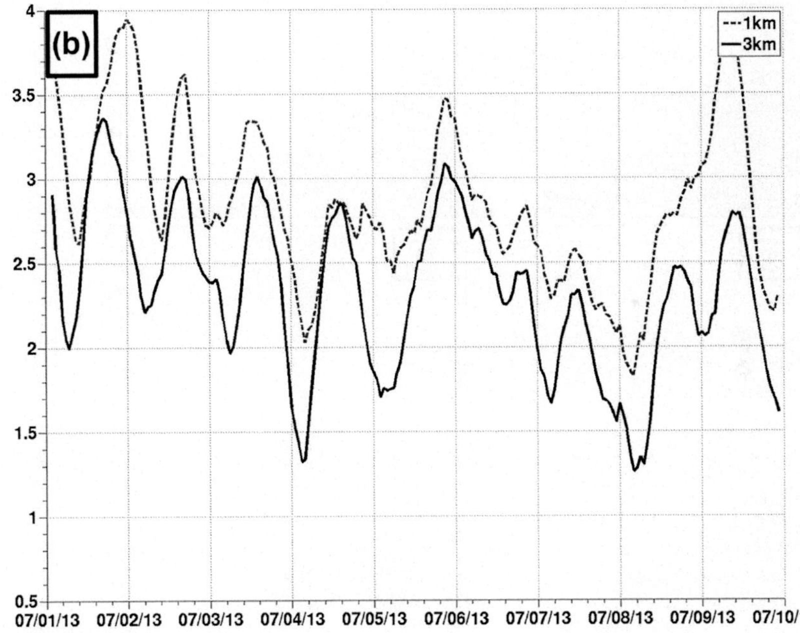

FIGURE 7

A comparison of the J_{FIT} values from the analyses (a) and the first guesses (b) from the 1-km (dashed) and 3-km (solid) model resolutions.

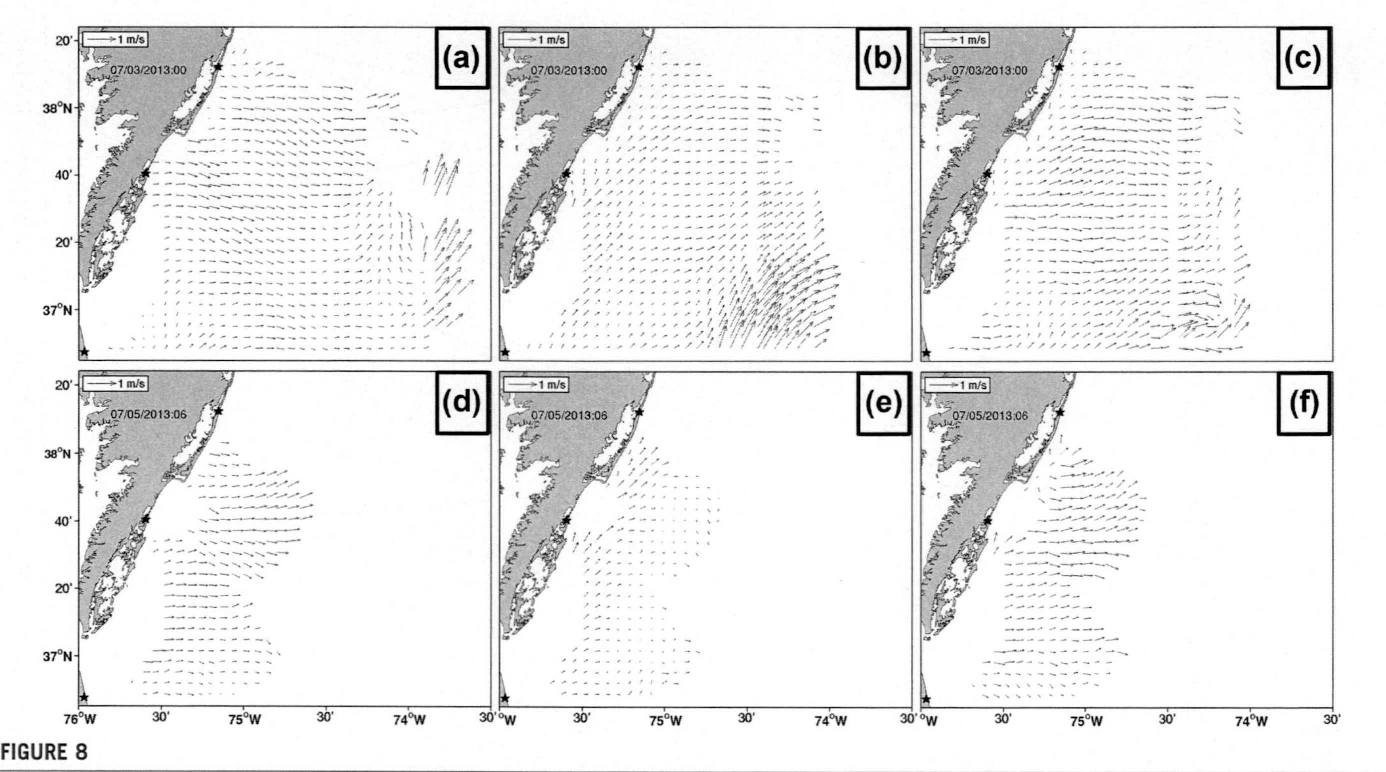

FIGURE 8

Similar to Figure 4, surface velocities from the observations (a, d), the first guess (b, e), and the analysis (c, f) on July 3 and July 5, respectively.

procedure corrects the first guess significantly to produce an analysis that fits (looks like) the observations, e.g., the Gulf Stream current is in better agreement with the observations, albeit not perfectly. However, the analysis on July 3 still has some patterns of the first guess, indicating that the background and model errors prescribed for the assimilation may still be too small. The analysis on July 5 shows a much better agreement with the observations (offshore circulation), even though the first guess was not (coastal circulation). Thus, even in the presence of an erroneous wind stress, the weak constraint assimilation fits the observations significantly better when the model's horizontal resolution is closer to that of the observations coverage.

7. CONCLUSION

Surface velocity observations from HF radar are a valuable data set for monitoring the coastal circulation, and they can be assimilated into a coastal circulation ocean model using the NCOM-4DVAR system, provided adequate model resolution, initialization, boundary conditions, and atmospheric forcing. It was shown in the experiments presented here that the assimilation cannot accurately fit the observations with rather small initial conditions and model errors. However, the biggest challenge for the assimilation system consists in an erroneous wind stress that consistently steers the model in a completely different direction than the observed surface velocities. Although the assimilation was able to reduce a noticeable portion of the model's discrepancy to the observations, those gains were quickly lost in the forecast stage for the following assimilation window, primarily due to the wrong wind stress and the high resolution of the model. Reducing the model resolution to be closer to that of the observations significantly improved the accuracy of the analysis and the forecast. The ability of the assimilation system to accurately fit the observations can also be improved by prescribing higher model errors. However, we suggest that instead of increasing the error levels in the assimilation system, which can be justified for the case at hand, the primary source of the errors must be addressed first, i.e., providing a wind stress that drives the model to be in more agreement with the observations. This can be achieved by a local nest of the atmospheric model, or better yet, a fully coupled ocean–atmosphere model.

ACKNOWLEDGMENTS

This work was sponsored by the Office of Naval Research Program Element 0601153N as part of "A multiscale Approach for Assessing Predictability of ASW environment" and "Rapid Transition NCOM-4DVAR" projects. The authors would like to thank the two anonymous reviewers whose constructive comments helped to improve the quality of this paper. The authors would also like to acknowledge Teresa Updyke at the Center for Coastal Physical Oceanography at Old Dominion University for freely providing the HF radar data used in this study and Rich Pawlowicz for providing the freely available M_MAP Matlab toolbox used here. This paper is NRL contribution number NRL/BC/7320-14-1231-4004.

REFERENCES

1. Oke PR, Allen JS, Miller RN, Egbert GD, Kosro PM. Assimilation of surface velocity data into a primitive equation coastal ocean model. *J Geophys Res* 2002;**107**:3122. http://dx.doi.org/10.1029/2000JC000511.
2. Paduan JD, Shulman I. HF radar data assimilation in the Monterey Bay area. *J Geophys Res* 2004;**109**:C07S09. http://dx.doi.org/10.1029/2003JC001949.
3. Barth A, Alvera-Azcárate A, Weisberg RH. Assimilation of high-frequency radar currents in a nested model of the West Florida Shelf. *J Geophys Res* 2008;**113**:C08033.
4. Li Z. A multi-scale three-dimensional variational data assimilation scheme and its application to coastal oceans. In: *Proceedings of the 9th workshop on adjoint model applications in dynamic meteorology, Cefalu, Sicily, Italy*; 2011. http://gmao.gsfc.nasa.gov/events/adjoint_workshop-9/presentations/Li.pdf.
5. Hoteit I, Cornuelle B, Kim SY, Forget G, Köhl A, Terrill E. Assessing 4D-VAR for dynamical mapping of coastal high-frequency radar in San Diego. *Dyn Atmos Oceans* 2009;**48**:175–97.
6. Zhang WG, Wilkin JL, Arango HG. Towards an integrated observation and modeling system in the New York Bight using variational methods. Part I: 4DVAR data assimilation. *Ocean Modell* 2010;**35**(3):119–33.
7. Yu P, Kurapov A, Egbert GD, Allen JS, Kosro PM. Variational assimilation of HF radar surface currents in a coastal ocean model off Oregon. *Ocean Modell* 2012;**49–50**:86–104. http://dx.doi.org/10.1016/j.ocemod.2012.03.001.
8. Ngodock HE, Carrier MJ. A 4D-Var assimilation system for the Navy coastal ocean model part I: system description and assimilation of synthetic observations in the Monterey Bay. *Mon Weather Rev* 2014:2085–107. http://dx.doi.org/10.1175/MWR-D-13-00221.1.
9. Carrier MJ, Ngodock HE, Smith SC, Jacobs GA, Muscarella PA, Ozgokmen T, et al. Impact of assimilating ocean velocity observations inferred from Lagrangian drifter data using the NCOM-4dvar. *Mon Weather Rev* 2014;**142**(4):1509–24.
10. Roarty H, Glenn S, Kohut J, Gong D, Handel E, Rivera E, et al. Operation and application of a regional high-frequency radar network in the Mid-Atlantic Bight. *Mar Technol Soc J* 2010;**55**(6).
11. Lipa BJ, Barrick DE. Least-squares methods for the extraction of surface currents from CODAR cross-loop data: application at ARSLOE. *IEEE J Oceanic Eng* 1983;**OE-8**:226–53. http://dx.doi.org/10.1109/JOE.1983.1145578.
12. Paduan JD, Rosenfeld LK. Remotely sensed surface currents in Monterry Bay from shore-based HF radar (coastal ocean dynamics application radar). *J Geophys Res* 1996;**101**:20669–86.
13. Martin P. *Description of the navy coastal ocean model version 1.0*. 2000. NRL report NRL/FR/7322—00-9961.
14. Barron CN, Kara AB, Martin PJ, Rhodes RC, Smedstad LF. Formulation, implementation and examination of vertical coordinate choices in the global navy coastal ocean model (NCOM). *Ocean Modell* 2006;**11**:347–75.
15. Goerss JS, Phoebus PA. The Navy's operational atmospheric analysis. *Weather Forecast* 1992;**7**:232–49.
16. Rosmond TE. The design and testing of the navy operational global atmospheric prediction system. *Weather Forecast* 1992;**7**:262–72.

17. Rosmond TE, Teixeria J, Peng M, Hogan TF, Pauley R. Navy operational global prediction system (NOGAPS): forcing for ocean models. *Oceanography* 2002;**15**:99–106. http://dx.doi.org/10.5670/oceanog.2002.40.
18. Bennett AF. *Inverse methods in physical oceanography.* New York: Cambridge University Press; 1992. 347 pp.
19. Bennett AF. *Inverse modeling of the ocean and atmosphere*. Cambridge University Press; 2002.
20. Ngodock HE, Chua BS, Bennett AF. Generalized inversion of a reduced gravity primitive equation ocean model and tropical atmosphere ocean data. *Mon Weather Rev* 2000;**128**: 1757–77.
21. Ngodock HE, Carrier MJ. A weak constraint 4D-Var assimilation system for the navy coastal ocean model using the representer method. In: Park SK, Xu L, editors. *Data assimilation for atmospheric, oceanic and hydrologic applications*, vol. II. Springer-Verlag Berlin Heidelberg; 2013. http://dx.doi.org/10.1007/978-3-642-35088-7_15.

CHAPTER

21 System-Wide Monitoring Program of the National Estuarine Research Reserve System: Research and Monitoring to Address Coastal Management Issues

Edward J. Buskey[1,*], Marie Bundy[2], Matthew C. Ferner[3], Dwayne E. Porter[4,5], William G. Reay[6], Erik Smith[5,7], Dwight Trueblood[8]

Marine Science Institute, University of Texas, Port Aransas, TX, USA[1]; Office for Coastal Management/NOS/NOAA, Silver Spring, MD, USA[2]; San Francisco Bay NERR, San Francisco State University, Tiburon, CA, USA[3]; Arnold School of Public Health[4]; The Baruch Institute for Marine and Coastal Sciences, University of South Carolina, Columbia, SC, USA[5]; Virginia Institute of Marine Science, Gloucester Point, VA, USA[6]; North Inlet Winyah Bay NERR, Georgetown, SC, USA[7]; Office for Coastal Management/NOS/NOAA, University of New Hampshire, Durham, NH, USA[8]

**Corresponding author: E-mail: ed.buskey@utexas.edu*

CHAPTER OUTLINE

1. INTRODUCTION TO THE NERRS

The National Estuarine Research Reserve System (NERRS) is a network of 28 estuarine reserves (Figure 1) established in 1972 by the Coastal Zone Management Act (as amended in 1990). The NERRS protects over 1.3 million acres of coastal land and waters of the United States and Puerto Rico, and it operates as a partnership between the National Oceanic and Atmospheric Administration (NOAA) and the coastal states. NOAA provides funding, national guidance, and technical assistance, while the states provide staff, infrastructure, and management for the operation of the reserves. The NERRS goals and priorities are focused on long-term research, monitoring, education, and stewardship to support coastal management.

The NERRS is characterized by a diverse set of coastal systems that represent different biogeographic regions and estuarine types. The selection process for designation of NERRS sites must be approved by NOAA, and sites must contribute to the biogeographic and typological balance of estuaries within the system. These sites must also be considered representative estuarine ecosystems and should, to the maximum extent possible, be estuarine ecosystems minimally affected by human activity or influence. Reserve boundaries are designed to encompass two areas: a core area composed of "key land and water areas" and a buffer zone. Core areas are considered those regions vital to the function of the estuarine ecosystem, and buffer zones are areas adjacent to the core areas that are essential to estuarine ecosystem integrity. The sites must also be suitable for long-term estuarine research that is compatible with existing and potential land and water uses in contiguous areas, and that provides important opportunities for education and outreach. These estuaries are among the most productive coastal ecosystems in the United States and provide a suite of valuable ecosystem services, including sediment trapping, shoreline stabilization, and contaminant and nutrient remediation. They also serve as critical nursery, feeding, and spawning areas for commercially and recreationally important marine and coastal species. Reserve staff work with state and local agencies and communities to provide a better understanding of how natural resource management affects coastal environmental issues such as nonpoint source pollution, habitat restoration, invasive species, and climate change. Through integrated research, education, training, and stewardship programs, reserves help communities develop meaningful strategies to successfully address a variety of coastal resource management challenges.

2. INTRODUCTION TO THE NERRS SYSTEM-WIDE MONITORING PROGRAM

Environmental observation and monitoring has been an important component of the NERRS research program from its inception and is a primary mechanism of assessing environmental impacts of climate change and other stressors on coastal

FIGURE 1

The national estuarine research reserve system.

Map provided by NOAA.

resources. In support of this effort, the NERRS System-Wide Monitoring Program (SWMP) was formally established in 1995 and has since become a robust, long-term observing system with the capacity to comprehensively address existing and emerging coastal management issues, improve the scientific understanding of estuarine processes, and inform management decisions. The mission is to develop quantitative measurements of short-term variability and long-term changes in the water quality, biologic systems, and land use and land cover characteristics of estuaries and estuarine ecosystems for the purpose of informing effective coastal zone management (NERRS SWMP Plan 2011; http://nerrs.noaa.gov/Doc/PDF/Research/2011SWMPPlan.pdf). In support of this mission, SWMP was designed to address three fundamental questions:

- How do environmental conditions vary through space and time within the network of reserve sites?
- How does ecosystem function vary through space and time within critical reserve habitats?
- To what extent are changes in estuarine ecosystems represented by the NERRS attributable to natural variability versus anthropogenic activity?

The locations of SWMP stations at each NERRS site were initially chosen to be both characteristic of the estuary while also trying to compare sites that were impacted by nonpoint source pollutants to sites that were not impacted or less impacted. When the SWMP program expanded from a minimum of two to four monitoring sites per reserve, SWMP stations were additionally located to address gradients in salinity, land use, or habitat. The use of standardized approaches, instrumentation, and protocols, and the high temporal resolution of the data across all 28 reserves makes SWMP unique among coastal observing systems. At every reserve, a suite of water quality and meteorologic data are made available in near real-time through satellite based telemetry. In addition, an integrated adaptive monitoring framework allows data to be useful on national, regional, and local scales. SWMP has been periodically evaluated by external scientists and coastal managers to ensure that the program is operationally efficient and cost-effective, and reserve staff are dedicated to interpreting and communicating SWMP data and products to a wide variety of users including scientists, educators, coastal decision makers, and the public.

This chapter focuses on four core aspects of SWMP: abiotic data collection, biologic observations, habitat and land use mapping, and sentinel site monitoring. A rigorous approach to data collection and management using standardized protocols, parameters, and approaches to data quality assurance/quality control (QA/QC) have been developed for each aspect of the program. The standardized approaches and resulting long-term data sets provide the capacity to address environmental issues at different temporal scales, and to conduct data analyses using observations from more than one reserve to answer regional and national questions.

3. ABIOTIC SWMP COMPONENTS

Abiotic monitoring includes high temporal resolution water quality and meteorologic data collected at 15-min intervals by automated data sondes and weather stations placed in field locations at all National Estuarine Research Reserves. Each reserve maintains at least one meteorologic station to quantify atmospheric conditions by measuring air temperature, relative humidity, barometric pressure, wind speed, wind direction, rainfall, and photosynthetically active radiation (PAR). Water quality is measured by data sondes at a minimum of four stations in each reserve, with station locations designed to characterize gradients in estuarine environmental conditions. All reserves use automated data sondes manufactured by Yellow Springs Instrument Co. and use standardized calibration and QA/QC protocols. Core measurements include water temperature, conductivity (salinity), dissolved oxygen, pH, turbidity, and pressure (water depth). Because the methods used for abiotic monitoring are independent of the geologic or biologic characteristics of the estuary, the same protocols and instruments can be used at all reserves, producing a high-quality data set that can be readily compared between locations. Data are available for 150 continuous water quality stations, with 113 of these still active; 144 nutrient and chlorophyll *a* water sampling stations, with 130 still active; and 32 meteorologic stations, with 29 still active. These numbers do not include supplemental stations at many reserves, for which data are not submitted to the NERRS Centralized Data Management Office (CDMO) to be archived.

The core abiotic elements of SWMP also include monthly water samples collected at the water quality stations where data sondes are located. These water samples are processed for measurements of inorganic nutrients and extracted chlorophyll *a*. Nutrient measurements include nitrate, nitrite, ammonium, and orthophosphate. In addition to these monthly samples, a diel series of water samples is also collected at one location per reserve at 2-h intervals over a 24-h tidal cycle. These diel samples are analyzed for the same suite of inorganic nutrients and chlorophyll *a*, providing information on the flux of nutrients and phytoplankton biomass during tidal exchange within the estuary. Additional measurements of water chemistry would provide a more complete picture of nutrient dynamics in the NERRS, and several reserves currently collect data on total dissolved nitrogen, total dissolved phosphorous, and total alkalinity as optional parameters for their water quality monitoring, with further expansion being planned.

The core abiotic elements of SWMP help establish baseline indicators of estuarine habitat condition and environmental quality, and they provide useful information to assess human use impacts and risks related to human health issues. Measurements of inorganic nutrients help assess the availability of nutrients that drive primary production in the estuary, and they allow for the determination of excess nutrients that can lead to eutrophication or harmful algal blooms. Measurement of chlorophyll *a* is an indicator of phytoplankton biomass and a useful indicator of water quality related to nutrient load and eutrophication, and dissolved oxygen measurements can indicate hypoxic conditions that can lead to the formation of "dead zones."

In addition to the parameters required to be measured by all reserves' SWMP programs, additional elective measures are made based on research and management issues at various reserves. Examples of some of these elective abiotic measures include deposition of atmospheric nitrogen and mercury, automated measurement of chlorophyll fluorescence by data sondes, and additional water chemistry parameters such as silicate, total suspended solids, dissolved and particulate organic carbon, pCO_2, and others.

4. BIOLOGIC SWMP COMPONENTS

Biologic monitoring in the NERRS is custom-designed to allow each reserve to address specific management issues and to provide indicators of the composition and diversity of the biologic communities within the reserve. Because NERRS estuaries encompass diverse habitats such as freshwater estuaries on the Great Lakes, subarctic fjords in Alaska, and mangrove dominated estuaries in Florida and Puerto Rico, it is not possible to standardize biologic monitoring across all reserves as has been done with abiotic monitoring. Therefore, the NERRS has adopted a "toolbox" approach by which biologic monitoring protocols are developed for various monitoring elements such as submerged aquatic vegetation (SAV), emergent vegetation (EV), and mangroves. Standard protocols and approaches have also been developed, or are under development, for other biologic components including phytoplankton and zooplankton, benthic invertebrate communities, nekton communities, and marsh shorebirds. A link to the mandatory and elective monitoring occurring at each NERRS site can be found at http://cdmo.baruch.sc.edu/data/availableTwo.cfm.

Periodic monitoring of SAV and EV has been the most widely implemented biologic component of SWMP and is currently applied across most reserves within the system. Periodic assessments of the ecological characteristics and areal extent of vegetated communities can provide a sensitive indicator of estuarine ecosystem structure and functionality. Standard protocols have been developed to assess changes in biomass, growth, and species abundance in SAV and EV estuarine habitats.[1] Sites for vegetation monitoring within reserves are chosen by their proximity to abiotic monitoring stations and, when possible, are located in areas where high-resolution mapping is available to ground-truth larger scale efforts to characterize vegetative communities within the reserve and its watershed. The protocols adopted for SWMP vegetation monitoring are in close alignment with protocols used by other agencies such as the National Park Service[2] and the global SeagrassNet monitoring program.[3]

5. HABITAT MAPPING AND CHANGE ANALYSIS

Habitat and land-use classification and change analysis have been long recognized as core elements of SWMP.[4,5] From a NERRS perspective and broadly defined, habitat refers to an ecological or environmental area that supports a community of plants and animals. Examples of estuarine habitats include salt marshes, submerged sea grass beds, mangrove forests, and others. The NERRS Habitat Mapping and Change (HMC) Plan[6] establishes a structured framework for mapping reserve habitats and boundaries with the primary goals of (1) characterizing short-term variability (1–10 years) and long-term trends (10+ years) in the spatial extent of reserve habitats at various scales (site, regional, and national) and (2) examining the impact of climate change, effects of land use within adjacent watersheds, and changes in local water level and inundation patterns on reserve habitats.

The development of a hierarchical land use/land cover classification that is reserve-wide, comprehensive, and multilevel was required for implementation of the NERRS Habitat Mapping and Change element. Using the U.S. Fish and Wildlife Service's national wetland mapping standards as a foundation, the NERRS modified and expanded the mapping standards to include upland and frozen habitat and cultural cover types[7] (Figure 2). Justifications for various aspects of the classification

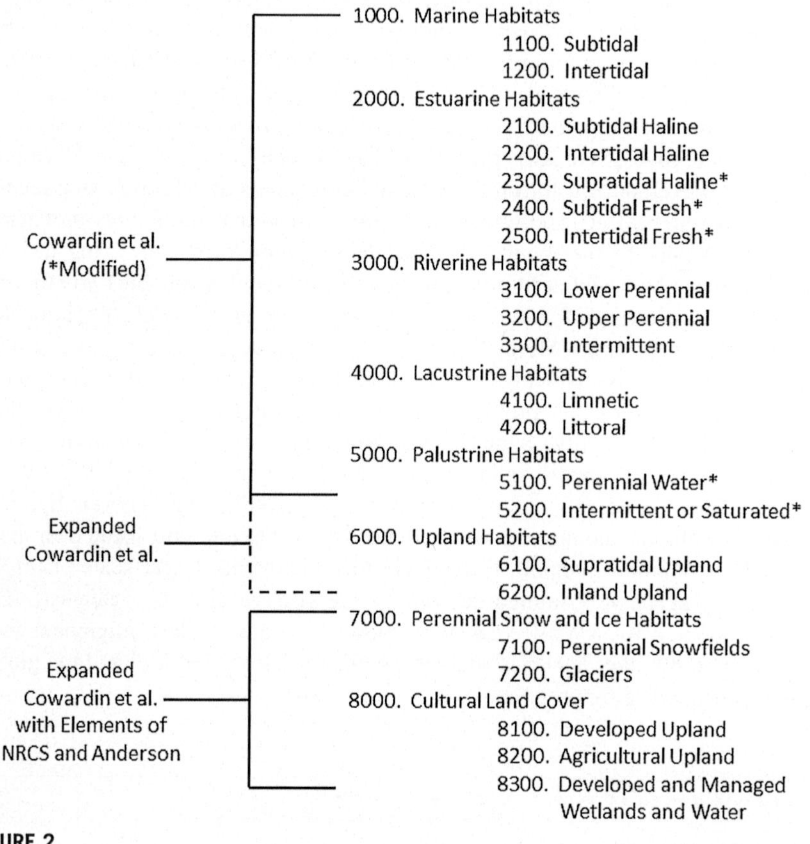

FIGURE 2

The structure and content of levels one and two of the NERRS classification scheme.[7]

are detailed in Kutcher et al.[8] Utilizing the developed classification scheme, the NERRS has implemented a two-level approach to characterize land use/land cover within reserve boundaries and associated watersheds. High-resolution imagery (1–3 m) is required to classify land use/land cover within reserve boundaries, whereas moderate resolution imagery (e.g., Landsat TM 30-m resolution data available through NOAA's Coastal Change Analysis Program) is required for watershed-level land use/land cover analysis. As part of the HMC Plan, habitat change analysis should be conducted on a 10-year basis for priority habitats that have been identified in each reserve. Although formal NERRS guidelines for change analysis have yet to be developed, individual reserves are currently using well-documented, peer-reviewed methodologies (e.g., pixel-based classification, object-oriented image analysis). The HMC Plan implementation protocols and standard operating procedures[9] have been established and data management and archiving are provided by NOAA/NERRS CDMO.

Habitat distribution and change analysis maps have been used extensively for a variety of special area management applications including basic documentation of existing habitat types and distributions for reserve site profiles, management plans, and disaster response plans. Additional stewardship and research uses include monitoring restoration project success, tracking native and invasive species distributions,[10] investigating vegetative impacts due to anthropogenic hydrologic modifications,[11] delineating and monitoring of marsh shoreline erosion and marsh-upland ecotone retreat, and predicting changes in habitat distributions resulting from climate change.[12]

6. SENTINEL SITES PROGRAM FOR EVALUATING CLIMATE CHANGE IMPACTS

An initial assessment of climate sensitivity in the NERRS has shown that critical factors to consider when assessing a reserve's vulnerability to climate change include social as well as biophysical factors.[13] The NERRS Sentinel Sites Program is designed to address the biophysical component of climate vulnerability and constitutes an early detection network for coastal impacts due to climate change. The ultimate goal of the program is to help determine reserve biophysical vulnerabilities to climate change and to translate that understanding to coastal communities and natural resource managers. Predictions of accelerating sea level rise[14] initially led the NERRS to prioritize evaluation of the effects of changing water levels on key coastal habitats. Specifically, the NERRS is linking long-term SWMP data to a network of specialized geodetically referenced infrastructure, such as survey monuments and deep-rod Surface Elevation Tables (SETs) that allow precise vertical measurement of effects of changing coastal water levels on marshes, mangroves, and submerged aquatic vegetation. The NERRS sentinel sites exemplify the value of integrated, long-term, place-based environmental monitoring. The program will be expanded to address effects of other climate-related stressors on coastal systems as resources allow.

Initial questions answered by the NERRS Sentinel Sites Program require fundamental data on changing water levels and inundation patterns, and include the following:

- What are the current distributions of vegetation communities with respect to elevation and tidal range, and how sensitive are spatial distributions and community composition to inter-annual variability?
- What are the responses, in terms of spatial distributions and community composition of vegetation communities, to long-term changes in local water levels and inundation patterns (e.g., changes in mean water level, tidal amplitude, and storm frequency and intensity)?
- What are the responses of sediment elevation of vegetated habitats (with respect to changes in sediment deposition, accretion, and/or subsidence) to discrete episodic inundation events as well as long-term changes in local water levels and inundation patterns?

To address these questions, SWMP vegetation monitoring protocols are used to quantify horizontal rates of marsh migration and upland transgression, changes in species composition, and changes in the areal extent of vegetated habitats. In those same areas, SETs and other methods are used to measure sediment elevation and elevation change as drivers of coastal vegetation patterns. Additionally, working with NOAA's National Geodetic Survey and Center for Operational Oceanographic Products and Services, the NERRS sentinel sites are equipped with local geodetic control networks to establish and accurately track spatial relationships between the observed environmental parameters.

Standardized integration of these monitoring approaches is the basis of the NERRS Sentinel Site Program, yet the societal value of the program lies in determining reserve vulnerabilities and translating that understanding to coastal resource managers and other audiences. An emerging strategy for data translation involves the use of previously developed numeric simulations of coastal wetland change.[15,16] These ecological forecasting approaches combine regionally downscaled estimates of future sea level rise with data from sentinel site monitoring and other SWMP activities to predict how the distribution of coastal habitats will change over time. Maps of model results (Figure 3) allow resource managers to prioritize conservation areas that are predicted to be resilient and restoration areas where management action is needed, inform related studies of green infrastructure and carbon sequestration, and facilitate climate communication among educators, decision makers, and the public. Model results also provide numerous testable hypotheses about incremental effects of sea level rise that should be evaluated with ongoing data collection through the NERRS Sentinel Sites Program.

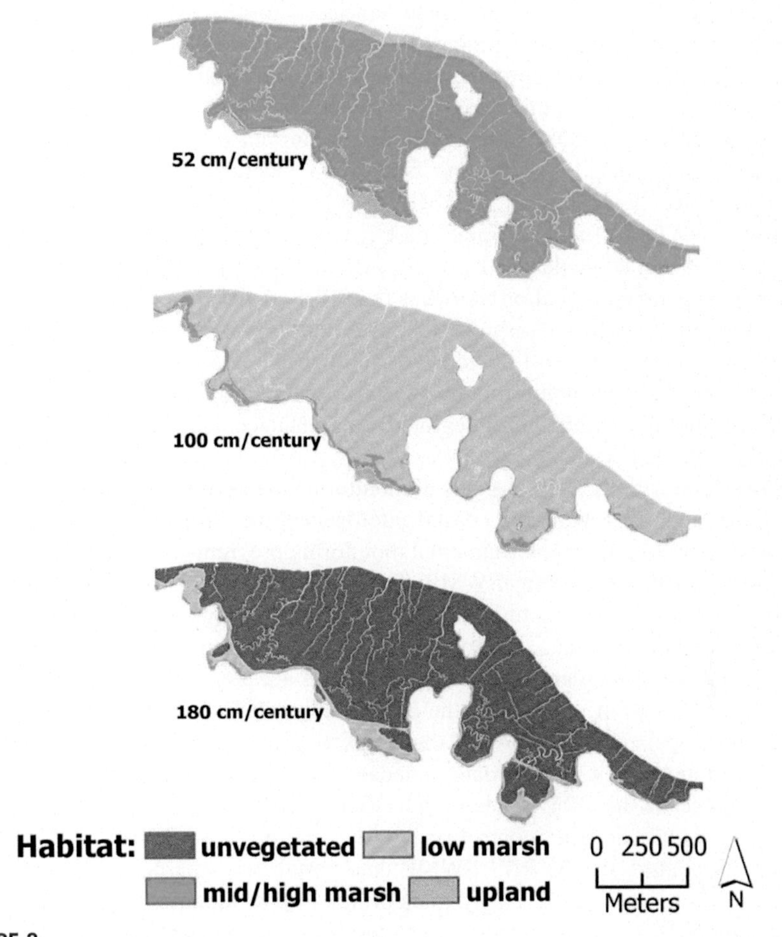

FIGURE 3

Predicted habitat distributions at a San Francisco Bay NERR salt marsh using three possible scenarios of century sea level rise in the Marsh Equilibrium Model, based on conditions in 2010. Data from the NERRS Sentinel Sites Program will be used to evaluate and refine model predictions over time.

7. NERRS SWMP DATA MANAGEMENT

Implementation and maintenance of robust data management and communications are critical challenges for development of successful collaborative scientific initiatives. The development of the data and information management components of coastal observing systems must address both core and cooperating programs' data and information exchange while also meeting the needs of the end users.

A goal of NOAA's National Ocean Service is to increase *coastal intelligence* with a commitment to integrating scientifically defensible data, models, and decision-support tools to improve the ability of decision-makers scaling from federal agencies to the private individual. To that end, the NOAA-led Integrated Ocean Observing System (IOOS) initiative is developing an integrated national system for the United States coastal zone built upon federal monitoring efforts (e.g., National Data Buoy Center (NDBC) and National Water Level Observation Network (NWLON)) and regional coastal ocean observing systems that monitor the state and characteristics of the oceans, Great Lakes, and estuaries of the United States. An operational goal of NOAA is the timely delivery of comprehensive data and information on the nation's coastal and estuarine waters and the open waters of the coastal ocean. This will be achieved through the coordinated development of national and regional monitoring and systems meeting the data and informational needs of multiple users via integrated data management on global, national, regional, and local scales.

Consistent with the NOAA goals of monitoring the health of our nation's coastal waters in support of increased coastal intelligence, the NERRS acknowledges the importance of long-term environmental monitoring programs, data, and information dissemination by supporting SWMP. The purpose of the NERRS CDMO is to manage the basic infrastructure and data protocols to support the assimilation and exchange of data, metadata, and information within the framework of NERRS sites.

The quality of the data ingested by and made accessible from the CDMO data portal is critical to the success of the SWMP effort. SWMP data must adhere to a documented system of data QA/QC that helps to identify spurious data and/or interesting data anomalies. Appropriate metadata are developed and maintained for all data available via the CDMO data portal. After SWMP data and associated metadata undergo a rigorous quality control review, they are made available via a variety of methods including the NERRS SWMP data portal (www.nerrsdata.org) and web services. These data can be easily transformed to pass along to appropriate nodes in existing NOAA data networks such as the National Oceanographic Data Center, NDBC, and IOOS. In addition, the NERRS has implemented Geostationary Operational Environmental Satellite-based telemetry systems for water quality and meteorologic stations to provide for near real-time access to provisional SWMP data. Other programs that provide near real-time access to data include U.S. Geological Survey stream gauge stations, National Weather Service weather stations, National Water Level Observing Network water level stations, and National Data Buoy Center ocean buoys, among others.

8. CONDITIONS ACROSS THE NERRS

The 28 reserves that currently make up the NERRS represent all major coastal regions of the United States, including the Great Lakes, as well as almost every established climate zone and biogeographic province within these coastal regions.

The estuaries represented by the NERRS also include the full suite of estuarine geomorphic types, including the river-to-lake freshwater estuaries of the Great Lakes. The NERRS encompasses the full range of estuarine salinities, with annual mean salinities across the reserves ranging from 0 to 37 psu and mean daily salinity range within site varying from 0 to 19 psu.[17] Although some reserves comprise just a single ecotype, many others encompass the full salinity gradient of their estuary. The specific habitats monitored by the SWMP are mainly located in the littoral zone of the estuaries or tidal creeks (mean SWMP site water depth is generally <2 m). As such, the temporal dynamics of water quality conditions monitored by SWMP largely represent conditions that occur in the shallow-water estuarine environments throughout the nation. Across the diversity of NERR sites and habitats, water quality conditions vary substantially. The concentration of chlorophyll *a* (Chl) is, perhaps, the single most widely used index of trophic condition in aquatic environments. Based on data collected from 2007 to 2012, the median Chl concentrations ranged over an order of magnitude across the NERRS (Figure 4).

The majority (65%) of sites had a median Chl of <5 μg/L, the threshold defining "good" water quality in the U.S. Environmental Protection Agency's Coastal Conditions Assessment.[18] Nine NERRS sites had median Chl concentrations indicative of "fair" water quality (5–20 μg/L), while just one marine site exceeded the threshold indicative of "poor" water quality (>20 μg/L). All sites exhibited substantial temporal variability in Chl, which was often similar in magnitude to that observed across sites. This within-site variability underscores the importance of high-frequency monitoring in determining average conditions within and across estuaries. A similar pattern in the distribution of water quality conditions, relative to the EPA's numeric criteria, could be seen using median dissolved inorganic nitrogen concentrations, which varied by close to four orders of magnitude across the NERRS, or dissolved inorganic phosphorus concentrations, which varied by over two orders of magnitude across the NERRS (data not shown). The lack of degraded water quality in many of the reserves is consistent with an analysis of hypoxia occurrence across SWMP sites. Though hypoxia occurred frequently at most SWMP sites, these events were generally of short duration with only a very few number of sites experiencing hypoxic conditions for periods >24 h.[17]

Previous attempts at characterizing water quality conditions across the NERRS using a variety of multivariate statistical approaches, specifically principle component analysis and nonlinear multidimensional scaling,[19,20] have shown salinity and temperature to be the primary factors determining the grouping of reserves based on similarities in environmental conditions. The strong co-variation of salinity and other factors (e.g., nitrate) in these analyses suggest that it is not salinity alone, but the delivery of nutrients, primarily nitrogen, associated with freshwater inflow that drives much of variability across the SWMP data set. Interestingly, these analyses all roughly grouped NERRS sites according to biogeographic region, suggesting that despite local-scale differences in the degree of anthropogenic

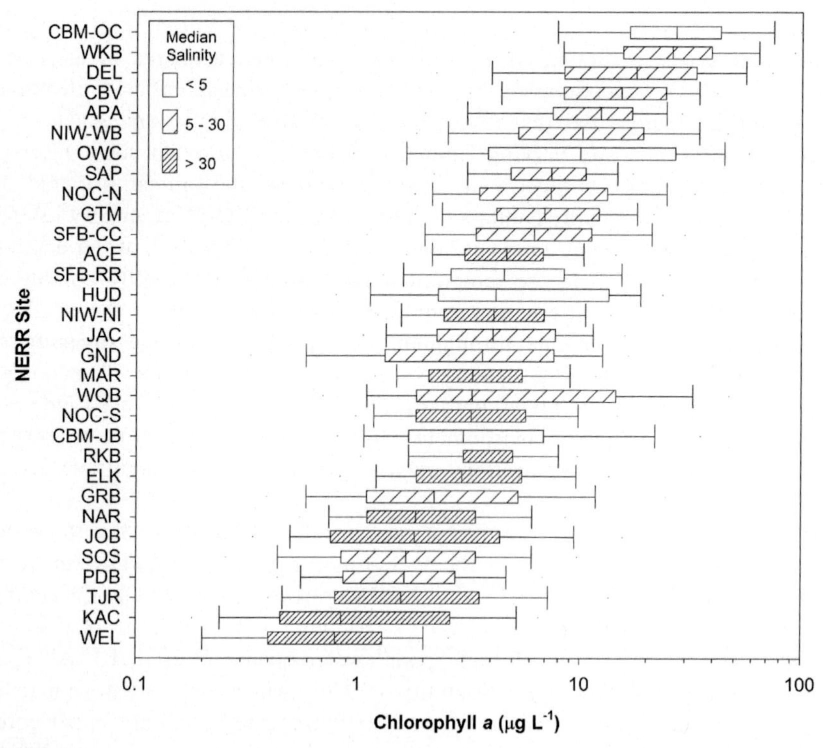

FIGURE 4

Box plots of chlorophyll *a* distributions across NERRS sites from 2007 to 2012. For reserves with geographically discontinuous components that had significantly different chlorophyll *a* distributions, component distributions are plotted separately. Site IDs, from highest to lowest median value are as follows: Chesapeake, Maryland, Otter Creek; Weeks Bay; Delaware Bay; Chesapeake, Virginia; Apalachicola Bay; North Inlet - Winyah Bay, Winyah Bay; Old Woman Creek; Sapelo Island; North Carolina, Northern Component; Guana Tolomato Matanzas; San Francisco Bay, China Camp; ACE Basin; San Francisco Bay, Rush Ranch; Hudson Bay; North Inlet—Winyah Bay, North Inlet; Jacques Cousteau; Grand Bay; Mission-Aransas; Waquoit Bay; North Carolina, Southern Component; Chesapeake, Maryland, Jug Bay; Rookery Bay; Elkhorn Slough; Great Bay; Narragansett Bay; Jobos Bay; South Slough; Padilla Bay; Tijuana River; Kachemak Bay; Wells.

influence on individual NERRS sites, regional-scale climatology and biogeography exert a strong influence on the water chemistry of shallow water estuarine ecosystems.

Although SWMP does not currently collect any direct biologic rate measurements, the dissolved oxygen time series generated by SWMP has successfully been used to generate system-scale and regional estimates of gross primary production and total respiration.[21–23] Net ecosystem metabolism (gross production − total respiration) has been shown to be another useful index of estuarine condition as it integrates many of the complex processes and trophic interactions that occur in estuaries.[24] Of the SWMP sites analyzed by this approach, most (23 of 27) were significantly net heterotrophic (production < total respiration on an annual basis). All the sites adjacent to salt marshes or mangroves were net heterotrophic, many often substantially so. Only sites adjacent to SAV or macroalgal beds were either metabolically balanced or net autotrophic (production > respiration). The fact that most sites were net heterotrophic reinforces the concept that organic inputs to coastal systems are important drivers of estuarine dynamics, despite the high primary productivity generally associated with estuarine environments.

The consistent and comparable data collected by SWMP across the diverse biogeography and physiography represented by the NERRS provide an ideal opportunity to increase our understanding of how environmental factors influence variability in estuarine water conditions and their temporal dynamics. This is essential for separating natural variation from that induced by human activities. The value of SWMP is, thus, both in the data it provides to inform better management of critical estuarine habitats as well as the unparalleled cross-system platform it provides for leveraging further research on the temporal dynamics regulating estuarine processes and conditions. Many examples of the use of SWMP in targeted research within and across the NERRS can specifically be found in the two special issues of the Journal of Coastal Research (in 2004 and 2008) dedicated to research and monitoring within the NERRS.

9. DATA APPLICATIONS: WATER QUALITY ASSESSMENT, PUBLIC HEALTH

The major water-quality issues facing coastal managers and public health officials in coastal environments include nutrient enrichment and dissolved oxygen depletion (hypoxia), harmful algal blooms, and beach water quality. Human activity in the United States has increased nutrient fluxes to the coast substantially,[25] and nearly two-thirds of the nation's coastal waters are degraded through nutrient loading and eutrophication.[26] A general consensus has emerged that eutrophication in estuaries is primarily a function of inorganic nitrogen loading, although other factors such as light limitation, residence time, and activity level of grazers may make the relationship more complex.[27,28] Excess nitrogen load can lead to high phytoplankton biomass, and when phytoplankton die and sink to the bottom, the high oxygen demand of bacterial decomposition can lead to low oxygen and hypoxia. One of the most common causes of fish kills in estuaries is low concentrations of dissolved oxygen. This can be a completely natural occurrence due to decomposition of organic matter near the bottom in a stratified water column that reduces mixing by surface waves, or it can be exacerbated by human activities such as excess nutrient loads entering an estuary from municipal wastewater treatment plants or agricultural and urban nutrient runoff.

Oxygen concentrations of 4 to 5 ppm are often used to set minimum water quality standards, whereas 2 ppm is considered the minimum level to support most animal life and is used as a threshold for hypoxia.[29] There are typically daily variations in dissolved oxygen concentrations due to daytime photosynthesis of phytoplankton and aquatic plants, and nighttime respiration of all living organisms, including photosynthetic plants, animals, and bacteria that decompose detritus and other nonliving organic matter. Hypoxic conditions may occur only at night, when respiration rapidly reduces oxygen concentrations, but they may persist throughout the day if water column stratification is created by warmer, fresher, less dense water from rivers forming a layer over cooler, saltier water at depth. Exposure to hypoxic and anoxic conditions can be a major factor in determining the distribution and abundance of aquatic organisms, and there is considerable evidence indicating that the duration and extent of hypoxic events is increasing worldwide.[30,31] Data sondes used in the NERRS SWMP monitoring program are placed within a meter of the bottom along an estuarine gradient, collecting dissolved oxygen, temperature, and salinity data at 15-min intervals. Investigating dissolved oxygen dynamics in estuarine systems, Wenner et al.[17] characterized spatial and temporal trends in dissolved oxygen within the NERRS in a 2004 study (at 55 continuous monitoring stations within 22 reserves). General findings indicated hypoxic conditions occurred over a broad geographic range, 95 percent of all hypoxic events were relatively short-lived (<12 hrs; very few sites experienced hypoxic conditions >24 h), and most events occurred in the summer when water temperatures were elevated. The authors suggest that the brief periods of hypoxia were within tolerance limits of aquatic animals that effectively take up and use oxygen, and in general, water quality within the reserves has not experienced significant persistent problems often associated with more populated areas of the country. A follow-up study would be useful to determine if oxygen dynamics at the reserves have changed since 2005. SWMP measurements of dissolved oxygen provide data on the frequency and duration of hypoxic events to help determine if remediation efforts to reduce nutrient loads to the estuary are needed.

Reduced oxygen level is the most commonly cited cause of fish kills, particularly in eutrophic waters exhibiting elevated water temperature,[31] although winter fish kills due to cold stress have also been documented. In January 2014, regional (North Carolina to Maryland) die-offs of speckled trout (*Cynoscion nebulosus*) were noted and attributed to recurrent outbreaks of cold weather where water temperatures declined past reported tolerance limits of some important commercial finfish. Based on available information, including SWMP contributions that documented the magnitude and duration of cold water extreme temperatures, state agencies responsible for fisheries management were able to enact emergency management actions to protect the spawning stock in the following season (Figure 5).

Nutrient loading of inorganic nutrients can also lead to changes in phytoplankton community structure, and it has been implicated in the formation of Harmful Algal Blooms (HABs).[32] It has further been suggested that human alteration of nutrient ratios in coastal waters can alter natural phytoplankton

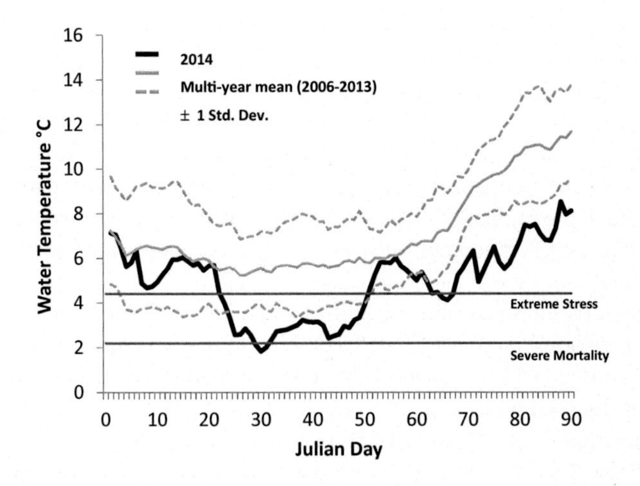

FIGURE 5

Daily water temperature variation between January 1, 2014 and March 3, 2014 at Gloucester Point, Virginia. Instantaneous measured low temperature was 1.2 °C. Low temperatures near the end of January resulted in local cold stun fish mortality.

Data provided by Chesapeake Bay NERR in Virginia.

community composition and possibly favor harmful or toxic species.[33] SWMP abiotic monitoring allows for examination of the relationships between inorganic nutrient concentrations and overall phytoplankton biomass (via chlorophyll *a*), but biotic monitoring of planktonic organisms, including HAB species, is not part of the biologic monitoring at most reserves. In some cases, plankton monitoring is accomplished through partnerships with other organizations. The Mission-Aransas reserve in Texas carries out continuous real-time monitoring for HAB species at their SWMP station in the Aransas Ship Channel in cooperation with Dr Lisa Campbell at Texas A&M University using an automated imaging flow cytometer,[34] and this program has been instrumental in providing warning to health officials of the initiation of HAB events that could impact public health through the consumption of oysters and other seafood.[35] In addition, monthly microplankton samples from each SWMP station in the Mission-Aransas reserve are analyzed for the presence of HAB species using an automated particle imaging system (the FlowCAM), which has been instrumental in documenting HABs such as *Karenia brevis* blooms.[36]

Recreational beach waters are also susceptible to contamination from potential pathogens. Often, these sources are from a mix of point and nonpoint discharges that include urban and suburban runoff, storm water effluent, and sewer system overflows that can transport pathogens known to cause illnesses in humans ranging from ear infections to cryptosporidiosis, an intestinal illness caused by the parasite *Cryptosporidium*. Standard sampling protocols rely on easily detectible bacteria, for example fecal coliforms, to indicate the possible presence of pathogens likely to cause human illness. Present methods used to determine bacterial concentrations require long incubation times (e.g., 24-hour), in which case beach swimmers and waders could be exposed to harmful pathogens before results are available and advisories can be posted. By enhancing the utility of existing coastal and ocean monitoring programs, improved predictive capabilities are desired to reduce the impact of unnecessary advisories (false positives) and failures to issue advisories when needed (false negatives) that sometimes occur because of delayed results.

In the state of South Carolina, an effort to address the need for timelier beach swimming advisories due to elevated bacteria levels was initiated in 2004.[37,38] With support from NOAA and the EPA, a collaboration including the University of South Carolina, University of Maryland, South Carolina Department of Health and Environmental Control, the Southeast Coastal Ocean Observing Regional Association, and the NERRS CDMO has produced a decision-support tool to provide timely assessments of local water quality conditions. To improve the information derived from the forecast models, a series of statistical analyses were performed to identify those coastal and ocean observing and monitoring system data that provide the best predictive capabilities of the presence of fecal coliforms. A number of parameters used to describe environmental conditions in the test area were included, such as air and water temperature, salinity, tidal stage, winds, and precipitation amounts. These near real-time monitoring data are available from a variety of sources including the NOAA NERRS and IOOS monitoring activities. In addition to spatially explicit rainfall data, measured through either a network of rain gauges or using rainfall estimates from the Next Generation Radar (NEXRAD) system, water temperature and salinity were found to improve the predictability of fecal coliform concentrations. These model parameters are related to processes known to regulate bacterial input and survival.

To make the daily forecast information available, an online portal and a mobile app with water quality predictions have been created to aid coastal resource managers, public health officials, and individuals. The outcomes of these efforts, made in conjunction with state and local resource managers and public health officials, have been more accurate and timely beach swimming advisory forecasts, with reduced false positives and lower indirect community costs by eliminating unwarranted advisories.

10. DATA APPLICATIONS: STORM SURGE

Storm surge can be defined as the abnormal rise in water level produced by a storm and is calculated as the difference between storm tide and the predicted astronomical tide. Storm surge is caused by high and sustained winds blowing across a water body creating a long wave that eventually piles up as it moves into shallower waters and by the storm's lower central air pressure, the former being the primary contributor. In addition to storm intensity and central pressure, other storm (size, speed, angle of approach to land mass), coastal (shape of the coast line, width and slope of ocean bottom), and local (barrier islands, bays, rivers, etc.) features influence storm surge.[39] The most severe storm surges impacting the nation are associated with large-scale tropical cyclone systems, such as tropical storms and hurricanes that generally occur in the Atlantic and Eastern Pacific basins from mid-May through November, and extra-tropical cyclone systems such as Nor'easters that impact the mid- and northeast Atlantic coasts typically during winter months.

The NERRS SWMP, with both its meteorological and water quality components, is suited to characterize storms and associated impacts on water levels and water quality at multiple spatial and temporal scales. Hurricane/Post Tropical Cyclone Sandy (also referred to as Superstorm Sandy), the largest Atlantic hurricane on record (with high wind fields >65 km/h (18 m/s) extending 1800 km) provided an opportunity to highlight SWMP geographic coverage and capacity of a single event. This storm system impacted much of the eastern seaboard, including 14 reserves (Figure 6). Barometric pressure varied from 946 to 1000 mbars across the East Coast reserve network with minimum values reported at the Delaware (958 mbar) and

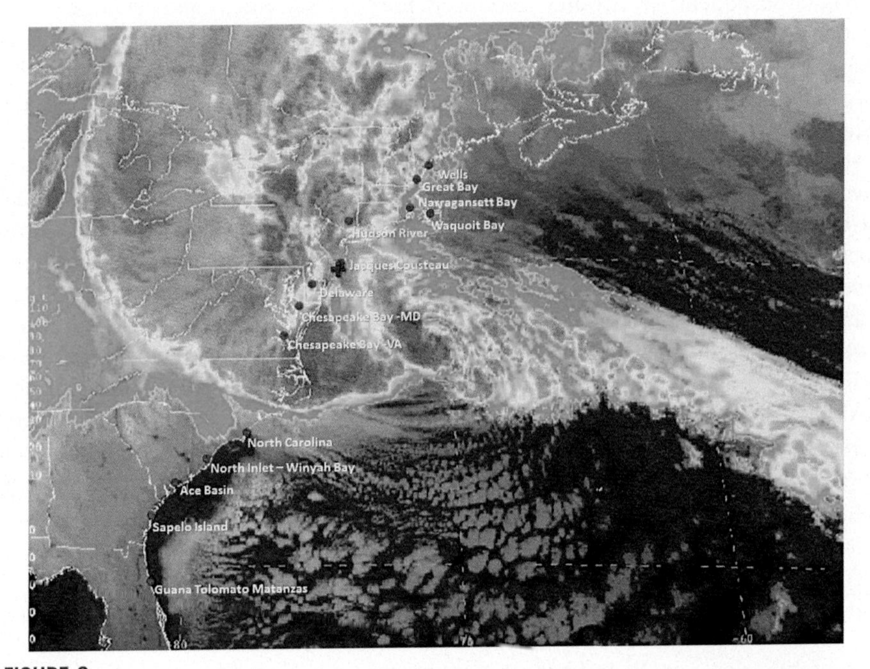

FIGURE 6

NOAA GOES-E infrared satellite image of Hurricane Sandy (12:15 UTC October 29, 2012) along with locations of East Coast NERRs and point of storm landfall (denoted by +).

Jacques Cousteau (946 mbar) reserve located on either side of landfall (Brigantine, New Jersey); measurements at Jacques Cousteau reserve matched the storm's lowest recorded measurement at Atlantic City, New Jersey. Wind gusts over 25 m/s (60 mph) were measured at several reserves in the mid-Atlantic (Jacques Cousteau reserve, 25.6 m/s) and Northeast regions (Narragansett Bay and Wells, 29.1 and 25.3 m/s, respectively). The storm surge gradient varied from 0.3 m (1.0 ft) to 2.0 m (6.6 ft) along the East Coast reserves, with maximum values occurring immediately adjacent and just north of landfall at the Jacques Cousteau and Hudson River reserves. For comparison, maximum recorded storm surge for Sandy was 3.86 m (12.7 ft) at Kings Point, New York.[40]

Reay and Moore[41] described both moderate (>week) and short-term (hours to days) impacts of tropical cyclone Isabel on the York River estuary, the Chesapeake Bay's fifth largest tributary where four Virginia NERRS components are located. Isabel produced a storm surge of 1.7 m near the mouth of the estuary and 2.0 m in the upper tidal freshwater regions, resulting in a thirty-fold increase in net salt flux during the surge.[42] Short-term (12–36 h) elevated salinity levels, greater than 10 psu above pre-storm conditions, were observed within oligohaline reaches of the York River system. Elevated turbidity, in some cases extreme (1000 NTUs), were in direct response to the storm surge and wave action; turbidity in large portions of the York River returned to pre-storm conditions within 24 to 36 h. Subsequent freshwater inflow depressed salinities and changed the estuary from partially mixed to a very strongly stratified estuary for a prolonged period. Low dissolved oxygen levels (3–4 mg/L) coincided with increased freshwater inflow and persisted (~10 days) following the passage of this large-scale storm.

Dix et al.[43] investigated the effects of extreme wind and rainfall conditions from sequential tropical storms on water quality within a tidal creek system of the Guana Tolomato Matanza reserve in Florida and compared results to longer term non-storm periods. Although short-lived, the storms drove high salinity marine water beyond the range of tidal exchange followed by extended periods of salinity depression and disruption of normal diel tidal fluctuations driven by elevated watershed freshwater discharges from high storm rainfall totals. Episodic and elevated nitrogen pulses were associated with storm events in contrast to phosphorus that did not exhibit a difference between periods of high and low watershed outflow. The authors discuss how significant changes in key water quality parameters from multiple storm events, both in terms of magnitude and timing, can influence coastal ecosystem structure and function.

11. DATA APPLICATIONS: EDUCATION

Coastal residents enjoy and benefit from resources provided by our nation's estuaries, but when asked in a survey about the health of coastal waters, over a quarter of the respondents reported that they did not know enough about these areas to give an opinion, and only a small percentage of the public understood that climate change is a threat to the ocean.[44] Consequently, educators, students, coastal resource managers, and other coastal residents are all important audiences for SWMP data. Case studies using SWMP data and observations provide tangible opportunities for both formal and informal education activities that even people living far from any coastline can appreciate. The destructive impact of Hurricane Sandy is an example. In late October 2012, as the storm was waning, the Jacques Cousteau reserve in southeastern New Jersey began an online dialogue on Facebook that included presentations of SWMP data demonstrating the hurricane's impact on local weather, flooding, and water quality. Outreach about the hurricane's impacts through translation of SWMP data then continued in the form of public presentations, teacher training workshops, and naturalist blogs that were spearheaded by reserve staff. The Jacques Cousteau NERR staff's interpretation of the hurricane increased the scientific literacy of participants and allowed for more thoughtful, science-based discussions of storm impacts in the coastal zone. Although it is difficult to quantify the benefits of such education, it is reasonable to expect that these efforts result in improved coastal stewardship through more informed personal actions such as voting and grassroots support for long-term research and monitoring programs. In this way, SWMP data serve as a vehicle and a mechanism for engaging people in coastal science and the NERRS research and monitoring programs.

Even when catastrophic events are not occurring, NERRS educators routinely translate research and teach using SWMP data as a heuristic tool. Discrete data sets are woven into stories that teach about estuarine science, such as in the NERRS Estuaries 101 (http://estuaries.noaa.gov/Teachers/Home.aspx) lesson plans that teach basic science concepts using examples from past research and data from NERRS sites around the country. For example, the last activity in the physical science section of the curriculum, "Human Impacts on Estuaries: A Terrible Spill in Grand Bay," uses SWMP water quality and nutrient data to explore the implications of a pollution spill in ways that foster students' abilities to inspect and interpret data, identify relationships among parameters, and think critically about the implications of those relationships. Many reserve sites offer professional development programs for teachers such as Teachers on the Estuary (TOTE; http://estuaries.noaa.gov/Teachers/Default.aspx?ID=387) to further build teachers knowledge of SWMP data and the nature of science, so that resources like Estuaries 101 lessons can lead to more substantial learning that comes from student-led investigations of data.

The foundations of educators' abilities to engage diverse audiences with SWMP are the web-based graphing tools (http://cdmo.baruch.sc.edu/get/landing.cfm) that allow nonscientists to quickly graph, compare, and interpret SWMP data. These tools are essential to data use in education. In San Francisco Bay, for example, real-time SWMP data are used by owners of duck hunting clubs to check salinity and water levels as they manage water exchange in their wetlands, and by swimmers to check the temperature of the bay water. The graphing tools go beyond reporting real-time data. For example, high school students have used the graphing interface at www.estuaries.noaa.gov to make complex comparisons of SWMP data over time and to develop projections of climate change impacts to San Francisco Bay.

Teachers, students, and coastal residents around the country have experienced the value of SWMP data by learning about short-term phenomena (like Hurricane Sandy) and by approaching long-term challenges such as the need to teach students how to think critically. Making SWMP data accessible and understandable to a variety of audiences is a hallmark of the NERRS that directly supports the shared mission of improving science literacy, stewardship, and understanding of the complex issues facing our nation's estuaries and coasts.

12. SUMMARY, CONCLUSIONS, AND CHALLENGES

The NERRS SWMP program is proving to be a powerful tool for assessing environmental change in the nation's estuaries. Coordinated monitoring and rigorous data management is supporting research and management applications ranging from climate change detection to water quality management. As the SWMP dataset grows, it becomes more valuable for addressing longer term research and management questions, but more effort and resources are needed for data analysis and comparative studies of SWMP data from across the NERRS. Challenges that the NERRS SWMP has overcome include establishing metadata standards and data comparability across diverse habitats, common quality assurance and control protocols, data archiving, making data available in real or near real time, and establishing long-term water quality and weather datasets to detect trends. However, challenges that remain include consistent and predictable funding to support collection of ecological data, the ability to respond to and efficiently take advantage of new technology while maintaining consistency in existing data streams and data standards, and the need for tightly linked partner networks that share common protocols and data standards to increase geographic coverage and spatial resolution of nearshore to offshore observing systems.

There is also a critical need to integrate terrestrial, estuarine/coastal, and offshore environmental monitoring systems. There are several other federally and state-supported coastal observing systems that are networked to provide data on water quality, weather, or coastal habitat, but none adequately link terrestrial and aquatic ecosystems across an upland-to-nearshore-to-offshore continuum. The NERRS is building this capacity by installing a suite of sentinel sites for understanding the impacts of climate on coastal ecosystems. The current focus is on sea level rise and vegetated habitat, but there is a need to tie into other terrestrial, nearshore, and offshore observing systems to create a seamless array of data and information that can inform management responses to episodic events and long-term climate change.

REFERENCES

1. Moore KA. *Long-term monitoring of estuarine submersed and emergent vegetation communities. National Estuarine Research Reserve Technical Report Series*. Silver Spring, MD: NOAA/NOS/OCRM/ERD; 2009.
2. Roman CT, Janes-Piri MJ, Heltshe JF. *Monitoring salt marsh vegetation. Long-term coastal ecosystem monitoring program*. Wellfleet, MA: Cape Cod National Seashore; 2001. 47 pp.
3. Short FT, McKenzie LJ, Coles RG, Vidler KP. *SeagrassNet manual for scientific monitoring of seagrass habitat*. University of New Hampshire; 2002. 55 pp.
4. Wenner EL. *The National Estuarine Research Reserve's system-wide monitoring program (SWMP): a scientific framework and plan for detection of short-term variability and long-term change in estuaries and coastal habitats of the United States*. MD: Silver Spring; 2002. Unpublished report to NOAA/NOS OCRM. 48 pp.
5. Nieder C, Porter D, Rumrill S, Wasson K, Wenner E, Stevenson B, et al. *Land use and land cover change analysis in the National Estuarine Research Reserve System*. White Paper; 2002.
6. NOAA/NERRS. *SWMP Phase III. Land use, land cover, and habitat change plan for the National Estuarine Research Reserve system. Report for the Estuarine Reserves Division*. Silver Spring, MD: NOAA/NOS/OCRM; 2009. 44 pp.
7. Kutcher TE. *Habitat and land cover classification scheme for the National Estuarine Research Reserve System*. Silver Spring, MD: NOAA/NOS/OCRM/ERD; 2008. 42 pp.
8. Kutcher TE, Garfield NH, Raposa KB. *A recommendation for a comprehensive habitat and land use classification system for the National Estuarine Research Reserve System. Report for the Estuarine Reserves Division*. Silver Spring, MD: NOAA/NOS/OCRM; 2005. 26 pp.
9. Garfield N, Madden K, Shull S, Upchurch S, Harold N, Ferner M, et al. *Mapping land use and habitat change in the NERRS: revised standard operating procedures*. Silver Spring, MD: NOAA Estuarine Research Reserve Division; 2012. 27 pp.
10. Bulthuis DA. Distribution of seagrasses in a North Puget Sound estuary: Padilla Bay, Washington, USA. *Aquat Bot* 1995;**50**:99–105.
11. Dyke EV, Wasson K. Historical ecology of a Central California estuary: 150 years of habitat change. *Estuaries* 2005;**28**:173–89.
12. Kairis PA, Rybczyk JM. Sea level rise and eelgrass (*Z. marina*) production: a spatially explicit relative elevation model for padilla Bay, WA. *Ecol Model* 2009;**221**:1005–16.
13. Robinson P, Leight AK, Trueblood DD, Wood B. *Climate sensitivity of the National Estuarine Research Reserve System*. Report to NOAA's Climate Program Office; 2013. p. 79.
14. Intergovernmental Panel on Climate Change. *Summary for policymakers*. Cambridge, UK and New York, USA: Cambridge University Press; 2007.
15. Fagherazzi S, Kirwan ML, Mudd SM, Guntenspergen GG, Temmerman S, D'Alpaos A, et al. Numerical models of salt marsh evolution: ecological, geomorphic, and climatic factors. *Rev Geophys* 2012;**50**:RG1002.
16. Schile LM, Callaway JC, Morris JT, Stralberg D, Parker VT, Kelly M. Modeling tidal marsh distribution with sea-level rise: evaluating the role of vegetation, sediment, and upland habitat in marsh resiliency. *PloS One* 2014;**9**:e88760.
17. Wenner EL, Sanger DM, Arendt MD, Holland AF, Chen Y. Variability in dissolved oxygen and other water quality variables within the National Estuarine Research Reserve System. *J Coast Res* 2004;**45**:17–38.

18. US Environmental Protection Agency. *National coastal condition report IV. US EPA office of water report # EPA-842-R-10-003*. 2012.
19. Sanger DM, Arendt MD, Chen Y, Wenner EL, Holland AF, Edwards D, et al. A synthesis of water quality data: national estuarine research reserve system-wide monitoring program (1995–2000). National Estuarine Research Reserve Technical Report Series 2002: 3. South Carolina department of natural resources. *Mar Resour Div Contrib* 2002;**500**:135.
20. Apple JK, Smith EM, Boyd TJ. Temperature, salinity, nutrients and the covariation of bacterial production and chlorophyll a in estuarine ecosystems. *J Coast Res* 2008;**55**: 59–75.
21. Caffrey JM. Production, respiration and net ecosystem metabolism in US estuaries. *Environ Monit Assess* 2003;**81**:207–19.
22. Caffrey JM. Factors controlling net ecosystem metabolism in US estuaries. *Estuaries* 2004;**27**:90–101.
23. Caffrey JM, Murrell MC, Amacker KS, Harper JW, Phipps S, Woodrey MS. Seasonal and inter-annual patterns in primary production, respiration, and net ecosystem metabolism in three estuaries in the northeast Gulf of Mexico. *Estuaries Coasts* 2013; **37**(S1):222–41.
24. Kemp WM, Smith EM, Marvin-DiPasquale M, Boynton WR. Organic carbon balance and net ecosystem metabolism in Chesapeake Bay. *Mar Ecol Prog Ser* 1997;**150**: 229–48.
25. Howarth RW, Boyer W, Pabich WJ, Galloway JN. Nitrogen use in the United States from 1961–2000 and potential future trends. *Ambio* 2002;**31**:88–96.
26. Bricker SB, Clement CG, Purhall DE, Orlando SP, Farrow DRG. *National estuarine eutrophication assessment: a summary of conditions, historical trends, and future outlooks*. 1999. Special Projects Office and National Centers for Coastal Ocean Sciences National Ocean Service, National Oceanic and Atmospheric Administration.
27. Howarth RW, Anderson D, Cloern J, Elfing C, Hopkinson C, Lapointe B, et al. Nutrient pollution of coastal rivers, bays and seas. *Issues Ecol* 2000;**7**:1–15.
28. Howarth RW, Marino R. Nitrogen as the limiting nutrient for eutrophication in coastal marine ecosystems: evolving views over three decades. *Limnol Oceanogr* 2006;**51**:364–76.
29. Chapman D. *Water quality assessments*. New York, NY: Chapman and Hall; 1992.
30. Diaz RJ, Rosenberg R. Marine benthic hypoxia: a review of its ecological effects and the behavioral responses of benthic macrofauna. *Oceanogr Mar Biol Annu Rev* 1995;**33**: 245–303.
31. Breitburg D. Effects of hypoxia, and the balance between hypoxia and enrichment, on coastal fishes and fisheries. *Estuaries* 2002;**25**:767–81.
32. Anderson DM, Glibert PM, Burkholder JM. Harmful algal blooms and eutrophication: nutrient sources, composition and consequences. *Estuaries* 2002;**25**:704–26.
33. Smayda TJ. Harmful algal blooms: their ecophysiology and general relevance to phytoplankton blooms in the sea. *Limnol Oceanogr* 1997;**42**:1137–53.
34. Olson RJ, Sosik HM. A submersible imaging-in-flow instrument to analyze nano- and microplankton: imaging FlowCytobot. *Limnol Oceanogr Methods* 2007;**5**:195–203.
35. Campbell L, Olson RJ, Sosik HM, Abraham A, Henrichs DW, Hyatt CJ, et al. First harmful *Dinophysis* (Dinophyceae, Dinphysiales) bloom in the U.S. is revealed by automated imaging flow cytometry. *J Phycol* 2010;**46**:66–75.
36. Campbell J. (2012). *The role of protozoan grazers in harmful algal bloom dynamics* [Ph.D. dissertation]. The University of Texas at Austin.

37. Fletcher M, Pournelle J, Ramage D, Porter D, Shervette V, Kelsey R. A southeast regional testbed for integrating complex coastal and ocean information systems. *Oceans* 2009;**09**.
38. Kelsey RH, Scott GI, Porter DE, Siewicki TC, Edwards DG. Improvements to shellfish harvest area closure decision making using GIS, remote sensing and predictive models. *Estuaries Coasts* 2010;**33**:712–22.
39. NOAA/National Weather Service. *Introduction to storm surge*. Link: http://www.nws.noaa.gov/om/hurricane/resources/surge_intro.pdf.
40. NOAA. *Hurricane sandy: NOAA water level and meteorological data report*. Silver Spring, MD: NOAA/NOS/Center for Operational Oceanographic Products and Services; 2013.
41. Reay W, Moore KA. Impacts of tropical cyclone Isabel on shallow water quality of the York River estuary. In: *Proceedings, Hurricane Isabel in perspective: developing an understanding of how storm events affect Chesapeake Bay Region*. CRC Press; 2005.
42. Gong W, Shen J, Reay WG. The hydrodynamic response of the York River estuary to tropical cyclone Isabel, 2003. *Estuar Coast Shelf Sci* 2007;**73**:695–710.
43. Dix NG, Philips EJ, Gleeson RA. Water quality changes in the Guana Tolomato Matanzas National Estuarine Research Reserve, Florida, associated with four tropical storms. *J Coast Res* 2008;**55**:26–37.
44. The Ocean Project. *America and the ocean: annual update*. 2011. Available at, http://theoceanproject.org/wpcontent/uploads/2011/12/TOP_AmericaOceansUpdate2011_online.pdf [accessed 7.10.14].

CHAPTER

22 Integrating Environmental Monitoring and Observing Systems in Support of Science to Inform Decision-Making: Case Studies for the Southeast

Dwayne E. Porter[1,*], Jennifer Dorton[2], Lynn Leonard[3], Heath Kelsey[4], Dan Ramage[1], Jeremy Cothran[1], Adrian Jones[4], Charlton Galvarino[5], Vembu Subramanian[6], Debra Hernandez[6]

Arnold School of Public Health and the Baruch Institute for Marine and Coastal Sciences, University of South Carolina, Columbia, SC, USA[1]; Center for Marine Science, University of North Carolina—Wilmington, Wilmington, NC, USA[2]; Department of Geography and Geology, University of North Carolina—Wilmington, Wilmington, NC, USA[3]; University of Maryland Center for Environmental Science, Cambridge, MD, USA[4]; Second Creek Consulting, Columbia, SC, USA[5]; Southeast Coastal Ocean Observing Regional Association, Charleston, SC, USA[6]

**Corresponding author: E-mail: porter@sc.edu*

CHAPTER OUTLINE

1. INTRODUCTION

The Southeast Coastal Ocean Observing Regional Association's (SECOORA) geographic footprint encompasses the coastal and ocean waters of North Carolina (NC), South Carolina (SC), Georgia (GA), and Florida (FL) to include the southeast

Atlantic seaboard, the Straits of Florida, as well as the eastern portion of the Gulf of Mexico. Like the other 11 Regional Associations (RAs) supported by the US Integrated Ocean Observing Systems (IOOS®), SECOORA continuously works to meet the needs of a variety of decision-making stakeholders in a region with disparate coastal and oceanographic environments. The Southeast region domain ranges from mangrove and saltmarsh estuaries to tropical coral reefs to subtropical whale calving grounds to mid-latitude barrier islands, and it includes the only western boundary current in the IOOS footprint. This region, frequently impacted by tropical storms, has one of the largest recreational and commercial fishing and boating industries in the United States, numerous major ports and military bases, and over 50% of the nation's cruise industry.

A challenging opportunity faced in the Southeast is the integration of multiple local, subregional, regional, and national coastal and ocean observing systems that span inshore waters, estuaries, and nearshore and open waters of the Atlantic Ocean and Gulf of Mexico.[1] The experiences (successes and/or failures) of these diverse systems, both individually and amassed, can be used to identify ongoing challenges and issues related to end-user requirements. Moreover, they serve as lessons learned when developing strategies to meet current and future user needs. The purpose of this chapter is to review the state of coastal and ocean monitoring and observing system efforts in the Southeast; to present case studies demonstrating the value of integrating data from monitoring and observing systems to support marine safety, water quality, and ecosystem management decision-making; and finally, to present straightforward recommendations for the future.

2. ROLE OF MONITORING AND OBSERVING SYSTEMS IN THE SOUTHEAST

The SECOORA is implementing a cohesive Regional Coastal Ocean Observing System (RCOOS) for the southeast United States as a regional 501(c)(3) partner in the US IOOS. It is building an RCOOS that leverages, integrates, and augments existing observational, modeling, data management, education, and scientific assets within the region. Finally, SECOORA is creating customized decision-support products to address the IOOS four thematic areas: marine operations; coastal hazards; ecosystems, fisheries, and water quality; and climate variability and change.

The observing system provides the basis for the RCOOS by supporting and integrating existing assets and observations specific to the development of products addressing the four IOOS theme areas. The observing system consists of a suite of coastal and offshore moored platforms, autonomous underwater gliders, satellite data receivers, and high-frequency radar (HFR) surface current installations. Table 1 lists observing system assets currently funded by SECOORA and the variables measured. Each observing platform collects and provides near real-time data to a variety of end users and stakeholders. SECOORA also has established a robust data management subsystem that integrates data from SECOORA and non-SECOORA

Table 1 SECOORA-Funded Monitoring Efforts

In-situ moored and coastal stations	*Observing systems*: • University of South Florida Coastal Ocean Monitoring and Prediction System (COMPS) • University of North Carolina Wilmington Coastal Ocean Research and Monitoring Program (CORMP) • University of Georgia–Gray's Reef National Marine Sanctuary buoy *Variables measured*: • Meteorologic: Air temperature, relative humidity, barometric pressure, wind speed and direction, short- and long-wave radiation • Oceanographic: In-water velocity and temperature, salinity, water level and waves, pCO_2, pH, dissolved oxygen, water temperature
High-frequency radar (HF radar) Stations–Seasonde CODAR and Helzel WERA systems	*HFR stations*: • University of South Florida West Florida Shelf CODAR systems • University of Miami WERA systems • Skidaway Institute of Oceanography WERA systems • University of South Carolina WERA systems • University of North Carolina, Chapel Hill CODAR systems *Variables measured*: • Surface currents and waves

platforms to facilitate delivery of coastal and ocean water and weather conditions and other information products via SECOORA's Website and data portal (http://secoora.org/maps).

3. THE ROLE OF DATA MANAGEMENT TO SUPPORT COLLABORATION AND INTEGRATION

Data management and communications (DMAC) infrastructure is recognized as a critical and prioritized component of the coastal and ocean observing systems.[2,3] SECOORA, in collaboration with its data providers and member partners in the region, has established a DMAC infrastructure to integrate high-quality, real-time and historical environmental data in support of science-based decision-making. SECOORA's member institutions and partners have extensive experience in development and implementation of Open Geospatial Consortium (OCG) recommended standards and technologies to promote aggregation, archival, visualization, dissemination, and interoperability of distributed federal, state, and nonfederal data sets.[4–8] These data management activities are central to a number of local, subregional, regional, and national programs, including local beach water quality monitoring programs, the Carolinas Regional Coastal Ocean Observing System (Carolinas

RCOOS), SECOORA, and the NOAA National Estuarine Reserve System's System-wide Monitoring Program (SWMP). All of these programs rely on real-time and delayed-mode data on environmental conditions, which are aggregated and managed to support diverse data uses and products. The capacities established during the development of these programs include the following:

- Infrastructure required for reliable data access, maintenance, and archival
- Ability to access a variety of data types (e.g., satellite, HF radar, glider, and buoy) and import them into sophisticated centralized data hubs
- A relational database structure that allows for rapid and customizable data queries to meet specific user needs
- Quality assurance and quality control protocols that identify spurious data and/or data anomalies prior to permanent archival
- Web-based tools, including map-based products that enable spatial visualization of data and derived information, and tailored alerts that notify the end user when user-specified thresholds are exceeded
- Workforce training and development of a community of data management experts in the region

Since 2002, significant federal, state, and private funds have been spent to collect a variety of environmental data for a range of purposes, such as weather forecasts, beach contamination monitoring, and fisheries resource management. The implementation of coordinated data management processes and protocols has increased the value of these observations. Though each data set still serves its original purpose, the data management system has enabled the sharing of this information, thereby expanding data interoperability and integration beyond coastal systems to other environmental programs.

4. CASE STUDIES

The true value of any observing system is only achieved when the data and derived information are used to advance science and also repurposed to meet applied needs that include stakeholder decision-making. By capitalizing on the value of collaboration across monitoring efforts to integrate data, opportunities exist to develop tools that support decision-making over scales from the individual family as related to public health, to multi-agency, regional resource planning entities. In this section, we present three case studies representing marine safety, water quality, and ecosystem management that highlight the role of integrated coastal and ocean monitoring and observing efforts in support of science-based decision-making for multiple stakeholders.

4.1 MARINE SAFETY

Regional observing initiatives have increased availability of real-time meteorologic and oceanographic data through the deployment of estuarine, coastal, and offshore mooring platforms. In 2007, the University of North Carolina Wilmington (UNCW) and the University of South Carolina (USC) partnered to operate moorings within the North Carolina and South Carolina coastal region under the Carolinas Regional Coastal Ocean Observing System (Carolinas RCOOS). The Carolinas RCOOS project personnel cultivated relationships with local NOAA National Weather Service (NWS) Weather Forecast Offices (WFOs) in Wilmington, NC, and Charleston, SC, which provided project staff with a better understanding of NWS online products, such as maps and text forecasts, used by the commercial and recreational marine communities.

At that time, each NWS WFOs had their own marine weather webpage with little commonality between them. In an effort to create a standardized marine weather presence for the Wilmington and Charleston WFOs, data management personnel from USC and community outreach personnel from UNCW created a small pilot project titled the "Carolinas Coast." The goal was to create a one-stop shop for marine weather forecasts and observations for the two-state region. The project team then asked a focus group, comprised of NWS forecasters and commercial and recreational marine users, to review the Carolinas Coast site. The simple design of the site appealed to a wide range of web users who reiterated the need for a map-based Website that included easily accessible marine forecasts and hazard alerts as well as map overlays (sea surface temperature, radar, radar loop, air pressure, and bathymetry). Project partners increased awareness of the site by speaking at trade shows, to local civic groups, and at other informal settings in both North Carolina and South Carolina. Ultimately, the Carolinas Coast Website was well received by the marine audience and the NWS.[9]

Based on this success, the NWS identified a need for a standardized marine weather portal for the eastern United States and the Gulf of Mexico. The Carolinas RCOOS project team expanded to also include the University of South Florida, the NWS Eastern Region Headquarters, the NWS Southern Region Headquarters, the NWS Office of the Chief Information Officer (OCIO), and a private business partner, Second Creek Consulting. In 2007, this team launched an initiative to expand the Carolinas Coast product through creation of the Marine Weather Portal (MWP), an effort that included 15 NWS WFOs and a coverage area extending from the Carolinas to Texas.[9] The goal of the MWP was to provide 24/7 access to critical marine weather information for commercial and recreational marine communities. Moreover, by integrating across adjacent NWS WFO domains, users could, for the first time, easily navigate across multiple WFO coverage areas (Figure 1).

The MWP is now hosted by SECOORA and available at http://secoora.org/data/marineweatherportal. It is one of the most popular products currently on the SECOORA Website. Critical to the success of the MWP is the use of familiar interfaces (e.g., Google maps and text-based NWS forecasts) and incorporation of lessons learned from the localized Carolinas Coast initiative. Stakeholders were engaged continuously during the portal development to ensure their comments and feedback resulted in the desired final product.

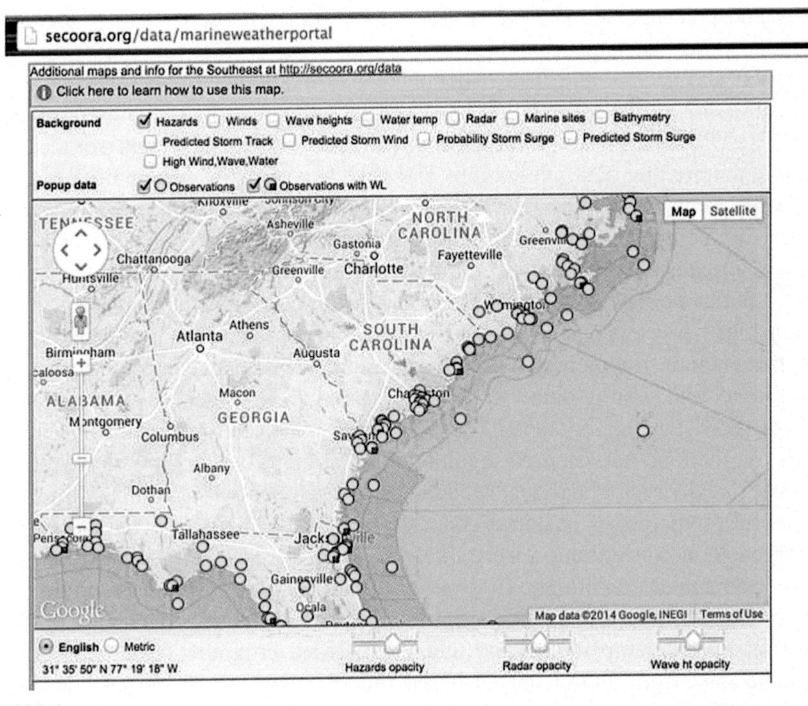

FIGURE 1

Screen shot of Marine Weather Portal (http://secoora.org/data/marineweatherportal). Users can toggle between types of data displayed and map overlays. In this example, the marine hazards layer has been activated and observations are shown as yellow dots. Note that this product spans multiple NWS domains in the SECOORA region. Product coverage also extends into the Gulf of Mexico Coastal Ocean Observing System region.

4.2 WATER QUALITY

Currently, decisions regarding bacterial water quality at many shellfish harvest areas and beach recreation areas are based on poorly supported assumptions. For instance, many states regulate periodic closures of shellfish harvest areas using only precipitation data,[10] and many beach recreation areas issue swimming advisories based on bacterial sampling results that may be several days old.[11] These decision-making methods use presumptive relationships that are often poorly defined or provide limited predictive ability.

An interdisciplinary team, comprised of researchers from USC and the University of Maryland Center for Environmental Science (UMCES), worked with the South Carolina Department of Health and Environmental Control (SCDHEC) and SECOORA to develop new tools using data integrated from multiple sources to support decision-making for water quality applications. An automated system was created that integrates necessary data to produce daily predictions of bacteria water quality and presents results in a user-friendly format for use by beach goers and managers (http://howsthebeach.org). USC was responsible for providing data integration and access, and it provided statistical modeling support. UMCES was responsible for statistical model development and creating the presentation of results. SCDHEC and SECOORA provided access to needed data sources and advice in automating data integration. The objective of the water quality modeling effort was to enable improved decision-making (e.g., public health warnings, shellfish harvest openings/closures) by creating decision-support tools that produce more reliable estimates of bacteria concentration and notifications of exceedance of water quality standards. The tools are based on a rigorous statistical evaluation of the relationships between concentration of indicator bacteria and environmental conditions that include remotely sensed precipitation, wind, tidal stage, water temperature, wave, and salinity data.

These new statistical models have been implemented in four South Carolina estuaries for shellfish harvest area regulation and at eight beach recreation areas for swimming advisory applications. Models were developed for each location to evaluate useful predictors of bacterial concentrations, recognizing that processes controlling bacteria concentration were likely different in areas with varying physical and environmental conditions. To develop the models, bacteria concentration and water quality data were integrated from the SCDHEC Shellfish Management (fecal coliform, salinity, temperature, and tides) and Beach Programs (enterococci) with additional data from NOAA (radar-based rainfall, wind, temperature, and tides), and SECOORA-supported data providers (water temperature, salinity, wave height, and wind speed and direction) for use as potential predictors. Models were evaluated for predictive capability using standard statistical methods, including cross-validation correlation coefficient and Receiver Operator Characteristic Curves (Figure 2).[10,12] Results indicate that these tools generate predictions that are improved over the traditional methods that rely on the previous day's bacteria level[11] and rainfall-only models (Figure 3).

This analysis yielded several recommendations to monitoring program managers. Development of the prediction tools suggests that radar-based rainfall data are more useful to bacteria prediction than rain gauge network data, which could reduce effort and cost in rainfall data acquisition.[10] Additionally, salinity was the single most important predictor of bacteria concentration in nearly every model developed.[10] Agencies that are reducing data acquisition costs by reducing the parameters sampled should consider retaining salinity in their monitoring programs or obtaining it from observing systems; it is inexpensive to measure and is the single most important driver of bacteria concentration.[10]

Automated access to multiple data streams continues to be challenging for implementation of the water quality modeling application. Interruption in delivery of necessary data streams, whether due to sensor failure, telemetry issue, or changes in data format by the data provider can interfere with or prevent the generation of daily predictions. Problematic data identified through standard quality assurance and quality control procedures may also trigger interruptions. Although some

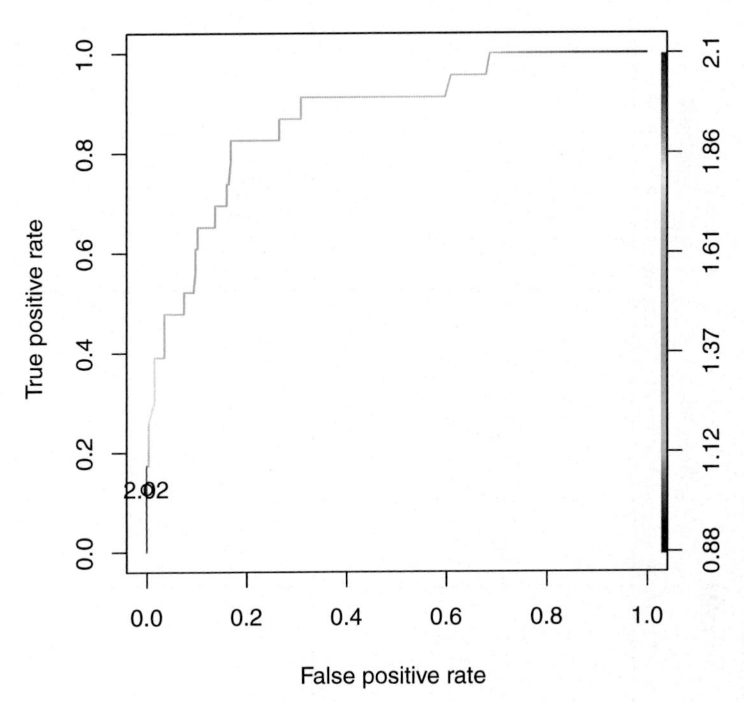

FIGURE 2

Evaluation of model false and true positive predictions from a model generated for the Myrtle Beach area using Receiver Operator Characteristic Curves. A perfect model would have a false positive rate = 0.0 and true positive rate = 1.0. Graphically, a line extending along the left *y*-axis and the top *x*-axis would represent that perfect model. Conversely, a straight line extending from the origin to the top right would represent a model with no utility. ROC curves are evaluated by the area under the curve; 1.0 is perfect, and 0.5 represents no utility.

interruptions are inevitable, planned activities for maintenance to observation system components could, for example, be scheduled to minimize interruptions during peak swimming seasons. To account for periodic interruptions (planned or unplanned), water-quality tools and products should be designed to incorporate contingency data, when needed, and to report the level of uncertainty associated with these predictions. These approaches allow state and local officials to evaluate the information to make informed decisions.

To increase use of these tools, a web-based mobile app (Figure 4, http://howsthebeach.org) has been developed. The mobile app (currently in the publication process for iOS and Android platforms) delivers daily predictions of bacteria concentration at eight beach recreation areas in the Myrtle Beach, SC, area. Results of past observations and the current official swimming advisory status are also

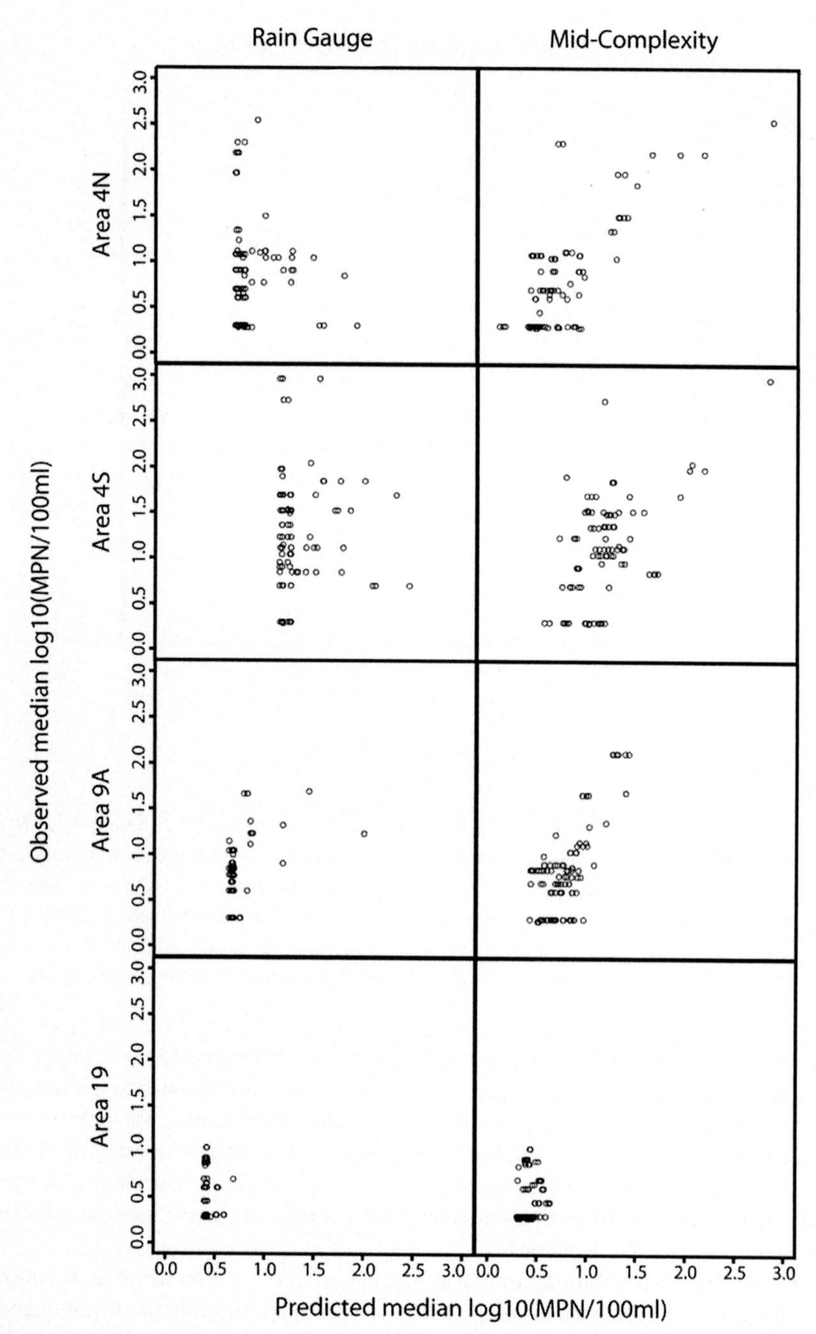

FIGURE 3

Cross-correlation results illustrate limited predictive capability of assumed relationships between rain gauge rainfall data (left column) and improved predictions from statistical models (right column) in four diverse shellfish areas evaluated in South Carolina.

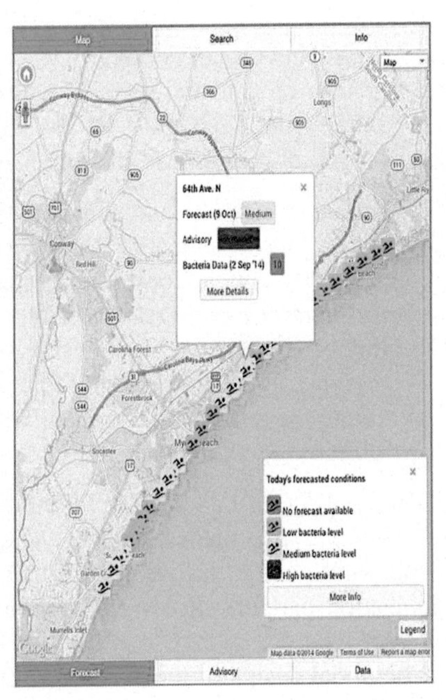

FIGURE 4

Screen shot of the Website and mobile app that presents results of daily predictions, last available data, and current advisory status.

presented. The success of the water quality modeling effort and mobile app has led to collaboration with the U.S. Environmental Protection Agency (EPA) to compare results and ease of use for a new beach bacteria modeling software package called Virtual Beach (http://www2.epa.gov/exposure-assessment-models/virtual-beach-vb), and to expand the app to additional geographies and to shellfish management applications. The team is currently developing similar products for marine recreation areas on Florida's Gulf Coast.

4.3 ECOSYSTEMS MANAGEMENT

Established in 2009, the Governors' South Atlantic Alliance (GSAA) is a partnership between the four southeastern states (NC, SC, GA, and FL) designed to increase collaboration around regional coastal and marine issues. Actions focus on four prioritized issue areas: healthy ecosystems, clean coastal and ocean waters, disaster-resilient communities, and working waterfronts. Common across all issue areas is the need for development of and access to regional data to provide baselines for measuring change, identify gaps in data and knowledge, and enable effective collaboration among the states, federal agencies, and other partners in the GSAA. With support from NOAA and the GSAA, a SECOORA-led interdisciplinary team consisting of data managers, geospatial analysts, resource managers, and outreach and education specialists developed a regional information management system, the GSAA Coast and Ocean Portal (www.gsaaportal.org), to provide member states access to a suite of state and regional data sets.

Leveraging the SECOORA DMAC infrastructure, GSAA Portal was designed to address regional priorities in the South Atlantic, including the following:

- Locations of offshore and nearshore sand resources to enable state and municipality managers to identify sediment resources for beach nourishment projects
- Locations of critical coastal and estuarine habitats and distributions of marine species to inform conservation efforts by state and federal agencies in support of healthy ecosystems and sustainable fishery populations
- Promotion of economic development, such as port expansion projects, by providing the critical marine data that will enable states to undertake activities with limited ecosystem disruption

The GSAA Coast and Ocean Portal development team also included representatives of state coastal management agencies, intended to be the primary user audience for the portal, and a technical team that included academic and nongovernmental organization partners. Roles and responsibilities of the partners are described in Table 2.

The initial design of the portal was driven by the needs of users. A user assessment was conducted to determine hardware and software preferences and limitations, priority data needs, and desired functionality. An additional component of the assessment involved a review of existing portals to identify the elements that appealed to the state managers and also met the technical requirements and preferences. These reviews were conducted via a series of webinars, one for each state (NC, SC, GA, and FL). Each webinar began with a state environmental manager providing an overview of their decision-making processes, environmental and biological data needs, and data analysis requirements. This was followed by a technical expert who demonstrated one or more existing portals to provide examples of possible layout, functionality, and decision-support-making capacities. This two-step approach allowed the project team to identify potential attributes and characteristics that were both desirable and realistic for a regional-based product. This information was used to develop a prototype portal. The portal was reviewed by the state environmental managers at a series of training sessions whose feedback was used to refine the portal.

The existing GSAA Coast and Ocean Portal is an online tool kit and resource center that consolidates state, regional, and federal geographic information system (GIS)-based datasets in one location. Only data sets that address management concerns identified by a user are included. The portal is not designed as a data warehouse and relies on data archived and served by a data provider. The portal accesses state and federal data sources and makes them readily available and easily accessible via a data catalog and map viewer. The site currently includes 108 spatial data layers and associated metadata; the majority are from existing sources, such as the Multipurpose Marine Cadastre and South Atlantic Fisheries Management

Table 2 Collaborators and Associated Roles and Responsibilities for the GSAA Data Portal in Support of Ecosystem and Fisheries Management

PIs/Partners	Roles	Tasks
SECOORA	Project manager	Manage funding to partners; lead grant reporting; provide overall project coordination and leadership
	Stakeholder engagement coordinator	Organize/staff regional stakeholder engagement team (RSET)
	Technical requirements team lead	Organize and support regional CMSP visioning technical team (RVTT)
NCDENR/NERR CTP	NC stakeholder engagement facilitator	
SCDHEC OCRM	SC Policy lead	Participate on RVTT; lead SC stakeholder engagement
SCDNR	SC technical lead	Participate on RVTT; lead SC data management (DM) team
SC NERRs CTP	SC stakeholder engagement coordinator	Participate on RSET; facilitate & support SC stakeholder education & assessment
GADNR CZM	GA Policy lead	Participate on RVTT; lead GA stakeholder engagement
GA Tech	GA technical lead; technical expert	Participate on RVTT, lead GA DM team; serve on decision-support tools (DST) development team
GA NERR CTP	GA stakeholder engagement facilitator	Participate on RSET; facilitate & support GA stakeholder education & assessment
FL NERR CTP	FL stakeholder engagement facilitator	Participate on RSET; facilitate & support FL stakeholder education & assessment
FL FWC FWRI	Technical expert	Participate on RVTT
SE TNC	Regional dataset lead; technical expert; DST developer	Serve on the RVTT as a technical advisor; develop regional-scale datasets; develop spatial analysis tools
Duke Marine Geospatial Ecology Lab	Regional dataset developer, technical expert; DST developer	Serve on the RVTT as a technical advisor; develop regional-scale datasets; develop spatial analysis tools
USC	Technical expert; DST developer	Serve on the RVTT as a technical advisor; develop spatial analysis tools
NOAA CSC	Technical expert	Participate in RVTT

Council. It also provides GIS data sets that were developed as part of this project by the SC Department of Natural Resources that includes fishery independent survey data, regional in-water sea turtle surveys, sediment analyses, and sea turtle nesting locations.

Increased data compatibility will be required so that each of the four states provide similar types of data (e.g., water quality, wildlife, and sediment sources) to facilitate access and product development at a regional scale. The portal team will continue to work with the GSAA's technical teams to support and distribute additional spatial data and associated metadata, for estuarine habitats, working waterfronts, coastal vulnerability, and military uses.

5. CONCLUSIONS

Though this chapter and presented case studies are specific to the Southeast, the lessons learned and associated challenges are not. Because each of the IOOS RAs shares a common mission, SECOORA's experiences have national relevance. Therefore, to further enhance the utility of integrating coastal and ocean observing and monitoring systems in support of science to inform decision-making, we recommend the following:

- Effective product development only occurs when users are fully engaged in defining requirements and stay engaged throughout the development process. This approach will ensure that the end product is relevant and utilized.
- Broad-based user input should be considered when prioritizing sensor location, sampling frequency, and variable type (e.g., water temperature, waves, salinity, and meteorologic data). This input will facilitate greater use of data provided by ocean observing systems.
- Small, localized pilot projects are an ideal first step in product development with the lessons learned transferred up to subregional or regional scales. Additionally, an iterative approach to product development can be cost-effective and allows for the inclusion of lessons learned so that the final product meets the needs of the target audience.
- End user satisfaction and product credibility depends on a long-term sustainability plan. Product maintenance is a long-term commitment that requires dedicated financial and personnel resources. It also requires that data sources used in the developed product (e.g., observing platforms, GIS databases, etc.) are maintained.

REFERENCES

1. DeVoe MR, Buckley E, Dorton J, Fletcher M, Leonard L, Lumpkin P, et al. Regional coastal IOOS development in the South Atlantic Bight. *Mar Technol Soc J* 2007; **40**(4):110–7.

2. Fletcher M, Cleary J, Cothran J, Porter DE. Southeast Atlantic coastal ocean observation system (SEACOOS) information management: evolution of a distributed community system. *Mar Technol Soc J* 2008;**42**(3):28–34.
3. Porter DE, Small T, White D, Fletcher M, Norman A, Swain D, et al. Data management in support of environmental monitoring and coastal management. *J Coast Res* 2004;**45**: 9–16.
4. Botts M. *OGC implementation specification 07-000: openGIS sensor model language (SensorML)*. Wayland (MA): Open Geospatial Consortium; 2007.
5. CF Metadata Group. *NetCDF climate and forecast (CF) metadata convention and CF standard name table*. 2012. Version 19, April 28, 2012. http://cf-pcmdi.llnl.gov/documents/cf-standard-names [accessed 30.08.12].
6. Rueda C, Bermudez L, Fredericks J. The MMI ontology registry and repository: a portal for marine metadata interoperability. In: *Proceedings of the MTS/IEEE oceans 09 conference*; 2009.
7. Neiswender C, Isenor A, Montgomery E, Bermudez L, Miller SP. Vocabularies: dictionaries, ontologies, and more. In: *MMI guid navigating world mar metadata*; 2011. http://marinemetadata.org/guides/vocabs [accessed 30.08.12].
8. Haines S, Subramanian V, Mayorga E, Snowden D, Ragsdale R, Rueda C, et al. IOOS vocabulary and ontology strategy for observed properties. In: *Proceedings of the MTS/IEEE oceans 12 conference*; 2012.
9. Dorton J, Subramanian V, Galvarina C, Porter DE. Ocean observing: linking observations to the NWS and marine community. In: *Proceedings of the MTS/IEEE oceans 09 conference*; 2009.
10. Kelsey RH, Scott GI, Porter DE, Siewicki TC, Edwards DG. Improvements to shellfish closure decision-making using GIS, remote sensing, and predictive models. *Estuaries Coasts* 2010;**33**:712–22.
11. Olyphant GA, Whitman RL. Elements of a predictive model for determining beach closures on a real time basis: the case of 63rd Street Beach, Chicago. *Environ Monit Assess* 2003;**98**:175–90.
12. Morrison MM, Coughlin K, Shine JP, Coull BA, Rex AC. Receiver operator characteristic curve analysis of beach water quality indicator variables. *Appl Environ Microbiol* 2003;**69**(11):6405–11.

CHAPTER

23 One System, Many Societal Benefits: Building an Efficient, Cost-Effective Ocean Observing System for the Gulf of Mexico

Christina Simoniello[1,*], Stephanie Watson[2], Barbara Kirkpatrick[3,6], Michael Spranger[4], Ann E. Jochens[5], Shinichi Kobara[5], Matthew K. Howard[5]

Texas A&M University, Based at University of South Florida, College of Marine Science, St. Petersburg, FL, USA[1]; Gulf of Mexico Coastal Ocean Observing System Consultant, Based at Stennis, MS, USA[2]; Gulf of Mexico Coastal Ocean Observing System, College Station, TX, USA[3]; University of Florida, Gainesville, FL, USA[4]; Texas A&M University, College Station, TX, USA[5]; Mote Marine Laboratory, Sarasota, FL, USA[6]

**Corresponding author: E-mail: chris.simoniello@gcoos.org*

CHAPTER OUTLINE

1. ORIGIN OF THE GCOOS "SYSTEM OF SYSTEMS" CONSTRUCT

The United States has a long history of using environmental data to benefit the citizenry of this country. The National Weather Service, dating back to the late 1800s, is a prime example. Just as President Ulysses S. Grant signed a joint resolution "to secure the greatest promptness, regularity, and accuracy in required weather observations" more than a century ago, President Barack Obama signed the Integrated Coastal and Ocean Observation System (ICOOS) Act of 2009, to systematically acquire "ocean weather" in a sustained manner for the benefit of society.

Although the technology has changed dramatically, the name given to the early weather program by Brigadier General Albert J. Myer, *The Division of Telegrams and Reports for the Benefit of Commerce*, largely applies to today's efforts to establish a comparable ocean observing system—we strive to acquire and communicate data and information in a timely way to benefit society. Unlike the weather service of old, today's system is intricately linked globally via the Global Ocean Observing System (GOOS), and the larger, all-encompassing Global Earth Observation System of Systems (GEOSS).

Within the GEOSS "System of Systems" construct are nested earth observations from the land, sea, and atmosphere, to measure the health of flora and fauna, including humans. The GOOS is the coastal and oceanic component of GEOSS, coordinated by United Nation agencies and involving the participation of some 100 ocean nations. This end-to-end system of observations, data management, and production and delivery of products has two modules: the global module designed to monitor, predict, and understand marine conditions and climate variability; and the coastal module designed to sustain healthy marine ecosystems, ensure human health, promote safe and efficient marine transportation, enhance national security, and predict and mitigate coastal hazards. The United States makes several contributions to this global effort, under the umbrella of the U.S. Integrated Earth Observation System. Included here are the National Science Foundation's Ocean Observatory Initiative, the National Oceanic and Atmospheric Administration's (NOAA) Climate Observation Program, and the U.S. Integrated Ocean Observing System (IOOS). The U.S. IOOS is currently comprised of 17 federal entities and 11 Regional Associations (RAs). The Gulf of Mexico Coastal Ocean Observing System Regional Association (GCOOS-RA), the focus of this chapter, is one of these RAs and encompasses the waters of the Gulf of Mexico, from the U.S. Exclusive Economic Zone inland to the end of tidal effects in estuaries, including nearly 17,000 miles of bay, estuary, and coastal shorelines from Texas to Florida.

2. THE GULF OF MEXICO: NATIONAL TREASURE AND ECONOMIC DRIVER

The area within the footprint of the GCOOS-RA is a national ecological and economic treasure critical to the lives and livelihoods of approximately 50 million residents.[1] The five Gulf states, Texas, Louisiana, Mississippi, Alabama, and Florida, account for 17% of the US gross domestic product,[2,3] and if considered a country, their gross domestic product of more than $2.4 trillion would rank as the seventh largest economy in the world.[4] Much of the revenue is generated from diverse natural resources. For example, the region supports 38% of US petroleum reserves, 48% of the natural gas reserves,[5] and 93% of the nation's offshore oil and gas production.[6] Commercial fish and shellfish exceed 1.6 billion pounds annually in the Gulf, valued at approximately $754 million, with shrimp landings and oyster poundage accounting for approximately 60% and 33% of the national total, respectively.[7] Likewise, recreational fishing accounts for nearly one third of the nation's fishing expenditures and supports approximately one quarter of US jobs related to recreational saltwater fishing.[8] Intricately linked to this is the region's tourism industry, which supports tens of thousands of jobs worth more than $20 billion annually.[9] The Gulf is also home to 13 of the nation's top 20 ports in terms of tonnage or cargo value, two of these are among the top seven globally.[10] Clearly, there is a need to better understand the Gulf system if our reliance on it is to be sustainable.

3. A COMPREHENSIVE BLUEPRINT FOR MONITORING IN THE GULF OF MEXICO

Perhaps seemingly disparate, the many demands on Gulf resources described previously have in common similar data and information requirements to monitor and mitigate their impacts on the ecosystem and communities throughout the region. Recognizing and implementing strategies to gain economies of scale is an underlying strength of the Gulf of Mexico Coastal Ocean Observing System (GCOOS), which is being built as one comprehensive, end-to-end system, carefully planned and operated, and serving data for multiple, regionally driven societal needs. The long-term plan for a comprehensive GCOOS, described in the GCOOS Build-out Plan,[11] is based on 10 years of identifying broad stakeholder needs, identifying common priorities, and cross-referencing priorities from more than 90 existing Gulf plans. More than 17 workshops over seven years, with nearly 700 individuals representing 297 unique organizations, have been held to collect stakeholder input into the plan. A summary of the GCOOS stakeholder workshops held is given in Table 1. A summary of priorities held in common across stakeholder groups is given in Table 2.

Stakeholder sectors include diverse industries, educational institutions, governmental agencies, recreational boaters, and the general public for which societal benefits are targeted. There are 19 elements described in this long-term vision of a Gulf regional observing and monitoring system, listed as follows:

1. Aircraft observations and unmanned aerial systems
2. Autonomous meteorologic measurement network
3. Bathymetry and topography mapping
4. Circulation modeling/physical modeling
5. Data management and communication subsystem
6. Ecosystem modeling
7. Ecosystem monitoring
8. Enhanced physical oceanography real-time systems
9. Fixed mooring network
10. Gliders and autonomous underwater and surface vehicles network
11. Harmful algal bloom integrated observing system
12. Integrated water quality monitoring network and beach quality monitoring

Table 1 Stakeholder Workshops Hosted by the GCOOS-RA from 2003 to 2014

Name	Dates	Location
The NVODS workshop for managers of coastal observing system activities in the Gulf of Mexico	14–15 January 2003	Stennis Space Center, MS
A workshop to explore private sector interests and roles in the U.S. integrated ocean observing system; focus on the Southeastern United States and Gulf of Mexico	2–4 March 2004	Marathon Oil Company, Houston, TX
The HABSOS-GCOOS workshop	13–15 April 2004	St. Petersburg, FL
The GCOOS and the private sector: oil and gas and related industry workshop	2–4 November 2005	Houston, TX
The GCOOS-SECOORA-NOAA CSC storm surge & inundation workshop	24–26 January 2007	New Orleans, LA
First GCOOS-GOMA workshop on a harmful algal bloom observing system plan for the Gulf of Mexico	14–16 November 2007	New Orleans, LA
The Eastern Gulf of Mexico recreational boaters workshop	4–5 February 2009	St. Petersburg, FL
Second GCOOS-GOMA workshop for a harmful algal bloom integrated observing system workshop	21–23 April 2009	St. Petersburg, FL
The Western Gulf of Mexico GCOOS educator GPS workshop	23–24 April 2009	Corpus Christi, TX
The Eastern Gulf of Mexico GCOOS educator GPS workshop	30 April–1 May 2009	Dauphin Island, AL
The Western Gulf of Mexico recreational boaters workshop	28–29 May 2009	Clear Lake, TX
GCOOS-GOMA-SECOORA ecosystem modeling workshop	14–16 October 2009	St. Petersburg, FL
Third GCOOS-GOMA HABIOS workshop	26–28 March 2012	Pensacola, FL
Southwest Florida potential water quality providers workshop	28 June 2012	Sanibel, FL
Integrated water quality network meeting	12 March 2013	New Orleans, LA
Ecosystem modeling workshop	7–8 April 2014	Houston, TX
Integrated tracking of aquatic animals in the Gulf of Mexico workshop	29–30 May 2014	St. Petersburg, FL
GCOOS workshop with Nongovernmental organizations	10–11 June 2014	Houston, TX

Table 2 Priorities Held in Common across GCOOS-RA Stakeholder Groups

Priority Product or Data	Stakeholder Sectors
Obtain accurate bathymetry and topography with consistent vertical control between data sets in the coastal zone, including locations of shorelines	Emergency managers, surge modelers, recreational boaters, urban planners and developers, insurance industry, oil and gas, marine transportation, NGOs, iTAG community
Improve coverage of real-time currents in the coastal zone and navigable estuaries using HF radars as primary technique	Marine transportation, recreational boaters, oil and gas sector, Coast Guard SAR, iTAG community
Improve real-time, offshore meteorology measurements (V, P, T, H)	Oil and gas sector, Coast Guard SAR, surge modelers, HABs monitoring, recreational boaters
Improve forecasts and nowcast models of sea level, winds, and waves; this requires added real-time measurements	Recreational boaters, oil and gas sector, Coast Guard SAR, storm surge modelers, emergency managers
Improve hurricane severity forecasts	Emergency managers, oil and gas sector, recreational boaters
Improve forecasts and nowcasts of surface currents offshore	HABs tracking, oil and gas sector, Coast Guard SAR, marine transportation, iTAG community
Improve severe weather monitoring, forecasting, and dissemination	Oil and gas sector, recreational boaters, HABs tracking and fate, emergency managers
Enhance measurements of water quality parameters	Oil and gas sector, recreational boaters, HABs detection and fate, NGOs
Implement a modern, real-time current and water level observing system in all major ports	Marine transportation, recreational boaters, NGOs
Establish coastal storm surge/inundation maps for mitigation planning (not real time)	Oil and gas sector, insurance, real estate, planners, emergency manager, NGOs
Improve information on and forecasts of visibility	Coast Guard SAR, recreational boaters, marine transportation
Produce upper ocean profiles of temperature, salinity, and currents	Oil and gas sector, recreational boaters (near artificial reefs and major diving locations)
Produce reliable forecast maps of three-dimensional currents offshore	Oil and gas sector, iTAG community
Improve real-time forecasts of coastal inundation	Emergency managers, general public, NGOs
Increase number of stations monitoring HABs	Public and animal health officials, HABs monitoring network, NGOs
Improve data and product dissemination techniques taking into account the sophistication of the user	Requirement of all sectors

13. Hypoxia monitoring
14. Monitoring of river discharge to the Gulf
15. Mooring network
16. Outreach and education subsystem
17. Satellite observations and products
18. Surface currents and waves network
19. Water level observation network

In addition to the 19 elements listed, the GCOOS Build-out Plan includes detailed plans for the subsystems designated as Governance and Management, the Role of Research and Development in GCOOS, and Budget and Funding. The latter includes cost estimates for each of the described elements and subsystems.

The GCOOS-RA has worked diligently to ensure that observing system priorities evolve with regional priorities. For this reason, the Build-out Plan, originally drafted in 2009, was extensively reviewed and updated following the 2012 Resources and Ecosystems Sustainability, Tourist Opportunities, and Revived Economies of the Gulf Coast States (RESTORE) Act, the National Academy of Sciences Gulf Research Program's Ecosystem Services Priorities, and priorities identified by experts participating in the 2014 Gulf of Mexico Research Initiative (GOMRI) Oil Spill and Ecosystem Science Conference Session on Ecosystem Monitoring. Examples of elements of the Build-out Plan, matched to stakeholder-based needs identified in these documents, are provided in Tables 3–5.

Table 3 Match of Stakeholder-Based Elements in GCOOS Build-out Plan to RESTORE Priorities

RESTORE Priorities–Common Themes across S. 1603, 1604, 1605	GCOOS Build-out Plan Element Examples
Restoration and protection of fish, wildlife, and natural resources	Integrated water quality monitoring network, ecosystem monitoring, ecosystem modeling, hypoxia monitoring, data management, O/E
Restoration and protection of marine and coastal resources, including barrier islands, beaches, and wetlands	Bathymetry and topography, river discharge monitoring, enhanced water level network, PORTS, ecosystem monitoring, surface currents and waves network, data management, O/E
Restoration and protection of ecosystems	Ecosystem monitoring, ecosystem modeling
Observing and monitoring	Observing system (14 elements)
Restoration and protection of economy, sustainable development and sustainable technology	PORTS, research and development, circulation modeling, beach monitoring

Table 4 Match of Stakeholder-Based Elements in GCOOS Build-out Plan to National Academy of Sciences Ecosystem (NAS) Services Priorities

Ecosystem Services (NAS, 2013)	GCOOS BOP V.2.0 Section Examples
Provisioning services (e.g., material goods such as food, feed, fuel, and fiber)	Fisheries monitoring, physical and ecosystem modeling
Regulating services (e.g., climate regulation, flood control, water purification)	River discharge to the gulf, Enhanced Water Level Network, Integrated Water Quality Network, autonomous meteorological stations, harmful algal bloom monitoring
Cultural services (e.g., recreational, spiritual, aesthetic)	Beach quality monitoring, surface currents and waves network
Supporting services (e.g., nutrient cycling, primary productions, soil formation)	Integrated water quality monitoring, hypoxia monitoring, plankton monitoring

Table 5 Match of GCOOS activities to General Recommendations from the 2014 GOMRI Oil Spill and Ecosystem Science Conference Session on Ecosystem Monitoring

GoMRI Session Recommendations	GCOOS/IOOS Activities
Develop a business model	Business model is under revision and development plan is in draft form
Highlight advanced technologies	Technology R&D are included in the Gulf of Mexico observing system plan (http://gcoos.tamu.edu/BuildOut/BuildOutPlan-V2.pdf)
Quantify economic value of an observing system	GCOOS-RA board identified this need and member, BOEM, funded an ongoing 3-year study in partnership with Louisiana State University (LSU)
Building consensus and vision	20 stakeholder workshops informed the Gulf of Mexico observing system plan, stakeholder-focused organizational structure (board with private, governmental, academic, outreach/education sector representatives; stakeholder-based councils, committees, and task teams)
Gap assessment and analysis	Undertaken for moorings, high-frequency radar, hypoxia, HABS, and more as part of the system plan
Improved communication of monitoring products	GCOOS data products page (http://gcoos.org/products/) and additional work identified in the system plan
Single location for accessing data	GCOOS data portal (http://data.gcoos.org) and additional improvements identified in the system plan
"Community of practice" standards	Quality assurance for Real-Time Oceanographic Data (QARTOD) QA/QC standards for water level and in situ parameters (e.g., temperature, salinity, dissolved oxygen, waves, currents) (http://www.ioos.noaa.gov/qartod/welcome.html); GOMA water quality PIT recommendations; IOOS Data Management and Communications (DMAC)

Table 5 Match of GCOOS activities to General Recommendations from the 2014 GOMRI Oil Spill and Ecosystem Science Conference Session on Ecosystem Monitoring—cont'd

GoMRI Session Recommendations	GCOOS/IOOS Activities
	standards http://www.ioos.noaa.gov/data/contribute_data.html; Interagency Ocean Observations Committee (IOOC) core variables list http://www.iooc.us/ocean-observations/variables/
"System of systems" approach	GCOOS has a partnership model based on existing systems
Inventory of assets	IOOS data catalog and asset viewer http://www.ioos.noaa.gov/catalog/welcome.html
Improve communication of benefits	GCOOS communications team with strategy and long-term plan, BOEM/LSU study, more details in the system plan
Data management requirements	GCOOS DMAC–portal, products, tools, and technical assistance for data providers, more identified in the system plan; IOOS DMAC-compliant

4. CHALLENGES QUANTIFYING THE RETURN ON INVESTMENT OF A GULF OBSERVING SYSTEM

Rather than duplicating efforts, the GCOOS system of systems is being developed on existing subsystems and capacities. GCOOS subsystem partners include the following:

- Center for Ocean-Atmospheric Prediction Studies (COAPS), Florida
- Central Gulf of Mexico Ocean Observing System (CenGOOS), Mississippi
- Coastal Ocean Monitoring and Prediction System (COMPS), Florida
- Dauphin Island Sea Lab (DISL), Alabama
- Florida Fish and Wildlife Research Institute (FWRI)
- Louisiana Universities Marine Consortium (LUMCON)
- Mote Marine Laboratory, Florida
- National Estuarine Research Reserve System, Gulf-wide
- NOAA National Data Buoy Center (NDBC), Gulf-wide
- NOAA National Water Level Observing Network (NWLON), Gulf-wide
- NOAA Physical Oceanographic Real-Time System (PORTS)
- National Ocean Service Sites, Gulf-wide
- Oil and Gas Industry Acoustic Doppler Current Profile data, Gulf-wide
- Sanibel Captiva Conservation Foundation (SCCF), Florida
- Texas Coastal Ocean Observation Network (TCOON)
- Texas Automated Buoy System (TABS)
- Wave Current Surge Information System (WAVCIS), Louisiana

These partnerships are integral to the RA's infrastructure. For a decade, this multi-institutional, interdisciplinary structure has enabled GCOOS to ingest data and information from diverse providers to be seamlessly aggregated, integrated, and served to meet numerous regional information priorities. Though common sense suggests regional sharing is good practice, quantifying the value of this sharing has proved difficult. However, recent events have enabled the GCOOS and other RAs to showcase benefits in response to natural (e.g., Superstorm Sandy) and anthropogenic (e.g., the Deepwater Horizon oil spill) events, and as a result, on the whole, the IOOS community has been doing a better job quantifying the value of ocean observing systems. While limited, there are several economic studies relevant to the Gulf of Mexico that have been or are in the process of being conducted.

Some of the early economic studies, such as those of Adams et al., in 2000[12] and Kaiser and Pulsipher in 2004,[13] were undertaken to look at the link between improved ocean observations and better short- and long-term weather and climate forecasts across a wide range of private and public economic sectors and activities. The conclusions were that national[12] and Gulf of Mexico-specific[13] observing system benefits would exceed costs significantly and that Federal support should be given to achieve the full benefits. In a 2008 special edition of *Coastal Management*,[14] several articles focused on estimates of the economic benefits of regional observing systems around the country. Annual economic benefits ranged from $33 million for the Gulf of Maine region to $381 million for the Gulf of Mexico.[14–16] Also relevant to GCOOS was an estimate of $170 million return on investment for the Southeast Atlantic region, including the states of Florida, Georgia, South Carolina, and North Carolina.[17]

In most of these studies, researchers identify five major categories where there are large economic benefits. These include maritime commerce transportation, commercial fishing, recreational boating and fishing, search and rescue operations, and oil spill prevention. Other benefit categories of importance include beach recreation and tourism, coastal erosion prevention, marine hazards and hurricane prediction, and offshore energy and power generation. All of these are highly dependent on the utilization of coastal resources and are relevant to the Gulf of Mexico.

Of significant importance to the Gulf of Mexico is a three-year cooperative project, *The Economic Benefits of the Gulf of Mexico Coastal Ocean Observing System [GCOOS]*, recently undertaken by GCOOS, the Bureau of Ocean Energy Management (BOEM), and Louisiana State University (LSU) to quantify the value of ocean monitoring systems, primarily the GCOOS, to create an economic model and assess the return on investment. The study, currently in its second year, is being led by Dr Rex Caffey, LSU Professor of Agricultural Economics and Agribusiness, and Resource Economics specialist, and it is anticipated to cost approximately $750,000. Compared to earlier studies that did not quantify actual benefits, but rather made estimates of what could be achieved if the quality and quantity of information could be increased,[15,17,18] the current study is assessing the value of GCOOS both during catastrophic events (e.g., hurricanes Ike, Isaac, and Katrina; Superstorm Sandy; the Deepwater Horizon oil spill) and for routine operations of

US Gulf of Mexico activities. Caffey stated, "Having fully aggregated, integrated, and searchable ocean and atmospheric information available prior to, during, and following catastrophic events has significant implications for the forecasts directed at saving lives, property, and ecosystems. From an economic perspective, the nominal investment made to date in observing systems has demonstrated great value to society. Yet the value has not been accurately quantified" (personal communication). Results of the study will surely equip advocates of regional ocean observing systems with data to justify to Congressional leaders that ocean observing systems are sound economic investments.

A fully developed observing system is not only sound business, it supports responsible management of natural resources. There exists a delicate balance between environmental protection and the high economic demands for resources. The rich biodiversity of the Gulf, including many endangered, threatened, and protected species, combined with the dense coastal human population, dictate that sustained monitoring take place to understand and protect the wealth of natural resources, human life, and property. Complicating the tenuous balance between sustainable resource use and economic vitality are the myriad stressors negatively impacting the Gulf. Natural hazards such as hurricanes in summer and extratropical cyclones in winter threaten all strategic activities. The position of the Gulf as a major drainage basin for 33 of the 48 contiguous states contributes significantly to nutrient loading, pollution, hypoxia (low oxygen levels detrimental to the wildlife), and other problems unique to its ecosystems.

5. MYRIAD GULF ISSUES, ONE COMPREHENSIVE SYSTEM

Among the most notable successes achieved during the decade of GCOOS program development, and possibly the most difficult to quantify, is the change in the fundamental way the business of ocean monitoring is conducted in the Gulf of Mexico. There has been a shift from stand-alone systems, designed to acquire proprietary data, to interconnected and coordinated systems that publicly share data in real time. This shift has numerous positive implications to those who live, work, and recreate in the Gulf region, and to the nation.

Following are descriptions of some of the major natural and anthropogenic issues within the Gulf region and corresponding examples of successes resulting from GCOOS-RA engagement with stakeholders.

Issue: Water Quality, Harmful Algal Blooms, and Waterborne Pathogens

Numerous waterborne pathogens exist in Gulf of Mexico waters including *Salmonella, Escherichia coli,* and *Vibrio* spp.[19] The primary vector for illness is through the consumption of undercooked or raw bivalves such as clams and oysters. Although few in number, deaths due to *Vibrio* through skin breaks causing septicemia have occurred. Expanding urbanization along our coastal waters and concomitant increases in storm water runoff, combined with environmental alterations due to climate change and future oil/pollutant spills, are cause for concern that these waterborne pathogens will also increase.[20]

Harmful algal blooms (HABs) occur worldwide, and the Gulf of Mexico waters are not exempt. The most frequent and widespread HAB in the Gulf is caused by the organism *Karenia brevis*, and the Gulf has almost annual blooms of this toxic dinoflagellate.[21] One of the unique features of *Karenia* blooms is that the toxin, brevetoxin, is not only carried in bivalves, causing neurotoxic shellfish poisoning if consumed, but it also becomes part of the marine aerosol that, when inhaled, causes respiratory illness in people, particularly those with chronic lung diseases such as asthma. In addition to *Karenia* blooms, blooms of *Dinophysis* have impacted human health with diarrheic shellfish poisoning. Although not as prevalent as *Karenia* blooms, *Dinophysis* has the potential for significant, acute human health effects through the consumption of shellfish.[22] Another emerging HAB in the Gulf of Mexico is *Gambierdiscus*, associated with ciguatera fish poisoning. Traditionally, these species were associated with the Bahamas and the Caribbean as reef phenomena. In recent years, with the installation of artificial reefs in the Gulf, ciguatoxic fish have been detected. Between impacts from climate change and possible introduction of new HAB species through ship ballast water, the Gulf of Mexico has challenges. The identification and tracking of these harmful species can be greatly enhanced through an integrated observing system, as evidenced by preliminary work using in situ sensors in ongoing monitoring and characterization studies.[23]

Examples of Success: Water Quality, Harmful Algal Blooms, and Waterborne Pathogens

Given the Gulf's propensity for harmful algal blooms, significant effort has been made to bring new technologies and integrate existing data to advance understanding of this phenomenon. In 2007, the GCOOS-RA began working with partners from the Florida Department of Health and the Gulf of Mexico Alliance to coordinate the development of a Harmful Algal Bloom Integrated Observing System Plan for the Gulf of Mexico. The collaboration resulted in the widely popular *A Primer on Gulf of Mexico Harmful Algal Blooms*[24] and led to engagement in several pilot projects to advance understanding of HABs and waterborne pathogens.

- Researchers at Texas A&M University (TAMU), the University of Texas, and Woods Hole Oceanographic Institution are collaborating on development of the Texas Observatory for Algal Succession Time Series (TOAST). GCOOS provides support to operate the Imaging Flow CytoBot (IFCB), the primary instrument of TOAST, and hosts data on the GCOOS data portal. The IFCB is an imaging-in-flow instrument that combines high-resolution video and flow cytometer technology to capture high-resolution images of plankton that range from approximately 10 to greater than 100 μm. The purpose of the IFCB time series is to provide sustained monitoring for early warning of toxic phytoplankton responsible for HABs along the Texas coast. The real-time and continuous operation of the IFCB has provided early warning for six toxic HABs (2008–2013) and has provided insight into the seasonal dynamics of phytoplankton along the coast of Texas.[23] It also has contributed to the discovery of a species of dinoflagellate new to the Gulf, and subsequent prevention of human illness. In 2008, images from the IFCB provided evidence for first *Dinophysis* bloom in the Gulf of Mexico.[25] This dinoflagellate produces okadaic acid, a toxin that accumulates in oysters and other filter-feeding shellfish and results in the syndrome known as Diarrhetic Shellfish Poisoning (DSP). Because of the early warning, oyster harvesting was closed, contaminated oysters recalled, and no instances of DSP or other human illness were reported.

- To promote easy access to diverse HAB information, GCOOS incorporated two separate human health and water quality information tools, originally reported on separate websites, into the GCOOS data portal. The Beach Conditions Reporting System, created in 2006 and operated by GCOOS member Mote Marine Laboratory, reports daily beach parameters on 33 beaches in Florida. The site uses beach sentinels to file real-time reports to the Website: http://coolcloud.mote.org/bcrs/. Parameters include impacts from Florida's dominant red tide species *Karenia brevis*, such as amount of dead fish on the beach and human response to the toxic marine aerosols. The Florida Department of Health maintains the Healthy Beaches program monitoring for bacterial levels (fecal coliforms and enterococci bacteria) at Florida beaches (http://www.floridahealth.gov/environmental-health/beach-water-quality/). By assimilating these two data sets into the GCOOS data portal, multiple water quality parameters can be linked and investigated.
- Some of the most interesting advances in our understanding of HAB dynamics have resulted from joint GCOOS efforts with Mote Marine Lab and the University of South Florida. By conducting coordinated autonomous underwater vehicle missions along the West Florida Shelf (WFS) during *Karenia* blooms, new insight into the three-dimensional bloom structure has emerged.[26,27] An important goal of the missions has been to combine glider data (e.g., conductivity temperature depth, chlorophyll, color dissolved organic matter, dissolved oxygen, optical properties, and passive acoustic readings) with other fixed-platform and satellite-based observations and model simulations to determine and predict water property distributions on the WFS. A significant outcome of this work is GCOOS progress toward development of mission glider viewing tools to display glider data in near real-time.[28] Water property distributions and currents are important because these influence many issues such as hazardous material tracking and search and rescue operations, as well as HAB monitoring. In addition to elucidating HAB dynamics, the joint missions are testing the boundaries of the gliders' capabilities, important details that will be of value as the regional glider fleet, a critical element of the fully developed observing system in the Gulf, is developed.
- The CenGOOS, supported in part by the GCOOS-RA and operated by the University of Southern Mississippi, is an important subsystem of GCOOS with a focus on the Mississippi Bight in the northern Gulf. In addition to monitoring the development of stratification, eutrophication, seasonal hypoxia, and effects

and efficacy of restoration projects, including expansion of the Port of Gulfport, CenGOOS has been a test bed for new approaches to measuring ocean acidification. In ongoing work to bring industry and academia together, GCOOS worked with CenGOOS and Liquid Robotics to test a new wave glider platform equipped with a CO_2 measuring system developed and built by the NOAA Pacific Marine and Environmental Laboratory. The project aimed to determine the feasibility of increasing the spatial coverage of sampling by ground truthing the data collected from the wave glider with long-standing measurements made with similar instruments mounted on buoys. GCOOS worked to make the different data streams interoperable and collaborated with Liquid Robotics on communications to raise awareness.

Issue: Oil and Gas Industry Operations and Spill Response

The oil and gas industry in the Gulf of Mexico is a vital energy resource not only for Gulf residents but for the entire nation. More than 40% of total US petroleum refining capacity is located along the Gulf coast, as well as 30% of total US natural gas processing plant capacity.[5] There are more than 4000 oil and gas platforms in the region. In addition to these, if placed end to end, existing Gulf oil and gas pipelines could wrap around the Earth's equator, exceeding a distance of more than 40,000 km.[29]

The Deepwater Horizon (DWH) oil spill of 2010 is certainly the first Gulf of Mexico oil spill that most people remember. However, over the years there have been six other significant Gulf events. Ixtoc occurred in 1979 and released 10,000 to 30,000 barrels per day for 9 months.[30] During hurricanes Katrina and Rita, more than 146 oil spills were recorded, 113 offshore platforms destroyed, and 457 pipelines damaged.[31,32] Because of the massive destruction and human suffering associated with Katrina, minimal attention was given to the oil incidents. Three other events involving ships also occurred. The *Burmah Agate* in 1979, the *Alvenus* in 1984, and *Ocean 255*, which involved three ships in Tampa Bay in 1993, all released significant amounts of oil into Gulf waters.[33,34] Finally, a lightning strike caused a massive fire on the *Megaborg* in 1990, releasing an oil slick into the environment. Given the history of spills and the increasing number of permits being issued by the BOEM to explore in deep waters of the Gulf's Outer Continental Shelf, the question is this: Will we be able to more effectively mitigate the next spill. The GCOOS-RA has worked closely with partners from the oil and gas industry, state and federal government, and academia to increase knowledge and predictive capabilities of ocean weather, particularly ocean currents needed for safe operations, and to provide accurate forecasting when mitigation is required.

Examples of Success: Oil and Gas Industry Operations and Spill Response

Paving the way for many of the ocean sciences advances in the Gulf were the early discussions among GCOOS partners, including the U.S. Minerals Management Service (now BOEM) that resulted in ocean current and meteorologic data collected by the oil and gas industry being freely served in real time to the public via the NOAA NDBC and the GCOOS data portal. These data have been used to develop theoretical models that provide insight into how Gulf currents move. The models have proven valuable for both improving oil spill response and for understanding the development and characteristics of hurricanes and extratropical storms.[35–37]

- At the suggestion of GCOOS-RA Board members, the GCOOS-RA was a major participant in the Gulf of Mexico Pilot Prediction Project (GOMEX PPP), initiated in 2010 as a 2.5-year, $1.56 million project to evaluate and demonstrate a computer modeling system, testing both individual models and multi-model ensembles, for the operational prediction of the circulation of the Gulf of Mexico. The project was sponsored by the Department of Energy via the Research Partnership to Secure Energy for America (RPSEA), a consortium of several dozen universities and energy companies, and the Climatology and Simulation of Eddies/Eddy Joint Industry Project, a consortium of several offshore oil and gas companies. GCOOS staff and members from Texas A&M University, the Naval Oceanographic Office, the Naval Research Laboratory, and the NOAA National Ocean Service participated.
- On April 20, 2010, the DWH oil spill in the Gulf of Mexico began with a tragic explosion. Specific GCOOS contributions to, and participation in, the response to the oil spill were numerous and included the following: (1) pre-spill establishment of the system providing immediate access to the available real-time and near real-time data through NDBC and the GCOOS data portal (http://data.gcoos.org); (2) ready access to legacy data through the portal; (3) generation of new data layers for the Emergency Response Management Application and the General NOAA Ocean Modeling Environment models; (4) rapid reinstallation of high-frequency radar in the at-risk locations for oiling; (5) compiled oil spill resources; (6) Gulf-wide education and outreach efforts to support accurate communications about the spill; and (7) support to partners deploying gliders for subsurface oil reconnaissance.
- Also in support of oil spill response, GCOOS staff and members from the Texas General Land Office (TGLO), the state agency responsible for oil spill response and mitigation, and TAMU's Geochemical and Environmental Research Group, worked to develop a rapidly deployable buoy network, capable of real-time oceanographic and meteorologic measurements. The buoy network and companion modeling system can forecast ocean currents 72 h into the future.[38,39] The information has assisted TGLO mitigation efforts during more than 60 spill events. For the Deepwater Horizon response, at the request of the NOAA Office of Response and Restoration, GCOOS worked with TGLO and TAMU to extend circulation forecasts from the Regional Ocean Model for ingestion into their General NOAA Operational Modeling Environment (GNOME).

Issue: Hurricanes and Extra-tropical Storms in the Gulf of Mexico

Over a span of 120 years (1890–2010), 395 named storms formed in the Atlantic Basin.[40] Several of the most intense of these made landfall along the US Gulf coast. For example, the Florida Keys Labor Day Hurricane (1935), and hurricanes Camille

(1969), Gilbert (1988), Rita, Katrina, and Wilma (2005) are among the top 10 storms with the lowest Atlantic hurricane central pressures on record. The human, environmental, and economic toll of such storms has changed over the years. The 1900 Galveston hurricane remains the deadliest natural disaster in US history, killing between 6000 and 12,000 people and causing $30 million (1900-dollar value) in damage. Hurricane Camille claimed an estimated 150 lives and brought a 20–25-foot storm surge to southeast Louisiana that contributed to $1.421 billion in damage. Hurricane Katrina claimed more than 1800 lives and became the most costly hurricane on record with over $75 billion in damage to southern Louisiana, southern Mississippi, and southern Alabama.

The importance of early warning systems from an integrated ocean observing system and potential life and economic savings cannot be underestimated. Hurricane Rita provides a compelling argument for the need for improved forecasts to help avoid deaths and injury. This 2005 storm caused one of the largest evacuations in US History, with severe traffic jams outside the Galveston/Houston, Texas, area. In addition to deaths and injuries directly related to the storm's passage over land, more than 20 deaths were indirectly attributed to the storm. Most of these were elderly evacuees from a nursing home, victims of a bus accident resulting from the perilous traffic. Despite the efforts of authorities to stagger evacuations, many people were stuck in clogged traffic or ran out of gas. Had Rita made landfall closer to Houston, significantly more people would have been in serious danger. An additional factor supporting the need for improved forecasting is that, although necessary to protect health and safety, mandatory evacuations result in substantial burdens on local governments. Communities that evacuated but were not hit by the storm were ineligible for Federal Emergency Management Agency (FEMA) funding (see Texas House Hearings, http://www.hro.house.state.tx.us/interim/int79-2.pdf).

Studies have shown that having a system in place that can better predict the path and intensities of hurricanes and tropical storms can save local economies millions of dollars. Reductions in the length of coastline that is under hurricane warnings with its associated evacuation costs and community preparedness have been estimated at a savings of $640,000 (2004-dollar value) per mile.[41] The costs of hurricane mitigation measures are significantly less than the costs of response.[42] When Hurricane Isaac entered the Gulf in 2012, nearly 35,000 offshore personnel were forced to evacuate, resulting in 23% of the nation's oil and gas exploration and production to shut down, at an estimated cost of $130 million a day, with ripple effects through the national economy.[43] Reducing the number of days this industry is shut down by increasing the availability of accurate forecasts can save lives and hundreds of millions of dollars for the country.[44]

Hurricane/Extra-tropical Cyclone Sandy, which made landfall in northern New Jersey in October 2012, is an excellent example of the importance of sustained ocean monitoring. It was the second-costliest hurricane in US history, second only to Hurricane Katrina, with more than $68 billion in damages. However, the presence of the existing U.S. IOOS is touted as saving countless lives and economic losses.

It has been documented that the early warning and predictions on the magnitude of the storm allowed the U.S. Navy to redeploy 80 of its ships out of Hampton Roads ports and shipyards three days prior to the onset of the storm. Estimates are that this prevented more than $500 million in damage to the ships.[45] Likewise, major shipping companies were able to divert their containers filled with goods for the approaching holiday season to other ports. More than 23,000 cargo vans were safely delivered and trucked or shipped via rail to the key markets during the 2012 Christmas shopping season. It is estimated that there were cost savings (i.e., reduced economic losses) of more than $1 billion.[45]

Examples of Success: Hurricanes and Extra-tropical Storms in the Gulf of Mexico

Increasing coastal populations and investment in waterfront development, combined with the growing body of literature supporting a link between global warming trends and storm intensification,[46] means improvements in accuracy of forecasts for storm intensities and trajectories are needed. The GCOOS-RA has been working with partners, particularly through the GCOOS Modeling Task Team, to advance forecasting capabilities in the Gulf.

- The earliest demonstration reinforcing the benefits of interoperable Gulf observing data and products came in 2005, shortly after the GCOOS-RA began aggregating and integrating data from partners. When the power of Hurricane Katrina rendered the NDBC web system inoperable for three days following the storm, GCOOS, through established partnerships, was able to take the lead in offering data and products to assist in the response effort. Because the GCOOS subsystems are diverse, distributed, and with a dispersed suite of observational capabilities, they were able to maintain access to the needed data streams, reinforcing the need for backup access to federal systems during catastrophic events.
- Also during Hurricane Katrina, when Navy wind products became unavailable due to failures in telecommunications equipment, NOAA Hazardous Materials (HAZMAT) personnel relied on data streams from the GCOOS network to drive their circulation models that predict the trajectories of debris and pollutants. These fields were available because of front-end work to make data ingested into the GCOOS meet the requirements recommended by the Ocean.US Data Management and Communications (DMAC) team.[47] The same interoperable data sets also enabled collaborations among GCOOS regional partners such as the University of South Florida, Louisiana State University, and Roffer's Ocean Fish Forecasting Service, Inc., to generate forecast models, plume trajectories, and tracking of the actual contaminant plumes that helped to coordinate post-hurricane cruise surveys.
- The GCOOS-RA-supported CenGOOS system, described previously, is also a valuable source of data for the National Hurricane Center and National Weather Service. During Hurricane Katrina, sensors from a CenGOOS buoy survived winds in excess of 280 km/h (170 mph) and transmitted a rare continuous record of meteorologic and oceanographic data as the historic storm passed.[48] Data sets

such as these are tremendously valuable because they are used in hindcast models and enable progress to be made in subsequent forecasts.

- GCOOS partners at the TAMU Conrad Blucher Institute have developed several specialized products using data from the TCOON, a data node partially funded by the GCOOS-RA. The first is software that enables the local National Weather Service Office in Corpus Christi, Texas, to seamlessly integrate near real-time, IOOS DMAC-compliant observations from TCOON into its weather forecasting system to more accurately predict the conditions of oncoming storms. Two other applications for mobile devices provide ship captains and harbor pilots operating in the Houston Ship Channel with real-time conditions of ocean water levels and currents (http://cbi-apps.tamucc.edu/transit), and coastal users with real-time wind measurements (http://cbi-apps.tamucc.edu/tcw).

Issue: Maritime Commerce Safety

Maritime commerce, responsible for more than 75% of the US overseas trade,[49] is an excellent example of the return on investment potential of an integrated observing system. There are 175 major seaports located in the United States, and it is expected that the volume of maritime commerce will double by 2021, and double again by 2030.[50] The gross domestic product from this maritime commerce has been estimated at more than $740 billion, with more than 13 million jobs supported by this sector.[51] With so much at stake, it is not surprising that this industry was one of the early adopters of ocean observing systems.

One of the first observing systems established in the United States was the Physical Oceanographic Real-Time System (PORTS) in 1991, as a response to a horrible accident in Tampa, Florida. In 1980, a 200,000-ton freighter lost its radar and steered out of the navigation channel, destroying the main support pier of the Sunshine Skyway Bridge, collapsing the road and killing 35 people. In addition to the irreplaceable loss of lives, the catastrophic event cost millions of dollars in port and road closures, and property and environmental damages. The PORTS program provides real-time observations and predictions on such parameters as water levels, currents, waves, clearance to bridge heights, and visibility, and it has greatly improved both the safety and efficiency of maritime commerce.[49]

In addition to challenges regarding maritime commerce safety at sea, inundation and loss of land mass pose real threats to the industry, particularly the low-lying lands of Louisiana and southwest Florida. Indeed, in several instances, we are past the "planning for" and well into the "mitigation of" submerged land. In addition to increasing rates of global sea level rise from 4 to 8 in over the past century to 0.13 in/year in the last two decades, the situation is exacerbated by having some of the world's most extreme subsidence rates, particularly in southeast Louisiana.[52] By 2050, much of the land in southeast Louisiana will be submerged, including the nation's largest energy port, Port Fourchon. Over 1.5 million barrels of oil per day pass through Port Fourchon, the land terminus of most oil and gas pipelines in the northern Gulf of Mexico.[53] Exacerbating the issue is the ever-increasing size of vessels pushing channel depth limits and the high volume of hazardous materials being transported.

Examples of Success: Maritime Commerce Safety
The National Ocean Service (NOS) is responsible for providing navigation products to promote safe and efficient navigation via the PORTS. While the system has documented benefits from 1996 through 2000, more than 12,000 commercial vessel collisions and groundings were reported by the National Transportation Safety Board[54] during this time. The issue is that despite the documented benefits, only about one third of the nation's ports are currently equipped with full PORTS arrays. The Gulf of Mexico falls within this average, having approximately 30 commercial ports and only seven PORTS to aid navigation.

- The Tampa Bay PORTS is the first NOS decision support tool developed to improve the safety and efficiency of maritime commerce through the integration of real-time environmental observations, forecasts, and geospatial information. Since its development in the early 1990s, six other systems have been implemented throughout the Gulf of Mexico. Today, the buoy observations, model forecasts, bathymetry data, and satellite products derived from GCOOS-funded partners, staff, and regional stakeholders are critical to the observations and predictions of water levels, currents, salinity, winds, atmospheric pressure, and air and water temperatures that mariners need to navigate safely.
- In 2012, the Gulf established the first of its kind Mobile Bay Storm Surge Monitoring Network (MBSSMN), consisting of five hurricane-hardened water level stations. The system, led by the Mobile County Commission and NOAA's Center for Oceanographic Products and Services (CO-OPS), combines data provided by multiple stations in CO-OPS existing Mobile PORTS, including meteorologic, water level, and ocean current data funded and made available by GCOOS. CO-OPS, an IOOS federal partner, has an enhanced data partnership with NDBC, one that enables CO-OPS to automatically ingest quality-controlled regional data sent to NDBC. In addition to strengthening the ability of emergency management officials to alert and prepare coastal communities to storm surges and other dangerous water level conditions, these data can be used for long-term climate assessments and validation of hydrodynamic model performance.

6. SUMMARY

In summary, increased population; poor water quality; risks associated with oil and gas exploration, extraction, and transport; severe weather; and loss of land mass are some of the many challenges the Gulf of Mexico faces now and into the future. Addressing the summative effects of these threats is a challenge that requires rigorous monitoring of the Gulf, from its estuaries and coastal wetlands to the deep water beyond the Outer Continental Shelf. During the building of the GCOOS enterprise, there has been convincing evidence that an integrated ocean observing "system of systems" is an effective, efficient system that can address these challenges and provide many societal benefits.

The Gulf of Mexico has come a long way toward establishing a regional ocean observing system in spite of the lack of needed new financial investment by government. The GCOOS has been built mainly because it has been developed with Gulf stakeholders voluntarily contributing substantial observing assets to be incorporated into the GCOOS, including states, universities, tribes, industries, nongovernmental organizations, federal partners, and the public. Moving forward, the Gulf community has an unprecedented opportunity via the DWH Clean Water Act penalty monies from responsible parties (e.g., British Petroleum and TransOcean) to expand the existing infrastructure into a comprehensive, fully developed system that can address the myriad issues facing the region. The blueprint for this fully developed system, with cost estimates, is clearly articulated in the GCOOS Build-out Plan. Rather than reinventing a new system, enhanced observations can fill known gaps while avoiding duplication and, when combined with legacy data, can generate meaningful ecosystem indicators that inform the restoration process, ascertain the success of restoration efforts, and build on the extensive multi-institutional, interdisciplinary relationships that have a long, trusted history. In this manner, the many societal benefits created by an efficient, cost-effective system can be realized.

ACKNOWLEDGMENTS

Major support for the projects summarized was from NOAA Cooperative Agreements NA11NOS0120024 (2011–2016) and NA08NOS4730411 (2008–2013) for development of the GCOOS System of Systems and Data Portal through the NOAA IOOS Program Office. The authors thank the many dedicated researchers, technicians, graduate students, and RA members who keep critical information flowing and account for all program success. Special thanks to Angie Jones, Eckerd College Office of Advancement, for generously sharing her time and expertise reviewing the manuscript.

REFERENCES

1. Wilson SG, Fischetti TR. Coastline population trends in the United States, 1960 to 2008. In: *U.S. Census Bureau, P25-1139*; 2010. http://www.census.gov/prod/2010pubs/p25-1139.pdf.
2. Beck MW, Odaya M, Bachant JJ, Bergan J, Keller B, Martin R, et al. *Identification of priority sites for conservation on the Northern Gulf of Mexico: an ecoregional plan*. Arlington (VA): The Nature Conservancy; 2000. http://www.conserveonline.org/2001/02/b/en/gulf.pdf.
3. Center for coastal studies: the first 25 Years, 1984–2009. In: Tunnell Jr JW, editor. *Center for Coastal Studies, College of Science and Technology*. Corpus Christi (Texas): Texas A&M University-Corpus Christi; 2009. p. 172.
4. NOAA, NOS. *The Gulf of Mexico at a glance: a second glance*. Washington (DC): U.S. Department of Commerce; 2011.
5. Energy Information Administration (EIA). 2014. http://www.eia.gov/special/gulf_of_mexico/.

6. Bureau of Ocean Energy Management. *Gulf of Mexico OCS region*. 2014. http://www.boem.gov/Gulf-of-Mexico-Region/.
7. NOAA, Office of Science and Technology, NMFS. *Fishing industry statistics*. 2012. http://www.st.nmfs.noaa.gov/Assets/economics/documents/feus/2011/FEUS2011%20-%20Gulf%20of%20Mexico.pdf.
8. NOAA NMFS. *Fisheries economics of the United States 2011*. 2011.
9. Stokes S, Lowe M. Wildlife tourism and the gulf coast economy. In: *Data report by Environmental Defense Fund and Walton Family Foundation*; 2013. http://www.daturesearch.com/wp-content/uploads/GulfWildlifeTourismReport_Datu-Research_07-09-2013.pdf.
10. American Association of Port Authorities. *Port industry Statistics*. 2013. http://www.aapaports.org/Industry/content.cfm?ItemNumber=900.
11. GCOOS. *A sustained, integrated ocean observing system for the Gulf of Mexico (GCOOS): infrastructure for decision-making*. 2014. http://gcoos.tamu.edu/BuildOut/BuildOutPlan-V2.pdf.
12. Adams R, Brown M, Colgan C, Flemming N, Kite-Powell H, McCarl B, et al. *The economics of sustained ocean observation: benefits and rationale for public funding*. Washington: National Oceanic and Atmospheric Administration and Office of Naval Research; 2000.
13. Kaiser MJ, Pulsipher AG. The potential value of improved ocean observation systems in the gulf of Mexico. *Mar Policy* 2004;**28**(6):469–89.
14. Taylor, Francis, editors. *Special edition coastal management*. ISSN 0892-0753, vol. 36(2); 2008.
15. Kite-Powell H, Colgan C, Weiher R. Estimating the economic benefits of regional ocean observing systems. *Coast Manag* 2008;**36**(2):125–45.
16. Kite-Powell H. Benefits of maritime Commerce from ocean surface vector wind observations and forecasts. *Ocean Surf Vector Winds Marit Commer*. 2008. http://manati.star.nesdis.noaa.gov/SVW_nextgen/QuikSCAT_maritime_report_final.pdf.
17. Dumas CF, Whitehead JC. The potential economic benefits of coastal ocean observing systems: the southeast region. *Coast Manag* 2008;**36**(2).
18. Kite-Powell H, Colgan C. *The potential economic benefits of coastal ocean observing systems: the gulf of Maine*. Joint publication of NOAA, Office of Naval Research, and Woods Hole Oceanographic Institution; 2001. http://www.publicaffairs.noaa.gov/worldsummit/pdfs/mainereport.pdf.
19. Girones R, Bofill-Mas S, Furones MD, Rodgers C. Food Borne Infectious diseases and monitoring of Marine Food resources. In: Walsh PJ, Smith SL, Fleming LE, Solo-Gabrielle HM, Gerwick WH, editors. *Oceans and human health: risks and remedies from the sea*. Burlington (MA): Academic Press; 2008. p. 359–80.
20. Solo-Gabriele HM. Infectious mircrobes in coastal waters. In: Walsh PJ, Smith SL, Fleming LE, Solo-Gabriele HM, Gerwick WH, editors. *Oceans and human health: risks and remedies from the sea*. New York: Elsevier Science Publishers; 2008. p. 331–6.
21. Walsh JJ, Jolliff JK, Darrow BP, Lenes JM, Milroy SP, Remsen A, et al. Red tides in the Gulf of Mexico: where, when, and why? *J Geophys Res* 2007;**111**:C11003. http://dx.doi.org/10.1029/2004JC002813.
22. Steidinger KA, Landsberg JH, Flewelling LE, Kirkpatrick BA. Toxic dinoflagellates. In. Walsh PJ, Smith SL, Fleming LE, Solo-Gabrielle HM, Gerwick HW, editors. *Oceans and human health: risks and remedies from the sea*. Burlington (MA): Academic Press; 2008. p. 239–56.

23. Campbell L, Henrichs DW, Olson RJ, Sosik HM. Continuous automated imaging-in-flow cytometry for detection and early warning of *Karenia brevis* blooms in the Gulf of Mexico. *Environ Sci Pollut Res* 2013. http://dx.doi.org/10.1007/s11356-012-1437-4.
24. GCOOS and GOMA. *A primer on gulf of Mexico harmful algal blooms*. 2013. http://gcoos.tamu.edu/documents/HabPrimer-10162013.pdf.
25. Harred LB, Campbell L. Predicting harmful algal blooms: a case study with *Dinophysis ovum* in the Gulf of Mexico. *J Plankton Res* 2014;**36**(6):1434–45. http://dx.doi.org/10.1093/plankt/fbu070.
26. Zhao J, Hu C, Lenes JM, Weisberg RH, Lembke C, English D, et al. Three- dimensional structure of a *Karenia brevis* bloom: observations from gliders, satellites, field measurements, and numerical models. *Harmful Algae* 2013;**29**:22–30.
27. Milroy SP, Dieterle DA, He R, Kirkpatrick GJ, Lester KM, Steidinger KA, et al. A three-dimensional biophysical model of *Karenia brevis* dynamics on the west Florida shelf: a look at physical transport and zooplankton grazing controls. *Cont Shelf Res* 2008;**28**:112–36.
28. Kobara S, Simoniello C, Mullins-Perry R, Jochens AE, Howard MK, Watson SM, Howden S. Ch. 16: Near real-time oceanic glider Mission Viewers. In: Wright D, editor. *Earth solutions, ocean solutions*. California: ESRI Press; 2015.
29. Minerals Management Service. *Geographic mapping data in digital format: pipelines*. 2008. http://www.gomr.mms.gov/homepg/pubinfo/repcat/arcinfo/index.html.
30. *Oil spill intelligence report*. Aspen; 1979–1980.
31. Veritas DN. *Minerals management service technical report. pipeline damage assessment from katrina and rita in the Gulf of Mexico*. 2007. Report No. 448 14183, http://www.bsee.gov/Research-and-Training/Technology-Assessment-and-Research/tarprojects/500-599/581AA/.
32. Minerals Management Service. *MMS updates hurricanes katrina and rita damage*. 2006. http://www.boem.gov/boem-newsroom/press-releases/2006/press0501.aspx.
33. NOAA Emergency Response Division, Office of Response and Restoration, National Ocean Service. Retrieved. Ixtoc I. In: *IncidentNews*. National Oceanic and Atmospheric Administration, US Department of Commerce; 2010.
34. NOAA. 2012. http://sero.nmfs.noaa.gov/deepwater_horizon/documents/pdfs/fact_sheets/historical_spills_gulf_of_mexico.pdf.
35. Li C, Rouse L. *Analysis of ocean current data from Gulf of Mexico oil and gas platforms*. Report to BOEM Environmental Studies Program. 2014., http://www.data.boem.gov/PI/PDFImages/ESPIS/5/5366.pdf.
36. Hetland RD, DiMarco SF. Skill assessment of a hydrodynamic model of circulation over the TX-LA continental shelf. *Ocean Model* 2012;**43–44**:64–76. http://dx.doi.org/10.1016/j.ocemod.2011.11.009.
37. Zhang X, Marta-Almeida M, Hetland RD. A high-resolution pre-operational forecast model of circulation on the Texas-Louisiana continental shelf and slope. *J Oper Oceanogr* 2012;**5**(1):19–34.
38. Wade TL, Sweet ST, Walpert JN, Sericano JL, Singer JJ, Guinasso Jr NL. Evaluation of possible inputs of oil from the Deepwater Horizon spill to the Loop Current and associated eddies in the Gulf of Mexico. In: Liu Y, et al., editors. *Monitoring and modeling the Deepwater horizon oil spill: a record-breaking enterprise. Geophys. monogr. ser.*, vol. 195. Washington (DC): AGU; 2011. p. 83–90. http://dx.doi.org/10.1029/2011GM001095.

39. Walpert JN, Guinasso NL. TABS responder: a quick-response buoy for oil spill applications. *Sea Technol Mag* October 2010;**51**(10):10–3.
40. NOAA. *National Weather Service, National Hurricane Center Hurricanes in History.* 2012. http://www.nhc.noaa.gov/outreach/history/.
41. Kite-Powell HK, Colgan CS, Kaiser MJ, Luger M, Pelsoci T, Pendleton L, et al. *Estimating the economic benefits of regional ocean observing systems.* A report prepared for the National Oceanographic Partnership Program. Marine Policy Center, Woods Hole Oceanographic Institution; 2004.
42. National Research Council. *Reducing coastal risks on the East and Gulf Coasts.* Washington (DC): The National Academies Press; 2014.
43. Oil and Gas Journal. *Hurricane Isaac Shuts in oil, gas production from the Gulf.* 2013. http://www.ogj.com/articles/print/vol-110/issue-9/general-interest/hurricane-isaac-shuts-in-oil.html.
44. Bureau of Safety and Environmental Enforcement. June 27, 2013. http://www.bsee.gov/Hurricanes/Offshore-Hurricane-Preparedness-Response-Forum/.
45. Marine Technology Society. *National Ocean Service/MTS co-hosted Tech Surge Event with 25 subject matter experts to develop a statement on the value the US IOOS had on preparations and damage avoidance from Superstorm Sandy.* December 20, 2012.
46. Emanuel K. Increasing destructiveness of tropical cyclones over the past 30 years. *Nature* 2005;**436**:686–8.
47. Hankin S, The DMAC Steering Committee. *Data management and communications plan for research and operational integrated ocean observing systems: I. Interoperable data discovery, access, and Archive.* Arlington (VA): Ocean.US; 2005. http://dmac.ocean.us/dacsc/imp_plan.jsp. 304.
48. Howden SD, Gilhousen D, Guinasso N, Walpert J, Sturgeon M, Bender L. Hurricane Katrina winds measured by a buoy mounted sonic anemometer. *J Atmos Ocean Tech* 2007;**25**(4):607–16.
49. NOAA Technical Report. *NOS CO-OPS 031, national physical oceanographic real-time systems (PORTSTM) management report.* 2000. http://tidesandcurrents.noaa.gov/publications/techrpt31.pdf.
50. American Association of Port Authorities. 2013. http://www.aapaports.org/Industry/content.cfm?ItemNumber=900.
51. Wolfe K, MacFarland E, MacFarland D. *An assessment of the value of the physical real-time system to the U.S. Economy (PORTS®).* National Oceanic and Atmospheric Administration; 2013. Available from: http://tidesandcurrents.noaa.gov/pub.html.
52. Barras J, Beville S, Britsch D, Hartley S, Hawes S, Johnston J, et al. *Historical and projected Louisiana coastal changes: 1978–2050.* 2004. USGS Open File Report OFR 03-334, http://lacoast.gov/LandLoss/newhistoricalland.pdf.
53. Port Fourchon: Port Facts: http://www.portfourchon.com/explore.cfm/aboutus/portfacts/.
54. National Transportation Safety Board Marine Accident Reports 1988–2013. https://www.ntsb.gov/investigations/reports_marine.html.

Index

Note: Page numbers followed by "f" and "t" indicate figures and tables respectively.

O

P

Y

Z

Printed in the United States
By Bookmasters